Computer and Network Security Essentials

Kevin Daimi

Editor

Computer and Network Security Essentials

 Springer

Editor
Kevin Daimi
University of Detroit Mercy
Detroit, MI, USA

Associate Editors
Guillermo Francia
Jacksonville State University, USA

Luis Hernandez Encinas
Institute of Physical and Information
 Technologies (ITEFI), Spain

Levent Ertaul
California State University East Bay
USA

Eman El-Sheikh
University of West Florida, USA

ISBN 978-3-319-86404-4   ISBN 978-3-319-58424-9   (eBook)
DOI 10.1007/978-3-319-58424-9

Printed on acid-free paper

This Springer imprint is published by Springer Nature
The registered company is Springer International Publishing AG
The registered company address is: Gewerbestrasse 11, 6330 Cham, Switzerland

Preface

The constantly increasing trend of cyber-attacks and global terrorism makes it vital for any organization to protect and secure its network and computing infrastructure. With the continuous progress the Internet is facing, companies need to keep up by creating and implementing various software products and by utilizing advanced network and system equipment that need to be protected against various attacks. Data stored in our computers can also be subject to unauthorized access. Attackers can modify our data, steal our critical information including personal information, read and alter our e-mail messages, change program code, and possibly mess with our photos including using them for wicked purposes. Intruders can also employ our computers to attack other computers, websites, and networks without our knowledge. By enforcing security of networks and other computing infrastructure, the possibility of losing important data, privacy intrusion, and identity theft can be countermeasured. Many professionals working in computer technology consider security as an afterthought. They only take it seriously when a security problem occurs. It is imperative that society should start accepting security as the new norm.

Computer and Network Security Essentials will introduce the readers to the topics that they need to be aware of to be able to protect their IT resources and communicate with security specialists in their own language when there is a security problem. It introduces IT security to the public at large to improve their security knowledge and perception. The book covers a wide range of security topics including computer security, network security, cryptographic technologies, biometrics and forensics, hardware security, security applications, and security management. It introduces the concepts, techniques, methods, approaches, and trends needed by security specialists to improve their security skills and capabilities. Further, it provides a glimpse of future directions where security techniques, policies, applications, and theories are headed. The book is a rich collection of carefully selected and reviewed manuscripts written by diverse security experts in the listed fields and edited by prominent security researchers.

University of Detroit Mercy, USA   Kevin Daimi

Acknowledgments

We would like to thank the following faculty and researchers for the generous time and effort they invested in reviewing the chapters of this book. We would also like to thank Mary James, Zoe Kennedy, Brinda Megasyamalan, Brian Halm, and Sasireka Kuppan at Springer for their kindness, courtesy, and professionalism.

Nashwa AbdelBaki, Nile University, Egypt
Hanaa Ahmed, University of Technology, Iraq
Ahmed Ali Ahmed Al-Gburi, Western Michigan University, USA
Abduljaleel Mohamad Mageed Al-Hasnawi, Western Michigan University, USA
Rita Michelle Barrios, University of Detroit Mercy, USA
Pascal Birnstill, Fraunhofer IOSB, Germany
Aisha Bushager, University of Bahrain, Bahrain
Ángel Martín del Rey, University of Salamanca, Spain
Alberto Peinado Domínguez, Universidad de Málaga, Spain
Xiujuan Du, Qinghai Normal University, China
Luis Hernandez Encinas, Spanish National Research Council (CSIC), Spain
Patricia Takako Endo, University of Pernambuco, Brazil
Jason Ernst, Left™, Canada
Levent Ertaul, California State University, East Bay, USA
Ken Ferens, University of Manitoba, Canada
José María De Fuentes, Universidad Carlos III de Madrid, Spain
Alejandro Sánchez Gómez, Universidad Autónoma de Madrid, Spain
Arturo Ribagorda Grupo, Universidad Carlos III de Madrid, Spain
David Arroyo Guardeño, Universidad Autónoma de Madrid, Spain
Hisham Hallal, Fahad Bin Sultan University, Saudi Arabia
Tarfa Hamed, University of Guelph, Canada
Zubair Ahmad Khattak, ISACA, USA
Irene Kopaliani, Georgian Technical University, Georgia
Stefan C. Kremer, University of Guelph, Canada
Gregory Laidlaw, University of Detroit Mercy, USA
Arash Habibi Lashkari, University of New Brunswick, Canada

Leszek T. Lilien, Western Michigan University, USA
Lorena González Manzano, Universidad Carlos III de Madrid, Spain
Victor Gayoso Martínez, Spanish National Research Council (CSIC), Spain
Natarajan Meghanathan, Jackson State University, USA
Agustín Martín Muñoz, Spanish National Research Council (CSIC), Spain
Mais W. Nijim, Texas A&M University–Kingsville, USA
Kennedy Okokpujie, Covenant University, Nigeria
Saibal Pal, Defense R&D Organization, India
Ioannis Papakonstantinou, University of Patras, Greece
Keyur Parmar, Indian Institute of Information Technology, INDIA
Bryson R. Payne, University of North Georgia, USA
Slobodan Petrovic, Norwegian University of Science and Technology (NTNU),
    Norway
Thiago Gomes Rodrigues, GPRT, Brazil
Gokay Saldamli, San Jose State University, USA
Jibran Saleem, Manchester Metropolitan University, UK
Narasimha Shashidhar, Sam Houston State University, USA
Sana Siddiqui, University of Manitoba, Canada
Nicolas Sklavos, University of Patras, Greece
Polyxeni Spanaki, University of Patras, Greece
Tyrone Toland, University of South Carolina Upstate, USA
Jesús Díaz Vico, BEEVA, Spain

Contents

About the Editors

 **Kevin Daimi** received his Ph.D. from the University of Cranfield, England. He has a long mixture of academia and industry experience. His industry experience includes working as senior programmer/systems analyst, computer specialist, and computer consultant. He is currently professor and director of computer science and software engineering programs at the University of Detroit Mercy. His research interests include computer and network security with emphasis on vehicle network security, software engineering, data mining, and computer science and software engineering education. Two of his publications received the Best Paper Award from two international conferences. He has been chairing the annual International Conference on Security and Management (SAM) since 2012. Kevin is a senior member of the Association for Computing Machinery (ACM), a senior member of the Institute of Electrical and Electronic Engineers (IEEE), and a fellow of the British Computer Society (BCS). He served as a program committee member for many international conferences and chaired some of them. In 2013, he received the Faculty Excellence Award from the University of Detroit Mercy. He is also the recipient of the Outstanding Achievement Award in Recognition and Appreciation of his Leadership, Service and Research Contributions to the Field of Network Security, from the 2010 World Congress in Computer Science, Computer Engineering, and Applied Computing (WORLDCOMP'10).

Guillermo Francia received his B.S. degree in mechanical engineering from Mapua Tech in 1978. His Ph.D. in computer science is from New Mexico Tech. Before joining Jacksonville State University (JSU), he was the chairman of the Computer Science Department at Kansas Wesleyan University. Dr. Francia is a recipient of numerous grants and awards. His projects have been funded by prestigious institutions such as the National Science Foundation, Eisenhower Foundation, Department of Education, Department of Defense, National Security Agency, and Microsoft Corporation. Dr. Francia served as a Fulbright scholar to Malta in 2007 and is among the first cohort of cyber security scholars awarded by the UK Fulbright Commission for the 2016–2017 academic year. He has published articles and book chapters on numerous subjects such as computer security, digital forensics, regulatory compliance, educational technology, expert systems, computer networking, software testing, and parallel processing. Currently, Dr. Francia holds a distinguished professor position and is the director of the Center for Information Security and Assurance at JSU.

Levent Ertaul is a full professor at the California State University, East Bay, USA. He received a Ph.D. degree from Sussex University, UK, in 1994. He specializes in network security. He has more than 75 refereed papers published in the cyber security, network security, wireless security, and cryptography areas. He also delivered more than 40 seminars and talks and participated in various panel discussions related to cyber security. In the last couple of years, Dr. Ertaul has given privacy and cyber security speeches at US universities and several US organizations. He received 4 awards for his contributions to network security from WORLDCOMP. He also received a fellowship to work at the Lawrence Livermore National Laboratories (LLNL) in the cyber defenders program for the last 4 years. He has more than 25 years of teaching experience in network security and cyber security. He participated in several hacking competitions nationwide. His current research interests are wireless hacking techniques, wireless security, and security of IoTs.

Luis Hernandez Encinas is a researcher at the Department of Information Processing and Cryptography (DTIC) at the Institute of Physical and Information Technologies (ITEFI), Spanish National Research Council (CSIC) in Madrid (Spain). He obtained his Ph.D. in mathematics from the University of Salamanca (Spain) in 1992. He has participated in more than 30 research projects. He is the author of 9 books, 9 patents, and more than 150 papers. He has more than 100 contributions to workshops and conferences. He has delivered more than 50 seminars and lectures. Luis is a member of several international committees on cybersecurity. His current research interests include cryptography and cryptanalysis of public key cryptosystems (RSA, ElGamal, and Chor-Rivest), cryptosystems based on elliptic and hyper elliptic curves, graphic cryptography, pseudorandom number generators, digital signature schemes, authentication and identification protocols, crypto-biometry, secret sharing protocols, side channel attacks, and number theory problems.

Eman El-Sheikh is director of the Center for Cybersecurity and professor of computer science at the University of West Florida. She teaches and conducts research related to the development and evaluation of artificial intelligence and machine learning for cybersecurity, education, software architectures, and robotics. She has published over 70 peer-reviewed articles and given over 90 research presentations and invited talks. Dr. El-Sheikh received several awards related to cybersecurity education and diversity and several grants to enhance cybersecurity education and training for precollegiate and college students that emphasize increasing the participation of women and underrepresented groups in cybersecurity. She leads the UWF ADVANCE Program, an NSF-funded grant aimed at enhancing the culture for recruiting, retaining, and advancing women in STEM. She enjoys giving presentations related to cybersecurity education and workforce development and mentoring students. El-Sheikh holds a Ph.D. in computer science from Michigan State University.

Part I
Computer Security

Chapter 1
Computer Security

Jeffrey L. Duffany

1.1 Introduction

Computer security can be viewed as a set of mechanisms that protect computer systems from unauthorized access, theft, damage and disruption of the services they provide. It includes protection from both internal and external threats. Internal threats can be flaws in a software program or operating system. External threats are unauthorized access or human error. Much of computer security is based on the principle of separation which states that one thing cannot affect another if they are suitably separated [1]. The main mechanisms for achieving separation are physical, temporal, logical and cryptographic [1]. Each of these four basic techniques is in widespread use today and security by separation is one of the fundamental principles of computer security. From an implementation standpoint, however, computer security is usually attained by a suitable set of mechanisms to provide confidentiality, integrity and availability of systems and data [1, 2] (see Fig. 1.1).

1.1.1 Confidentiality

Confidentiality is the principle that information is not disclosed unless intended [1]. One of the primary techniques to achieve confidentiality is through the use of cryptography [2]. Cryptographic techniques involve scrambling information so it becomes unreadable by anyone who does not possess the encryption key. For

J.L. Duffany (✉)
Universidad del Turabo, Gurabo, Puerto Rico
e-mail: jeduffany@suagm.edu

© Springer International Publishing AG 2018
K. Daimi (ed.), *Computer and Network Security Essentials*,
DOI 10.1007/978-3-319-58424-9_1

3

Fig. 1.1 Security at the
intersection of confidentiality,
integrity and availability

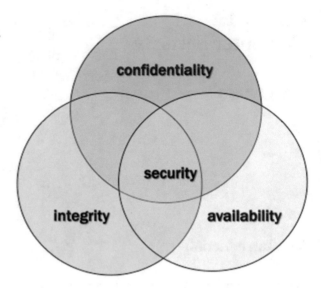

example, hard drives can be encrypted so that information is not compromised in the
event of theft or loss. Trusted parties who possess the encryption key can decipher
the encrypted data while others cannot.

1.1.2 Integrity

Integrity is assuring the accuracy and completeness of data over its entire life cycle.
This means that data cannot be modified in an unauthorized or undetected manner.
The mechanism to ensure integrity often involves the use of a hash function, a
one-way mathematical function that provides a digital signature of the data to be
protected [2].

1.1.3 Availability

For any information system to serve its purpose the stored data must be available
when it is needed [1]. High availability systems are designed to remain available at
all times avoiding service disruptions due to power outages, hardware failures and
system upgrades. Ensuring availability also includes the ability to handle denial-of-
service attacks which send a flood of messages to a target system in an attempt to
shut it down or block access [1].

1.1.4 Vulnerabilities and Attacks

A vulnerability is a system susceptibility or flaw in the design of the hardware or software and can be exploited to gain unauthorized access. A desktop computer faces different threats as compared to a computer system used in a government or military network. Desktop computers and laptops are commonly infected with malware designed to steal passwords or financial account information or to construct a botnet [1]. Smart phones, tablet computers and other mobile devices have also become targets. Many of these mobile devices have cameras, microphones and Global Positioning System (GPS) information which could potentially be exploited. Some kind of application security is provided on most mobile devices. However, applications of unknown or untrusted origin could result in a security compromise as a malicious attacker could embed malware into applications or games such as Angry Birds.

Government and military networks and large corporations are also common targets of attack. A recent report has provided evidence that governments of other countries may be behind at least some of these attacks [3]. Software and communication protocols such as Supervisory Control and Data Acquisition (SCADA) [4] are used by many utilities including the power grid and other types of critical infrastructure such as the water distribution system. Web sites that store credit card numbers and bank account information are targets because of the potential for using the information to make purchases or transfer funds. Credit card numbers can also be sold on the black market thereby transferring the risk of using them to others. In-store payment systems and ATMs have been exploited in order to obtain Personal Identification Numbers (PINs), credit card numbers and user account information.

1.2 Historical Background

Computing as we know it today had its origins in the late 1930s and 1940s during World War II when computers were developed by England and the United States to break the German Enigma cipher [2]. However computers did not find widespread government, commercial and military use in the United States until the decade of the 1960s. At that time the threatspace was rather limited and the emphasis was on functionality and getting things to work. Computing in the 1960s was carried out using large mainframe computers where users had to share the same memory space at the same time which leads to computer security issues. One program could affect another although this could be intentional or unintentional. This leads to the principle of separation as a primary means of implementing security. Physical separation was not always practical because of the expense, however, temporal and logical separation was widely employed in early mainframe computers even though it leads to somewhat inefficient use of resources. Temporal separation required programs to run sequentially while logical separation was used to give a virtual machine address space to each program.

The 1970s saw the migration toward smaller more affordable minicomputers and the rise of the Unix operating system. One minicomputer cost only a small fraction of what it cost to purchase and maintain a mainframe computer and could support dozens of users. These systems were highly scalable simply by adding more machines connected by networking equipment. Individual machines were often given fanciful names such as harpo, zeppo, chico, (the Marx brothers) or precious stones (diamond, emerald, etc.). Each user had one or more accounts on one or more machines and after logging on to their account were given a command line interface very similar to the Linux systems of today. Basic networking and electronic mail was supported. Each file or folder was given a set of read, write and execute (rwx) permissions to the owner and other users designated by the owner. Toward the end of the 1970s the first personal computers began to emerge from companies such as Apple and IBM.

The 1980s continued the revolution of the personal computer first beginning with the desktop and then laptop computers. Personal computers in the early 1980s typically had hard drives in the range of 40 MB, 64 K of RAM, 8 bit processors and command line user interfaces. As the command line interface was boring to many people one of the main uses of personal computers at that time was video games such as Space Invaders and PacMan (Fig. 1.2). Laptop computers were relatively expensive in the 1980s and became a prime target for theft. The first computer viruses (Fig. 1.3) also began emerging during the 1980s [5]. Floppy disks were used to boot and to share files. The first cybercrimes started making their way into the courtroom and as a result the Computer Fraud and Abuse Act (CFAA) (1984) was passed [1]. On 2 November 1988 Robert Morris released the first computer worm onto the internet and was subsequently found guilty of violating the new CFAA-related statutes [1]. During the mid-1980s Microsoft started developing the NTFS as a replacement for the outdated and severely limited File Allocation Table (FAT) filing system. The US Government issued the TCSEC Trusted Computer System Evaluation Criteria as a means of letting vendors know what they needed to do to make their operating systems more secure [1, 6]. Early adopters started subscribing to online services such as AOL and Compuserve which gave them access to electronic mail, chatrooms and bulletin boards. A member of the Chaos Computer Club in Germany accessed several US government military computer networks [7].

By the 1990s many companies had provided their employees with desktop or laptop computers running the latest version of Microsoft Windows. Many individuals owned their own desktop or laptop computers which were continuously adding new technological features while steadily reducing in price. The 1990s also saw the meteoric rise of the internet and web browsers. E-commerce was enabled by web browsers that supported secure connections such as Netscape [2]. Computer viruses continued to wreak havoc (Fig. 1.3) and the early 1990s saw the rise of many individual antivirus companies that were bought out by their rivals

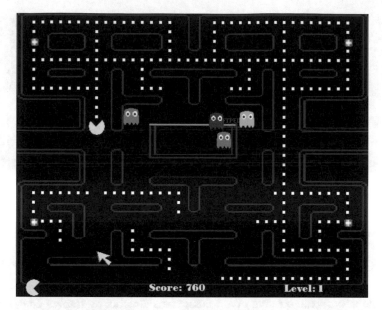

Fig. 1.2 PacMan game screen capture from early 1980s personal computer

consolidating down to a few major competitors. Cellular phones started becoming more affordable to the masses. The Data Encryption Standard (DES) [8] was broken by the Electronic Frontier Foundation [9]. Meanwhile wireless networks and the Wired Equivalent Privacy (WEP) standard emerged that used RC4 stream coding [10]. The Digital Millennium Copyright Act anticipated the potential abuse of copying information in digital form [1].

The decade of 2000 saw increasingly widespread use of the internet and social networking (Facebook, Twitter, etc.). Google introduced their electronic mail system called gmail (2004). Many privacy issues emerged especially after the Patriot Act (2001) gave the US government expanded powers of surveillance of anyone who might be suspected of terrorism. The Advanced Encryption Standard (AES) [11] officially replaced the Data Encryption Standard (DES) [8] in 2001. The US government began accelerating efforts to secure cyberspace and critical infrastructure while developing countermeasures against cyberterrorism and the threat of cyberwarfare [12, 13]. A continuing series of government, military and corporate data breaches made news headlines on a regular basis. Many individuals became victims of various forms of internet fraud including phishing attacks designed to get their passwords or other personal information through electronic mail.

The decade of 2010 continued to see major corporate and government security breaches. The Office of Personnel Management (OPM) had social security numbers and data of millions of persons (e.g., social security numbers) stolen. The decade also brought with it the concept of cloud computing and the Internet of Things (IoT) both of which presented new security and privacy challenges. Evidence emerged

Fig. 1.3 Spread of computer virus by electronic mail

about the widespread hacking of US computer networks by foreign countries [3]. Software for exploiting computer security vulnerabilities such as Metasploit [14] and Kali Linux continued to increase in popularity [14]. A plethora of computer-security-related conferences (such as DefCon) and websites arose which allowed people to share information about and learn about exploiting computer vulnerabilities. Evidence released by whistleblower NSA contractor Edward Snowden indicated that the US government was working with companies such as Microsoft, Google and Apple and Facebook to access personal information about their clients. Information warfare on a large scale seemed to play a more dominant role in deciding the outcome of US presidential elections than ever before.

1.3 Computer Security Vulnerabilities and Threats

The main goals of computer security are to protect the computer from itself, the owner and anything external to the computer system and its owner. This includes mainly forces of nature (earthquakes, hurricanes, etc.) and individuals known as intruders or attackers. Probably the single biggest threat to computer system security are the individuals (i.e., attackers) who employ a variety of mechanisms to obtain data or resources of a computer system without the proper authorization. A standard part of threat modelling for any system is to identify what might motivate an attack on that system and who might be motivated to attack it. This section includes an overview of the major computer security threats being faced today by computer systems and their users. This includes intrusion by various means, physical access, social engineering, password attacks, computer viruses, malware, botnets and denial-of-service attacks.

1.3.1   The Attacker (Intruder)

An intruder is someone who seeks to breach defenses and exploit weaknesses in a computer system or network. Attackers may be motivated by a multitude of reasons such as profit, protest, challenge or recreation. With origins in the 1960s anti-authority counterculture and the microcomputer bulletin board scene of the 1980s many of these attackers are inspired by documented exploits that are found on alt.2600 newsgroup and Internet Relay Chat (IRC). The subculture that has evolved around this type of individual is often referred to as the computer underground. Attackers may use a wide variety of tools and techniques to access computer systems [14, 15]. If the intruder can gain physical access to a computer, then a direct access attack is possible. If that is not the case, then the intruder will likely attack across a network, often hiding behind a proxy server, vpn tunnel or onion router/tor browser [16].

1.3.2   Physical Access

An unauthorized user gaining physical access to a computer is most likely able to directly copy data from it. Even when the system is protected by standard security measures such as the user account and password it is often possible to bypass these mechanisms by booting another operating system or using a tool from a CD-ROM to reset the administrator password to the null string (e.g., Hiren Boot disk). Disk encryption [17] and Trusted Platform Module [18] are designed to prevent these kinds of attacks.

1.3.3   Social Engineering and Phishing

Social engineering involves manipulation of people into performing actions or giving out confidential information [15]. For example, an attacker may call an employee of a company and ask for information pretending to be someone from the IT department. Phishing is the attempt to acquire sensitive information such as usernames, passwords and credit card details directly from users [15]. Phishing is typically carried out by email spoofing and it often directs users to enter details at a fake website whose look and feel are almost identical to the legitimate one. As it involves preying on a victim's trust phishing can be classified as a form of social engineering [15].

1.3.4   Attacker Software Tools

To gain access the attacker must either break an authentication scheme or exploit some vulnerability. One of the most commonly used tools by attackers is Nmap [14].

Nmap (Network Mapper) is a security scanner used to discover hosts and services on a computer network thus creating a "map" of the network. Nmap sends specially crafted packets to the target host and then analyses the responses. Nmap can provide a wealth of information on targets including open port numbers, application name and version number, device types and MAC addresses.

Once a target host and open ports are identified the attacker then typically tries using an exploit to gain access through that port. One of the most powerful tools is Metasploit [14] which has already made code to inject to perform the exploit. Metasploit also takes advantage of other operating system vulnerabilities such as stack or buffer overflow and can also perform privilege escalation. Metasploit can also perform SQL injection [1, 14] which is a technique where SQL statements are inserted into an entry field for execution. SQL injection exploits a security vulnerability that takes advantage of incorrectly filtered or misinterpreted user input.

1.3.5   Botnets

The word botnet is a combination of the words robot and network. A botnet is a number of Internet-connected computers under control of an attacker that are typically used to send spam email or participate in distributed denial-of-service attacks [1] (Fig. 1.4). Botnets can contain hundreds of thousands or even millions of computers. Botnets can be rented out to other attackers for a fee that can be untraceable if paid, for example, in bitcoins [19]. Phishing emails or other techniques are used to install program code in the target computer also known as zombies. The attacker takes great care to ensure that the control messages cannot easily be traced back to them.

1.3.6   Denial-of-Service Attack

Denial-of-service (DoS) attacks [1] are designed to make a machine or network resource unavailable to its intended users. Attackers can deny service to individual victims such as by deliberately entering a wrong password enough consecutive times to cause the victim account to be locked. Or they may overload the capabilities of a machine or network and block all users at once. While a network attack from a single IP address can be blocked by adding a new firewall rule many forms of denial-of-service attacks are possible. When the attack comes from a large number of points such as in the case of a distributed denial-of-service attack (DDOS) and defending is much more difficult. Such attacks can originate from the zombie computers of a botnet, but a range of other techniques are possible including reflection and amplification attacks, where innocent systems are fooled into sending traffic to the

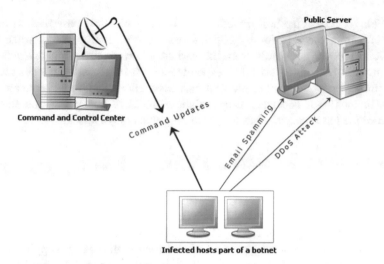

Fig. 1.4 Anatomy of a typical botnet

victim. Denial-of-service attacks are often used in an attempt to cause economic loss to the victim (usually a competitor) and to damage their reputation by making the outage appear to be their fault.

1.3.7 Password Cracking

Perhaps the easiest way to find out a user's password is through social engineering [15]. For example, some people write down their password on a yellow sticky pad and then post it on the wall next to their desk in case they forget it. If direct access or social engineering is not possible, the attacker can attempt to use widely available tools to attempt to guess the passwords. These tools work by dictionary attack of likely passwords and variations of those passwords possibly incorporating user's personal information such as birthdate or the name of their dog. Password cracking tools can also operate by brute force (i.e., trying every possible combination of characters). Lists of possible passwords in many languages are widely available on the Internet. Password cracking tools allow attackers to guess poorly chosen passwords. In particular, attackers can quickly recover passwords that are short, dictionary words, simple variations on dictionary words or that use easy to guess patterns.

Computer systems normally do not store user passwords instead it stores a hash of the password. A hash is a one-way mathematical function. If you know the password, you can easily compute the hash. However, if you only know the hash, you cannot easily compute the password. In some cases it might be possible to copy the entire file of hashed passwords from a system. Normally it is computationally infeasible to reverse the hash function to recover a plaintext password. However, there is a time space trade-off [20] that can be used that might in some cases be able to recover passwords from the hashed password file. Rainbow tables are precomputed hash tables that allow expedited search for a password since the time consuming step of computing the hash has been eliminated. Attackers can spend weeks or months if necessary using rainbow tables to find passwords since the password file has no mechanism for preventing this type of attack.

1.3.8 Malware

One of the most common and well-known threats to computer systems is "malware" which includes computer viruses [21]. A computer virus is a software program that installs itself without the user's consent then replicates by copying its own source code infecting other computer programs or the operating system itself (e.g., a boot virus). A computer virus often spreads itself by electronic mail (Fig. 1.3.) and attachments to the email that can contain executable code. Malicious software or "malware" includes computer viruses along with many other forms of malicious software such as computer worms, ransomware, trojan horses, keyloggers, rootkits, spyware, adware and other malicious software. Malware often performs some type of harmful activity on infected host computers such as accessing private information, corrupting data, logging keystrokes, creating botnets or providing a backdoor for future access.

The majority of viruses target systems running Microsoft Windows employing a variety of mechanisms to infect new hosts and using anti-detection strategies to evade antivirus software. Motives for creating viruses can include financial gain or simply a sociopathic desire to harm large numbers of people. The Virus Creation Laboratory (VCL) was one of the earliest attempts to provide a virus creation tool so that individuals with little to no programming expertise could create computer viruses. A hacker dubbed "Nowhere Man", of the NuKE hacker group, released it in July 1992.

1.3.9 Software Piracy

Software piracy is a major computer security issue for organizations that develop proprietary software products. It relates mainly to violation of copyright laws where individuals download software from the internet and make use of that software without compensating the software developer. The cost of software products ranges from free to several hundreds of dollars or more. Peer-to-peer networks are often used to circumvent copyright laws [1] and allow distribution of copyrighted materials and proprietary software to unauthorized individuals. Countermeasures usually involve some type of product code that is needed to activate the software.

Perhaps the most well-known example of this is the product key and activation process that is necessary to install and use many Microsoft operating systems and proprietary software products. Intruders often use reverse engineering techniques such as decompiling the machine language code to circumvent the various software protection mechanisms [22].

1.4 Countermeasures

There are many different ways of gaining unauthorized access into computers and computer systems. It can be done through a network, system, Wi-Fi connection or physical access. Computer systems can be protected by properly designed software and hardware that can help and prevent security failure and loss of data. To secure a computer system it is important to understand the attacks that can be made against it. One of the main techniques used in computer security is the separation of the intruders from the computer or data and this separation can be typically either physical, logical, cryptographic or temporal [1].

In computer security a countermeasure is a technique that reduces a threat, a vulnerability or an attack by eliminating or preventing it or by minimizing the harm it can cause or by discovering and reporting it so that corrective action can be taken. The countermeasures will vary depending on the system to be secured. A risk analysis can also help to determine appropriate countermeasures. Not all security breaches can be detected as they occur so some type of auditing should be included as an integral part of computer security. Audit trails track system activity so that when a security breach occurs the mechanism and extent of the breach can be determined. Storing audit trails remotely can help to prevent intruders from covering their tracks by preventing them from modifying the audit log files.

1.4.1 Authentication

Authentication is the act of verifying a claim of identity and is one of the primary techniques of separation used in computer security [23]. Across the internet you cannot see the person who is trying to access a website. If the person provides the proper credential, they are allowed access. This is one of the areas of computer security of most vulnerability. Passwords are by far the most predominant means of authentication in use today because of the ease of implementation and low cost. Biometric authentication [24] (for example, fingerprints, face recognition, hand geometry, retinal scan, voice recognition) is also in limited use. Strong authentication requires providing more than one type of authentication information (for example, two-factor authentication requires two independent security credentials).

A password is a string of characters used for user authentication to prove identity to gain access to a resource. User names and passwords are commonly used by people during a log in process that controls access to desktop or laptop computers, mobile phones, automated teller machines (ATMs), etc. A typical computer user has many passwords for email, bank account and online e-commerce. Most organizations specify a password policy that sets requirements for the composition and usage of passwords typically dictating minimum length, type of characters (e.g., upper and lower case, numbers, and special characters) and prohibited strings (the person's name, date of birth, address, telephone number). Some passwords are formed from multiple words and may more accurately be called a passphrase. The terms passcode and passkey are sometimes used when the secret information is purely numeric, such as the personal identification number (PIN) commonly used for ATM access.

1.4.2 Data and Operating System Backup

It is not always possible to forsee or prevent security incidents which involve loss of data or damage to data integrity. However, it is possible to be more resilient by having all important data backed up on a regular basis which allows for a faster recovery. Backups are a way of securing information and as such represent one of the main security mechanisms for ensuring the availability of data [1]. Data backups are a duplicate copy of all the important computer files that are kept in another separate location [1]. These files are kept on hard disks, CD-Rs, CD-RWs, tapes and more recently on the cloud. Operating systems should also be backed up so they can be restored to a known working version in case of a virus or malware infection. Suggested locations for backups are a fireproof, waterproof and heat proof safe, or in a separate, offsite location in which the original files are contained. There is another option which involves using one of the file hosting services that backs up files over the Internet for both business and individuals also known as the cloud. Natural disasters such as earthquakes, hurricanes or tornados may strike the building where the computer is located. There needs to be a recent backup at an alternate secure location in case of such kind of disaster. Having recovery site in the same region of the country as the main site leads to vulnerabilities in terms of natural disasters. Backup media should be moved between sites in a secure manner in order to prevent it from being stolen.

1.4.3 Firewalls and Intrusion Detection Systems

Firewalls [2] are an important method for control and security on the Internet and other networks. Firewalls shield access to internal network services, and block certain kinds of attacks through packet filtering. Firewalls can be either hardware or software-based. A firewall serves as a gatekeeper functionality that protects intranets and other computer networks from intrusion by providing a filter and safe transfer point for access to and from the Internet and other networks.

Intrusion detection systems [2] are designed to detect network attacks in-progress and assist in post-attack forensics. Intrusion detection systems can scan a network for people that are on the network but who should not be there or are doing things that they should not be doing, for example, trying a lot of passwords to gain access to the network. Honey pots are computers that are intentionally left vulnerable to attackers. They can be used to find out if an intruder is accessing a system and possibly even the techniques being used to do so.

1.4.4 Antivirus and Protection Against Malware

Computer viruses are reputed to be responsible for billions of dollars worth of economic damage each year due to system failures, wasted computer resources, corrupting data and increasing maintenance costs. It is estimated that perhaps 30 million computer viruses are released each year and this appears to be on an increasing trend. Many times a clean installation is necessary to remove all traces of a computer virus as the virus makes many changes throughout the system, for example, the registry in the case of Microsoft Windows systems. In response to the widespread existence and persistent threat of computer viruses an industry of antivirus [25] software has arisen selling or freely distributing virus protection to users of various operating systems. Antivirus scanners search for virus signatures or use algorithmic detection methods to identify known viruses. When a virus is found it removes or quarantines it. No existing antivirus software is able to identify and discover all computer viruses on a computer system.

1.4.5 General Purpose Operating System Security

Most general purpose operating system security is based on the principle of separation by controlling who has access to what and this information is kept in an access control list (ACL). The ACL is modifiable to some extent according to the rules of mandatory access control and discretionary access control [1]. The ACL itself must be secure and tamperproof otherwise an attacker can change the ACL and get whatever access they want.

1.4.5.1 NTFS Security

New Technology File System (NTFS) is a proprietary file system developed by Microsoft. It has replaced FAT and DOS in the late 1990s and has been the default filing system for all Microsoft Windows systems since then. NTFS has a number of improvements over the File Allocation Table (FAT) filing system it superceded such as improved support for metadata and advanced data structures to improve performance, reliability and disk space use. Additional improvements include security based on access control lists (ACLs) and file system journaling. In NTFS, each file or folder is assigned a security descriptor that defines its owner and contains two access control lists (ACLs). The first ACL, called discretionary access control list (DACL), defines exactly what type of interactions (e.g., reading, writing, executing or deleting) are allowed or forbidden by which user or groups of users. The second ACL, called system access control list (SACL), defines which interactions with the file or folder are to be audited and whether they should be logged when the activity is successful or failed.

1.4.5.2   MAC OSX and Linux Security

MAC OSX and Linux have their roots in the UNIX operating system and derive most of their security features from UNIX. A core security feature in these systems is the permissions system. All files in a typical Unix-style file system have permissions set enabling different access to a file which includes "read", "write" and "execute" (rwx). Permissions on a file are commonly set using the "chmod" command and seen through the "ls" (list) command. Unix permissions permit different users access to a file. Different user groups have different permissions on a file. More advanced Unix file systems include the access control list concept which allows permissions to be granted to additional individual users or groups.

1.4.5.3   Security Enhanced Linux (SE Linux)

NSA security-enhanced Linux [26] is a set of patches to the Linux kernel and some utilities to incorporate a mandatory access control (MAC) architecture into the major subsystems of the kernel. It provides an enhanced mechanism to enforce the separation of information based on confidentiality and integrity requirements which allows threats of tampering and bypassing of application security mechanisms to be addressed and enables the confinement of damage that can be caused by malicious or flawed applications. A Linux kernel integrating SE Linux enforces mandatory access control policies that confine user programs and system server access to files and network resources. Limiting privilege to the minimum required reduces or eliminates the ability of these programs to cause harm if faulty or compromised. This confinement mechanism operates independently of the discretionary access control mechanisms.

1.4.6   Program Security and Secure Coding

Program security reflects measures taken throughout the Software Development Life Cycle (SDLC) [27] to prevent flaws in computer code or operating system vulnerabilities introduced during the design, development or deployment of an application. Programmer reviews of an application's source code can be accomplished manually in a line-by-line code inspection. Given the common size of individual programs it is not always practical to manually execute a data flow analysis needed in order to check all paths of execution to find vulnerability points. Automated analysis tools can trace paths through a compiled code base to find potential vulnerabilities. Reverse engineering techniques [27] can also be used to identify software vulnerabilities that attackers might use and allow software developers to implement countermeasures on a more proactive basis, for example, to thwart software piracy [27].

Securing coding [28] is the practice of developing computer software in a way that guards against the introduction of security vulnerabilities. Defects, bugs and logic flaws are often the cause of commonly exploited software vulnerabilities. Through the analysis of large numbers of reported vulnerabilities security professionals have discovered that most vulnerabilities stem from a relatively small number of common software programming errors. By identifying coding practices that lead to these errors and educating developers on secure alternatives, organizations can take proactive steps to help significantly reduce vulnerabilities in software before deployment.

1.4.7 CyberLaw and Computer Security Incidents

It is very important to bring cybercriminals to justice since the inability to do so will inevitably inspire even more cybercrimes. Responding to attempted security breaches is often very difficult for a variety of reasons. One problem is that digital information can be copied without the owner of the data being aware of the security breach. Identifying attackers is often difficult as they are frequently operating in a different jurisdiction than the systems they attempt to breach. In addition they often operate through proxies and employ other anonymizing techniques which make identification difficult. Intruders are often able to delete logs to cover their tracks. Various law enforcement agencies may be involved including local, state, the Federal Bureau of Investigation (FBI) and international (Interpol). Very rarely is anyone ever arrested or convicted of initiating the spread of a computer virus on the internet [29].

Application of existing laws to the cyberspace has become a major challenge to Law Enforcement Agencies (LEA). Some of the main challenges are the difficulties involved in enforcing cyberlaws and bringing cybercriminals to justice. International legal issues of cyber attacks are complicated in nature. Even if a Law Enforcement Agency locates the cybercriminal behind the perpetration of a cybercrime it does not guarantee they can even be prosecuted. Often the local authorities cannot take action due to lack of laws under which to prosecute. Many of the laws we have today were written hundred of years ago before computers were invented and information in digital form did not exist. Identification of perpetrators of cyber crimes and cyber attacks is a major problem for law enforcement agencies.

1.5   Summary and Future Trends

The future of computer security appears to be that of a never-ending arms race between the attackers and the computer system users and administrators, designers and developers of hardware, software and operating systems. The average computer system user does not have extensive security training but nonetheless has to face the reality of computer security threats on a daily basis. For example, most people have to deal with a large number of passwords for different devices and websites. For that reason it can be expected that we will see a trend toward greater usability in security, for example, a trend toward password manager software [30] or perhaps the elimination of passwords altogether (https://techcrunch.com/2016/05/23/google-plans-to-bring-password-free-logins-to-android-apps-by-year-end/). One way that this could be done is to use the built-in signature of individual behaviours to act as an inexpensive biometric authentication (https://techcrunch.com/2016/05/23/google-plans-to-bring-password-free-logins-to-android-apps-by-year-end/) or by putting authentication into a computer chip [23].

The average person is relatively unsophisticated and is likely to be unaware of computer system vulnerabilities and even if they were they probably would not know how to deal with them. Therefore we can expect to see a trend toward building security into computing systems especially moving it from software into hardware where it is more difficult to compromise. The Next Generation Secure Computing Base initiative and the Trusted Platform Module [18] represent a step in that direction, however, it is not clear how long it will take before that type of technology reaches the consumer market. Secure coding practices [28] are likely to lead to incremental improvements in program and web application security as time goes on.

An overall sense of complacency seems to prevail currently for both computer users and manufacturers. The goal of a secure cyberspace seems to be replaced with a lesser goal of not allowing the situation to get any worse and simply trying to manage the security issues as best as possible as they arise. The current state of security complacency also appears to have become somewhat institutionalized. The number of computer viruses increases each year but no one is ever arrested or convicted as a result [29]. Manufacturers have little motivation to improve security as customers are more focused on features. Critical infrastructure is being increasingly controlled via computer programs that expose new vulnerabilities. Vulnerabilities will continue to be discovered and operating systems will continue to be patched, however, the operating systems in use now have not significantly improved from a security perspective since they were developed in the 1970s and 1980s. Improvements in computer security are not likely to occur proactively rather reactively as a result of cyberwarfare or cyberterroristic events [12, 13].

References

1. Pfleeger, C. P., & Pfleeger, S. L. (2015). *Security in computing* (5th ed.). Upper Saddle River, NJ: Prentice Hall. ISBN:978-0134085043.
2. Stallings, W. (2016). *Cryptography and network security: Principles and practice* (7th ed.). London: Pearson. ISBN:978-013444284.
3. Clarke, R. A. (2011). *Cyber war: The next threat to national security and what to do about it.* Manhattan, NY: Ecco Publishing. ISBN 978-0061962240.
4. Boyer, S. A. (2010). *SCADA supervisory control and data acquisition* (p. 179). Research Triangle Park, NC: ISA-International Society of Automation. ISBN:978-1-936007-09-7.
5. Cohen, F. (1987). Computer viruses. *Computers & Security, 6*(1), 22–35. doi:10.1016/0167-4048(87)90122-2.
6. Caddy, T., & Bleumer, G. (2005). Security evaluation criteria. In H. C. A. van Tilborg (Ed.), *Encyclopedia of cryptography and security* (p. 552). New York: Springer.
7. Stoll, C. (1988). Stalking the wily hacker. *Communications of the ACM, 31*(5), 484–497.
8. *FIPS 46-3: Data encryption standard.* csrc.nist.gov/publications/fips/fips46-3/fips46-3.pdf
9. Loukides, M., & Gilmore, J. (1998). *Cracking DES: Secrets of encryption research, wiretap politics, and chip design* (pp. 800–822). San Francisco, CA: Electronic Frontier Foundation.
10. Benton, K. (2010). *The evolution of 802.11 wireless security.* Las Vegas, NV: University of Nevada.
11. Daemen, J., & Rijmen, V. (2002). *The design of Rijndael: AES – the advanced encryption standard.* Berlin: Springer. ISBN 3-540-42580-2.
12. Singer, P. W., & Friedman, A. (2014). *Cybersecurity: What everyone needs to know.* Oxford, UK: Oxford University Press. ISBN:978-0199918199.
13. Clarke, R. A. (2011). *Cyber war: The next threat to national security and what to do about it.* Manhattan, NY: Ecco Publishing. ISBN 978-0061962240.
14. Kennedy, D. (2011). *Metasploit: The penetration tester's guide.* San Francisco, CA: No Starch Press. ISBN:978-1-59327-288-3.
15. Conheady, S. (2014). *Social engineering in IT security: Tools, tactics and techniques.* New York City, NY: McGraw-Hill. ISBN:978-00071818464. (ISO/IEC 15408).
16. Smith, J. (2016). *Tor and the dark net: Remain anonymous and evade NSA spying.*, ISBN:978-00071818464978-0692674444. New Delhi: Pinnacle Publishers.
17. Fruhwirth, C. (2005). *New methods in hard disk encryption. Institute for computer languages: Theory and logic group (PDF).* Vienna: Vienna University of Technology. ISBN: 978-00071818464978-0596002428.
18. England, P., Lampson, B., Manferdelli, J., Peinado, M., & Willman, B. (2003). A trusted open platform (PDF). *Computer, 36*(7), 55–62.
19. Nakamoto, S. (2009). *Bitcoin: A peer-to-peer electronic cash system* (PDF). Retrieved February 20, 2017, from https://bitcoin.org/bitcoin.pdf
20. Hellman, M. E. (1980). A cryptanalytic time-memory trade-off. *IEEE Transactions on Information Theory, 26*(4), 401–406. doi:10.1109/TIT.1980.1056220
21. Aycock, J. (2006). *Computer viruses and malware* (p. 14). New York: Springer. ISBN: 978-00071818464.
22. Eilam, E. (2005). *Reversing: Secrets of reverse engineering.* Indianapolis, IN: Wiley Publishing. ISBN:978-0007181846413-978-0-7645-7481-8.

23. Richard E. S. (2001), Authentication: From passwords to public keys., ISBN: 978-00071818464978-0201615999.
24. Jain, A., Hong, L., & Pankanti, S. (2000). Biometric identification. *Communications of the ACM, 43*(2), 91–98. doi:10.1145/328236.328110
25. Szor, P. (2005). *The art of computer virus research and defense*. Boston: Addison-Wesley Professional. ASIN 0321304543.
26. *National Security Agency shares security enhancements to linux*. NSA Press Release. Fort George G. Meade, Maryland: National Security Agency Central Security Service. 2001-01-02.
27. Sommerville, I. (2015), Software engineering., ISBN:978-0133943030.
28. Graff, M. G., & van Wyk, K. R. (2003). *Secure coding: Principles and practices*. Sebastopol, CA: O'Reilly Media, Inc.
29. *List of computer criminals*. https://en.wikipedia.org/wiki/List_of_computer_criminals
30. Li, Z., He, W., Akhawe, D., & Song, D. (2014). *The emperor's new password manager: Security analysis of web-based password managers (PDF)*. Usenix.

Chapter 2
A Survey and Taxonomy of Classifiers of Intrusion Detection Systems

Tarfa Hamed, Jason B. Ernst, and Stefan C. Kremer

2.1 Introduction

Nowadays, the Internet is experiencing many attacks of various kinds that put its information under risk. Therefore, information security is currently under real threat as a result of network attacks [40]. Therefore, to overcome the network attacks, intrusion detection systems (IDS) have been developed to detect attacks and notify network administrators [16]. The IDSs are now being studied widely to provide the defense-in-depth to network security framework. The IDSs are usually categorized into two types: *anomaly detection* and *signature-based detection* [40]. Anomaly detection utilizes a classifier that classifies the given data into normal and abnormal data [34]. Signature-based detection depends on an up-to-date database of known attacks' signatures to detect the incoming attacks [40]. Network Intrusion Detection Systems (*NIDS*) are considered as classification problems and are also characterized by large amount of data and numbers of features [44].

In recent years, Internet users have suffered from many types of attacks. These cyber attacks are sometimes so damaging and cost billions of dollars every year [28]. Some of these attacks were able to access sensitive information and reveal credit cards numbers, delete entire domains, or even prevent legitimate users from being served by servers such as in the case of denial-of-service (DoS) attacks. The most common type of Internet attack is intrusion. These days, the most popular Internet services are being attacked by many intrusion attempts every day. Therefore,

T. Hamed (✉) • S.C. Kremer
School of Computer Science, University of Guelph, Guelph, ON, Canada
e-mail: tarafayaseen@gmail.com; skremer@uoguelph.ca

J.B. Ernst
Left Inc., Vancouver, BC, Canada
e-mail: jason@left.io

© Springer International Publishing AG 2018
K. Daimi (ed.), *Computer and Network Security Essentials*,
DOI 10.1007/978-3-319-58424-9_2

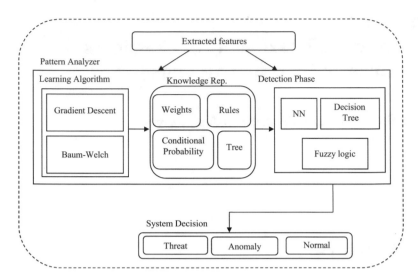

Fig. 2.1 The IDS components covered in this chapter

designing a robust system to detect against cyber attacks has become a necessity that needs the collaborations from all individuals.

The chapter is organized as follows: in Sect. 2.2 we explain the extracted features that result from the pre-processing phase. Next, the different IDS pattern analyzers are presented in detail in Sect. 2.3 with the knowledge representation used by those learning algorithms and the classifier systems. In Sect. 2.4 we present the decision making component of the IDS. The algorithms used in the detection phase produce different system decisions and are explained in this section. The system decision component with some details is presented in Sect. 2.5. The remaining parts of the IDS framework are beyond the scope of this chapter. Section 2.6 presents the conclusions of the chapter in addition to the open issues. We also provided a useful comparison and some critiques at the end of each component. The IDS components covered in this chapter are shown in Fig. 2.1.

2.2   Extracted Features

The pre-processing phase produces some patterns (sets of features) that would be suitable for the pattern analyzer, and the classification phase. These patterns are of different types (integer, float, symbols, etc.) according to the learning algorithm used. In [32], the resulting features are the statistical properties of packet sequences after converting them into statistical properties related to the transitions of the state machine.

In [27], the produced patterns represent the signature generated from the pre-processing phase. The algorithm presented in [24] calculates the empirical probability of a token appearing in a sample (whether it is *malicious* or *normal*). In [6], the extracted features included normal behaviors, audit data from ten users which have been collected for users who performed several types of actions such as programming, navigating Web pages, and transferring FTP data over the course of 1 month.

Now, having explained the extracted features resulting from the pre-processing phase and their types, we will explain the pattern analyzer of the system in the next section.

2.3   Pattern Analyzer

The next step is to use a suitable classifier to categorize the resulting extracted features from the pre-processing phase into one of threat, anomaly or normal data. In this section, the aim is to explain how the pattern analyzers work. In later sections, when discussing the classifiers, comparisons are given between the performances of various approaches after defining the typical metrics used to evaluate them. Some classifiers of intrusion detection systems deal with a user profile and behavior, therefore they use machine learning to learn the user profile in order to compare this profile later with the observed behavior to detect anomalies [22].

However, some other intrusion detection classifiers do not use any learning algorithm in making the final decision [43].

In general, a classification system can be viewed as consisting of three major components:

1. A decision making component, which ultimately classifies the data coming from the preceding phase,
2. A knowledge representation component, which incorporates information gathered from example data and informs the decision making component, and
3. An optional learning algorithm which is used to generate the knowledge representation of the previous component.

However, the chronological order of the above components is just the opposite, but we want here to start with the decision making component since the main objective of this phase is the detection process which is done by the decision making component. In addition, the decision making component needs a knowledge representation to make its decision, and to generate the knowledge representation, a learning algorithm is required to perform this task. The next sections will explain each part in details.

2.3.1   Learning Algorithms

In order to utilize the resulting features from the pre-processing phase for detecting intrusions, it is desirable to use a learning algorithm to learn from this data and later to use it to detect the intrusions. Learning algorithms are different in terms of the used input data whether they are labeled, un-labeled, and the type of the features. Some datasets like KDD Cup 99 contain labeled data either normal or attack (with only one specific attack type) for training and testing purposes, while some other datasets do not label their data. Researchers have been using several kinds of learning algorithms for intrusion detection purposes. In this context, several learning algorithms are discussed: *gradient descent, Baum–Welch algorithm, learning statistical properties, Genetic Network Programming, and some other machine learning algorithms*.

2.3.1.1   Gradient Descent Algorithm

Neural networks are one of the active approaches in building a learning system for detecting intrusions. In [22], the researcher has used back-propagation as a learning algorithm to train the network on the input data and use it to classify the test data. Back-Propagation (BP) is an algorithm used to train multi-layer, feed-forward, and supervised neural network. In this approach, the network is trained on different types of attacks and normal data to make it able to detect different attacks. Finding the optimal weights of the networks is accomplished by applying conjugate gradient descent algorithm. The host-based intrusion detection system is another type of intrusion detection system which collects input data from the host being monitored. The model proposed in [17] was used to detect both anomaly and misuse intrusions by incorporating two approaches: log file analysis and (BP) neural network. The researcher proposed a host-based intrusion detection system using a (BP) neural network to detect anomaly and misuse intrusions. The BP network was trained on the mentioned values to construct a user profile using a multi-layer neural network in anomaly detection [17].

2.3.1.2   Baum–Welch Algorithm

The Hidden Markov Model (HMM) is another technique used in intrusion detection. In [6], an HMM is used to model sequence information regarding system tasks, in order to minimize the false-positive rate and maximize the detection rate for anomaly detection. Usually, to estimate the parameters for an HMM, a standard Baum–Welch algorithm with the maximum-likelihood estimation (ML) criterion is used. The researcher used the Baum–Welch algorithm for HMMs since it is simple, well-defined, and stable [6].

2.3.1.3 Learning Statistical Properties

This approach focuses on unusual behavior to detect anomalies, so the approach needs to learn the frequency of making a transition from a state representing normal behavior to a state representing abnormal behavior. In this approach, the researchers used frequency distributions to represent network phenomena. Frequency distributions are used for type 1 properties (when there is a specific transition on the state machine) while for type 2 properties (the value of a specific state variable or a packet field when a trace traverses a transition), distribution of values for the state variable of interest are applied [32].

2.3.1.4 Genetic Network Programming (GNP)

Genetic Network Programming (GNP) is another approach for detecting intrusions of both types: anomaly and misuse. In [9], a learning algorithm starts with rule mining, which uses GNP to check the attribute values and compute the measurements of association rules using processing nodes.

In order to obtain the distribution of the average matching, the average matching degree between normal connection data and the rules in the normal rule pool is calculated. The matching degrees will be used later in the classification phase (detection phase) to make the system's decision.

2.3.1.5 Some Other Machine Learning Algorithms

In [15], where the researcher uses machine learning for detecting anomalies, the detection phase consisted of two steps: computing sequence similarity and classifying user behavior. In step one: the system calculates a numerical similarity measure which results from the number of adjacent matches between two sequences. Higher score of this measure means higher similarity [15].

The second step of the detection phase is classifying user behavior. This step processes the stream, token by token, and indicates at each point whether the user is a normal or an anomalous user. This determination is called classification of users. The classification is achieved according to a threshold value. If the mean value of the current window is greater than the threshold, then the current window is classified as normal, otherwise the window is classified as abnormal [15].

In [35], which employs a machine learning algorithm for anomaly detection, the empirical detection phase consists of three sub-steps: packet filtering, field selection, and packet profiling. Each sub-step is explained as follows [35]:

a. Packet filtering: The goal of packet filtering is to eliminate malformed packets from raw traffic.

b. The field selection scheme is performed using a Genetic Algorithm (GA). Preliminary tests are done using the typical genetic parameter values to find acceptable genetic parameters.
c. For packet profiling, a Self-Organized Feature Map (SOFM) neural network is used to create different packet clusters. The prepared raw packets are 60,000 raw packets from two different sources with 30,000 each. One source was for normal data and the other was for different types of packets aggregated from the internet.
d. Comparisons among the three SVMs and cross-validation tests: This step involves testing the three SVMs: soft margin SVM as a supervised method, one-class SVM as an unsupervised method, and the proposed enhanced SVM. The test for all of them was concluded using four different kinds of SVM kernel functions

In [45], the learning phase is divided into two steps: rule growing and rule pruning. In the rule growing step (GrowRule), the rule growing algorithm is used to handle each feature attribute in a growing set and decide the best split condition. During the learning process, the network is trained on normal and attacking data. The rule learning algorithm (FILMID) is utilized to perform inductive learning and construct a double-profile detection model from labeled network connection records. Besides FILMID, another two algorithms (RIPPER and C4.5) have been used in the training for four attack classes.

From the above learning algorithms used in pattern analysis phase, several comparisons can be drawn. Using neural networks helps in constructing a user profile or to train on a training data and test on testing data to detect both anomaly and misuse intrusions [17, 22], while the HMM is used to model normal behavior only from normal audit data [4]. Learning statistical properties was used in detecting anomalies only by learning the frequency distribution of the network to detect unusual behavior [32]. GNP was used by rule mining in checking the attribute values and computing the measurements of association rules using processing nodes to detect both anomaly and misuse intrusions [9]. Anomalies only were detected using machine learning in [15] by comparing the sequence similarities of the observed behavior and the stored behavior and then classifying user behavior to know whether the user is normal or anomalous. The POMDP learning algorithm was used in [14] in both anomaly and misuse detection. The learning involved the model parameters using an EM algorithm. Machine learning was used also in [35] for anomaly detection only. The detection phase of the approach involved packet filtering, field selection, and packet profiling to achieve detecting intrusions. The model comprised of building a double profile based on inductive learning to take the advantages of both anomaly and misuse detection techniques.

Some learning algorithms produce intermediate data which can be used later for classifier decision making during the detection phase. Some common forms of the generated knowledge representations are explained in the next section.

2.3.2 *Knowledge Representation*

In the intrusion detection problem, the knowledge representation can be one of the following types: weights resulting from training a neural network, rules resulting from fuzzy logic, conditional probabilities resulting from applying Hidden Markov Models, a cost function from POMDP, events from a log monitor, decision trees, or signature rules. Each of the aforementioned knowledge representation types is explained in the next sections.

a. Weights

   The result of the gradient descent learning algorithm represents the values of connection weights between the neurons which are normally organized as matrix and called a weight matrix. As an example of using the neural networks in IDS is the model presented in [22], where the conjugate gradient descent algorithm has been used to train a feed-forward neural network on both normal and attack data. In [10], the same concept was used but on two neural networks: Multi-Layer Perceptron (MLP) and Self-Organizing Maps (SOM). The used approach utilized the SOM network first to cluster the traffic intensity into clusters and then trained the MLP network to make the decision.

b. Rules

   Fuzzy rules are another form of knowledge representation that is used to provide effective learning. In [33], fuzzy rules consisted of numerical variables which represent the IF part and a class label which is represented by THEN part. Fuzzy rules are obtained automatically by "fuzzifying" the numerical variable of the definite rules (IF part) while the THEN part is the same as the resultant part of the definite rules [33].

c. Conditional probabilities

   The Baum–Welch learning algorithm produces a conditional probability which can be used later in the detection phase to check the status of the system if it is under attack or not. In [6], after providing an input sequence, the HMM performs the modelling for this sequence with its own probability parameters using the Markov process. After finishing building the model, then evaluating the probability with which a given sequence is generated from the model is performed [6].

d. Cost Function in POMDP

   The model presented in [14] is based on representing both the attacker and the legitimate user as unobservable, homogeneous Markov random variables by $A_t \in \{a_1,..,a_n\}$ and $U_t \in \{u_1,...,u_n\}$, respectively. At time t the computer state is called X_t which is generated by either an intruder (attacker) or a user and is controlled by a decision variable $D_t \in \{USER, ATTACKER\}$. The system is considered under intrusion when the captured data is produced by intruder, i.e., when $D_t = ATTACKER$. The next step is the action selection when an additional variable $C_t \in \{ALARM, NOALARM\}$ is used to model intrusion detection system actions [14].

e. Events from log monitor

In [17], the log file is monitored by the log monitor and events are sent to the log analyzer in case of a log change. In addition, system resources are also monitored by the systems resource monitor and their status is sent to the system resources analyzer during each time unit [17]. Finally, the active response unit, which receives the events from the log analyzer and system resources analyzer, is responsible for choosing the appropriate response to that situation which can be: notifying users, auditing, disconnecting from the network, etc. [17].

f. Decision trees

Pattern analyzers can also utilize decision trees in building an intrusion detection model. Decision trees have a learning process that results into the knowledge representation (the tree itself) that can be used in the detection phase. The main goal of decision tree classifier is to repeatedly separate the given dataset into subsets so that all elements in each final subset belong to the same class [12]. Three models of decision trees were used in [12] in the classification process: ID3, C4.5, and C5.0 algorithms. Another type of decision trees is called NBTree which is a hybrid between decision trees and NaïveBayes. The knowledge representation that results from NBTree is a tree whose leaves are NaïveBayes classifiers [8]. In intrusion detection problem, the decision tree classifier can be used to identify network data as malicious, benign, scanning, or any other category utilizing information like source/destination ports, IP addresses, and the number of bytes sent during a connection [12].

g. Signature Rules

One of the effective techniques in detecting intrusions is to use signature rules. In [19], several firewall rules were generated from network information such as packet source address, packet destination address, port from where packet is received, and packet type (protocol). The generated rules (knowledge representation) are dynamically modified based on the network requirement [19]. Behavior rule is another kind of rules that can be used to detect intrusions such as the model proposed in [20]. The knowledge representation of the model was based on behavior rules for defining acceptable behaviors of medical devices [20].

2.4 Decision Making Component (Detection Phase)

The second phase of the intrusion detection systems is the actual process of detecting the intrusions. Different detection algorithms need different steps to achieve this goal. Some of them need training and some do not, while others need rule generation as shown in some of the following examples.

2.4.1   Neural Networks

In [22], after a network was trained on two classes of data: normal and attack, the network now is ready for the testing phase. The three networks have shown detection accuracy of about 99%. The limitation of this approach is that it did not take into account a specific kind of attack and it dealt with only two classes of data: normal and attack. Some of the new datasets now differentiate between the attacks as the reaction of the IDS would be different against each type of attack.

2.4.2   Decision Tree

Decision trees have been successfully used in many applications due to its effectiveness. In [8], the researcher used an approach for network intrusion detection based on classifiers, decision trees, and decision rules. The detection phase in this work consisted of multiple steps and used multiple classifier algorithms and decision trees. For the classification algorithms, J48 (C4.5 Decision Tree Revision 8) was used. Next, the NnaïveBayes Tree classification algorithm was applied, and then decision table was used to evaluate feature subsets using a best-first search algorithm. The last classification algorithm was OneR, which was used for generating a one-level decision tree with a set of rules representation [8]. However, the approach did not involve calculating the False Alarm Rate (FAR), which is an important metric in evaluating an IDS.

2.4.3   Fuzzy Logic

The model presented in [33] was used to detect anomaly intrusions on the network. The researcher applied the model on the KDD cup99 dataset. Since the KDD cup99 dataset is very large to deal with, only 10% of the whole dataset is selected for training and testing and the data is selected from normal and attack data. The detection phase which uses fuzzy logic to detect intrusions consists of two sub-steps: a fuzzy decision module and finding an appropriate classification for a test input. The first step is used to select the most suitable attribute for a record's classification (normal or attack). This selection is performed by applying the deviation method [33] which uses the mined 1-length frequent items from each attribute and stores them in a vector.

The rule base is a knowledge base consisting of a set of rules acquired from the definite rules. The result of the inference engine would be selected from the set{Low, High}. Then, the "defuzzifier" transforms that output into useful values.

These useful values vary between 0 and 1, where 0 indicates normal data and 1 indicates pure attack data [33].

2.4.4 Genetic Network Programming

After calculating the matching degree in the learning phase, the class of a new connection data d needs to be recognized. The detection phase involves entering into a set of IF-THEN-ELSE statements to predict from the mentioned calculations the class of the current connection data whether it is normal, a known intrusion or an unknown intrusion or [9]. However, the limitation of this approach is that it did not give better accuracy than 90% which is not considered that high compared to the recent approaches.

2.4.5 Support Vector Machine

The model proposed in [42] depends on using a support vector machine (SVM) approach in detecting network intrusions. The proposed model was tested against four intrusion types: DoS, R2L, U2R, and probing attack. The intrusion detection system consists of three parts: an acquisition module of data packets, an intrusion detection agent, and a management agent. The intrusion detection agent is responsible for detecting illegal network activity (i.e., an attack). This agent uses a support vector machine to identify intrusions. The management agent—the third part—is responsible for organizing the performance of the intrusion detection agents and maintaining the whole system.

A possible drawback of this could be its lack in applying cross-validation in evaluating the results of the SVM classifiers to obtain more reliable results.

2.4.6 Some Other Decision Making Approaches

For space restriction, we are providing here some of other decision making approaches that we encourage readers to explore such as specification-based approach [32], mobile agent approach [7], situation awareness [13], malware detection [27], fast inductive learning [45], and negative selection [29]. Table 2.1 gives a brief summary on the detection approaches and their benefits discussed above.

Table 2.1 Detection approaches of IDS and their benefits

Reference	Detection approach	Benefits
[22]	Neural networks	The ability of neural networks to learn and generalize to detect attacks in a testing dataset
[8]	Decision tree	Accurate classification results for the input patterns
[33]	Fuzzy logic	The Fuzzifier was used to convert input attributes to linguistic variables and the Defuzzifier was used to transform the output of the inference engine to useful values (0 for normal and 1 for attack)
[32]	Specification-based method	Detecting anomalies when the observed statistics were so different from what was learnt
[27]	Malware detection	Efficient malware detection which can discover if there is any malware from the tokens of the signature
[7, 43]	Mobile agent	Efficient intrusion detector which was based on comparing the information collected by mobile agents with intrusion patterns
[9]	Genetic network programming (GNP)	Predicts the current connection's class whether it is normal or, known, or an unknown intrusion
[45]	Fast inductive learning	Used double profile to decrease the false positive and false negative in the classification results
[13]	Situation awareness	Distinguished attacks by maintaining a network security situation awareness graph and updating it periodically to detect attacks
[42]	Support vector machine (SVM)	Four SVMs were used as the kernel of an IDS to detect normal data and DoS, R2L, and U2R attacks
[29]	Negative selection	The detectors were able to reduce the detection speed by 50% in anomaly detection

2.5   Classifier's Decision

Generally, the detection phase should give a decision about what was discovered from the detection algorithm used. In some works like [17], the decision is made as a report to the administrator and called an auditing report. This report may involve notifying users, auditing or disconnecting from the network. The process of intrusion detection and the attack type are recorded by the audit database to be

32   T. Hamed et al.

used in the future [17]. Generally, a system decision can be one of the following three forms: threat, anomaly, or normal.

Different papers have been surveyed in this chapter with different types of decisions. Some of them just give a decision whether the data was an anomaly or normal such as [2, 15, 32, 38]. Some other papers limited their decisions to one of three options: anomaly, misuse, or normal. The coming sections explain the IDS decisions in more details.

2.5.1 Threat

Computer networks are the targets of many kinds of attacks and they are exposed to many new kinds of threats through the internet every day. In this section, four fundamental classes of attacks [18] are explained and illustrated with their subclasses in Fig. 2.2. The four fundamental classes are explaind in detail as follows:

a. *Access*

When an attacker tries to obtain information that the attacker is not allowed to access. Information may be exposed to this kind of attack while residing or during transmission. This type of attack puts the confidentiality of the information at risk. In general, access attacks are divided into three subclasses [18]:

1. Snooping: Snooping examines information files in order to find useful information.
2. Eavesdropping: Is listening on a conversation by a person who is not part of it. This typically occurs when an unauthorized person occupies a location where the information is expected to pass by as is shown in Fig. 2.3.

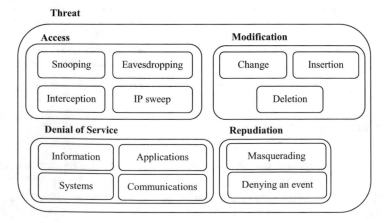

Fig. 2.2 Threat types with their subclasses

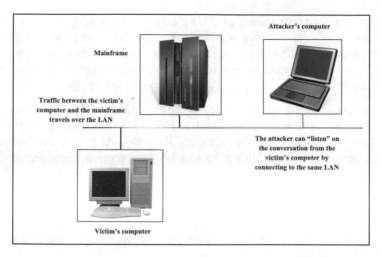

Fig. 2.3 Eavesdropping

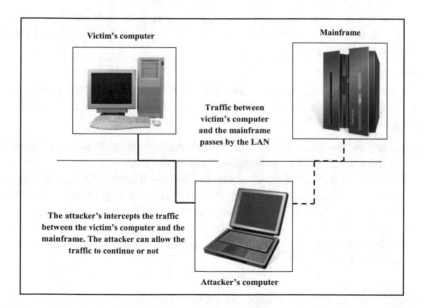

Fig. 2.4 Interception

3. Interception: Interception is considered more serious than eavesdropping to the system. That is because the attacker intercepts information, inserts himself/herself in the path of the information and captures it before it reaches its destination. After analyzing the information, the attacker has the choice to let the information continue to its destination or not as shown in Fig. 2.4.

4. IP sweep (Network scanning): This kind of attack is designed to identify the range of IP addresses that are mapped to live hosts on a target network by sending many ping requests to the full IP range and waiting for the reply. The purpose behind this technique is that it helps the attacker to know legitimate IP addresses in the target domain at the time of attack [32].

A probing attack is another kind of network scanning attack [36]. This attack occurs when an attacker performs a comprehensive scan on a network to collect information or find known vulnerabilities. Port scanning is a technique used to determine what ports are open, and that can inform the attacker what potential services running on a system are available to the attacker. There are two sides to port scanning. The first one is that the result can be utilized by network and system administrators as a part of network security audits for network maintenance. The second face is that it can be utilized by attackers who aim to compromise the system by exploiting a known vulnerability for one of the detected running services on its open port. Port scanning has some additional applications such as [36]:

– Identifying which hosts are active on a network.
– Identifying other network topological information, such as IP addressing, MAC addressing, router and gateway filtering, firewall rules, and IP-based trust relationships.

b. *Modification*
   A modification attack is when the attacker tries to alter information that the attacker is not authorized to. Generally, modification attacks are divided into three subclasses [18]:

1. Changes: This kind of attack involves changing existing sensitive information by the attacker such as an employee's salary.
2. Insertion: This kind of attack involves inserting information that did not exist before. For example, in a banking system an attacker might insert a transaction that moves money from a customer's account to his own account.
3. Deletion: Basically, a deletion attack is the removal of existing information such as the removal of a historical record that represents a transaction record in a banking system.

c. *Denial of Service*
   Denial-of-service (DoS) attacks are attacks that forbid the use of resources to authentic users of the system [18]. DoS attacks usually target the availability goal of the security and sometimes they are called availability attacks [26]. DoS attacks have been considered as one of the most significant attacks to networks in the last few years since they can cause a huge disorder to network functions. In addition, DoS attacks have been proven to be difficult to protect from [41].

d. *Repudiation*
   A repudiation attack is an attempt to give false information, or to deny that a real event or transaction has been performed by a particular entity [18]. Preventing

an entity from denying that it has taken part in a message exchange is called non-repudiation [23].

Usually, repudiation attacks are divided into two subclasses [18].

1. Masquerading
   Masquerading means the attacker attempts to imitate or impersonate someone else or some other system. This attack can threaten personal communication, transactions, or system-to-system communications.
2. Denying an Event
   Denying an event is the rejection of an event such as denying a bill of some purchase or denying cash withdrawal from a bank account.

2.5.2 Anomaly

If the intrusion detection system was designed to detect anomalies in the network, then it should be able to distinguish these events from those that it has seen previously in a training phase. Usually, this kind of IDS considers any deviation from the normal behavior of that network as an anomaly. However, this kind of IDS tends to suffer from false-positive classification.

Anomaly intrusions are of different types and a remote to local (R2L) attack is a class of anomaly attacks where an attacker tries to exploit the machine's vulnerability to illegally gain local access (becomes a local user) to a machine by sending packets to a machine over a network. It can happen when an attacker sends packets over a network to a machine which the attacker does not have an account on and exploits some vulnerability on that machine to acquire local access as a legitimate user of that machine. The Dictionary, FTP-Write, Guest, and Xsnoop attacks are types of R2L attack which all attempt to take advantage of any weaknesses or shortcoming in the configuration of system security policies [36].

User to root (U2R) attacks form another class of anomaly attacks when an attacker starts with access to a normal user account on the system then tries to acquire root access to the system by exploiting any vulnerability on the system. Usually, the attacker starts by accessing the system using a normal user account—which might have been obtained by some techniques like: sniffing passwords, a dictionary attack, or social engineering—and moves to exploiting some vulnerability to achieve the goal (gaining root access to the system). Buffer overflow attacks are the best known type of U2R attacks which come in many forms. Buffer overflows happen when a program copies a huge amount of data into a static buffer without checking the space availability in that buffer [36].

In [2], an approach for reusing information from a different layer for intrusion detection was adopted. WSN was divided into several layers and each layer ran a different protocol. The proposed technique used different information from different layers for ID as shown in Table 2.2.

Table 2.2 Summary of the layer information taken for ID

Layer	Protocols/techniques for anomaly detection	Use
Physical	RSSI value	Detects masquerade
Routing	Maintain neighbor lists, MAC layer transmission schedules are also used	Guarantees information authentication
MAC	TDMA: Check if adversary follows TDMA schedule	Keeps track of TDMA schedules of other nodes
	S-MAC: Check if sender is supposed to be sleeping	Keep track of sleep-wake up schedules of other nodes
Application	Use triangulation to detect intrusions	Detects masquerade
	Round trip time	Detects masquerade

Table 2.3 Types of decisions for each kind of attack and the references used

Type of attack	References
Access	[1, 8, 11, 14, 18, 21, 22, 30, 32, 33, 37, 39, 42]
DoS	[8, 9, 11, 18, 22, 30, 32, 33, 37, 39, 42, 43, 45]
Repudiation	[31, 45]
Anomaly	[2, 9, 14, 15, 17, 25, 27, 32, 35, 37, 38]

2.5.3 Normal

When the data is neither a threat nor an anomaly then it is considered normal data. This normal data represents the regular network traffic for that network or user. From the above, we can summarize the types of decisions used in this chapter for each class of data, as shown above in Table 2.3.

2.6 Conclusion and Open Issues

In this chapter and taxonomy of the IDSs, we have explored a wide range of pattern analyzers (classifiers) used in the IDSs and presented the taxonomy of the knowledge base that is produced as intermediate step. We also presented different techniques that have been utilized in the actual detection phase of the IDSs. We also explored the taxonomy of the classifiers' decision and explained each subcategory of these decisions.

As a matter of fact, the intrusion detection will keep developing as long as there are new attacks on the computer networks every day. In the last years, the Internet witnessed severe attacks that led to catastrophic consequences on multiple levels of computer users (i.e., end-users, governments, companies, etc.). Having said that, the world will still need to find new techniques to defend the computer networks and provide the ultimate security to the users from these attacks.

One of the major open issues is that since pattern classifiers have to work in adversarial environments (where the classifier needs to discriminate between normal and hostile patterns such as spam filtering, intrusion detection, and biometric identity verification), these classifiers need to deal with the attacks that try to avoid detection or force a classifier to generate many false alarms [4]. These days the attacks are being more sophisticated such that the input data can be intentionally tampered by skilful adversary to overcome the classifiers. According to [5], now this is considered as an arm race between adversary and classifier designers. The procedure of classifier designer could be either "reactive" or "proactive" arm race between the adversary and the classifier designer. The "reactive" procedure starts after an adversary analyzes the classifier defenses and formulates an attack strategy to defeat them. The designer reacts to the attack by analyzing the attack's effects and devising countermeasures. The "proactive" arm race involves the designer's attempt to anticipate the adversary by mimicking possible attacks, evaluating their effects, and developing countermeasures if necessary [5]. To improve the robustness of a classifier, different techniques have been used in the literature. One of the early efforts was proposing multiple classifiers systems (bagging and random subspace method) to improve the robustness of linear classifiers to adversarial data manipulation [3].

References

1. Bergadano, F., Gunetti, D., & Picardi, C. (2003). Identity verification through dynamic keystroke analysis. *Intelligence Data Analaysis, 7*(5), 469–496.
2. Bhuse, V., & Gupta, A. (2006). Anomaly intrusion detection in wireless sensor networks. *Journal of High Speed Networks, 15*(1), 33–51.
3. Biggio, B., Fumera, G., & Roli, F. (2010). Multiple classifier systems for robust classifier design in adversarial environments. *International Journal of Machine Learning and Cybernetics, 1*(1), 27–41. doi:10.1007/s13042-010-0007-7
4. Biggio, B., Fumera, G., & Roli, F. (2011). Design of robust classifiers for adversarial environments. In *IEEE international conference on systems, man, and cybernetics (SMC)* (pp. 977–982). IEEE.
5. Biggio, B., Fumera, G., & Roli, F. (2014). Security evaluation of pattern classifiers under attack. *IEEE Transactions on Knowledge and Data Engineering, 26*(4), 984–996. doi:10.1109/TKDE.2013.57
6. Cho, S. B., & Park, H. J. (2003). Efficient anomaly detection by modeling privilege flows using hidden markov model. *Computers & Security, 22*(1), 45–55. doi:10.1016/S0167-4048(03)00112-3
7. Dastjerdi, A. V., & Bakar, K. A. (2008). A novel hybrid mobile agent based distributed intrusion detection system. *Proceedings of World Academy of Science, Engineering and Technology, 35*, 116–119.
8. Gandhi, G. M., Appavoo, K., & Srivatsa, S. (2010). Effective network intrusion detection using classifiers decision trees and decision rules. *International Journal of Advanced Networking and Applications, 2*(3), 686–692.
9. Gong, Y., Mabu, S., Chen, C., Wang, Y., & Hirasawa, K. (2009). Intrusion detection system combining misuse detection and anomaly detection using genetic network programming. In *ICCAS-SICE, 2009*, (pp. 3463–3467).

10. Haidar, G. A., & Boustany, C. (2015). High perception intrusion detection system using neural networks. In *2015 ninth international conference on complex, intelligent, and software intensive systems* (pp. 497–501). doi:10.1109/CISIS.2015.73

11. Jalil, K. A., Kamarudin, M. H., & Masrek, M. N. (2010) Comparison of machine learning algorithms performance in detecting network intrusion. In *2010 international conference on networking and information technology* (pp. 221–226). doi:10.1109/ICNIT.2010.5508526

12. Kumar, M., Hanumanthappa, M., & Kumar, T. V. S. (2012). Intrusion detection system using decision tree algorithm. In *2012 IEEE 14th international conference on communication technology* (pp. 629–634). doi:10.1109/ICCT.2012.6511281

13. Lan, F., Chunlei, W., & Guoqing, M. (2010). A framework for network security situation awareness based on knowledge discovery. In *2010 2nd international conference on computer engineering and technology* (Vol. 1, pp. V1–226–V1–231). doi:10.1109/ICCET.2010.5486194.

14. Lane, T. (2006). A decision-theoritic, semi-supervised model for intrusion detection. In *Machine learning and data mining for computer security* (pp. 157–177). London: Springer.

15. Lane, T., & Brodley, C. E. (1997). An application of machine learning to anomaly detection. In *Proceedings of the 20th national information systems security conference* (Vol. 377, pp. 366–380).

16. Lin, W. C., Ke, S. W., & Tsai, C. F. (2015). Cann: An intrusion detection system based on combining cluster centers and nearest neighbors. *Knowledge-Based Systems, 78*, 13–21. doi:10.1016/j.knosys.2015.01.009

17. Lin, Y., Zhang, Y., & Ou, Y-J (2010). The design and implementation of host-based intrusion detection system. In *2010 third international symposium on intelligent information technology and security informatics* (pp. 595–598). doi:10.1109/IITSI.2010.127

18. Maiwald, E. (2001). *Network security: A beginner's guide*. New York, NY: New York Osborne/McGraw-Hill. http://openlibary.org./books/OL3967503M

19. Mantur, B., Desai, A., & Nagegowda, K. S. (2015). *Centralized control signature-based firewall and statistical-based network intrusion detection system (NIDS) in software defined networks (SDN)* (pp. 497–506). New Delhi: Springer. doi:10.1007/978-81-322-2550-8_48

20. Mitchell, R., & Chen, I. R. (2015). Behavior rule specification-based intrusion detection for safety critical medical cyber physical systems. *IEEE Transactions on Dependable and Secure Computing, 12*(1), 16–30. doi:10.1109/TDSC.2014.2312327

21. Mo, Y., Ma, Y., & Xu, L. (2008). Design and implementation of intrusion detection based on mobile agents. In *2008 IEEE international symposium on IT in medicine and education* (pp. 278–281). doi:10.1109/ITME.2008.4743870

22. Mukkamala, S., Janoski, G., & Sung, A. (2002). Intrusion detection: Support vector machines and neural networks. *IEEE International Joint Conference on Neural Networks (ANNIE), 2*, 1702–1707.

23. Muntean, C., Dojen, R., & Coffey, T. (2009). Establishing and preventing a new replay attackon a non-repudiation protocol. In *IEEE 5th international conference on intelligent computer communication and processing, ICCP 2009* (pp. 283–290). IEEE.

24. Newsome, J., Karp, B., & Song D. (2005). Polygraph: Automatically generating signatures for polymorphic worms. In *2005 IEEE symposium on security and privacy (S&P'05)* (pp. 226–241). IEEE.

25. Pannell, G., & Ashman, H. (2010). Anomaly detection over user profiles for intrusion detection. In *Proceedings of the 8th Australian information security management conference* (pp. 81–94). Perth, Western Australia: School of Computer and Information Science, Edith Cowan University.

26. Pfleeger, C. P., & Pfleeger, S. L. (2006). *Security in computing* (4th ed.). Upper Saddle River, NJ: Prentice Hall PTR.

27. Rieck, K., Schwenk, G., Limmer, T., Holz, T., & Laskov, P. (2010). Botzilla: Detecting the phoning home of malicious software. In *Proceedings of the 2010 ACM symposium on applied computing* (pp. 1978–1984). ACM.

28. Di Pietro, R., & Mancini, L. V. (2008). *Intrusion detection systems* (Vol. 38). New York, NY: Springer Science & Business Media.

29. Sadeghi, Z., & Bahrami, A. S. (2013). Improving the speed of the network intrusion detection. In *The 5th conference on information and knowledge technology* (pp. 88–91). doi:10.1109/IKT.2013.6620044

30. Sarvari, H., & Keikha, M. M. (2010). Improving the accuracy of intrusion detection systems by using the combination of machine learning approaches. In *2010 international conference of soft computing and pattern recognition* (pp. 334–337). doi:10.1109/SOCPAR.2010.5686163

31. Schonlau, M., DuMouchel, W., Ju, W. H., Karr, A. F., Theus, M., & Vardi, Y. (2001). Computer intrusion: Detecting masquerades. *Statistical Science, 16*(1), 58–74.

32. Sekar, R., Gupta, A., Frullo, J., Shanbhag, T., Tiwari, A., Yang, H., & Zhou, S. (2002). Specification-based anomaly detection: A new approach for detecting network intrusions. In *Proceedings of the 9th ACM conference on computer and communications security, CCS '02* (pp. 265–274). New York, NY: ACM. doi:10.1145/586110.586146

33. Shanmugavadivu, R., & Nagarajan, N. (2011). Network intrusion detection system using fuzzy logic. *Indian Journal of Computer Science and Engineering (IJCSE), 2*(1), 101–111.

34. Sheng Gan, X., Shun Duanmu, J., Fu Wang, J., & Cong, W. (2013). Anomaly intrusion detection based on {PLS} feature extraction and core vector machine. *Knowledge-Based Systems, 40*, 1–6. doi:10.1016/j.knosys.2012.09.004

35. Shon, T., & Moon, J. (2007). A hybrid machine learning approach to network anomaly detection. *Information Sciences, 177*(18), 3799–3821. doi:10.1016/j.ins.2007.03.025

36. Singh, S., & Silakari, S. (2009). A survey of cyber attack detection systems. *IJCSNS International Journal of Computer Science and Network Security, 9*(5), 1–10.

37. Terry, S., & Chow, B. J. (2005). *An assessment of the DARPA IDS evaluation dataset using snort* (Technical report, UC Davis Technical Report).

38. Trinius, P., Willems, C., Rieck, K., & Holz, T. (2009). *A malware instruction set for behavior-based analysis* (Technical Report TR-2009-07). University of Mannheim.

39. Vasudevan, A., Harshini, E., & Selvakumar, S. (2011). Ssenet-2011: a network intrusion detection system dataset and its comparison with kdd cup 99 dataset. In *2011 second asian himalayas international conference on internet (AH-ICI)* (pp. 1–5). IEEE.

40. Wang, W., Guyet, T., Quiniou, R., Cordier, M. O., Masseglia, F., & Zhang, X. (2014). Autonomic intrusion detection: Adaptively detecting anomalies over unlabeled audit data streams in computer networks. *Knowledge-Based Systems, 70*, 103–117. doi:10.1016/j.knosys.2014.06.018

41. Wang, Y., Lin, C., Li, Q. L., & Fang, Y. (2007). A queueing analysis for the denial of service (dos) attacks in computer networks. *Computer Networks, 51*(12), 3564–3573.

42. Xiaoqing, G., Hebin, G., & Luyi, C. (2010). Network intrusion detection method based on agent and svm. In *2010 2nd IEEE international conference on information management and engineering* (pp. 399–402). doi:10.1109/ICIME.2010.5477694

43. Xu, J., & Wu, S. (2010). Intrusion detection model of mobile agent based on aglets. In *2010 international conference on computer application and system modeling (ICCASM 2010)* (Vol. 4, pp. V4-347–V4-350). doi:10.1109/ICCASM.2010.5620189

44. Xue-qin, Z., Chun-hua, G., & Jia-jun, L. (2006). Intrusion detection system based on feature selection and support vector machine. In *2006 first international conference on communications and networking in China* (pp. 1–5). doi:10.1109/CHINACOM.2006.344739

45. Yang, W., Wan, W., Guo, L., & Zhang, L. J. (2007). An efficient intrusion detection model based on fast inductive learning. In *2007 international conference on machine learning and cybernetics*, (Vol. 6, pp. 3249–3254). doi:10.1109/ICMLC.2007.4370708

Chapter 3
A Technology for Detection of Advanced Persistent Threat in Networks and Systems Using a Finite Angular State Velocity Machine and Vector Mathematics

Gregory Vert, Ann Leslie Claesson-Vert, Jesse Roberts, and Erica Bott

3.1 Identification and Significance of the Problem or Opportunity

3.1.1 Introduction

Computers are an integral part of our society. Computer security efforts are engaged in an asymmetric fight against an enemy comprised of thousands of both independent and interrelated actors on an incredible number of fronts across a grand surface area [1]. Given the asymmetry, it is inefficient and impractical for analysts to manually investigate each new attack [2]. While this is true for the general case of known attacks, it is especially true for Advanced Persistent Threats (APTs). Advanced Persistent Threats are cyber-crimes that occur when an unauthorized person accesses a network and remains for a prolonged time period to steal information as opposed to compromise the organization itself [2, 3]. APTs are used to target financial institutions, military defense, aerospace, healthcare, manufacturing industries, technologies, public utilities, or political entities [2–5]. In discussing potential automation of parts of the process or the entirety of it all, it behooves us to observe the process first. When a new attack occurs, detection is the first step. Once detected, the offending code and its effects are isolated from background noise and studied. That is, its function must be determined, as well as potential relationships,

G. Vert (✉) • J. Roberts • E. Bott
College of Security and Intelligence, Embry-Riddle Aeronautical University, Prescott, AZ, USA
e-mail: gvert@erau.edu; roberj49@my.erau.edu; botte@my.erau.edu

A.L. Claesson-Vert
School of Nursing, College of Health and Human Services, Northern Arizona University, Flagstaff, AZ, USA
e-mail: Ann.Claesson@nau.edu

© Springer International Publishing AG 2018   41
K. Daimi (ed.), *Computer and Network Security Essentials*,
DOI 10.1007/978-3-319-58424-9_3

if any, to preexisting attacks. While it may present with a previously undiscovered and unique signature, that signature may still bear some semblance to prior attacks. Function refers to the goal of the attack; e.g., disruption of the service. This relationship can both be familial, as in a new variant of an existing exploit, or typical as in a broad type (i.e., category) of attack. Lastly, the design and implementation of preventative and ameliorative efforts to remove the existing infection are crucial. It is in these prevention and amelioration tasks that the previously determined factors such as type, family, function, and code become important.

While automation of the entirety of the process is attractive, this technology focuses on the automation of the identification and classification stages of the process [2]. The major focus is on detection, though a certain degree of type classification is possible and desirable as well. Additionally, this classification will assist both external efforts in amelioration as well as serve to reduce false positive rates. This not only indicates unauthorized activity, but it matches patterns of behavior associated with a given broad category of Advanced Persistent Threat.

3.1.2 Background and Significance

Research has identified limitations of traditional Intrusion Detection Systems (IDS) in the areas of human error, cost, and high error-rates due to large volumes of data being processed simultaneously. Traditional intrusion detection requires processing large quantities of audit data, making it both computationally expensive and error-prone [2, 5–7]. Limitation of traditional IDS (Intrusion Detection System) techniques are as much a function of the ability of a human to process large amounts of information simultaneously as they are limitations of the techniques themselves [4–7]. Machines currently have a limited ability to recognize unusual states or attack states. Humans are more effective at properly recognizing anomalies, yet have limited ability to consistently and effectively sift through large amounts of data [4]. These findings support the need for alternative automated approaches to pattern recognition for better analysis of real or perceived APT attack [2, 3, 5–9]. New approaches to pattern recognition that simplify the process for human beings, such as FAST-VM, are beneficial in better analysis of attack data.

Data visualization and visual algebra is a potentially useful method for dealing with the limitations of tabular log-based data, due to the amount of information it can organize and present [10–16]. Additional research stressed methods to define and mediate APTs, malware, and their reduction of false positive rates. Self-organizing taxonomies that can detect and counter attack malware operations were examined by Vert et al. [17], application of variable fuzzy sets to reduce false alarms by Shuo et al. [18], and the application of genetic algorithms by Hoque et al. [19]. Another major issue facing traditional IDS techniques is they are either signature based and fail to detect brand new attacks, or are anomaly based and are prone to false positive rates far more than what human operators can handle. Since IDS operators are already overwhelmed by the volume of data, an excess of false positives would be problematic.

Prior research by Vert et al. resulted in the development of a conceptual mathematical model based on vector math and analytic visual algebra for detecting intrusion attempts on computers based on the concept of identification of normal versus abnormal fluctuation patterns of known vectors. This approach was part of a larger project called "Spicule" [2, 20]. Spicule is a visualization technique that builds on this work and uses vectors to display system activity. A Spicule is an easy way for a human to visualize a large amount of data in a short span of time. This is a potentially powerful approach to intrusion detection because it uses vectors that offer themselves to various mathematical operations. Using mathematical properties allows the same data set to take on different semantics. Furthermore, detection and classification become automated through the inherent algebra of Spicule [20]. Although not the primary focus of this work, it was anticipated that the visualization capabilities greatly assist the developmental efforts and may provide a future area for feature expansion. The focus of this work is on the utilization of extremely mathematically efficient algorithms as a detection mechanism.

Erbacher et al. built upon existing work by Vert et al. [20] and the Spicule concept to develop a visualization system that monitored a large-scale computer network [14]. Their system monitors computer activity using several glyphs in conjunction with time. Chandran, Hrudya, and Poornachandran referenced the FAST-VM model in their 2015 research on efficient classification roles for APT detection [21]. This research is an example of the broader applicability of this area of research.

3.1.3 Problems and Opportunities

The area of non-signature-based intrusion detection is a challenging, yet rich area for investigation. While APTs have greater access to novel and unique attacks, they do not have exclusive access. Rather, zero day exploits are a constant factor in computer security. As a result, all security companies are constantly scrambling to provide signatures to emergent attacks to protect their customers.

Any techniques capable of addressing APTs can be retooled to address general computer security. These techniques are practically guaranteed to be novel; as traditional signature-based detection mechanisms are not reactive enough to catch the novelty of APTs. To put it simply, research into APT detection and prevention systems has the potential to be game changing as evidenced by the existing research areas [1, 2, 4–6, 8, 14, 20].

This does not overshadow the challenge of addressing APTs. It is not sufficient to look for a signature of a specific piece of code, or exploit alone [1, 2]. A true understanding of the state of the machines is needed. If an attacker influences a machine, the attacker may change its state in subtle ways that are not easily predictable. It might be possible to classify attacks or types of state changes, however. This is hindered further by the sheer dimensionality of the data when the state of a machine is considered. One cannot simply rely on a human log-file type analysis to infer these state relationships; better detection mechanisms are necessary [2, 3, 5–9].

Another potential approach is that of simply trying to check for anomalous states—states that the machine does not normally enter. Unfortunately, the set of attacks seem to be a miniscule subset of the set of all anomalous states if current research is any indication. This leads to false positive rates that are simply intractable. Therefore, while it is important to try to predict expected "normal" states, it alone is insufficient.

Truthfully, it is likely any successful approach must take some combination of approaches, both predicting expected machine states and attempting to understand the interrelation between state change and attack [22–24]. This process is used to crosscheck anomalous states with states that indicate an attack. While humans alone cannot be relied upon to make this determination, there is certainly room to leverage human intuition in this process. Potential approaches vary, but include both computational techniques and visualization techniques. The implications of even partial functional success (as opposed to purely theoretical success) are many fold, and potentially ripple through many domains.

3.2 Concept

Our concept is to use vectors to model state changes affected by APTs. These vectors mitigate problems with the application of standard statistical methods for APT detection stemming from variation between hosts. Additionally, the algebra of the vectors, once decomposed to its base elements, is intuitive and easy to analyze for the security analyst or researcher [24]. This is of great importance currently in the developmental and research phases. Later this capability could extend into a general visualization to aid detection at a production level. There has been exploration conceptually and experimentally of the vector math and visualization. However, the domain of APTs is a new application area. Previous applications were extremely promising, but focused on simpler exploits.

There is no known upper bound on the number of state variable vectors that may be required for accurate detection. This is due to the subtle and sophisticated nature of APTs. The underlying algebra demonstrates experimentally to easily and efficiently handle thousands of state variable vectors without an appreciable degradation in execution time in large part due to being based upon integer calculations. This allows for the reduction of high order state variable vector spaces to only the "needles in the haystack," which are indicative of APT presence. Anything less efficient would be unacceptably computationally expensive for practical detection of threats.

We utilize expected state prediction to reduce false positives rates (FPRs). We refer to this capability as Jitter, which is adaptive over time. It uses sweep region statistical analysis and adaptive adjustment. It can also utilize Bayesian probabilities based on previous locations for a state variable vector. This allows the change from algorithm to remove vectors that are jittering and known versus truly new APT effects on vectors. Jitter is subject to future developments.

FAST-VM unifies the three major areas of IDS (anomaly, misuse, and specification) into a single model. FAST-VM is a signature of Spicules, the experimental visual form of the vector math, as they transition to new states, which can be measured by their velocity. Velocity is the rate of change in thousands of vectors over time to a new Spicule state. FAST-VM signatures or branches can generate in any direction in 3D space. When an APT starts to generate a FAST-VM signature branch, it can be compared to existing branches. This determines if it is similar to previous patterns of misuse. This comparison allows the system to mitigate APTs during instantiation, prior to activation.

3.2.1   Technical Objectives

Our goal was to create a preliminary prototype that models high order data spaces of state variables that could be affected by APT and reduce them to the essence of the operations of that APT over time. This is the Spicule vector mathematics model that integrates into the Finite Angular State Velocity Transition Machine (FAST-VM). The FAST-VM performs the key functions found in Table 3.1.

A real-time Spicule has been implemented as a prototype. The algebra has solid mathematical underpinnings as well as circumvents problems with application of statistical analysis methods. Further discussion of this appears later in this chapter.

The goals of this technology are:

1. Identify categories of APT attack and activities and collect or develop software tools to simulate APT activity in a networked environment.
2. Using current research, identify a large collection of state variables describing host and network operation. Implement and test FAST-VM against several categories of APT. Evaluate and refine its model including identification of state variables that are most sensitive and diagnostic of various categories of APT threat.
3. Implement and evaluate the adaptive Jitter methods previously discussed to measure and fine tune the FRR and FPR rates.

Table 3.1 Fast-VM capabilities

	Capability
1	*Find* "the needles in the haystack" representing the APT effects on the system
2	*Detect* a previously unknown APT
3	*Classify* state activity changes of unknown APTs as similar to known APTs
4	*Predict* what this category of APT attack on a system might look like so that it can be monitored for presence

3.3 Implementation

3.3.1 Overview

The broad strokes of the concept were summarized in the preceding concept section. The implementation for FAST-VM revisits each of those topics in detail before laying out a plan and series of tasks for moving forward. Integer-based vector mathematics is central to this concept and the approach and its justifications are discussed in detail. First vector mathematics is discussed, then the mechanisms used to represent threat on a system. This leads to a consideration of time effects, and how the complexity of APTs is captured through time. Finally, the mechanisms for counteracting the problems of FPRs and FRRs, which tend to plague non-invariant-based approaches such as anomaly detection, are defined.

Integer-based vector mathematics can be utilized to model state variables in a system. State variables are hardware and software attributes that change based on the systems operation, for instance, CPU usage. APTs require a high degree of stealth over a prolonged duration of operation to be successful.

Threats on a system or network can be detected as changes in state variables if the correct variables are modeled. Due to the persistence nature of an APT, vector modeling also needs to follow state variable changes in a temporal range. Additionally, APT attacks are generally mounted by sophisticated actors. This necessitates the need to model a high order data set of state variables and their changes over time. The FAST-VM has the capability to address these challenges. It can reduce the high order of state variable changes that have subtle changes in them over time to an easy to comprehend threat analysis.

FAST-VM is also integer based. Integer mathematics is one of the fastest ALU operations on most computers. In some past tests utilizing FAST-VM concepts to generate authentication signatures and comparing that to cryptographic methods, FAST-VM generated vector signatures approximately fifty times faster [2]. Basing FAST-VM model on this approach, it is possible that this technology could run in real time or near real time to detect APT presence.

Finally, it is important to address false positives and false negatives in any APT system. State variables are going to have normal ranges of operation that will change over time. A mechanism called adaptive jitter addresses this. The model allows a system to dynamically adapt its jitter function for the location of state variable vectors such that the false positive and false negative rates can automatically be reduced and perhaps eliminated. The user of FAST-VM can change jitter values as they see fit so they get the desired false positive rate.

3.3.2 Vector Mathematics Versus Other Methods

A few types of mathematical approaches could be utilized to potentially detect APT presence. Most likely would be statistical methods, producing state variable averages and standard deviations. There are a few reasons why this approach is not taken in the FAST-VM model.

The first of these is that APT attacks are a systemic and complicated attack. Statistical methods for state variables on a single host are fine for detection. A problem arises, however, when comparing, interpreting, and analyzing state variable statistics across multiple platforms. This is far too computationally intensive. Additionally, standard deviations, which might indicate APT presence, have different values among different hosts that APT may be in operation on. This leads to the question of how to compare systems and conduct a meaningful analysis among all systems. The binary properties of the vector approach are meant to display no vectors or few vectors if a system does not have APT present and large numbers of vectors if it is present. As noted below, this makes interpretation by an analyst much easier.

APT is also detected by monitoring many state variables and analyzing them simultaneously into a single aggregated picture for interpretation. The FAST-VM method has the capability to do this rapidly for any number of state variables at the same time. This could range from 10 to 10000 variables all representing some aspect of a host's operation in a network under APT attack. The analysis is done in a human intuitive fashion which makes training people to use the system easy. Similar sorts of capability using statistical methods get expensive computationally and are complicated to interpret.

There are other arguments that can be made for a vector-based approach, but in the final evaluation, a human analyst is required to interpret the data to determine the presence of an APT in a specified system of computers. Humans are highly skilled at fuzzy thinking. The vector method allows for an extremely intuitive method for analysis of the APT data from a system of hosts and the interpretation of such data thus aiding the human analyst. Subsequent sections further discuss this approach.

3.3.3   Vector Mathematics Background

Vectors have a variety of expressions usually denoted by a lower-case letter. They have magnitudes and point to locations in space indicating a precise value for a state variable or they point to a fixed location and grow in magnitude to indicate changes in state variable values. The FAST-VM uses a combination of these types of variables.

The vector-based approach of the FAST-VM model is simple and lends itself to an algebra that can detect state variable changes OR predict what a state variable will look like if an APT has affected its value. This allows the vectors to:

1. Detect change in a state variable if an APT has started operation on a system.
2. Predict what a state variable vector would look like if an APT is present and affects its value.

Looking at (1) first, given state variable vector v whose value at time t_0 is collected and w for the same state variable collected at time t_{0+1}.

The operation of subtraction has the following result if the values of the vectors have not changed

$$w-v = y$$

where v is the *Normal Form vector,* and w is the *Change Form vector*, what has changed since v was sampled.

If $y = 0$, no change to the state variable has been detected, suggesting no APT presence. This is referred to as a *Zero Form.*

If $y \neq 0$, the effect of APT presence on the state is indicated and is referred to as an *Observe Form.* If this specific state change has been detected previously and associated with a known attack, the change is referred to as an *Attack Form.*

If $w \neq v$, then $y \neq 0$, indicating the effect of APT on a given state variable. The *jitter* part of the model does address the notion of being able to say w and v are slightly different but essentially the same over time to reduce false readings.

Considering prediction of the effects of an unknown APT on what a state variable's value might be given a similar category of APT that has previously been detected, its algebra is defined as follows:

v - *Normal form*, state variable without APT present

z - Previously detected class of APT effects on the state variable, referred to as the *Attack Form*

p - *Predicted form* of an APT effects on a state variable previously detected, referred to as a predict form

o - Unknown APT affecting a given state variable.

$$v + z = p$$
Equation for predict form

If an unknown APT is similar to a previously seen APT, then:

$$o-p = q$$

q - If zero indicates the presence of APT, referred to as a *Zero Form.* If not equal to zero, indicates that the APT is new and previously unknown but has now been detected.

The usefulness of prediction is in application of previously developed mitigation methods rapidly versus having to develop new mitigation methods for a previously unknown APT. The FAST-VM model also allows rapid classification of an attack using algebra and logic like the above predict form calculation.

3.3.4 Previous Work and Example Approach

The vector mathematics and algebra previously presented can be extended to model a state variable environment consisting of thousands of variables that could be utilized to detect the subtle changes APT might have on a system of computers. Because of the binary property of differencing (if a future vector is the same as a past vector the resultant is zero) and the application of jitter control previously discussed, a model containing thousands of state variable vectors from the past and future can be differenced to reduce high order data to the essence of exactly what has changed. This can cull out the essence of APT effects on state variable and thus can be analyzed to determine for potential presence of APT as presented in the following sections about Spicule and FAST-VM.

Analytic visual mathematics can be used to redefine mathematics spatially [2]. This type of visual rendering is not diagrams or pictures, it has an algebra that can be utilized to analyze data. This is the concept behind the development of a 3D data representation of high order state variable vector data Spicule. Spicule does this by modeling variables describing a system's operation. It is possible to analyze up to tens of thousands of individual state variables and their change to determine APT presence. This is done by population of state variable vectors around the radius of a Spicule in as small a degree increment as required. Analysis for change, and thus APT presence, is almost instantaneous using the fastest computational operation on a computer, integer addition and subtraction of vectors data. Spicule's mathematical model and underpinning is based on a vector calculus. Its algebraic visual model can do the following:

1. *Detect changes* to a system instantly by only visualizing what has changed in the system (this form of Spicule is referred to as the *Change Form*). This facilitates human interpretation of the significance of the change and its potential threat. It also lends to automatic response and classification of malware activity.
2. *Predict* what a system will look like under attack (referred to as the *Predict Form*).
3. *Identify the essence* of how an attack changes a system (referred to as the *Attack Form*).
4. Determine if the states of a system have changed or not changed (referred to as the *Zero Form* or the *Ball Form*).

The Spicule interface is simple and intuitive for humans to interpret and requires very little training. It lends nicely to interpretation of events in a system facilitating human/fuzzy ways of reasoning and interpreting a possible APT attack such as "most likely APT," "no APT," "sort of similar to previous APT," or types of change analysis. The *Change Form* finds the "needles in the haystack" of a high order state variable data space and presents that alone for analysis of APT presence.

3.3.5 Visualization Work: Spicule

While the goal of this effort is not to produce a prototypical visualization system for APTs, it is self-evident from previous work that the visualization is a useful tool in the developmental and research phases. In short, it is a feature that can be later developed to enhance any resultant product once said product has been proven.

The Spicule is visualized as a sphere with two types of state variable vectors (Fig. 3.1). There can be an infinite number of these vectors representing thousands of state variable for a given host, or network of hosts. The two types of vectors are defined as:

1. Fixed vectors (green) that represent state variables ranging from 0 to infinity; for example, the number of users that are logged into the system.
2. Tracking vectors (blue) that range in value from 0 to 100% and track scalar state variables; for example, CPU usage.

Each vector is located at a degree location around the equator of the Spicule ball. Each vector represents a state variable that is being monitored for change. In a simple case, with tracking vectors ranging from 0 to 90° located 360° around the equator, and the tip of each tracking vector indicating a state the system is in, it is possible to model 32,400 (90 × 360) unique states at any given moment in time. This makes it possible to instantly analyze change between Spicules from two moments in time to see if malware is active (using the *Zero Form*). Subdivision of degree locations for the vectors around the equator leads mathematically to an almost infinite number of states that could be modeled. This is represented graphically in Figs. 3.1, 3.2, 3.3, and 3.4 below.

A *Zero Form* (Fig. 3.5), shown below as a round featureless ball, results when a Spicule at time T_1 is subtracted from a Spicule at time T_0 and no change has occurred in state variables being modeled by the tracking and fixed vectors. A *Zero Form* indicates that no malware is in operation.

Fig. 3.1 Equatorial view of Spicule, showing state variable vectors tracking normal or malware operation

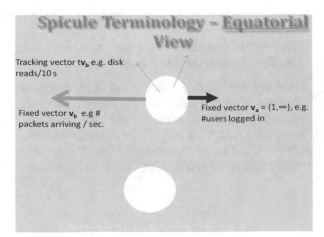

Fig. 3.2 Spicule showing port activity on a system (*Normal Form*)

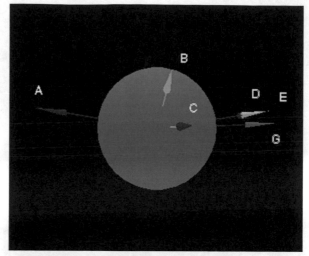

Fig. 3.3 Spicule showing a system under a SubSeven attack

Normal Form *Change Form* *Attack Form*

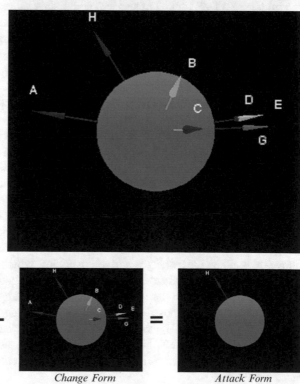

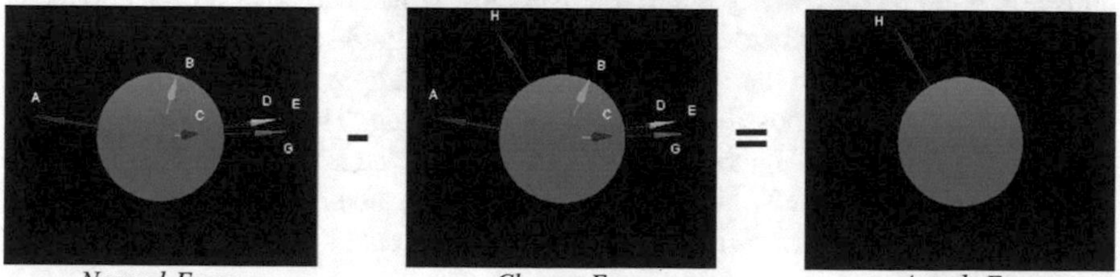

Fig. 3.4 The mathematics of calculating the *Attack Form*

Attack Form   *Observe Form*   *Zero Form*

Fig. 3.5 Algebra for the identification of an attack

The Spicule approach is to display system activity or state variables in the form of vectors that project from the center of a sphere as in Fig. 3.2. These vectors move or track as changes occur over time in a system. For example, a vector may represent CPU usage, which can range from 0 to 100%. A CPU usage vector would normally start out at the equator to denote low CPU usage; but if the system found itself in the middle of a DoS (denial of service) attack, that same vector would be translated to pointing out of the northern pole to denote high CPU usage (near 100%). Vectors can be mapped to represent any number of system state variables or log data that might be useful in detecting an attack.

3.3.5.1   Previous Work on Spicule Visualization Prototype

For the initial development of a working Spicule prototype, we chose to test the concept by monitoring ports. While this prototypical test is not directly targeted at the realm of APTs in specific, it does serve to illustrate the early concept and so is included. In testing, as any given port becomes opened, Spicule shows this by rendering a vector at the equator. As throughput increases on this port, the vector moves vertically up the sphere towards the northern pole. Figure 3.2 shows the purple-tipped vector pointing to the left; this vector is moving towards the northern pole as the activity on the SSH port (port 22) increases. Since it is just slightly above the equator, the activity is still relatively low which can be interpreted to mean that activity is characteristic of a system not under attack. In contrast, if the same vector were standing on the northern pole and pointing up, activity would be near maximum indicating a possibly dangerous system state. The Spicule in Fig. 3.2 is monitoring ports 22 (SSH, labeled A), 23 (Telnet, labeled B), 80 (HTTP, labeled C), 110 (POP3, labeled D), 137 NetBIOS (Name Service, labeled E), and 443 (HTTPS, labeled G). As the system's state changes, so will Spicule's. This generates a set of state variables.

To test this, a prototype used Backdoor SubSeven to simulate attack activity on specific ports. Backdoor SubSeven is a well-known Trojan. SubSeven works by opening an arbitrary port specified by the attacker. Most commonly, attacks happen to ports 1243, 6776, and 27374. Figure 3.3 shows the same system as before except that it is now under attack from SubSeven. The difference between these two

Spicules is this new purple-tipped vector (labeled H) which has appeared suddenly with a great deal of traffic on an otherwise reserved port (1243).

In Figure 3.4 above, a *Normal Form*, *Change Form*, and *Attack Form* are illustrated. Finally, the mathematics of calculating the *Attack Form* and the relative reduction of data and interpretation of change is discussed below.

3.3.5.2 Mathematical Properties and Visual Algebra

The Spicule model is comprised of six unique states: *Normal Form*, *Zero Form*, *Change Form*, *Attack Form*, *Observe Form*, and *Predict Form*. These forms are generated utilizing the previously discussed section on vector mathematics. The *Normal Form* is the state in which the system is operating normally and not under attack. Opposite to this is the *Change Form*, which is a representation of a system under attack. The *Attack Form* is a signature (or isolated) view of an attack in progress that is occurring inside the *Change Form*. *Attack Forms* could be stored in a database for later reference, in which case they become *Predict Forms*, which are predictions of future attacks. The *Observe Form* is a state in which may or may not be an attack signature. Through mathematical operations, an *Observe Form* can be compared to a *Predict Form*. Each one of these forms has a unique visual appearance and mathematical signature. The algebra for each of these forms in Fig. 3.4 is listed in Table 3.2 and discussed in more detail below.

Most operations to produce the above forms are accomplished by adding two forms (their state variable vectors) together or subtracting one from another. The algebra is performed by iterating through the vectors of each Spicule and performing individual vector operations depending on the algebraic function being calculated. For example, to isolate an attack and produce an *Attack Form*, simply subtract the *Normal Form* from the *Change Form* in Formula (1) above where S is the Spicule. The algorithm for this above process is listed below. Note that FOR EACH Vector (i) on the Spicule:

$$V_{Attack\ Form(i)} = V_{Normal\ Form(i)} - V_{Change\ Form(i)}$$

The visual representation of this algebra is presented in Fig. 3.4 above. Here, one can see the essence of the attack's visual characteristics in the *Attack Form*. This is vector H. Such forms can be potentially stored into a database as the *Attack Form*

Table 3.2 Mathematical operations per Spicule model

Formula	Spicule model	Mathematical operation
(1)	Attack Form	$S_{Attack\ Form} = S_{Normal\ Form} - S_{Change\ Form}$
(2)	Observe Form	$S_{Observe\ Form} = S_{Normal\ Form} - S_{Change\ Form}$
(3)	Zero Form	$S_{Zero\ Form} = S_{Attack\ Form} - S_{Observe\ Form}$
(4)	Predict Form	$S_{Predict\ Form} = S_{Normal\ Form} + S_{Attack\ Form}$

of SubSeven or the family of malware that operates similar to SubSeven. Once they become stored and classified, they become our *Predict Form*.

An *Attack Form* is created from pre-classification of attack families for the major families of malware. They can be stored and used for identification. They would be one phase of this research. They are subtracted with a *Change Form* to classify an attack and thus respond if a *Zero Form* results from the algebra. The *Attack Form* of Spicule is a classification of a type or family of attacks based on how they change the system over time. This may also be stored in a database library of attacks for future use.

An *Observe Form* may or may not be an *Attack Form*. It is generated by subtracting Spicules at different points in time to see if any change vectors appear. It can then be subtracted with an *Attack Form* stored in a database to classify the family of attack that is occurring on the system. It is created from pre-classification of attack families for the major families of malware. These are stored and used for identification. They are subtracted with a *Change Form* to classify an attack and thus respond if a *Zero Form* results from the algebra.

A *Change Form* is always Spicule at time T_1 that a *Normal Form* (at time T_0) is subtracted from to calculate the *Observe Form*. One can detect an attack by using Formula (2) where S is the Spicule.

The major difference between these two formulas is that the latter (2) is used to create an *Observe Form*, which is a *possible Attack Form*, whereas the former (1) is used when creating an *Attack Form* only. The reasoning behind this is that Formula (1) will be used to create a library of all attacks ever witnessed, and the result of (2) will be used to detect an attack underway against attacks stored in our library. Figure 3.6 shows the actual Spicules applied to Formula (2).

The *Observe Form* is potentially what an attack would look like while underway. It is compared against the *Attack Form*s to identify an attack. The method of performing this comparison is an algebraic subtraction as shown in Formula (3).

Here, S is the Spicule. Figure 3.6 below shows the actual Spicules applied to Formula (3). Note that this can easily be automated.

A *Predict Form* is meant to determine what a system might look like if a given attack from a family of malware is present on the system. It is one method of how to watch for such an event if it occurs. The *Predict Form* is created by the additive property of the algebra. It is calculated using Formula (4): *Predict Form = Normal Form + Attack Form*.

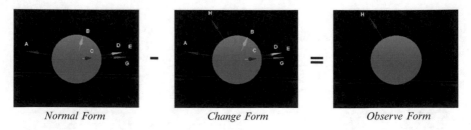

Normal Form   Change Form   Observe Form

Fig. 3.6 *Observe Form* algebra

This produces what we expect the attack to look like if it occurs. The subtraction operation then identifies and confirms that the malware exists via:

$$S_{Zero\ Form} = S_{Predict\ Form} - S_{Change\ Form}$$

If a *Zero Form* exists, then the attack has been identified, classified, and can be responded to. It is important to note that a Zero Form occurs with subtraction of one set of state variable vectors from another when they exactly match or jitter control has been applied.

In the above example, the mathematics of Spicule produces a featureless Spicule (*Zero Form*) if the *Observe Form* equals the *Predict Form*. This drastically simplifies and speeds the process of recognition. The potential gain in identification time has the possibility to extend Spicule methodology to real-time visual and/or automated detection. This illustrates the impact of the analytic visual algebra because a security officer can look for the *Zero Form*, which dramatically displays that the system might be under attack.

3.3.6   False Positive, False Negative Mitigation, and Jitter Control in FAST-VM Model

The goal of this part of the FAST-VM will be to minimize the false report rate of the individual vector activity in the model. Vector location during normal operation and over time for a state variable will fluctuate. The FAST-VM algebra discussed previously would detect this as potential APT presence. Because these fluctuations—referred to as jitter—change over time, an approach to jitter control needs to be adaptive by the system. This method must operate such that normal jitter is differentiated from abnormal (APT) jitter and not flagged as a threat. There are several methods that might be implemented for mitigation: (1) sweep region adaption and (2) Bayesian probabilistic methods.

Sweep region adaption argues that tracking vectors—the ones that range from 0 to 100%—will have a region that they characteristically like to settle into based on time of day. For instance, the "CPU usage" state variable for a given host may range from 40 to 60% over a 12-hour period. This is referred to as its characteristic sweep region. Additionally, statistical methods can be employed on the vector to determine where it characteristically tends to be found, for example, 51% with a standard deviation of ±2%. When conducting the FAST-VM Change Form analysis to detect state changes possibly due to APT, a vector for this state variable if falling within its typical sweep region would not be presented in the change form as an indicator of APT presence. The adaptive part to mitigate FPR and FRR is that if the sweep region is causing the vector to flag non-APT presence, or not to flag APT presence, the sweep region can be adjusted automatically as a variable in the algorithm generating the change form analysis.

Bayesian probabilistic methods can be utilized in a fashion akin to sweep region adaptation to predict and fine tune the probabilistic location of a state variable vector based on where it was identified in the past. This method can also be made adaptive, and further statistical analysis can be performed not unlike the sweep region method.

3.3.6.1 Finite Angular State Transition-Velocity Machine

The Finite Angular State Transition-Velocity Machine (FAST-VM) extends the Spicule concept into a state model. Unlike other state models, this state model also models the velocity of change in the system state over time to create very advanced capabilities to capture the complex state changes created as malware operates in a system. Current methods cannot capably model the high order of state complexity and change that FAST-VM handles very easily. Additionally, the FAST-VM is one of the first methods to integrate all three major methods of performing intrusion detection (anomaly detection, misuse detection, and specification detection) into a unified model. This unification develops powerful synergies for malware classification and identification that have not previously existed.

3.3.6.2 Fast-VM Operation

The FAST-VM consists of Spicules as they transition over time (the anomaly detection at a given moment in time for high order state variable vectors) combined into N-dimension state transitions. Each transition has a velocity. The velocity is the rate of change in Spicule *Change Forms over time* and an attribute of probabilistic confidence that denotes the transitioned to state as a recognized state (such as one would might fine when APT modifies the state of the system). In each state of the graph, the Spicule algebra can be applied for analysis. The model looks as shown in Fig. 3.7.

In this example (Fig. 3.7), a variety of characteristics is evident:

1. Spicules representing state changes at various points in time,
2. A velocity equation $|h|$ (magnitude of the transition) that describes the transition speed from T_0 to T_1,
3. A cumulative *Attack Form* describing the attack signature for APT summed over time at T_3,
4. A Bayesian probability P based on confidence that the attack signatures are known to be part of the transition attack profile for malware, where M_x is malware

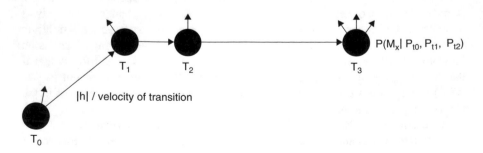

Fig. 3.7 A sample FAST-VM

x and P_n is the probability of having a known attack Spicule form at T_n, P is thought to deal with the issue of jitter, in that for a given malware family, Spicule *Attack Forms* at any given point in time should be similar but may not be identical.

The above diagram shows a FAST-VM creating a signature trail as APT is analyzed at each step in the process. This trail considers the rate of change to a new state over time and the high order of state variables that can evaluated as previously discussed. These attributes give FAST-VM a capacity to model and analyze very large amounts of data as it changes over time for the detection of the subtle changes in a system of hosts such as would be found with APT.

The above diagram also models a single category of APT as it is being analyzed. This becomes a known signature trail for this category. Notice the diagram is moving left to right in 2D space. FAST-VM is not limited to 20 signature trails. They could be created branching anywhere into 3D space creating an almost infinite modeling capacity. Unknown APT threat could be compared against signature branches shown above to rapidly classify the unknown APT into a category of previously seen APT without having to run the development of a full signature trail or branch. This offers the potential to stop an APT attack while it is instantiating and before it has taken hold.

The FAST-VM can do analysis at any given point in time using the Spicule mathematics and analyze Spicule *Attack Forms* in a time series sequence. Finally, it has a proposed method for dealing with jitter and classifications into families of malware using Bayesian probability, confidence, and velocity of transitions.

3.4 Application to Networks

FAST-VM is a powerful concept that can also be applied to entire networks of computers, not just a single host. There are two main strategies for applying FAST-VM to a network:

1. Consider each host's Spicule to be a vector on a larger "network spicule." This is useful for detecting attacks that affect many machines on a network at once.
2. Make a web of Spicules and analyze them all simultaneously. A standard network diagram can be adapted by replacing each system with a Spicule.
3. Deploy FAST-VM only on outward-facing gateways. This saves computational power because FAST-VM is not running on each host. It also reduces the workload for the admin because there are fewer Spicules to inspect. It would be effective at blocking external threats; however, it is not effective at detecting threats originating within the network.

Ultimately, it is up to the end user to decide where to deploy FAST-VM within their network, and there is no single correct way to do so. It will vary depending on the specific network. The basic FAST-VM algorithm stays the same when applied to networks. The only major difference between monitoring a single host and monitoring the network is the state variables and the methods of collecting them.

3.4.1 State Variables

FAST-VM relies on a list of state variables to create its Spicules and display information to the user. For FAST-VM to be useful, these state variables need to be well thought-out, specific attributes of a system that can be measured and analyzed. The following tables provide examples of the application of FAST-VM procedures in various industries and real-world scenarios.

In a practice application, variables are tailored based on specific function. Various systems can use the FAST-VM concepts including cars, medical devices, unmanned aircraft, and of course, personal computers and networks. In Table 3.3, the variables of intrusions detectable by FAST-VM are listed by device type. Table 3.4 lists FAST-VM intrusion detection for Automobile and Truck CAN ID systems and Table 3.5 defines intrusion detection capabilities for Network Protocols. Table 3.6 shows FAST-VM application in the health and medical industry listing Internet-Connected Medical Devices and types of intrusions detected. Table 3.7 defines

Table 3.3 Variables for individual computers or computer networks

Variable	Type of intrusion detected
Login frequency by day and time	Intruders may be likely to log in during off-hours
Frequency of login at different locations	Intruders may log in from a location that a specified user rarely or never uses
Time since last login	Break-in on "dead" account
Elapsed time per session	Significant deviations might indicate masquerader
Quantity of remote output	Excessive amounts of data transmitted to remote locations could signify leakage of sensitive data
Session resource utilization	Unusual processor or I/O levels could signal an intruder
Login failures	Attempted break-in by password guessing
Execution frequency	May detect intruders who are likely to use different commands, or a successful penetration by a legitimate user who has gained access to more privileged commands
Program resource utilization	An abnormal value might suggest injection of a virus or Trojan horse, which performs side-effects that increase I/O or processor utilization by a defined program.
Execution denials	May detect penetration attempt by individual user who seeks higher privileges
Read, write, create, delete frequency	Abnormalities for read and write access for individual users may signify masquerading or browsing
Records read, written	Abnormality could signify an attempt to obtain sensitive data by inference and aggregation
Failure count for read, write, create, delete	May detect users who persistently attempt to access sensitive data

Table 3.3 (continued)

CPU usage	DoS attack, malware activity
Open ports	Determine if a port being open is unusual
Metadata modification	Indicates an attacker is present on the system and could be injecting malware or doing other harmful work to the system.
Exhaustion of storage space	Denial of service attack; Malware might be present
Failure to receive SYN-ACK	The client's machine is sending the SYN packet to establish the TCP connection and the web server receives it, but does not respond with the SYN/ACK packet. Can indicate a stealth scan
Half-open connections	Denial of service attack; can also indicate stealth scan

Table 3.4 Variables for automobile and truck CAN ID systems

Variable	Type of intrusion detected
Time interval between messages	Messages normally are generated at a specific interval. Any interval besides the set one is likely an attack
Volume of messages	Helps detect DoS attacks
Frequency of diagnostic messages	These are rare and generated by critical component failure. Frequent diagnostic message will rarely happen except as part of an attack
Car movement status (driving/idle)	This variable is combined with the one above indicate an attack, since diagnostic messages usually only appear while the car is idle

From "Intrusion detection system based on the analysis of time intervals of CAN messages for in-vehicle network" by Song et al. [25]

Table 3.5 Variables for network protocols

Variable	Type of intrusion detected
Number of illegal field values	Illegal values are sometimes user generated, but not in high quantities, assuming the user identifies and corrects their mistake
Number of illegal commands used	Illegal commands are sometimes user generated, but not in high quantities, assuming the user identifies and corrects their mistake
Field lengths	Helps detect buffer overflow vulnerabilities
Protocol or service not matching standard port/purpose	Occasionally a legitimate user will set up a service on a non-standard port, but it is far more likely that malware is attempting to use the port instead
Volume of data from destination to source	Useful in detecting DoS attacks
Network service used on destination	Some services will stand out as unusual

From "A hybrid approach for real-time network intrusion detection systems" by Lee et al. [26] and "intrusion detection tools and techniques: a survey" by Karthikeyan and Indra [27]

Table 3.6 Variables for internet-connected medical devices

Variable	Type of intrusion detected
Number of requests for patient controlled analgesic (PCA)	An acceptable range can be set for this value. Any deviation from this range indicates a problem
Defibrillator status (on/off)	Can be combined with other variables such as pulse rate or requests for PCA. (Note: Unconscious patients are unable to press the PCA button)
Pacemaker setting	Compared with pulse rate to determine if pacemaker is working properly
Pulse rate	Compared with pacemaker setting to determine if pacemaker is working properly
Blood pressure, oxygen saturation, respiration rate, and temperature	An acceptable range can be set for this value. Any deviation from this range indicates a problem
Standard deviation of vital signs sensors	Multiple sensors are often used to gather vitals. If one sensor is attacked to give a false reading, but not another, it will result in an increased standard deviation between the two

From "Behavior rule specification-based intrusion detection for safety critical medical cyber physical systems" by Mitchell and Ing-Ray [28]

Table 3.7 Variables for armed unmanned aircraft systems (UAS)

Variable	Type of intrusion detected
Weapons ready (true/false)	Combined with location and status to be useful. For example, weapons should not be ready while taxiing
Location (target, airbase, non-target)	Combined with weapon status and flight destination
Thrust level	Combined with status. Each status should have a range of acceptable thrust levels so that fuel is not wasted
Status (taxi, transit, loiter, attack)	Combined with thrust. Each status should have a range of acceptable thrust levels so that fuel is not wasted
Landing gear status (up, down, error)	Combined with status and location; gear should not be up while taxiing or down while loitering over a target, for example
Flight destination (whitelisted, not whitelisted)	If the destination is set to a non-whitelisted location, it could be an operator error. Alternatively, it could be a third party trying to capture the UAV
Communication destination (whitelisted, not whitelisted)	If the comm's destination is set to a non-whitelisted location, it could be an operator error. Or it could be a third party trying to intercept UAV communications
Standard deviation of redundant flight sensors	Multiple redundant sensors are often used to gather flight information (airspeed, altitude, etc.). If one sensor is attacked to give a false reading, but not another, it will result in an increased standard deviation between the two

From "Specification based intrusion detection for unmanned aircraft systems" by Mitchell and Ing-Ray [29]

Table 3.8 Variables for disk drives/storage devices

Variable	Type of intrusion detected
Modification of specific files	There are system executables, configuration files, log files, and system header files that shouldn't be modified, per the admin's definition
Modification of metadata, timestamps, or file permissions	Rarely done for legitimate purpose
Active disk time	Excessive active time could be a result of malicious activity
Numbers of hidden files or empty files	Rarely done for legitimate purpose, can be a sign of a race condition exploit in progress

From "Slick," by Bacs et al. [30]

the application of FAST-VM with Unmanned Aircraft Systems (UAFs) as initially discussed in the introduction to this chapter. Finally, in Table 3.8, the use of FAST-VM for intrusion detection is presented for Disk Drives and Storage Devices.

3.5 Conclusion

In conclusion, computer security and the detection APTs is vital to strong e-commerce, military defense, aerospace, healthcare, financial institutions, and manufacturing industries [2–5]. In prior research, limitations of traditional Intrusion Detection Systems (IDS) were identified in the areas of human error, cost, and high error-rates due to large volumes of data being processed [2–9]. Current automated systems are restricted in the ability to recognize unusual states or attack states anomalies thus requiring a human analyst [4]. Yet humans have limited ability to consistently and effectively sift through large amounts of data which are the proficiency of computerized automated systems [4]. As cyber-crimes against business and society increase, automated systems to supplement human analysis are required to ensure safe secure networks and technologies [2, 3]. These findings and the information provided in this chapter support the need for alternative automated approaches to pattern recognition such as the FAST-VM for better analysis of real or perceived APT attack and present new technology [2, 3, 5–9].

The Finite Angular State Velocity Machine (FAST-VM) models and analyzes large amounts of state information over a temporal space. Prior development of the technology revealed capabilities of the FAST-VM to analyze 10,000,000 state variable vectors in around 24 ms. This demonstrates the application of "big data" to the area of cyber security. FAST-VM also unifies the three major areas of IDS (anomaly, misuse, and specification) into a single model. The FAST-VM mathematical analysis engine has shown great computational possibilities in prediction, classification, and detection but it has never been instrumented to a system's state variables. In this chapter, the ability of the FAST-VM to map the state variables in a UAS system to detect APT as well as practical application in industry was examined.

References

1. Turner, J. (2016, September). *Seeing the unseen—Detecting the advanced persistent threat* [Webcast]. Dell SecureWorks Insights. Retrieved from https://www.secureworks.com/resources/wc-detecting-the-advanced-persistent-threat
2. Vert, G., Gonen, B., & Brown, J. (2014). A theoretical model for detection of advanced persistent threat in networks and systems using a finite angular state velocity machine (FAST-VM). *International Journal of Computer Science and Application, 3*(2), 63.
3. Dell SecureWorks. (2016, September). *Advanced persistent threats: Learn the ABCs of APTs – Part I*. Dell SecureWorks Insights. Retrieved from https://www.secureworks.com/blog/advanced-persistent-threats-apt-a
4. Daly, M. K. (2009, November). *Advanced persistent threat (or informational force operations)*. Usenix.
5. Ramsey, J. R. (2016). *Who advanced persistent threat actors are targeting* [Video]. Dell SecureWorks Insights. Retrieved from https://www.secureworks.com/resources/vd-who-apt-actors-are-targeting
6. Scarfone, K., & Mell, P. (2012). *Guide to intrusion detection and prevention systems (IDPS)* (pp. 800–894). Computer Security and Resource Center, National Institute of Standards and Technology.
7. Kareev, Y., Fiedler, K., & Avrahami, J. (2009). Base rates, contingencies, and prediction behavior. *Journal of Experimental Psychology: Learning, Memory, and Cognition, 35*(2), 371–380.
8. MacDonald, N. (2010, May). *The future of information security is context aware and adaptive*. Stamford, CT: Gartner Research.
9. Othman, Z. A., Baker, A. A., & Estubal, I. (2010, December). Improving signature detection classification model using features selection based on customized features. In *2010 10th international conference on intelligent systems design and applications (ISDA)*. doi: 10.1109/ISDA.2010.5687051
10. Eick, S., & Wills, G. (1993, October). Navigating large networks with hierarchies, In *Proceedings Visualization Conference '93* (pp. 204–210), San Jose, CA.
11. Han, G., & Kagawa, K. (2012). Towards a web-based program visualization system using Web3D. In *ITHET conference*.
12. Bricken, J., & Bricken, W. (1992, September). A boundary notation for visual mathematics. In *Proceedings of the 1992 IEEE workshop on Visual Languages* (pp. 267–269).
13. Damballa, Inc. (2010). *What's an advanced persistent threat?* [White Paper.] Damballa, Inc. Retrieved from https://www.damballa.com/downloads/r_pubs/advanced-persistent-threat.pdf
14. Erbacher, R., Walker, K., & Frincke, D. (2002, February). Intrusion and misuse detection in large-scale systems. In *IEEE computer graphics and applications*.
15. Vert, G., & Frincke, D. (1996). Towards a mathematical model for intrusions. In *NISS conference*.
16. Vert, G., Frincke, D. A., & McConnell, J. (1998). A visual mathematical model for intrusion detection. In *Proceedings of the 21st NISSC conference*, Crystal City, VA.
17. Vert, G., Chennamaneni, A., & Iyengar, S. S. (2012, July). A theoretical model for probability based detection and mitigation of malware using self organizing taxonomies, In *SAM 2012*, Las Vegas, NV.

18. Shuo, L., Zhao, J., & Wang, X. (2011, May). An adaptive invasion detection based on the variable fuzzy set. In *2011 international conference on network computing and information security (NCIS)*.
19. Hoque, M. S., Mukit, A., & Bikas, A. N. (2012). An implementation of intrusion detection system using genetic algorithm. *International Journal of Network Security & ITS Applications (IJNSA), 4*(2), 109–120.
20. Vert, G., Gourd, J., & Iyengar, S. S. (2010, November). Application of context to fast contextually based spatial authentication utilizing the spicule and spatial autocorrelation. In: *Air force global strike symposium cyber research workshop*, Shreveport, LA.
21. Chandran, S., Hrudya, P., & Poornachandran, P. (2015). An efficient classification model for detecting advanced persistent threat. In *2015 international conference on advances in computing, communications and informatics (ICACCI)* (p. 2003). doi:10.1109/ICACCI.2015.7275911
22. Vert, G., & Triantaphyllou, E. (2009, July). Security level determination using branes for contextual based global processing: An architecture, In *SAM'09 The 2009 international conference on security and management*, Las Vegas, NV.
23. Vert, G., Harris, F., & Nasser, S. (2007). Modeling state changes in computer systems for security. *International Journal of Computer Science and Network Security, 7*(1), 267–274.
24. Vert, G., Harris, F., & Nasser, S. (2007). Spatial data authentication using mathematical visualization. *International Journal of Computer Science and Network Security, 7*(1), 267.
25. Song, H. M., Kim, H. R., & Kim, H. K. (2016). Intrusion detection system based on the analysis of time intervals of CAN messages for in-vehicle network. In *2016 international conference on information networking (ICOIN)*.
26. Lee, S. M., Kim, D. S., & Park, J. S. (2007). A hybrid approach for real-time network intrusion detection systems. In 2007 *international conference on computational intelligence and security (CIS 2007)*.
27. Karthikeyan, K., & Indra, A. (2010). Intrusion detection tools and techniques—A survey. *International Journal of Computer Theory and Engineering, 2*(6), 901–906.
28. Mitchell, R., & Ing-Ray, C. (2015). Behavior rule specification-based intrusion detection for safety critical medical cyber physical systems. *IEEE Transactions on Dependable and Secure Computing, 12*, 1.
29. Mitchell, R., & Ing-Ray, C. (2012). Specification based intrusion detection for unmanned aircraft systems. In *Proceedings of the first ACM MobiHoc workshop on airborne networks and communications—Airborne '12*.
30. Bacs, A., Giuffrida, C., Grill, B., & Bos, H. (2016). Slick. In *Proceedings of the 31ˢᵗannual ACM symposium on applied computing – SAC '16. Computer Science and Network Security, 7*(1), 293–295. January 2007.

Gregory Vert is a US citizen, who specializes in advanced security research in the areas of authentication, malware detection, classification, and modeling of state changes caused by malware in a system. He is the inventor of the contextual security model and Spicule state change model for malware detection. He has extensive experience in industry as a software engineer and extensive security training from a variety of places such as Black Hat, DEFCON, SANS Hackers Exploits, and Wireless Security as well as having earned a CISSP security certification. He has held two security clearances, one during his military service and one while working for Boeing. He has taught soldiers from Fort Hood who attend Texas A&M and recently published a book defining the new field of Contextual Processing. As a part of his work he has developed a new model for security based on context referred to as Pretty Good Security that has the potential to be faster and more computationally efficient than existing methods. He is currently teaching cyber security at College of Security and Intelligence at Embry-Riddle Aeronautical University in Prescott, Arizona.

Ann Leslie Claesson-Vert is an Associate Clinical Professor in the School of Nursing, College of Health and Human Services at Northern Arizona University in Flagstaff, Arizona. Her expertise lies in clinical research application, systems analysis, and development of innovative application of technology with a competency-based approach to practice in various industries. She also functions as an Assistant Professor at the Department of Medicine & Health Sciences George Washington University, Systems Analyst for the Higher Learning Commission, and Grant Peer Reviewers for the US Department of Health & Human Services, and Council for International Exchange of Scholars (CIES) – Fulbright Scholars program.

Jesse Roberts is an undergraduate student at Embry-Riddle Aeronautical University in the Cyber Intelligence and Security program. He assists Dr. Vert in researching and collecting state variables and developing the FAST-VM concept.

Erica Bott is an undergraduate student at Embry-Riddle Aeronautical in the Cyber Intelligence and Security program. She assists Dr. Vert in developing the FAST-VM concept and making it understandable for real people.

Chapter 4
Information-Theoretically Secure Privacy Preserving Approaches for Collaborative Association Rule Mining

Nirali R. Nanavati and Devesh C. Jinwala

4.1 Introduction

The massive proliferation of digital data is one of the results of modernization. The main reason behind this digital information explosion is the rising capabilities of the digital devices and their plummeting prices [1]. The multitude of data may be stored in different databases distributed throughout. Hence, the vastness of data makes it non-trivial to infer vital knowledge from the same [2]. Without due inference that could be put to use, the data effectively remains useless. Data mining and the tools used with it help inferring appropriate knowledge of use from these data. However, for due inference from data belonging to different owners, it is necessary that their data be exposed to the mining tools. Such exposure, on the other hand, obviously sacrifices the privacy of the data, at least in privacy sensitive applications [3, 4].

A data security policy is the means to the desired end, which is data privacy [5]. Giving an example of a company it is possible that their system is secure, but does not respect your privacy because they might be selling your data. A recent example of infringement of privacy is the National Security Agency's (NSA) PRISM program. Edward Snowden revealed the existence of this clandestine program in 2013. It was revealed that the PRISM program was mining the data of as many as 100 such companies without the end users knowing that their data was actually leaked to NSA [6].

There are two alternatives that we are left with in such situations wherein collaborative data mining needs to be undertaken (Fig. 4.1)—the first one being to

N.R. Nanavati (✉)
Sarvajanik College of Engineering and Technology (SCET), Surat, India
e-mail: nirali1111@gmail.com

D.C. Jinwala
S. V. National Institute of Technology (SVNIT), Surat, India
e-mail: dcjinwala@acm.org

© Springer International Publishing AG 2018
K. Daimi (ed.), *Computer and Network Security Essentials*,
DOI 10.1007/978-3-319-58424-9_4

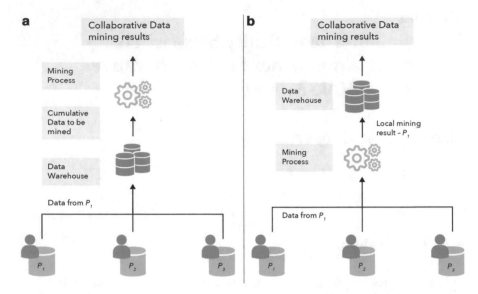

Fig. 4.1 Collaborative data mining methodologies. (**a**) Traditional collaborative data mining at data warehouse. (**b**) Distributed data mining by merging local mining results at the data warehouse

get all the data in one place like a data warehouse and then mine the entire data set (Fig. 4.1a). However, this solution is not preferred by many organizations mainly for loss of the privacy of the people involved with the organizations, which could comprise of customers, patients, employees, etc. Hence, the second approach of mining the data individually at each of the different sites and then merging the results is more likely to be used (Fig. 4.1b). This approach, known as Distributed Data Mining (DDM) [7] helps protect the privacy to a large extent among these competitors who wish to collaborate selectively.

However, these mining results, which are shared, could also comprise of sensitive information. Hence, Privacy Preservation techniques in Distributed Data Mining (PPDDM) are applied that help preserve the privacy while undertaking collaborative data mining.

PPDDM is a significant Secure Multiparty Computation (SMC) problem among other SMC problems like privacy preserving database query, privacy preserving intrusion detection and many others [8, 9]. SMC when applied to DDM helps in knowing how the competitors are performing without compromising on either party's privacy. The solution of SMC when applied to DDM is such that only the data mining results of each of the sites that satisfy a certain function f are known in the cumulative data. The confidential data (DB_1, $DB_2 \ldots DB_p$) of the collaborating parties ($P_1 \ldots P_p$) remains private as shown in Fig. 4.2.

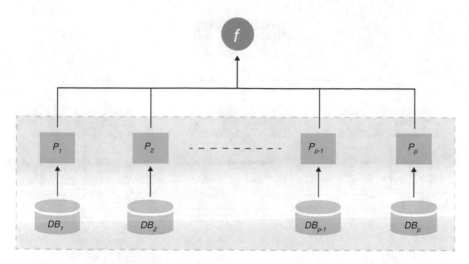

Fig. 4.2 Secure multi-party computation for distributed data mining

The primary motivation for SMC used for PPDDM is *benchmarking* [10] wherein multiple organizations undertake collaborative data mining to compare themselves with the best practice in their field of operation. This would in turn help them learn, plan and be a motivating factor for their businesses [10]. However, this has to be definitely done while preserving the privacy of the companies' private data.

In this chapter, we focus on the techniques proposed in literature that provide information-theoretic security (the highest level of security) to improve the state-of-the-art of the privacy preserving techniques for distributed data mining. In particular, we focus on Privacy Preservation in Distributed Frequent Itemset Mining (PPDFIM) or Privacy Preservation in Distributed Association Rule Mining (PPDARM) [11] which are case studies of the problem of PPDDM.

A number of approaches have been proposed in literature for the problem of PPDARM [11–22]. These algorithms are classified based on different issues that arise in a typical PPDARM scenario as shown in Fig. 4.3.

In Sect. 4.2, we first discuss the significance of information-theoretically secure schemes. Further, we discuss about schemes that provide information-theoretic security for each of the classification issues and the solutions therein for PPDARM. In Sect. 4.3, we discuss about the methodology of PPDARM and the schemes that can be applied to horizontally partitioned data and in Sect. 4.4, we discuss about the methodology and schemes applicable to vertically partitioned data for PPDARM.

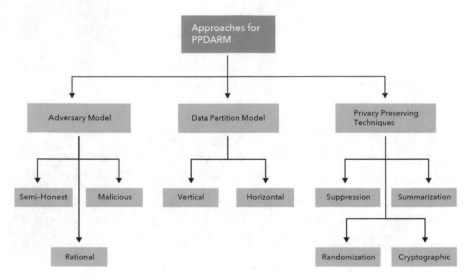

Fig. 4.3 Hierarchical division of approaches for PPDARM

4.2 Computational Security Versus Information-Theoretic Security

The cryptographic schemes for secure multi-party computation can be classified into two primary genres. These are computationally secure schemes and the ones that are information-theoretically secure.

Computationally secure schemes are based on computationally intractable problems and the assumption that the adversary is not omni-potent [23] (not all powerful in terms of computation power). Since these computationally secure public key schemes are based on computationally hard problems, they work with large numbers (in the order of 1000s of bits) [24]. Hence, they incur a high computation cost. The computationally secure schemes also need to be adapted continually so that they are not broken with advances in computer technology [25].

On the other hand, information—theoretically secure schemes are based on the concept of perfect security which is an ideal scenario. These schemes do not rely for their effectiveness on computationally hard problems [26]. Such schemes are not vulnerable to future developments in computation power. Shamir's secret sharing under the assumptions that the adversary is not omni-present (is unable to get hold of all the information) is an important example of an information—theoretically secure scheme [27].

Shamir's $[m, m]$ secret sharing scheme is additively homomorphic in nature, information-theoretic and has lower computational costs.

In Algorithm 1, we give the details of finding the sum of the secret information at each site using the Shamir's secret sharing technique [20, 27]. The notations are as follows: let P_i $(0 < i \leq p)$ represent the participating parties such that the minimum value of p is 3, V_i represent the secret values at P_i and $x_1, x_2 \ldots x_p$ are the set of publically available random values. The polynomial is chosen in the same field as the secret (generally 32 or 64 bits) [24].

Algorithm 1 Shamir's $[m, m]$ additive secret sharing algorithm [20, 27]

Require: The common random numbers $X = \{x_1 \cdots x_p\}$ are distinct publically available numbers in a finite field $\mathbb{F}$ of size P where P is a prime number and secrets $V_i < P$. $(0 < i \leq p)$. The coefficients $\{a_1 \cdots a_{p-1}\} < P$.

Ensure: *SecureSum* of the secret values V_i for p parties

1: **for** each party P_i, $(i = 1, 2, \cdots, p)$ **do**
2:     each party selects a random polynomial $q_i(x) = a_{p-1}x^{p-1} + \cdots + a_1 x^1 + V_i$
3:     compute the share of each party $P_y (y = 1, 2, \cdots, p)$, where share $(V_i, P_y) = q_i(x_y)$
4: **end for**
5: **for** each party P_i, $(i = 1, 2, \cdots, p)$ **do**
6:     **for** $y = 1$ to $p(i \neq y)$ **do**
7:         send share(V_i, P_y) to party P_y
8:         receive the shares share(V_i, P_y) from every party P_y.
9:     **end for**
10:    compute Sum$(x_i) = q_1(x_i) + q_2(x_i) + \cdots, +q_p(x_i)$
11: **end for**
12: **for** each party P_i, $(i = 1, 2, \cdots, p)$ **do**
13:    **for** $y = 1$ to $p(i \neq y)$ **do**
14:       send Sum(x_i) to party P_y
15:       receive the results Sum(x_i) from every party P_y
16:    **end for**
17:    solve the set of equations to find the sum of $\sum_{i=1}^{p} V_i$ secret values.
18: **end for**

4.3 PPDFIM Across Horizontally Partitioned Databases

In the seminal work [11], the authors explain that in horizontally partitioned databases, primarily two phases are required for PPDARM. The two phases are: discovering the candidate itemsets (those that are frequent at one or more sites) and finally finding the candidate itemsets that are globally frequent. The methodology for PPDFIM across horizontally partitioned data is shown in Fig. 4.4 and detailed in [11].

The first phase uses commutative encryption [11]. The sub-protocol used for this phase is the Secure Set Union. Each party encrypts its own frequent itemsets using its public key. These encrypted itemsets are passed to the other parties in a ring topology. Once all the itemsets of all the parties are encrypted, they are sent to a common semi-trusted party that eliminates the duplicates. Further, all the parties decrypt the data to finally get the candidate itemsets or itemsets that are frequent at one or more parties.

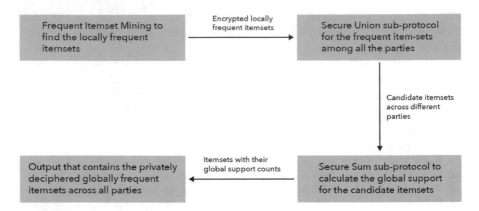

Fig. 4.4 Methodology of PPDFIM across horizontally partitioned data

In the second phase, each of the locally supported itemsets is tested to see if it is supported globally [11]. For example, the itemset {A,B,C} is found to be supported at one or more sites from Phase 1. Each party has computed their local support. Further, the secure sum sub-protocol is used to find the global support count of the candidate itemset {A,B,C}. Hence, it is possible to privately find if the itemset is frequent globally if the global support exceeds the global support threshold without knowing the private individual support counts at each of the parties. This is the actual goal of PPDFIM. The results of the same secure sum protocol are used to find the global confidence count and hence to find the globally frequent association rules. In this chapter we focus on the *secure sum* sub-protocol for the scenario with *homogeneous*(horizontally partitioned) databases.

In the horizontally partitioned data model, the proposed schemes must consider semi-honest, rational and malicious adversaries which exist in a realistic scenario. In the section below, we discuss the information-theoretically secure schemes for horizontally partitioned data for the three types of adversaries.

4.3.1 Information-Theoretically Secure Schemes for PPDFIM—Semi-Honest Model

Once the globally frequent itemsets are found using the *secure sum* sub-protocol for PPDFIM, the results of the *secure sum* sub-protocol are used to find the globally frequent association rules for the problem of PPDARM. Hence, we aim to find a privacy preserving approach that is suitable for a large number of parties and is information-theoretically secure in a PPDFIM setup.

The symmetric scheme proposed in [28] is similar to the one time pad. It is argued that for an equivalent level of security, asymmetric schemes are generally less efficient than symmetric ones. With proper key management, this scheme provides unconditional security and is highly efficient [29].

Hence, for the problem of undertaking secure sum in PPDFIM, the authors in [30] propose an efficient information-theoretically secure symmetric key based scheme based on [28] where the keys are generated using pseudo random functions in a semi-honest model. In [30], the authors further show a comparative analysis of this scheme with the:

* secure sum scheme [14] based on Paillier public key homomorphic scheme (provides computational security).
* information-theoretically secure Shamir's secret sharing scheme in the No Third Party (NoTP) model [20, 31].
* information-theoretically secure Shamir's secret sharing scheme in the Semi-honest Trusted Third Party (STTP) model [32].

The authors in [30] conclude that the Shamir's scheme is more efficient in terms of execution cost up to a certain number of parties after which the symmetric key-based scheme performs better. However, the symmetric key-based scheme still faces the issue of key management.

4.3.2 Game-Theoretic Privacy Preserving Schemes for PPDFIM: Rational and Malicious Model

Along with the malicious model, parties could also be rational in behaviour. In a co-opetitive setup, the rational participants will try to maximize their own benefit or utility and then prefer that the other agents have the least utility [33–35]. One of the goals of PPDDM is to ensure maximum participation from the contending participants. In order to do so, the scheme proposed must incorporate not only preventive, but also corrective measures. These measures aim at eliminating or correcting the negatively performing rational and malicious participants.

Incorporating these corrective and preventive measures in our scheme necessitates a game-theoretic approach. There have been approaches proposed for rational secret sharing which has rational participants using the game theory [33, 36–38]. Game theoretic concepts mainly aim at imposing punishments to attain the stable Nash equilibrium [12] state which is the optimum state for the setup wherein there is maximum possible participation of parties. However, none of these approaches discuss secret sharing among rational agents without mediators for secure sum in a repetitive PPDDM model.

Hence, for the rational party based model, the authors in [35] propose a game-theoretic secret sharing scheme that models Shamir's secret sharing in PPDDM as a repeated game without using mediators. $[m, m]$ secret sharing has been used in PPDDM by [20, 31, 39] to decipher the sum privately. The authors in [35] have further analysed this model by proposing three novel punishment policies for PPDFIM. They conclude that the rating-based punishment policy takes the least number of rounds to attain the stable Nash equilibrium state.

Further, the authors in [35] identify the problem that the schemes discussed until now would not be able to deal with a mixed model of rational, semi-honest and malicious parties in a game-theoretic setting that encourages maximum participation. Hence, they further propose a scheme [40] that works in such a mixed model.

The novel information-theoretic game-theoretic scheme that the authors in [35] propose and analyse for privacy preservation can be used or extended for different settings and privacy preserving techniques in a PPDDM model. These repetitive games would foster co-operation and corrective behaviour eventually.

4.4 PPDFIM Across Vertically Partitioned Databases: Semi-Honest Model

In [41], the authors first showed how secure association rule mining can be done for vertically partitioned data by extending the *Apriori* algorithm. Vertical partitioning implies that an itemset could be split between multiple sites. Most steps of the *Apriori* algorithm can be done locally at each of the sites. The crucial step involves finding the support count of an itemset as shown in Algorithm 2. If the support count of an itemset can be securely computed, one can check if the support is greater than the threshold, and decide whether the itemset is frequent. Using this, association rules can be easily mined securely. The methodology for PPDFIM is shown in Fig. 4.5.

There are three sub-protocols proposed in literature for PPDFIM across vertically partitioned data which are Secure Sum [20, 42], Set Intersection Cardinality [19] and Secure Binary Dot Product [17, 41] protocols. The *secure sum* protocol leads to leakage of private information and defeats the entire purpose of privacy preservation in a vertically partitioned PPDFIM scenario. It compromises the privacy of the parties leading to a protocol failure.

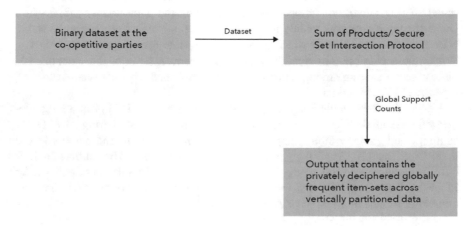

Fig. 4.5 Methodology of PPDFIM across vertically partitioned data

The key insight of [41] is that computing the support of an itemset is exactly the scalar product of the binary vectors representing the sub-itemsets with different parties. For the multi-party scenario, it is the sum of the component-wise multiplication of the vectors at each of the parties as shown in Step 11 of Algorithm 2. Thus, the entire secure association rule mining problem can be reduced to computing the sum of products of vectors in a privacy-preserving way. In [41], the authors also proposed an algebraic method to compute the scalar product for a two party scenario. It proposes the Set Intersection Cardinality for a multi party scenario of sparse datasets. For general datasets, Vaidya et al. [41] propose a public key scheme for the component-wise sum of product of vectors.

A strong point of the secure association rule mining protocol in a vertically partitioned setup is that it is not tied to any specific scalar product protocol. Indeed, there have been a number of secure scalar product protocols proposed [17, 41]. All of them have differing trade-offs of security, efficiency, and utility. Any of these could be used. In [14], the authors propose a secure protocol to compute the scalar product using homomorphic encryption.

Algorithm 2 Distributed frequent itemset mining algorithm across vertically partitioned data based on [41]

Require: p=number of parties, N= number of records
Ensure: Globally frequent itemsets
1: $L_1 = \{large1 - itemset\}$
2: **for** $k = 2; L_{k-1} \neq \phi; k + + $ **do**
3:    $C_k = apriori - gen(L_{k-1})$
4: **end for**
5: **for** all candidates $c \in C_k$ **do**
6:    **if** all the attributes in c are entirely in the same party **then**
7:        that party independently computes **c.count**
8:    **else**
9:        let P_i have l_i attributes
10:       construct vector $\mathbf{X_i}$ at P_i where $\mathbf{X_i} = \prod_{j=1}^{l_i} P_{ij}$
11:       collaboratively compute **c.count** $= \sum_{i=1}^{N} \mathbf{X_1} * \mathbf{X_2} \cdots \mathbf{X_p}$ (This is the sum of the component-wise product of p vectors which we refer to as the Sum-Product).
12:    **end if**
13:    $L_k = L_k \cup \{c | c.count \geq globalsup_{min}\}$
14: **end for**
15: return $L = U_k L_k$

In [17], the author mentions that multiple Secure Binary Dot Product (SBDP) protocols can be used for PPDFIM in a vertically partitioned setup. The work by Du-Atallah [43] has a two vector dot product protocol. In [44], the authors extend this efficient scheme to a cloud setup across three miners for two vectors using a share multiplication protocol. However, a PPDFIM setup will require multi-vector 'sum of products' protocol to find the sum of products of items that are split across different parties (Fig. 4.6). In [41], the authors propose a multi-vector 'Sum of Products' protocol based on the computationally expensive public key scheme. Hence, for

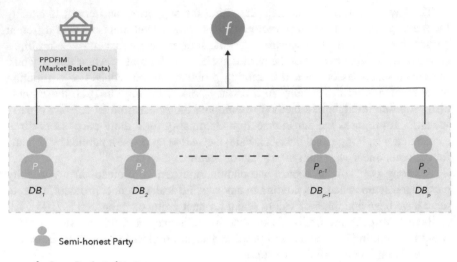

Semi-honest Party

f = Sum - Product of Vectors

Fig. 4.6 A semi-honest model for component-wise sum of product of vectors in PPDFIM across vertically partitioned data

an efficient scheme for this scenario, an extension of the two vector multiplication (non-public key based) protocol by Du-Atallah [43] to a multi-vector multi-party protocol [45] for a PPDFIM setup across vertically partitioned data is proposed. The instance of 3-vectors 3-party scenario is detailed in Algorithm 3 below which can further be extended to m-party m-vectors.

The scheme proposed by [45] is applicable to all the approaches where PPDFIM needs to be done across vertically partitioned data and is information-theoretically secure and efficient. In [45], the authors also give a comparative analysis and show that their scheme performs better in terms of execution cost than the public key based scheme proposed in [41].

The information-theoretically secure scheme proposed for the semi-honest model can further be extended to the malicious model as mentioned in [45].

4.5 Conclusion and Scope of Future Work

The increase in the volume and spread of data has inspired competitors to increasingly collaborate. These collaborations would provide an important value addition to the businesses of these competitors and are essential for *benchmarking* with the best practice in their field of operation.

These competitors that selectively collaborate in a co-opetitive setup, however, face the issue of privacy preservation of their sensitive data which is an important research direction.

Algorithm 3 Proposed sum of product algorithm instance for 3—Vectors 3 party PPDFIM scenario for vertically partitioned data based on [45]

Require: P_1 has $\mathbf{x_1} = (\mathbf{x_{11}} \ldots \mathbf{x_{1N}})$; P_2 has $\mathbf{x_2} = (\mathbf{x_{21}} \ldots \mathbf{x_{2N}})$; P_3 has $\mathbf{x_3} = (\mathbf{x_{31}} \ldots \mathbf{x_{3N}})$; Each party has N records; $p = 3$;

Ensure: Sum of product of 3-vectors across 3 parties.

1: Three parties P_1, P_2 and P_3 generate three random numbers r_1, r_2 and r_3.

2: P_1, P_2 and P_3 generate three random vectors $\mathbf{R_1}$, $\mathbf{R_2}$ and $\mathbf{R_3}$.

3: P_1 sends $(\mathbf{m} + \mathbf{R})$ and $(\mathbf{m'} + \mathbf{R'})$ to STTP where $\mathbf{m} = \mathbf{x_1} * \mathbf{R_2}$ and $\mathbf{m'} = \mathbf{x_1} * \mathbf{R_3}$ to the STTP. The STTP can hence calculate $s_1 = \sum_{i=1}^{N} \mathbf{x_{1i}} * \mathbf{x_{3i}} * \mathbf{R_{2i}}$ and $s_1' = \sum_{i=1}^{N} \mathbf{x_{1i}} * \mathbf{x_{2i}} * \mathbf{R_{3i}}$ using the 2-itemset Algorithm in [45]. (R and R' are random vectors generated privately between P_1 and P_3; P_1 and P_2 respectively)

4: P_1 calculates $\mathbf{w_1} = \mathbf{x_1} + \mathbf{R_1}$ and $s_1'' = \sum_{i=1}^{N} \mathbf{x_{1i}} * \mathbf{R_{2i}} * \mathbf{R_{3i}} + r_1$ and sends it to STTP.

5: P_2 sends $(\mathbf{m''} + \mathbf{R''})$ to the STTP where $\mathbf{m''} = \mathbf{x_2} * \mathbf{R_1}$. The STTP can hence calculate $s_2 = \sum_{i=1}^{N} \mathbf{x_{2i}} * \mathbf{x_{3i}} * \mathbf{R_{1i}}$ using the 2-itemset algorithm. (R'' is generated privately between P_2 and P_3 (as in Algorithm in [45]).)

6: P_2 calculates $\mathbf{w_2} = \mathbf{x_2} + \mathbf{R_2}$ and $s_2' = \sum_{i=1}^{N} \mathbf{x_{2i}} * \mathbf{R_{1i}} * \mathbf{R_{3i}} + r_2$ and sends it to STTP.

7: P_3 calculates $\mathbf{w_3} = \mathbf{x_3} + \mathbf{R_3}$ and $s_3 = \sum_{i=1}^{N} \mathbf{R_{1i}} * \mathbf{R_{2i}} * (\mathbf{x_{3i}} + \mathbf{R_{3i}}) + r_3$ and sends it to STTP.

8: STTP computes $v = \sum_{i=1}^{N} (\mathbf{w_{1i}} * \mathbf{w_{2i}} * \mathbf{w_{3i}}) - s_1 - s_1' - s_1'' - s_2 - s_2' - s_3$ and gets $v = \sum_{i=1}^{N} (\mathbf{x_{1i}} * \mathbf{x_{2i}} * \mathbf{x_{3i}}) - (r_1 + r_2 + r_3)$ and sends it to P_1, P_2 and P_3.

9: P_1, P_2 and P_3 get $\sum_{i=1}^{N} (\mathbf{x_{1i}} * \mathbf{x_{2i}} * \mathbf{x_{3i}}) = v + (r_1 + r_2 + r_3)$
//Component wise Sum of Product across three vectors for 3-itemset 3 party scenario. (All the vectors in the algorithm are shown in bold face which is the standard norm).

Hence, in this chapter, we detail the methodology of evaluation for PPDARM across horizontally and vertically partitioned data. Further, we discuss techniques for privacy preservation in collaborative association rule mining and improve the state-of-the-art of cryptographic techniques in PPDARM based on efficiency, types of adversaries handled and the security model applicable for different data partition models with the primary focus on information-theoretic security.

Privacy preserving data mining has the inherent challenge to balance between efficiency and security. Hence, we have discussed efficient and secure privacy preserving schemes that an application designer could choose from.

There is scope for further research in the area of PPDDM:

- Our focus in this chapter is on privacy preserving algorithms for collaborative association rule mining. However, the schemes proposed for sub-protocols like *secure sum* and *sum of products* of vectors could be analysed for other data mining methods, *viz. classification, clustering and support vector machines* wherein these sub-protocols are used.
- The extensions discussed could also be explored for investigating other privacy preserving functions for PPDDM scenarios, *viz. secure union, secure logarithm and secure polynomial evaluation.*

- It is necessary to investigate the trade-off between privacy and efficiency. Hence, optimum privacy preserving solutions for different application scenarios need to be explored that provide a balance between privacy and efficiency.
- Solutions that can deal with completely malicious behaviour without a trade-off of efficiency also need to be explored.

References

1. Data. "Data everywhere," The Economist, Feb 2010. [Online]. Available: http://www.economist.com/node/15557443. Accessed 13-January-2015.
2. Fan, W., & Bifet, A. (2013). Mining big data: current status, and forecast to the future. *ACM SIGKDD Explorations Newsletter, 14*(2), 1–5.
3. Seifert, J. W. (2013). *CRS report for congress: data mining and homeland security an overview August 27, 2008 - RL31798*. Mannheim, W. Germany, Germany: Bibliographisches Institut AG.
4. Aggarwal, C. C., & Yu, P. S. (2008). An introduction to privacy-preserving data mining. In *Privacy-Preserving Data Mining Models and Algorithms, ser. Advances in Database Systems* (vol. 34, pp. 1–9). New York: Springer US.
5. "9 important elements to corporate data security policies that protect data privacy," The Security Magazine, may 2016, [Online]. Available: http://www.securitymagazine.com/articles/. Accessed 18-February-2017.
6. Bachrach, D. G., & Rzeszut, E. J. (2014). Don't Let the Snoops In. In *10 Don'ts on Your Digital Devices*. Berkeley, CA: Apress.
7. Kantarcioglu, M., & Nix, R. (2010). Incentive compatible distributed data mining. In *Second International Conference on Social Computing (SocialCom)* (pp. 735–742). Minneapolis, Minnesota, USA: IEEE.
8. Du, W., & Atallah, M. J. (2001). Secure multi-party computation problems and their applications: a review and open problems. In *Proceedings of the 2001 Workshop on New Security Paradigms, ser. NSPW '01* (pp. 13–22). New York, NY, USA: ACM.
9. Lindell, Y., & Pinkas, B. (2000). Privacy preserving data mining. In *Proceedings of the 20th Annual International Cryptology Conference on Advances in Cryptology, ser. CRYPTO '00* (pp. 36–54). London, UK, UK: Springer-Verlag.
10. Bogetoft, P., Christensen, D., Damgård, I., Geisler, M., Jakobsen, T., Krøigaard, M., et al. (2009). Secure multiparty computation goes live. In *13th International Conference on Financial Cryptography and Data Security, ser. Lecture Notes in Computer Science* (vol. 5628, pp. 325–343). Accra Beach, Barbados: Springer/Berlin/Heidelberg.
11. Kantarcioglu, M., & Clifton, C. (2004). Privacy-preserving distributed mining of association rules on horizontally partitioned data. *IEEE Transactions on Knowledge and Data Engineering, 16*(9), 1026–1037.
12. Kargupta, H., Das, K., & Liu, K. (2007). Multi-party, privacy-preserving distributed data mining using a game theoretic framework. In *Proceedings of the 11th European conference on Principles and Practice of Knowledge Discovery in Databases, ser. PKDD* (pp. 523–531). Berlin/Heidelberg: Springer-Verlag.
13. Sekhavat, Y., & Fathian, M. (2010). Mining frequent itemsets in the presence of malicious participants. *IET Information Security, 4*, 80–92.
14. Kantarcioglu, M. (2008). A survey of privacy-preserving methods across horizontally partitioned data. In *Privacy-Preserving Data Mining, ser. Advances in Database Systems* (vol. 34, pp. 313–335). New York: Springer US.
15. Cheung, D. W., Han, J., Ng, V. T., Fu, A. W., & Fu, Y. (1996). A fast distributed algorithm for mining association rules. In *Proceedings of the Fourth International Conference on Parallel and Distributed Information Systems, ser. DIS '96* (pp. 31–43). Washington, DC, USA: IEEE Computer Society.

16. Wang, W., Deng, B., & Li, Z. (2007). Application of oblivious transfer protocol in distributed data mining with privacy-preserving. In *Proceedings of the The First International Symposium on Data, Privacy, and E-Commerce* (pp. 283–285). Washington, DC, USA: IEEE Computer Society.
17. Vaidya, J. (2008). A survey of privacy-preserving methods across vertically partitioned data. In *Privacy-Preserving Data Mining, ser. The Kluwer International Series on Advances in Database Systems* (vol. 34, pp. 337–358). New York: Springer US.
18. Samet, S., & Miri, A. (2009). Secure two and multi-party association rule mining. In *Proceedings of the Second IEEE International Conference on Computational Intelligence for Security and Defense Applications, ser. CISDA'09* (pp. 297–302). Piscataway, NJ, USA: IEEE Press.
19. Vaidya, J., & Clifton, C. (2005). Secure set intersection cardinality with application to association rule mining. *Journal of Computer Security, 13*(4), 593–622.
20. Ge, X., Yan, L., Zhu, J., & Shi, W. (2010). Privacy-preserving distributed association rule mining based on the secret sharing technique. In *2nd International Conference on Software Engineering and Data Mining (SEDM 2010)* (pp. 345–350). Chengdu: IEEE.
21. Evfimievski, A., & Grandison, T. (2007). *Privacy preserving data mining.* San Jose, California: IBM Almaden Research Center.
22. Aggarwal, C. C., & Yu, P. S. (2008). A general survey of privacy-preserving data mining models and algorithms. In *Privacy-Preserving Data Mining, ser. The Kluwer International Series on Advances in Database Systems* (vol. 34, pp. 11–52). New York: Springer US.
23. Barthe, G., Grégoire, B., Heraud, S., & Zanella Béguelin, S. (2009). Formal certification of ElGamal encryption—A gentle introduction to CertiCrypt. In *5th International Workshop on Formal Aspects in Security and Trust, (FAST 2008), ser. Lecture Notes in Computer Science* (vol. 5491, pp. 1–19). Malaga, Spain: Springer/Berlin/Heidelberg.
24. Pedersen, T. B., Saygin, Y., & Savas, E. (2007). Secret sharing vs. encryption-based techniques for privacy preserving data mining. *Sciences-New York*, 17–19.
25. Casey, E., & Rose, C. W. (2010). Chapter 2 - Forensic analysis. In *Handbook of Digital Forensics and Investigation* (pp. 21–47). San Diego: Academic Press.
26. Wikipedia. (2014). Information-theoretic security — Wikipedia, The Free Encyclopedia.
27. Shamir, A. (1979). How to share a secret. *Communication ACM, 22*, 612–613.
28. Castelluccia, C., Chan, A. C.-F., Mykletun, E., & Tsudik, G. (2009) Efficient and provably secure aggregation of encrypted data in wireless sensor networks. *ACM Transactions on Sensor Networks (TOSN), 5*(3), 20:1–20:36.
29. Vetter, B., Ugus, O., Westhoff, D., & Sorge, C. (2012). Homomorphic primitives for a privacy-friendly smart metering architecture. In *International Conference on Security and Cryptography (SECRYPT 2012)*, Rome, Itly (pp. 102–112).
30. Nanavati, N. R., Lalwani, P., & Jinwala, D. C. (2014). Analysis and evaluation of schemes for secure sum in collaborative frequent itemset mining across horizontally partitioned data. *Journal of Engineering, 2014*, p. 10.
31. Nanavati, N. R., & Jinwala, D. C. (2012). Privacy preserving approaches for global cycle detections for cyclic association rules in distributed databases. In *International Conference on Security and Cryptography (SECRYPT 2012)* (pp. 368–371). Rome, Italy: SciTePress.
32. Nanavati, N. R., Sen, N., & Jinwala, D. C. (2014). Analysis and evaluation of efficient privacy preserving techniques for finding global cycles in temporal association rules across distributed databases. *International Journal of Distributed Systems and Technologies (IJDST), 5*(3), 58–76.
33. Miyaji, A., & Rahman, M. (2011). Privacy-preserving data mining: a game-theoretic approach. In *Proceedings of the 25th Annual IFIP WG 11.3 Conference on Data and Applications Security and Privacy, ser. Lecture Notes in Computer Science* (vol. 6818, pp. 186–200). Richmond, VA, USA: Springer/Berlin/Heidelberg.
34. Nanavati, N. R., & Jinwala, D. C. (2013). A novel privacy preserving game theoretic repeated rational secret sharing scheme for distributed data mining. In *Security and Privacy Symposium*, IIT Kanpur, 2013. [Online]. Available: http://www.cse.iitk.ac.in/users/sps2013/submitting.html.

35. Nanavati, N. R., & Jinwala, D. C. (2013). A game theory based repeated rational secret sharing scheme for privacy preserving distributed data mining. In *10th International Conference on Security and Cryptography (SECRYPT)* (pp. 512–517), Reykjavik, Iceland. [Online]. Available: http://www.scitepress.org/DigitalLibrary/Index/DOI/10.5220/0004525205120517.

36. Abraham, I., Dolev, D., Gonen, R., & Halpern, J. (2006). Distributed computing meets game theory: robust mechanisms for rational secret sharing and multiparty computation. In *Proceedings of the Twenty-Fifth Annual ACM Symposium on Principles of Distributed Computing, ser. PODC '06* (pp. 53–62). New York, NY, USA: ACM.

37. Halpern, J., & Teague, V. (2004). Rational secret sharing and multiparty computation: extended abstract. In *Proceedings of the Thirty-Sixth Annual ACM Symposium on Theory of Computing, ser. STOC '04* (pp. 623–632). New York, NY, USA: ACM.

38. Maleka, S., Shareef, A., & Rangan, C. (2008). Rational secret sharing with repeated games. In *4th International Conference on Information Security Practice and Experience (ISPEC), ser. Lecture Notes in Computer Science* (vol. 4991, pp. 334–346). Sydney, Australia: Springer/Berlin/Heidelberg.

39. Nanavati, N. R., & Jinwala, D. C. (2012). Privacy preservation for global cyclic associations in distributed databases. *Procedia Technology, 6*(0), 962–969. In *2nd International Conference on Communication, Computing and Security [ICCCS-2012]*.

40. Nanavati, N. R., Lalwani, P., & Jinwala, D. C. (2014). Novel game theoretic privacy preserving construction for rational and malicious secret sharing models for collaborative frequent itemset mining. *Journal of Information Security and Applications (JISA)*. Submitted for consideration in Sep-2016.

41. Vaidya, J. S. (2004). Privacy preserving data mining over vertically partitioned data (Ph.D. dissertation, Centre for Education and Research in Information Assurance and Security, Purdue, West Lafayette, IN, USA, Aug 2004), aAI3154746. [Online]. Available: http://citeseerx.ist.psu.edu/viewdoc/summary?doi=10.1.1.2.4249.

42. Keshavamurthy, B. N., Khan, A., & Toshniwal, D. (2013). Privacy preserving association rule mining over distributed databases using genetic algorithm. *Neural Computing and Applications, 22*(Supplement-1), 351–364. [Online]. Available: http://dx.doi.org/10.1007/s00521-013-1343-9.

43. Du, W., & Atallah, M. (2001). Protocols for secure remote database access with approximate matching. In *E-Commerce Security and Privacy, ser. Advances in Information Security* (vol. 2, pp. 87–111). New York: Springer US.

44. Bogdanov, D., Jagomägis, R., & Laur, S. (2012). A universal toolkit for cryptographically secure privacy-preserving data mining. In *Proceedings of the 2012 Pacific Asia Conference on Intelligence and Security Informatics, ser. PAISI'12* (pp. 112–126). Berlin/Heidelberg: Springer-Verlag.

45. Nanavati, N. R., & Jinwala, D. C. (2015). A novel privacy-preserving scheme for collaborative frequent itemset mining across vertically partitioned data. *Security and Communication Networks, 8*(18), 4407–4420.

Chapter 5
A Postmortem Forensic Analysis for a JavaScript Based Attack

Sally Mosaad, Nashwa Abdelbaki, and Ahmed F. Shosha

5.1 An Overview of Web Browsers and Their Possible Attacks

Based on [23], Internet users spend more than 60 h per week surfing online contents. The web browsers, however, are complex software developed using various technologies and have to process different file formats and contents that may be vulnerable or contain malicious code. On the other hand, cybercriminals understand that the user is the weakest link in the security chain. Moreover, s/he is a higher possibility to a successful attack. That is why attackers are trying to exploit vulnerabilities in web browsers or luring users to visit malicious websites. By typing a web page URL into web browser, bunch of requests are created to get content from various web servers and resource directories. Since Web 2.0 revolution, new web technologies appear to provide dynamic web and active client-side content. One of these technologies is JavaScript. It is commonly used language to create interactive effects within web browsers and mobile applications. Each browser has its own JavaScript engine to interpret and execute the embedded JavaScript code in the visited web sites. Web based attacks target end users and their connected web devices. A user can be hacked because s/he has valuable information; will be used for launching a bigger attack such as Denial of Service attack (DoS), or just a curious script-kiddie takes the advantage of forgetting her/him to make a system update.

S. Mosaad (✉) • N. Abdelbaki • A.F. Shosha
Nile University, Cairo, Egypt
e-mail: smosaad@nu.edu.eg; nabdelbaki@nu.edu.eg; ashosha@nu.edu.eg

© Springer International Publishing AG 2018   79
K. Daimi (ed.), *Computer and Network Security Essentials*,
DOI 10.1007/978-3-319-58424-9_5

5.1.1 Drive-by-Download Attack

One of the most popular attacks is Drive-by-Download attack, which can be defined as a malicious content downloaded to a user's system using the web browser. This content may be in different file format like:

- A malicious Flash file or embedded action script code [22]
- Malicious PDF with embedded JavaScript code [15]
- Obfuscated JavaScript code in a web page [5] that exploits vulnerability in the user's system.

Drive-by-Download attack is known as a pull-based malware infection. This downloaded malware can be triggered by different actions such as opening, scrolling, or hovering a mouse cursor over a malicious web page or a hidden *iframe*. In a typical Drive-by-Download attack, an innocent user is redirected to a malicious web page. This page is commonly denoted as landing site, which is a web page that contains the shell code or a small binary payload [25]. This code can be either written in JavaScript or in VBScript. The code will then exploit a browser's vulnerability, browser's installed plug-ins, or insecurely designed APIs. If succeeded, it will download a malware from a malicious site into the victim's machine. Often a chain of redirection operations will take place before the user's browser gets to that malicious site, to make it more difficult to trace the attacker. A Drive-by-Download attack is developed for a specific vulnerability in a specific browser's version. A common initial activity in this attack vector is reconnaissance and fingerprinting the web browser meta-data. This means that the embedded script will attempt to collect information about the browser type, version, language, installed plug-ins, and the installed operating system. Based on the collected information a malicious shell code will download the appropriate exploit or it may behave in a completely benign manner if, for example, an analysis environment was detected.

Academic and professional researches are commonly focusing on the detection and prevention techniques of this attack vector [5, 11]. The currently proposed techniques are mainly based on either analyzing the properties of a malicious web page URL [25] or analyzing the code contained in the web page. The analysis for the code is done using static, dynamic analysis or a combination of both which is known as hybrid analysis.

- *Static analysis*: It uses a set of predefined features to determine that a malicious pattern or code exists in particular web page without code execution. Several machine-learning techniques and approaches may also be integrated to (1) define the set of features required for the analysis, (2) cluster, classify, and/or determine malicious web pages out of benign web pages [6, 11]. In this analysis approach, a low processing overhead may be required. However, the static analysis generally can be impeded if some obfuscation and/or encryption methods are employed [2].

• *Dynamic/semi dynamic analysis*: It uses a controlled environment, commonly called Sandboxing. In this analysis a subset or all of the possible execution paths for the embedded code are executed to detect the presence of a malicious behavior. This will help in case of obfuscated code. However, additional processing resources may be required. An attacker may also execute a legitimate code, suppress the execution of her/his malicious code, or attempt to self-delete it, if s/he suspects that s/he was detected [11, 21]. Moreover, malware can use extended sleep calls to void detection.

A combination of both techniques is also used for analyzing the embedded JavaScript code, and to avoid the drawbacks associated with each approach. Typically, a static analysis technique is used as an initial filter to define the web pages that require a dynamic analysis. Applying this hybrid analysis may guarantee accurate detection with minimum resources [11]. The implemented techniques are differing mainly in the type of code analysis performed and whether it is prevention or a detection only technique. Many tools have been proposed using static analysis technique such as ARROW [25], Prophiler [2], ZOZZLE [6], and PJScan [15]. Because static analysis is limited, tools such as JSAND [5], BLADE [17], JSGuard [9], ROZZLE [13], NOZZLE [21], and Shellzer [8] have been proposed using dynamic analysis. Revolver [12] and EVILSEED [10] are also proposed using a semi-dynamic way to take the advantages of using both techniques. This is by applying static analysis for identifying similarities between the analyzed web page and known malicious web pages. If similarities are found, a dynamic analysis using honey clients and sandboxing is used for taking the final decision. This is to minimize the time and resources needed by using dynamic analysis alone. A summary of these tools is given in Table 5.1. Other researches focus on analyzing exploit kits that are used to launch a Drive-by-Download attack [14]. *Exploit kit* is a malicious toolkit that exploits security flaws found in software applications. Using an exploit kit requires no proficiency or software development background. It is equipped with different detection/avoidance methods. In [7], the authors focus on the server side of a Drive-by-Download attack. They analyzed the source code for multiple exploit kits using Pexy. It is a system for bypassing the fingerprinting of an exploit kit and getting all of its possible exploits by extracting a list of possible URL parameters and user agents that can be used.

In a recent study presented in [24], the authors proposed a system using Chrome JavaScript Debugger to detect browser's extensions that inject malicious ads into a web page. The study revealed that 24% of ad network domains bring malicious ads. These ads will redirect the user to a landing page, which will finally download a malicious executable into the user's machine. Authors in [3] proposed the idea of extracting and validating Indicators of Compromise (IOCs) for web applications. They talked about how it is important to inspect not just the script content of a web page but also the context in which it is used. They recognized that attackers could use benign script code to perform malicious actions. They also noticed that most of JavaScript files used by attackers are not installed on the compromised hosts but

Table 5.1 Summary of Drive-by-Download prevention/detection tools

Tool	Type	Description	Limitations
ARROW	Static detection	Detects Drive-by-Download attack by investigating the URL and generating a set of regular expressions-based signatures. The implemented algorithm helps in detecting Malware Distribution Networks (MDNs) and blocks all landing pages [25]	It depends only on using URL features, which is not sufficient if analyzed alone
Prophiler	Static detection	Works as a front-end filter to reduce the number of pages to be analyzed. It uses a set of collected features, such as web page's content and URL combined with a machine-learning algorithm to classify the page as either malicious or benign [2]. The pages that are likely to be suspicious are further analyzed using a dynamic analysis tool (WEPAWET)	It depends on static detection, which cannot detect suspicious URLs with dynamic content such as obfuscated JavaScript
ZOZZLE	Static detection and prevention	Performs static analysis of the de-obfuscated JavaScript code in the browser, by hooking[a] into the JavaScript engine to get the final version of the code. Each code segment sent to the JavaScript engine for compilation is transferred into a JavaScript Abstract Syntax Tree (AST)[b]. Features are extracted from AST nodes. Bayesian classification is performed to predict maliciousness ([11]; [6])	Depending only on static analysis makes it impossible to firmly establish a complete and final analysis decision
PJSCAN	Static detection	Detects malicious JavaScript code in PDF documents using lexical analysis and One-Class Support Vector Machine (OCSVM) as a learning method [15]	Cannot detect de-obfuscated JavaScript code. Operates with a high false positive rate
JSAND/WEPAWET	Dynamic detection	Uses anomaly detection techniques and dynamic emulation to detect malicious JavaScript content. It uses ten collected features classified as necessary to characterize Drive-by-Download attack. It can characterize new exploits and generate exploit signature for signature-based tools (PhonyC)[c] ([5]; [11]). WEPAWET is its web service	Works when exploit is successfully executed. It fails in detecting new attacks whose signatures were not included during the learning phase
BLADE	Dynamic detection and prevention	Since Drive-by-Download attack is based on shell code injection and execution, BLADE prevents it by creating a non-executable sandbox to ensure that no downloaded file will be executed without explicit user acceptance [17]	Cannot detect malicious codes that are executed directly from memory, without writing the binary to disc [1]

JSGuard	Dynamic detection	Detects JS shell code using JS code execution environment information. It creates a virtual execution environment to monitor shell code behavior using malicious JS string detector and shell code analyzer [9]	Focuses on binary code analysis to detect heap spraying attack
ROZZLE	Dynamic detection	A Microsoft tool to detect Drive-by-Download attack. It executes both possibilities whenever it encounters control flow branching that is environment dependent [13]	Helpless in avoiding server-side cloaking
NOZZLE	Dynamic detection and prevention	Detects heap spraying attack by scanning the memory heap for NOP sled detection. By doing this, NOZZLE can prevent shell code execution ([11]; [21])	Made for a certain attack type (heap spraying)
Shellzer	Dynamic detection	Analyzes the shell code in both web-based malware and PDF documents. It uses single step instrumentation to improve the detection performance using a Trap Flag (TF) in EFlag register. This enables the tool to step through the execution and generates a complete list of all API functions directly called by the shell code [8]	It can be easily evaded by indirect API calls
Revolver	Hybrid detection	Detects evasion attempts in JavaScript code. It is not a detection tool in its own. It computes the similarities between scripts that were classified using an existing Drive-by-Download detection tool (Malicious, benign scripts). If there is similarity between two scripts classified differently, then one of these scripts is likely using evasion techniques [12]	Its output will depend on the efficiency and success of the detection tools used
EVILSEED	Hybrid detection	Improves detection of malicious pages that launch Drive-by-Download attack by searching the web more efficiently. It uses number of gadgets to find similarities between known malicious pages (seeds) to guide the search for additional malicious content (candidate URLs). Candidate pages are fully analyzed using WEPAWET [10]	Its effectiveness will depend on the quality and the diversity of the input seeds

[a] Hooking refers to intercepting code event or system call to monitor/debug/modify code behavior
[b] An Abstract Syntax Tree is a tree model to represent the source code for analysis
[c] PhonyC is a virtual honeyclient that impersonates a user browser to provide information on what happened during the attack and how

instead included from public URLs which facilitate the modification for these files. Attackers can also use popular library names like *jquery.js* to hide their malicious code.

5.1.2   Browser Forensics

Investigating the web browser of an infected or suspected machine is essential to understand the anatomy of the attack. Browser forensics is an emerging topic of the digital forensics science that refers to the process of extracting and analyzing the web browsers artifacts and the user's browsing activities for forensic investigation purposes [16]. Browsers store a significant amount of data about the user's activities over the Internet if it is used in its normal mode. Less data may also be collected if user opts to browse in the private browsing mode [20]. Private mode enables users to browse the web without storing local data. This means there will be no saved history or cached data. However, researches proved that even while using this mode there would be little traces for the user that can be collected.

On the forensics side, researches [4, 20] focused on private/portable browsing and how to collect the remaining evidences from the memory and the file system. In [19], the authors discussed the importance of making an integrated analysis for different browsers at the same time to understand what happened. They proposed a tool for constructing a timeline for the user's activities.

Although browsers store a lot of data about the user's activities over the Internet, still a digital forensic investigation process is required to reconstruct the browser activities. Knowing these activities is essential to understand what really happened. Many tools have been developed to extract the stored information and display it in a readable, user-friendly way. These tools help in knowing the history of a web browser usage if it was used in a normal browsing mode. This could be received/sent email, visited site, searches, stored cookies, or a downloaded file that the user voluntarily downloads, all of these actions were intentionally done by the user but what about things done without his knowledge?

None of these tools dealt with the browser memory. In this study we propose a methodology to acquire the browser's memory stack frames and extract the called functions with their parameters. We believe that this will help investigators to know the actual executed code when visiting a malicious URL. The investigator will be able to create a complete trace file that will assist in a postmortem investigation.

5.2   Proposed System Description

In this section, we are proposing a digital forensic methodology to forensically investigate a malicious web page. That malicious page is suspected to download and/or further execute malicious code within a web browser. A typical scenario

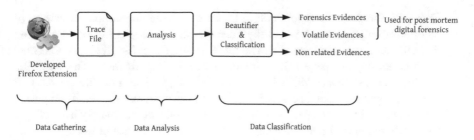

Fig. 5.1 A linear procedure for the proposed system

would be, a user noticing uncommon activities occurring in her/his system. This can be a suspicious/unrelated web advertisement appeared while surfing the web. Another scenario would be a network admin in corporate noticing unusual network traffic inbound or outbound from a system that might be compromised. The admin may also notice a connected machine to the corporate network visiting a black listed web server known to host malicious contents. In this case, a forensic analyst would perform an examination of the system to determine indications of compromise. This could be searching for a URL of the malicious web page in web-browsing history, a cookie file or a temp file in the Internet storage directory. If identified, it is crucial for the forensic investigation to determine what other resources have been downloaded and executed into the browser from this malicious website. To reconstruct the attack's executed events and analyze its actions, we develop a browser extension for postmortem forensic analysis (FEPFA) using Mozilla Debugger API [18]. We use FEPFA to monitor, log, and debug the details of an executed malicious JavaScript codes subject to investigation [7]. We create a virtual machine similar to the compromised system. The investigator will use the created virtual machine after installing FEPFA on it to access the same malicious URL. FEPFA will then create a data trace file with all the called and executed functions. After analyzing the code a list of digital forensics evidences are produced.

Our proposed methodology consists of the following sequential procedures as shown in Fig. 5.1.

- *Data Gathering*: It is a process of accessing the malicious URL in a setting similar to the compromised system. This is to lure the malicious URL to download the set of resources (content, code, and exploit payload) similar to those that have been downloaded in the system subject to investigation. Simulating the settings of a compromised system subject to investigation avoids downloading and executing code that has never been executed in the original system subject of the incident. We assume that the user was running a Firefox web browser. As such, our Firefox Browser Extension (FEPFA) monitors, logs, and debugs the downloaded resources after accessing a malicious web page with a particular attention to the executed embedded JavaScript code. In our Proof-of-Concept implementation we customize *Mozilla Debugger API*, which is a debugging interface provided by Mozilla JavaScript engine "*SpiderMonkey*."

This API enables JavaScript code to observe and manipulate the execution of other JavaScript codes. We use the debugger to develop a browser extension that outputs a detailed trace file. The trace file logs and lists the code executed from accessing the page subject to the forensic investigation. The trace file is generated in JSON, which is a JavaScript Object Notation file format that includes objects created/accessed/modified on the system with details about the stack frames of the executed code and the execution timestamps. By inspecting a browser's memory frame instance, we can find the script source code the frame was executing. Moreover, we proceed to older stack frames and find the lexical environment in which the execution is taking place. We select the most relevant properties from the extracted frames that can reveal important data about the executed JavaScript code. The generated JSON object from FEPFA contains data about the function whose application created the memory frame, the name/value pairs of the passed parameters to this function, the URL of the page in which the function has been called and the script source code being executed in this frame. We test FEPFA using over 200 real malicious URLs collected from public malware databases.[1] We access and analyze 103 malicious web sites after filtering the blocked and offline URLs. To filter these URLs and generate the required trace files, we access each of the 103 URLs separately using a virtual machine after installing FEPFA on it. Figure 5.2 demonstrates the categorization for the 103 analyzed URLs. FEPFA will then load the page, get the stack frames for the called and executed JavaScript functions from the browser memory, and create a JSON object with all the required data. After generating the JSON objects, FEPFA will create a file with the data gathered for that web site.

- *Data Analysis*: The analysis for the generated trace files requires a detailed examination for the extracted JavaScript code. We develop an analyzer using NodeJS, which is a JavaScript runtime built on Chrome's V8 JavaScript engine.

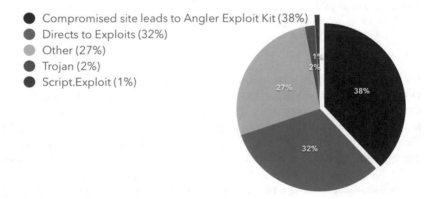

Fig. 5.2 Categorization for the analyzed URLs

---

[1] www.malwaredomainlist.com/mdl.php, http://www.malwareurl.com/.

```
function() {
var as = document.createElement('script');
as.src ="xxxd3lqbvlcthcecs.cloudfront.net/atrk.js ";
```

```
var tempImage = new Image();
tempImage.src = smf_prepareScriptUrl(smf_scripturl)+
'action=keepalive;time=' + curTime;
```

Fig. 5.3 Injecting a remote JavaScript code

We create number of regular expressions to be used by NodeJS analyzer. The developed analyzer is used to search for specific patterns. We search for patterns of obfuscation, encoding/decoding, checking for vulnerability, URL redirection, downloading external resources, and creating local files on the system. There are different well-known and commonly used techniques for cybercriminals to use a JavaScript code to perform malicious actions. For example, to download an external resource, an attacker may employ one of the following methods as shown in Fig. 5.3:

– Create a script tag and set the source attribute to the required downloadable file.
– Create an image tag with source to a malicious URL.

These different methods are taken into consideration during the analysis process. After running the analyzer on the generated data trace files, the output from the analyzer will show the number of occurrence or usage for each searched event. Following are the events we use:

– Vulnerability checking using ActiveXObject and Shockwave.
– Downloading external resources by assigning the source (*src*) attribute or by using the *iframe* tag.
– Created cookie files.
– Using of encoding function.
– Browser fingerprinting and URL redirection.

After this we transform the extracted code into a human readable format, this is by utilizing a web-based services named JavaScript beautifier.[2] The extracted code is further analyzed to get a closer look into each evidence and extract the common patterns between various URLs. A part of the generated trace files before beautifying is shown in Fig. 5.4. We only beautify the script part, which contains the JavaScript code.

• *Data Classification*: To avoid providing a forensic analyst with a significant amount of irrelevant information, data classification, and analysis procedure is a crucial activity for eliminating data that is not relevant to the case subject to investigation. If evidence is related to the examined attack type, it will then be classified as relative and is further classified into (1) volatile or (2) non-volatile forensic evidence. Other data will be considered as non-related evidences.

---

[2]http://jsbeautifier.org, http://codebeautify.org/jsviewer.

```
{"kind": "push",
    "code": {
    "type": "call",
    "class": "Window",
    "url": "xxx/show_afd_ads.js","parameters": [],
    "script": "(function(){var g=this,l=function(){},m=
    function(c){var a=typeof c;if(\"object\"==a)if(c){if
    (c instanceof Array)return\"array\";...
```

Fig. 5.4 Part of the generated trace file

Fig. 5.5 Creating cookie file

```
f.cookie = e + "=; expires=Thu, 01 Jan 1970 00:00:01 GMT;
path=/ " + (t ? "; domain= " + (l("msi ") ?" " : ". ") +
        "xxxaddthis.com " : " ")
```

- Volatile evidence: i.e., in memory shell code, encoding/encryption code.
- Non-volatile evidence: Such as created file on the system, downloaded resource, URL redirection with a trace of the URL in the browser history.

The main reason behind this classification is to ensure that a forensic analyst would know that a volatile data related to the investigation might exist but not necessarily can be recovered nor reconstructed. The identified non-volatile forensics can be further used to develop an attack signature. The forensic analyst can then use the generated attack signature to detect if there is an attack on other systems. Figure 5.5 is an example for one of the counted evidences. This code sample shows a cookie file created from one of the analyzed JavaScript files. The code contains many details like cookie name, expiry date, path, and/or domain. Most of the analyzed URLs are also checking vulnerabilities and trying to create an ActiveX object or shockwave flash. ActiveX controls are Internet Explorer's version of plug-ins. Creating an ActiveX object in JavaScript is used for getting a reference connected to another application or programming tool and it can be used for malicious purposes. The same with Shockwave Flash that is the Adobe's Flash Player built directly into the browser. In Fig. 5.6 the attacker is trying to create different versions from both ActiveX and Shockwave.

Attackers also attempt to use URL redirection when launching their attack. This is to redirect and forward the user to the malicious web page. An example for URL redirection is shown in Fig. 5.7. The attacker in this code uses a function named *smf_prepareScriptUrl*, which returns a string. This string is then concatenated with another string to form the link he needs. The attacker uses the generated link to change the *window.location* and redirect the user to another page. Another common behavior used is fingerprinting for the user browser. The most common way to do so is to get the *userAgent* property. This enables the attacker to detect the browser's JavaScript engine, browser version, and operating system. An example for it is shown in Fig. 5.8.

Fig. 5.6 Checking vulnerability

```
d = b[c], -1 < d[r][q]("Shockwave Flash ") &&
...
try {
    c = new ActiveXObject(d + ".7"),
...
```

Fig. 5.7 Redirection to another web page

```
onclick="window.location.href = \\'' +
smf_prepareScriptUrl(smf_scripturl) + '
```

Fig. 5.8 Fingerprinting for the user's browser

```
var is_opera5 = ua.indexOf('opera/5')!=-1||
...
var is_iphone = ua.indexOf('iphone')!=-1 ||
...
```

Table 5.2 Properties list example

Evidence name	Properties
URL redirection	Domain name, path
Vulnerability check	Branches depth, vulnerabilities name
String manipulation	String operations, string value, and length
Downloaded resource	Resource name, source URL, resource type
Created file	File name, file path, file type

After analyzing and classifying identified evidences, a list of properties is extracted based on the evidence type. Table 5.2 lists the possible properties for evidences based on their type. For example, when finding a trace for a URL redirection, it's important to get the domain name and path to know if it is a cross domain or not. Also if the attacker is using a string manipulation and evaluating a string using JS *eval* function, this can be an indication of heap spraying attempt. If a string manipulation exists in the extracted malicious code, the investigator will need to find the string values, length, and operations performed. This can be for spreading the shell code into the user's memory.

A typical Drive-by-Download attack will download an external resource into the system, so if this is detected we will have to get the resource name, type, and the link or URL leads to this resource. The forensics investigator will also have a list of all created files with the file name, type, and path.

5.3 Experiment and Findings

In this section, we present the results of the introduced web browser forensic analysis method. We collect the trace files generated from FEPFA. The generated files are then provided to the developed analyzer. The NodeJS analyzer searched for specific patterns and keywords related to our selected events and outputs the number

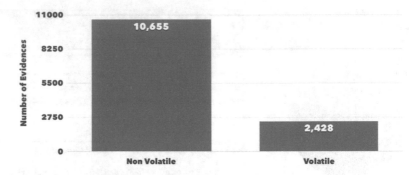

Fig. 5.9 Evidences distribution

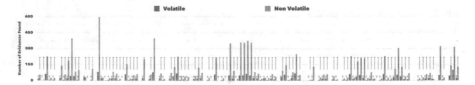

Fig. 5.10 The distribution for the searched events

of traces found for each searched event. The analysis shows that the percentage of non-volatile evidences is larger than the percentage of volatile evidences as shown in Fig. 5.9.

Based on these values, 81% of the identified evidences are artifacts that we believe it could assist forensic investigator to determine if web browser or a system subject to examination is compromised or not, and the indications of compromises. Figure 5.10 demonstrates the distribution for volatile/non-volatile evidences for each URL. The x-axis represents the analyzed URLs and the y-axis represents the number of volatile/non-volatile evidences. As shown in figure, the number of non-volatile evidences for each URL is higher than the number of volatile evidences. This proves that the majorities of the extracted evidences for each malicious page are beneficial and can be used in postmortem forensic analysis.

The distribution for both volatile and non-volatile traces as searched by our developed analyzer is shown in Fig. 5.11. 23% of the non-volatile evidences are downloaded files using *iframe* HTML tag and 12% are downloaded files by setting the source attribute for a script or image HTML tag. There are also 11% of the traces found for created cookie files.

By inspecting the extracted frames' type, we can count the number of usage for *eval* function. Figure 5.12 shows the URLs with the highest number of executed frames with type *eval*. This indicates a high possibility for a Drive-by-Download attack, as *eval* JavaScript function indicates evaluation for a JavaScript expression and in many times used to evaluate and execute shell codes. After beautifying the extracted code we get a closer look into the type of downloaded files and the way used to download it. Figure 5.13 demonstrates the distribution for the founded files.

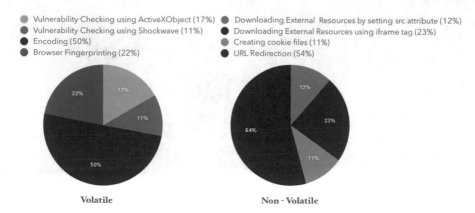

- ● Vulnerability Checking using ActiveXObject (17%)
- ● Vulnerability Checking using Shockwave (11%)
- ● Encoding (50%)
- ● Browser Fingerprinting (22%)

- ● Downloading External Resources by setting src attribute (12%)
- ● Downloading External Resources using iframe tag (23%)
- ● Creating cookie files (11%)
- ● URL Redirection (54%)

Fig. 5.11 Distribution for the volatile/non-volatile traces

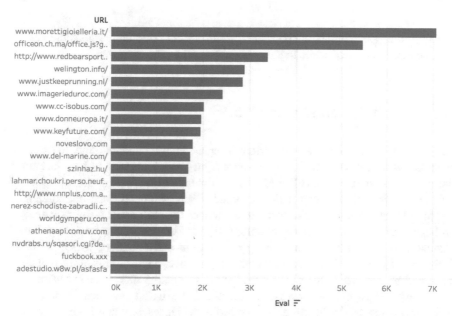

Fig. 5.12 URLs with the highest number of executed frames with type eval

Our experiment shows that we can get a detailed trace file for any executed malicious JavaScript code. If a JavaScript code is trying to execute a PHP file or load a malicious ad, the developed system is capable of identifying traces for that file. For example, if attacker creates *iframe* and sets the source attribute to a malicious web page as shown in Fig. 5.14, FEPFA extracts that code and the developed analyzer will show the usage of *iframe* and the downloaded resource name. FEPFA gets script files whose functions are loaded and executed in memory. Script files that have no executed functions and are not loaded in memory are not in our scope in this study.

Fig. 5.13 Distribution for the
downloaded resources

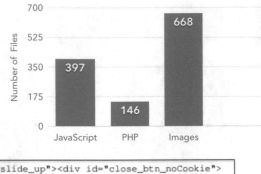

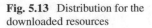

```
document.write('<div id="slide_up"><div id="close_btn_noCookie">
X</div><iframe id="su_frame" src="xxxpcash.imlive.c../releasese/..
```

Fig. 5.14 Security embedded iframe loads a malicious page

As such draggable.min.js, menu.min.js, mouse.min.js that are responsible for UI interactions if not used while loading the page will not be extracted.

5.4 Conclusion and Future Work

In this study we introduce a postmortem forensic analysis methodology to examine web browsers artifacts produced by accessing malicious URLs. We develop a Firefox Browser extension (FEPFA) to obtain detailed trace file for the downloaded malicious files and executed malicious code. Each generated trace file contains a set of volatile and non-volatile forensic evidences that will assist a forensic analyst in her/his investigation. Our methodology focuses on one of the JavaScript based attacks that are gaining an increasing momentum and attention, which is called Drive-by-Download. The proposed methodology gives a closer look at the real code executed by attacker at the client side. The developed system is tested using 103 malicious web pages and successfully identified the digital evidence of the attack. The percentage of the identified non-volatile evidences is much higher than the percentage of volatile evidences. This proves that most of the identified evidences can help in the postmortem forensic analysis.

In the future work, we can focus on other types of code such as PHP, Action Script, and VB Script. Moreover, we can investigate malicious code that exists into Adobe Flash Player as the numbers of vulnerabilities in Adobe plug-ins have grown significantly in recent years. From the server side, we can use NodeJS for examining application servers and extracting evidences from them.

References

1. Afonso, V. M., Grgio, A. R. A., Fernandes Filho, D. S., & de Geus, P. L. (2011). A hybrid system for analysis and detection of web-based client-side malicious code. In *Proceedings of the IADIS international conference www/internet* (Vol. 2011).
2. Canali, D., Cova, M., Vigna, G., & Kruegel, C. (2011, March). Prophiler: a fast filter for the large-scale detection of malicious web pages. In *Proceedings of the 20th international conference on world wide web* (pp. 197–206). ACM.
3. Catakoglu, O., Balduzzi, M., & Balzarotti, D. (2016, April). Automatic extraction of indicators of compromise for web applications. In *Proceedings of the 25th international conference on WorldWideWeb* (pp. 333–343). InternationalWorldWideWeb Conferences Steering Committee.
4. Choi, J. H., Lee, K. G., Park, J., Lee, C., & Lee, S. (2012). Analysis framework to detect artifacts of portable web browser. In *Information technology convergence, secure and trust computing, and data management* (pp. 207–214). Netherlands: Springer.
5. Cova, M., Kruegel, C., & Vigna, G. (2010, April). Detection and analysis of drive-by-download attacks and malicious JavaScript code. In *Proceedings of the 19th international conference on world wide web* (pp. 281–290). ACM.
6. Curtsinger, C., Livshits, B., Zorn, B. G., & Seifert, C. (2011, August). ZOZZLE: Fast and Precise In-Browser JavaScript Malware Detection. In *USENIX security symposium* (pp. 33–48).
7. De Maio, G., Kapravelos, A., Shoshitaishvili, Y., Kruegel, C., & Vigna, G. (2014, July). Pexy: The other side of exploit kits. In *International conference on detection of intrusions and malware, and vulnerability assessment* (pp. 132–151). Cham: Springer.
8. Fratantonio, Y., Kruegel, C., & Vigna, G. (2011, September). Shellzer: a tool for the dynamic analysis of malicious shellcode. In *International workshop on recent advances in intrusion detection* (pp. 61–80). Berlin Heidelberg: Springer.
9. Gu, B., Zhang, W., Bai, X., Champion, A. C., Qin, F., & Xuan, D. (2012, September). Jsguard: shellcode detection in JavaScript. In *International conference on security and privacy in communication systems* (pp. 112–130). Berlin Heidelberg: Springer.
10. Invernizzi, L., & Comparetti, P. M. (2012, May). Evilseed: A guided approach to finding malicious web pages. In *2012 IEEE symposium on security and privacy (SP)*, (pp. 428–442). IEEE.
11. Jayasinghe, G. K., Culpepper, J. S., & Bertok, P. (2014). Efficient and effective realtime prediction of drive-by download attacks. *Journal of Network and Computer Applications, 38*, 135–149.
12. Kapravelos, A., Shoshitaishvili, Y., Cova, M., Kruegel, C., & Vigna, G. (2013, August). Revolver: An automated approach to the detection of evasive web-based malware. In *USENIX security* (pp. 637–652).
13. Kolbitsch, C., Livshits, B., Zorn, B., & Seifert, C. (2012, May). Rozzle: De-cloaking internet malware. In *2012 IEEE symposium on security and privacy (SP)*, (pp. 443–457). IEEE.
14. Kotov, V., & Massacci, F. (2013, February). Anatomy of exploit kits. In *International symposium on engineering secure software and systems* (pp. 181–196). Berlin Heidelberg: Springer.
15. Laskov, P., & Šrndić, N. (2011, December). Static detection of malicious JavaScript-bearing PDF documents. In *Proceedings of the 27th annual computer security applications conference* (pp. 373–382). ACM.
16. Ligh, M., Adair, S., Hartstein, B., & Richard, M. (2010). *Malware analyst's cookbook and DVD: Tools and techniques for fighting malicious code*. Hoboken, NJ: Wiley.
17. Lu, L., Yegneswaran, V., Porras, P., & Lee, W. (2010, October). Blade: An attack-agnostic approach for preventing drive-by malware infections. In *Proceedings of the 17th ACM conference on computer and communications security* (pp. 440–450). ACM.

18. Mohamed, S. M., Abdelbaki, N., & Shosha, A. F. (2016, January). Digital forensic analysis of web-browser based attacks. In *Proceedings of the international conference on security and management (SAM)* (p. 237). The Steering Committee of the World Congress in Computer Science, Computer Engineering and Applied Computing (WorldComp).
19. Oh, J., Lee, S., & Lee, S. (2011). Advanced evidence collection and analysis of web browser activity. *Digital Investigation, 8*, S62–S70.
20. Ohana, D. J., & Shashidhar, N. (2013). Do private and portable web browsers leave incriminating evidence?: A forensic analysis of residual artifacts from private and portable web browsing sessions. *EURASIP Journal on Information Security, 2013*(1), 6.
21. Ratanaworabhan, P., Livshits, V. B., & Zorn, B. G. (2009, August). NOZZLE: A defense against heap-spraying code injection attacks. In *USENIX security symposium* (pp. 169–186).
22. Van Overveldt, T., Kruegel, C., & Vigna, G. (2012, September). FlashDetect: ActionScript 3 malware detection. In *International workshop on recent advances in intrusion detection* (pp. 274–293). Berlin Heidelberg: Springer.
23. Virvilis, N., Mylonas, A., Tsalis, N., & Gritzalis, D. (2015). Security busters: Web browser security vs. rogue sites. *Computers & Security, 52*, 90–105.
24. Xing, X., Meng, W., Lee, B., Weinsberg, U., Sheth, A., Perdisci, R., & Lee, W. (2015, May). Understanding malvertising through ad-injecting browser extensions. In *Proceedings of the 24th international conference on world wide web* (pp. 1286–1295). ACM.
25. Zhang, J., Seifert, C., Stokes, J. W., & Lee, W. (2011, March). Arrow: Generating signatures to detect drive-by downloads. In *Proceedings of the 20th international conference on world wide web* (pp. 187–196). ACM.

Part II
Network Security

Chapter 6
Malleable Cryptosystems and Their Applications in Wireless Sensor Networks

Keyur Parmar and Devesh C. Jinwala

6.1 Introduction

Sensor network's primary objectives are to sense the phenomena and transmit the sensed information towards the base station. The complex computation is expected to be carried out at the base station. However, the transmission of redundant data can impose an enormous communication overhead on sensor nodes that are close to the base station. In-network processing helps in reducing the redundant communication traffic. In-network processing performs en route aggregation of reverse multi-cast traffic in wireless sensor networks (WSNs) [10]. In addition, hostile and unattended deployments and unreliable communication environment pose security threats for communicated sensor readings.

In WSNs, aggregator nodes collect data from sensor nodes and aggregate them before sending the result towards the base station. If the data are encrypted, for the aggregation purpose, aggregator nodes need to decrypt data, perform an aggregation, and re-encrypt data before forwarding the result towards the next hop. Such hop-by-hop security increases the resource consumption in resource starved WSNs, and it is a risk against the privacy of individual sensor readings. The process often referred to as the hop-by-hop secure data aggregation in WSNs. In the hop-by-hop secure data aggregation, the compromised intermediate nodes become the bottleneck for the security of WSNs protocols [12]. Therefore, the need to ensure the privacy of sensor readings at intermediate nodes leads the development of end-to-end secure data aggregation protocols [12, 14].

K. Parmar (✉)
Indian Institute of Information Technology (IIIT), Vadodara, India
e-mail: keyur.mtech@gmail.com

D.C. Jinwala
S. V. National Institute of Technology (SVNIT), Surat, India
e-mail: dcjinwala@acm.org

© Springer International Publishing AG 2018   97
K. Daimi (ed.), *Computer and Network Security Essentials*,
DOI 10.1007/978-3-319-58424-9_6

Encrypted data processing has been studied extensively in the last few decades. A property, namely, privacy homomorphism presented by Rivest et al. [19], makes any cryptosystems malleable [4, 19] and has interesting applications in resource-constrained WSNs. However, privacy homomorphism has adversarial effects on the performance of other security metrics such as data integrity and data freshness. Moreover, traditional solutions do not comply with the end-to-end security model that supports en route aggregation. In this chapter, we analyzed the malleable cryptosystems adopted by different WSNs protocols. The proposed chapter helps in understanding different cryptosystems adopted by WSNs protocols to process the sensor readings at intermediate nodes. The discussion presented in the chapter not only helps to understand different cryptosystems but it also helps to understand numerous research articles presented in WSNs and adopts these cryptosystems.

The rest of the chapter is organized as follows. In Sect. 6.2, we discuss the impact of in-network processing on security requirements such as privacy, integrity, and freshness. In Sect. 6.3, we discuss privacy homomorphism. In Sect. 6.4, we present malleable cryptosystems that are based on symmetric-key cryptography and have been used in numerous WSNs protocols [17]. In Sect. 6.5, we present malleable cryptosystems that are based on asymmetric-key cryptography and adopted in WSNs protocols [12, 14]. Section 6.6 concludes the chapter by emphasizing our contributions.

6.2 Impact of In-Network Processing

WSNs are vulnerable to a wide range of attacks [8]. These attacks include eavesdropping, traffic analysis, integrity violation, replay attacks, physical attacks, denial-of-service attacks, etc. [21]. In addition, the denial-of-service attacks in WSNs include a wide variety of attacks ranging from simple jamming to more sophisticated attacks such as Sybil attack, wormhole attack, sinkhole attack, and flooding [18]. However, due to space constraints, we omit the discussion of different attacks and their countermeasures, but the same can be found in the relevant literature [8, 21].

In-network processing has a severe impact on the security of sensor network protocols. The conflicting requirements such as en route processing and end-to-end security cannot be realized using the traditional security mechanisms. In addition, as WSNs share similarity with conventional networks, such as wireless networks, security requirements of WSNs remain similar to those found in conventional networks [16]. However, the security requirements of WSNs have significant impact of in-network data aggregation [6, 10] and encrypted data processing [19]. In this section, we discuss the impact of in-network processing on vital security primitives such as privacy, integrity, and freshness.

6.2.1   Privacy

Deployments in hostile environments and lack of physical protection make sensor nodes vulnerable to the node capture attacks. Therefore, nodes that process sensor readings in a raw form are prime targets for attackers. If nodes closer to the base station are being captured, they can have a severe impact on the privacy of gathered sensor readings. One of the solutions that thwart node capture attacks is to process the encrypted sensor readings without decrypting them at intermediate aggregator nodes. Privacy homomorphism helps in processing the encrypted data using public parameters. Although privacy homomorphism protects against passive attackers, it makes sensor readings vulnerable to active attackers whose goal is to modify or inject fake data in the network.

6.2.2   Integrity

In-network processing modifies the original data en route. Hence, traditional mechanisms cannot provide end-to-end integrity verification in data-centric networks. End-to-end integrity verification has been considered as a formidable research issue [3]. However, Parmar and Jinwala [15] show the viability of end-to-end integrity protection in resource-constrained WSNs. In brief, the integrity verification in data-centric networks requires the integrity verification at intermediate nodes as well as at the base station, and it requires the integrity verification of unaggregated as well as aggregated sensor readings. The major obstacles in achieving these objectives are as follows: (1) The use of inherently malleable privacy homomorphism helps not only the genuine aggregator nodes, but also the attackers in modifying the encrypted data. (2) The en route aggregation changes the representation of original data. Hence, it becomes challenging to verify the correctness of aggregated data.

6.2.3   Freshness

The freshness plays a crucial role in the correctness of gathered sensor readings. The replay protection using counter or nonce, in a traditional way, only provides hop-by-hop replay protection. Such replay protection only considers outsider adversaries. However, sensor networks may have compromised intermediate nodes. The replay protection against compromised (or captured) nodes becomes imperative for the correctness of the collected information.

6.3 Privacy Homomorphism

The privacy homomorphism (or encrypted data processing) is a property of cryptosystems that supports processing of encrypted data without decryption. The property is utilized by different WSNs protocols to provide the privacy of sensor readings at vulnerable intermediate nodes.

As shown in Eq. (6.1), the encryption key and the decryption key can be the same for some cryptosystems such as the Domingo-Ferrer's cryptosystem.

$$D_k(E_k(x) + E_k(y)) \mod n = (x + y) \mod n \qquad (6.1)$$

As shown in Eq. (6.2), the encryption key and the decryption key can be different for some cryptosystems such as the Paillier's cryptosystem.

$$D_{k'}(E_k(x) \times E_k(y)) \mod n = (x + y) \mod n \qquad (6.2)$$

6.3.1 Privacy Homomorphism: Addition

As shown in Eq. (6.3), the CMT cryptosystem [1, 2] supports additive homomorphic operations over encrypted data. In addition, cryptosystems proposed by Koblitz [9], Okamoto and Uchiyama [11], Paillier [13], and Domingo-Ferrer [5] also support additive homomorphic operations over encrypted data.

$$D_k(E_k(x) + E_k(y)) \mod n = x + y \mod n \qquad (6.3)$$

6.3.2 Privacy Homomorphism: Multiplication

The RSA cryptosystem [20] supports multiplicative privacy homomorphism. The multiplicative privacy homomorphism enables the computation over encrypted data in such a way that the resultant data when decrypted yield the same result as the product of corresponding unencrypted data. As shown in Eq. (6.4), the decryption of the product of two ciphertexts yields the same result as the product of corresponding plaintexts.

$$D_{K'}(E_k(x) \times E_k(y)) \mod n = x \times y \mod n \qquad (6.4)$$

6.3.3 Privacy Homomorphism: Exclusive OR

The Goldwasser–Micali's cryptosystem [7] is homomorphic with respect to X-OR operations. As shown in Eq. (6.5), in order to compute the X-OR of plaintexts, Goldwasser–Micali's cryptosystem computes the product of corresponding ciphertexts. In Sect. 6.5.3, we discuss the Goldwasser–Micali's cryptosystem [7] and its applications in WSNs.

$$D_{K'}(E_k(x) \times E_k(y)) \mod n = x \oplus y \mod n \qquad (6.5)$$

6.4 Symmetric-Key Based Privacy Homomorphism

In this section, we discuss two symmetric-key based malleable cryptosystems, namely, Domingo-Ferrer's Cryptosystem and CMT Cryptosystem. The detailed discussion of these cryptosystems and their applications in WSNs is presented by Parmar and Jinwala [14]. Hence, in this section, we only present an elegant way to analyze these cryptosystems.

6.4.1 Domingo-Ferrer's Cryptosystem

Domingo-Ferrer's cryptosystem [5] supports encrypted data processing when data are encrypted using the same key. As shown in Fig. 6.1, we present the encryption and decryption operations of Domingo-Ferrer's cryptosystem. In Domingo-Ferrer's cryptosystem, the size of parameter d affects the size of ciphertext. Domingo-Ferrer's symmetric-key based cryptosystem uses a secret parameter r for encryption and computes r^{-1} for corresponding decryption.

As shown in Fig. 6.2, each plaintext is divided into d sub-plaintexts, and each sub-plaintext is encrypted using a secret parameter r and a public parameter n.

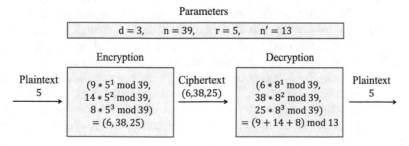

Fig. 6.1 Encryption and decryption using Domingo-Ferrer's cryptosystem

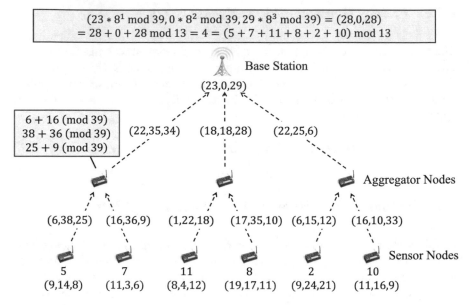

Fig. 6.2 Tree-based data aggregation using Domingo-Ferrer's cryptosystem

As shown in the tree-based data aggregation topology described in Fig. 6.2, the aggregator node performs computation over encrypted data using the public parameter n. The decryption is carried out at the base station, and it requires an inverse of the secret parameter r. In addition, the decryption operation requires the scalar product of r^{-1} and coordinates of the aggregated ciphertext.

6.4.2 CMT Cryptosystem

Castelluccia et al. [1, 2] adopted the well-known Vernam cipher to propose a provably secure additive aggregation scheme often referred to as the CMT cryptosystem. In the CMT cryptosystem, as shown in Fig. 6.3, the encryption operation is performed by the addition of plaintext and secret key, while the decryption operation is performed by subtracting the key from ciphertext. Although encryption and decryption are computationally efficient cryptographic operations, the generation of pseudo-random keys introduces significant computation overhead.

As shown in Fig. 6.4, each sensor node performs the addition of its plaintext and key to produce the ciphertext. The ciphertexts received at aggregator nodes are aggregated using the modular addition. The base station in CMT cryptosystem subtracts the aggregated key from the aggregated ciphertext to retrieve the aggregated plaintext.

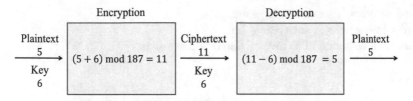

Fig. 6.3 Encryption and decryption using CMT cryptosystem

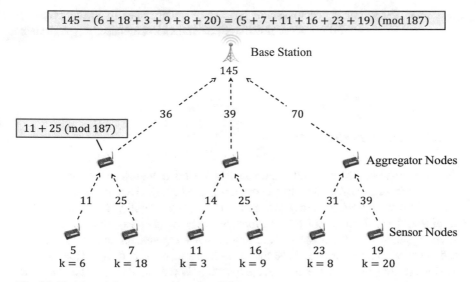

Fig. 6.4 Tree-based data aggregation using CMT cryptosystem

Table 6.1 Comparison of symmetric-key based malleable cryptosystems

Cryptosystem	Key management	Operation(s)	Message expansion
Castelluccia [2]	The base station shares a unique secret key with each node in the network	$\oplus \ \ominus \ \otimes_c$	1
Domingo-Ferrer [5]	The base station shares a global secret key across the network	$\oplus \ \ominus \ \otimes \ \otimes_c$	$\dfrac{d \times n}{n'}$

In Table 6.1, we present the comparison of symmetric-key based malleable cryptosystems.

- $\oplus$—Homomorphic addition
- $\ominus$—Homomorphic subtraction
- $\otimes$—Homomorphic multiplication
- $\otimes_c$—Homomorphic multiplication with a constant

- n—Randomly generated large integer
- n'—Integer $n' > 1$ such that $n' \mid n$
- d—Plaintext should be divided into $d > 2$ sub-parts

6.5 Asymmetric-Key Based Privacy Homomorphism

Asymmetric-key based cryptosystems have been widely used in WSNs protocols [12, 14]. The detailed discussion of different asymmetric-key based cryptosystems and their applications in WSNs protocols is presented by Parmar and Jinwala [14]. Hence, in this section, we only focus on an exemplary introduction of asymmetric-key based cryptosystems adopted by numerous WSNs protocols.

6.5.1 RSA Cryptosystem

In 1978, Rivest, Shamir, and Adleman introduced a method to implement the asymmetric-key cryptosystem that is often regarded as the RSA cryptosystem [20]. The security of RSA cryptosystem relies on an intractability of factoring large numbers. The advantage of RSA cryptosystem over other asymmetric-key based cryptosystems is that it does not have any message expansion, i.e., a plaintext and the corresponding ciphertext have the same block size ($m, c \in \mathbb{Z}_n$). However, the advantage that restricts the message expansion is due to the fact that the RSA cryptosystem does not use any random components during the encryption. Hence, the biggest advantage of not expanding the message in turn be the biggest drawback of the RSA cryptosystem. In addition, due to the deterministic nature of RSA cryptosystem, it remains semantically insecure [7].

The RSA cryptosystem supports multiplicative homomorphic operations over encrypted data. However, applications of concealed data aggregation require support for additive homomorphism. Therefore, concealed data aggregation protocols have not adopted the RSA cryptosystem to ensure the privacy of sensor readings at intermediate nodes. However, it is amongst the first asymmetric-key based cryptosystems that have been applied in WSNs to analyze the feasibility of asymmetric-key based cryptosystems in WSNs.

6.5.2 Example

In this section, we present an example of the RSA cryptosystem. Although for the ease of calculation, we use small parameters. However, the same can be extended for more realistic parameter settings. As shown in Fig. 6.5, a plaintext $m = 23$ is encrypted using the encryption key $e = 7$ and public parameter $n = 187$, while

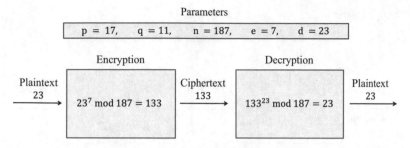

Parameters

$$p = 17, \quad q = 11, \quad n = 187, \quad e = 7, \quad d = 23$$

Encryption

Plaintext
23
$\rightarrow$ $23^7 \bmod 187 = 133$

Ciphertext
133
$\rightarrow$

Decryption

$133^{23} \bmod 187 = 23$

Plaintext
23
$\rightarrow$

Fig. 6.5 Encryption and decryption using RSA cryptosystem

$$(169^{23} \bmod 187 = 152 = 23 * 27 * 14 * 9 * 37 * 42 \bmod 187)$$

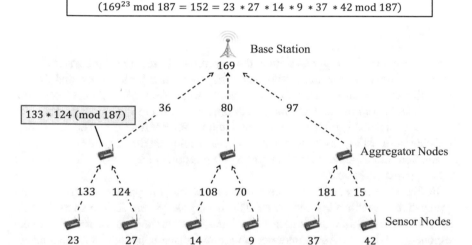

Fig. 6.6 Tree-based data aggregation using RSA cryptosystem

a ciphertext $c = 133$ is decrypted using the decryption key $d = 23$ and public parameter $n = 187$. Here, the encryption key e is publicly available, while the security of RSA cryptosystem relies on the secrecy of private key d. The example presents the deterministic nature of RSA cryptosystem where each plaintext m is converted into the same ciphertext c, if the parameters e and n remain the same.

The RSA cryptosystem enables multiplicative homomorphic operations over encrypted data. As shown in Fig. 6.6, leaf nodes encrypt the sensor readings using a public key e, while intermediate nodes compute the product of ciphertexts using a public parameter n. The base station decrypts the ciphertext using a private key d. The resultant ciphertext received at the base station, when decrypted, yields the same result as the aggregation of individual plaintexts.

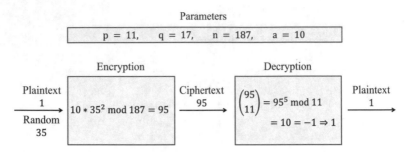

Fig. 6.7 Encryption and decryption using Goldwasser–Micali's cryptosystem

6.5.3 Goldwasser–Micali's Cryptosystem

Goldwasser–Micali's cryptosystem is the first provable secure cryptosystem based on intractability of the quadratic residuosity problem. Goldwasser and Micali formalize the notion of semantic security. In Goldwasser–Micali's cryptosystem, the plaintext is represented as a bit zero or one. As shown in Fig. 6.7, a plaintext bit 1 can be encrypted using the public key a and public parameter n. The decryption in Goldwasser–Micali's cryptosystem requires to compute the Legendre symbol $\left(\frac{c}{p}\right)$. In addition, Goldwasser–Micali's cryptosystem requires the secret primes p and q to decrypt the ciphertext.

In Fig. 6.8, an example is presented to show the encryption, decryption, and aggregation operations in Goldwasser–Micali's cryptosystem. As shown in Fig. 6.8, ciphertexts are multiplied at intermediate nodes to attain the X-OR effect on the corresponding plaintexts. Aggregator nodes require the public parameter n to aggregate the ciphertexts. The base station in Goldwasser–Micali's cryptosystem decrypts the aggregated ciphertext using the Legendre symbol and secret parameter p.

6.5.4 Okamoto–Uchiyama's Cryptosystem

Okamoto and Uchiyama [11] proposed a provably secure additive homomorphic cryptosystem. The proposed cryptosystem is semantically secure under the p-subgroup assumption. Security of Okamoto–Uchiyama's cryptosystem relies on the intractability of factoring $n = p^2q$. However, the fastest algorithm for factoring a composite number n is the number field sieve algorithm [11]. The running time of the number field sieve algorithm depends on the size of a composite number n. Hence, the parameters of Okamoto–Uchiyama's cryptosystem should be chosen such that the size of $n = p^2q$ remains the same as the size of $n = pq$ of the 1024-bit RSA cryptosystem [20]. As shown in Fig. 6.9, Okamoto–Uchiyama's probabilistic cryptosystem uses the random number r and plaintext p to produce the corresponding ciphertext c. The encryption is performed using the public keys

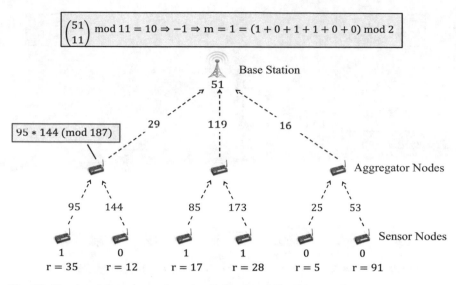

Fig. 6.8 Tree-based data aggregation using Goldwasser–Micali's cryptosystem

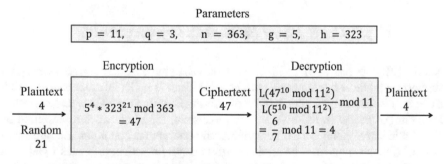

Fig. 6.9 Encryption and decryption using Okamoto–Uchiyama's cryptosystem

g and h, whereas the decryption is carried out using the private key p. The division operation in Okamoto–Uchiyama's cryptosystem requires the multiplicative inverse of denominator.

The encryption, decryption, and aggregation operations in a tree-based data aggregation topology are described in Fig. 6.10. In Okamoto–Uchiyama's cryptosystem, the ciphertexts are multiplied together to attain the addition effect on the corresponding plaintexts. The aggregator nodes use a public parameter n to aggregate the ciphertexts.

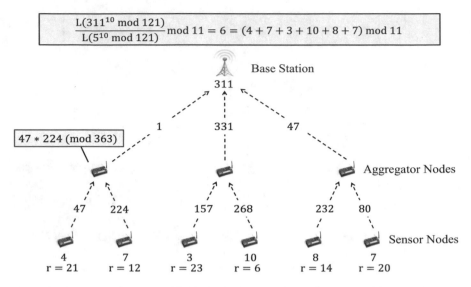

Fig. 6.10 Tree-based data aggregation using Okamoto–Uchiyama's cryptosystem

6.5.5   Elliptic Curve Based ElGamal's Cryptosystem

Koblitz [9] proposed the first elliptic curve based asymmetric-key cryptosystem with support for additive homomorphism. The elliptic curve ElGamal cryptosystem (EC-ElGamal) is based on an intractability of solving Elliptic Curve Discrete Logarithm Problem (ECDLP). In EC-ElGamal cryptosystem, the plaintext is represented as an elliptic curve point before performing the encryption operation. As shown in Fig. 6.11, the example uses an affine coordinate system with coordinates x and y. In the same way, a plaintext value can be represented using other coordinate systems such as projective coordinate system, Jacobian coordinate system, etc. For example, the projective coordinate system requires three coordinates x, y, and z to represent the plaintext as an elliptic curve point. The encryption in EC-ElGamal cryptosystem produces two ciphertext points c_1 and c_2 on the elliptic curve $E(\mathbb{F}_p)$. The decryption in EC-ElGamal cryptosystem produces a single point on the elliptic curve $E(\mathbb{F}_p)$. The elliptic curve point generated by the decryption operation needs to be mapped back to the corresponding plaintext value.

Figure 6.12 shows an example of encryption, decryption, and aggregation operations in a tree-based data aggregation topology. The reverse mapping function in EC-ElGamal cryptosystem is based on brute-force techniques. However, due to the resource-rich base station and limited message space, the EC-ElGamal cryptosystem becomes the widely adopted asymmetric-key cryptosystem for reverse multicast traffic of resource-constrained WSN.

In Table 6.2, we present the comparison of asymmetric-key based malleable cryptosystems. The RSA cryptosystem supports multiplicative homomorphic opera-

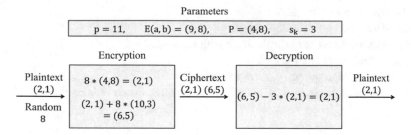

Fig. 6.11 Encryption and decryption using elliptic curve based ElGamal's cryptosystem

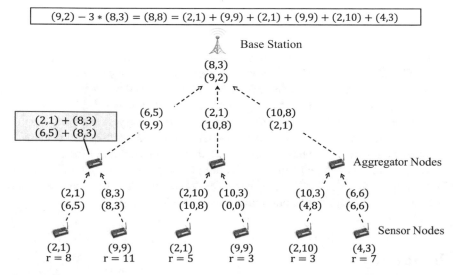

Fig. 6.12 Tree-based data aggregation using elliptic curve based ElGamal's cryptosystem

Table 6.2 Comparison of asymmetric-key based malleable cryptosystems

Cryptosystem	Security assumption(s)	Homomorphic operation(s)	Message expansion
RSA [20]	Integer factorization and RSA problem	$\otimes$	1
Goldwasser–Micali [7]	Quadratic residuosity problem	X-OR	n
Okamoto–Uchiyama [11]	p-Subgroup assumption	$\oplus \ \ominus \ \otimes_c$	$\dfrac{n}{2^{k-1}}$
EC-ElGamal [9]	ECDLP	$\oplus \ \ominus \ \otimes_c$	2 (+ 2-bit)

tions, Goldwasser–Micali's cryptosystem supports X-OR homomorphic operations, and Okamoto–Uchiyama cryptosystem and EC-ElGamal cryptosystem supports additive homomorphic operations over encrypted data.

- $\oplus$—Homomorphic addition
- $\otimes$—Homomorphic multiplication
- $\otimes_c$—Homomorphic multiplication with a constant
- $\ominus$—Homomorphic subtraction
- k—Size of large primes p (or q) in bits
- n—Size of ciphertext space such that the factorization of n is hard

6.6 Conclusion

A malleability property of cryptosystems has been often considered as an undesirable property due to its negative impact on the integrity of encrypted data. However, the malleability property has interesting applications in resource-constrained WSNs. The malleability property is used in WSNs to ensure the privacy of sensor readings at vulnerable intermediate nodes of WSNs. In addition, the integrity of sensor readings can also be maintained along with the privacy of sensor readings through different mechanisms. In this chapter, we discuss various encryption algorithms that have been used in WSNs to ensure the privacy of sensor readings at vulnerable intermediate nodes. To the best of our knowledge, the discussion of the algorithms presented in the chapter helps in understanding complex cryptographic algorithms. Cryptosystems discussed in the chapter can be applied to other research areas, such as Internet of Things, Cloud Computing, and Network Coding.

References

1. Castelluccia, C., Chan, A. C. F., Mykletun, E., & Tsudik, G. (2009). Efficient and provably secure aggregation of encrypted data in wireless sensor networks. *ACM Transactions on Sensor Networks (TOSN), 5*(3), 20:1–20:36. DOI 10.1145/1525856.1525858.
2. Castelluccia, C., Mykletun, E., & Tsudik, G. (2005). Efficient aggregation of encrypted data in wireless sensor networks. In *Proceedings of the 2nd Annual International Conference on Mobile and Ubiquitous Systems: Networking and Services, MOBIQUITOUS* (pp. 109–117). Washington, D.C., USA: IEEE. DOI 10.1109/MOBIQUITOUS.2005.25.
3. Chan, A. C. F., & Castelluccia, C. (2008). On the (im)possibility of aggregate message authentication codes. In *Proceedings of the International Symposium on Information Theory, ISIT* (pp. 235–239). Toronto, Canada: IEEE. DOI 10.1109/ISIT.2008.4594983.
4. Dolev, D., Dwork, C., & Naor, M. (1991). Non-malleable cryptography. In *Proceedings of the 23rd Annual Symposium on Theory of Computing, STOC* (pp. 542–552). New Orleans, USA: ACM. DOI 10.1145/103418.103474.
5. Domingo-Ferrer, J. (2002). A provably secure additive and multiplicative privacy homomorphism. In *Proceedings of the 5th International Conference on Information Security, ISC, Lecture Notes in Computer Science* (Vol. 2433, pp. 471–483). Sao Paulo, Brazil: Springer-Verlag. DOI 10.1007/3-540-45811-5_37.
6. Fasolo, E., Rossi, M., Widmer, J., & Zorzi, M. (2007). In-network aggregation techniques for wireless sensor networks: a survey. *Wireless Communications, 14*(2), 70–87. DOI 10.1109/MWC.2007.358967.

7. Goldwasser, S., & Micali, S. (1984). Probabilistic encryption. *Journal of Computer and System Sciences, 28*(2), 270–299. DOI 10.1016/0022-0000(84)90070-9.
8. Karlof, C., & Wagner, D. (2003). Secure routing in wireless sensor networks: attacks and countermeasures. *Ad Hoc Networks, 1*(2–3), 293–315. DOI 10.1016/S1570-8705(03)00008-8.
9. Koblitz, N. (1987). Elliptic curve cryptosystems. *Mathematics of Computation, 48*(177), 203–209. DOI 10.1090/S0025-5718-1987-0866109-5.
10. Krishnamachari, B., Estrin, D., & Wicker, S. (2002). The impact of data aggregation in wireless sensor networks. In *Proceedings of the 22nd International Conference on Distributed Computing Systems, ICDCSW* (pp. 575–578). Vienna, Austria: IEEE. DOI 10.1109/ICD-CSW.2002.1030829.
11. Okamoto, T., & Uchiyama, S. (1998). A new public-key cryptosystem as secure as factoring. In *Proceedings of the International Conference on the Theory and Application of Cryptographic Techniques, Advances in Cryptology, EUROCRYPT, Lecture Notes in Computer Science* (Vol. 1403, pp. 303–318). Espoo, Finland: Springer-Verlag. DOI 10.1007/BFb0054135.
12. Ozdemir, S., & Xiao, Y. (2009). Secure data aggregation in wireless sensor networks: a comprehensive overview. *Computer Networks, 53*(12), 2022–2037. DOI 10.1016/j.comnet.2009.02.023.
13. Paillier, P. (1999). Public-key cryptosystems based on composite degree residuosity classes. In *Proceedings of the 17th International Conference on Theory and Application of Cryptographic Techniques, EUROCRYPT, Lecture Notes in Computer Science* (Vol. 1592, pp. 223–238). Prague, Czech Republic: Springer-Verlag. DOI 10.1007/3-540-48910-X_16.
14. Parmar, K., & Jinwala, D. C. (2016). Concealed data aggregation in wireless sensor networks: A comprehensive survey. *Computer Networks, 103*(7), 207–227. DOI 10.1016/j.comnet.2016.04.013.
15. Parmar, K., & Jinwala, D. C. (2016). Malleability resilient concealed data aggregation in wireless sensor networks. *Wireless Personal Communications, 87*(3), 971–993. DOI 10.1007/s11277-015-2633-6.
16. Perrig, A., Szewczyk, R., Tygar, J. D., Wen, V., & Culler, D. E. (2002). SPINS: security protocols for sensor networks. *Wireless Networks, 8*(5), 521–534. DOI 10.1023/A:1016598314198.
17. Peter, S., Westhoff, D., & Castelluccia, C. (2010). A survey on the encryption of convergecast traffic with in-network processing. *IEEE Transactions on Dependable and Secure Computing, 7*(1), 20–34. DOI 10.1109/TDSC.2008.23.
18. Raymond, D. R., & Midkiff, S. F. (2008). Denial-of-service in wireless sensor networks: attacks and defenses. *IEEE Pervasive Computing, 7*(1), 74–81. DOI 10.1109/MPRV.2008.6.
19. Rivest, R. L., Adleman, L., & Dertouzos, M. L. (1978). On data banks and privacy homomorphisms. *Foundations of Secure Computation, 4*(11), 169–180.
20. Rivest, R. L., Shamir, A., & Adleman, L. (1978). A method for obtaining digital signatures and public-key cryptosystems. *Communications of the ACM, 21*(2), 120–126. DOI 10.1145/359340.359342.
21. Wang, Y., Attebury, G., & Ramamurthy, B. (2006). A survey of security issues in wireless sensor networks. *IEEE Communications Surveys & Tutorials, 8*(2), 2–23. DOI 10.1109/COMST.2006.315852.

Chapter 7
A Survey and Taxonomy on Data and Pre-processing Techniques of Intrusion Detection Systems

Tarfa Hamed, Jason B. Ernst, and Stefan C. Kremer

7.1 Introduction

Network security has become one of the most important fields of research in the area of digital communications in the last 10 years since the wide prevalence of the Internet and its applications over the world. Due to the huge amount of information on the Internet and its importance, the security has become an issue that needs to be solved. Therefore, according to [62], Intrusion Detection (ID) topic is being studied excessively in computer networks in the last years. An ID is responsible for detecting any inappropriate activity on a network. The idea of ID has been developed to a system now called Intrusion Detection System (IDS). The IDSs use particular technique(s) to detect attacks and alert network administrators [25]. An IDS aims to discover any violation to the confidentiality, availability, and integrity threatens the data over the network. Such a violation can range in severity from allowing an unauthorized intruder to read a small amount of data in a network to the breakdown of the entire network of interconnected devices.

In general there are two common types of IDSs according to their nature of detection: *Anomaly-based IDS* and *signature-based IDS* (or *misuse-based IDS*).

a- An *anomaly-based IDS* involves learning the baseline or the normal behavior of a system and any deviation in the observed behavior from the normal behavior will be considered an intrusion. The advantage of this type is its ability to detect

T. Hamed (✉) • S.C. Kremer
School of Computer Science, University of Guelph, Guelph, ON, Canada
e-mail: tyaseen@uoguelph.ca; skremer@uoguelph.ca

J.B. Ernst
Left Inc., Vancouver, BC, Canada
e-mail: jason@left.io

© Springer International Publishing AG 2018
K. Daimi (ed.), *Computer and Network Security Essentials*,
DOI 10.1007/978-3-319-58424-9_7

novel attacks (because it depends on comparing to normal behavior) while its disadvantage is that it suffers from a high false positive rate [16].

b- A *misuse-based IDS*, on the other hand, uses attack signatures of known intrusions, compares them with the observed signatures, and considers any match between them as an *intrusion*. The advantage of this type is that it has high detection rate but its disadvantage is that it cannot detect novel attacks (because it depends on comparing to known signatures) [51].

Many algorithms have been developed to address the problem of identifying intrusions on the network to keep up with many novel attacks. These attacks are also being developed and are using more sophisticated methods than in the past. In this chapter, the focus is on the components that most intrusion detection approaches utilize to achieve the goal of detecting the intrusions. Rather than just limiting the review to the intrusion detection system per se, the chapter also looks at techniques used to build and train the systems, and datasets used, thereby reviewing the broader field of study, as opposed to just the proposed solutions. We call this approach a *components-based approach* to distinguish it from a *paper-by-paper* approach to surveying a field, as was used in, for example, [19]. This unique approach gives the reader a different view of the field since it details the constituent components of intrusion detection systems. This enables an easier comparison of methods, and highlights commonalities in the overall approaches.

It has been discovered through this chapter that most intrusion detection systems employ three major phases: **pre-processing, detection, and empirical evaluation**. Some algorithms are implemented just for anomaly detection, while others are implemented for both anomaly and misuse detection. Only a few algorithms are used to deal with malicious software; these are also discussed in this chapter. This chapter covers how to collect the data, how to prepare this data for different types of processing, and a wide variety of pre-processing techniques employed in IDSs. We opted for a *component-by-component* organization for this chapter, rather than a *paper-by-paper* organization, since we believe this will give the reader a broader perspective about the process of constructing an IDS. The main components of an intrusion detection system covered in this chapter are shown in Fig. 7.1.

We think that the reader can acquire an extensive knowledge about the IDS by selecting required components and following components of the diagram presented in Fig. 7.1. In this way, a new IDS may be constructed through the novel combination of previously proposed components. This chapter is devoted and connected to the sequence of the diagram, so that each box in the diagram (in the dashed areas only) is explained in a dedicated section or subsection of the chapter.

Two types of rectangles are used in drawing Fig. 7.1: the rounded rectangle which refers to data (either input data or output data), and the normal rectangle which represents a process. In both cases, when a rectangle (normal or rounded) contains other rectangles, the outer rectangle is a category or classification of a process or data. The inner rectangles represent particular examples of these general categories. It can be noticed that after each processing step there is a rounded rectangle which refers to the data resulting from that processing.

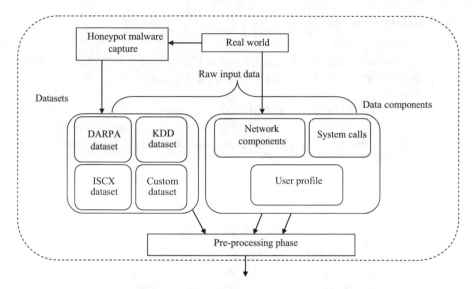

Fig. 7.1 The IDS components covered in this chapter

The rest of the chapter is organized, as per Fig. 7.1, Sect. 7.2 lists the real-world resources that serve as data for some IDSs. Section 7.3 explains Honeypots, their types, and roles in IDSs. Section 7.4 describes each type of the raw input data of IDSs in more detail. Section 7.5 presents the pre-processing phase on the raw input data and a wide range of pre-processing techniques.

7.2   Real World

To begin to understand the network intrusion problem, a researcher needs to deal with the context of the intrusion to know where the vulnerabilities are, and how they can be fixed. In light of this, it is started with discussing the types of real-world attacks. Next, details on how information on these attacks may be collected are provided. This section is concluded with a classification of the types of raw data the IDS may encounter. The context (real world) is the focus of the first section of this chapter, detailed below. The real-world environment of IDS is a computer network (either private or public network) and the incoming and outgoing packets. The IDS needs to examine the network packets and extract the required information for detecting any kind of intrusions that might be occurring in the network. Computer networks might be exposed to many kinds of malwares. However, a malware (which stands for malicious software) is any code added, changed, or discarded from a software system that leads to or damages the desired function of the system [28, 60]. Malwares can be one of the following forms [28]:

A- **Direct malicious programs**

   a. Viruses
     A virus is a harmful code that has the ability to bind itself to host programs
     and reproduce when executing an infected program.
   b. Worms
     Worms are malicious programs that target networked computers. Worms
     differ from viruses in strategy; they do not attach themselves to a host
     program. However, worms accomplish planned attacks to jump from machine
     to machine across the network.
   c. Trojan Horses
     Trojan horses are also malicious programs that hide in host programs and
     pretend to be innocent, but in fact have an alternate intent, like capturing
     passwords.
   d. Attack scripts
     Attack scripts usually exploit any security vulnerability across the network,
     to carry out an attack. Exploiting a buffer overflow is an example of an attack
     script which is usually performed by "smashing the stack."

B- **Malicious programs' assistance**

   a. Java Attack Applets
     Are attack programs that completely clone a website to provide an entry to
     the system through a web browser.
   b. Dangerous ActiveX controls
     Are program components that help a malicious program to exploit appli-
     cations or the operating system through a web browser by downloading a
     malicious program onto the user's computer.

The second group of enemies of computer networks are called intrusions which
represent series of actions that violate the network security goals (*Confidentiality,
Integrity, and Availability*) [52]. In order to capture the patterns of malware,
Honeypots are often used. We will discuss Honeypots in more detail in the next
section.

7.3 Honeypots

Honeypots are systems that can emulate real vulnerabilities, or weaknesses, such
as an easily guessable SSH password, in order to attract and record activity of
attackers and network worms for the purpose of studying their strategies [24].
Malware capturing involves using network Honeypots to emulate weak points of
JYSK network services to collect attack vectors in a controlled environment [43].
This controlled environment is called Sandnet or a Honeynet. A Sandnet is usually
adjusted during each repetition by changing the date, the network environment, the
host operating system, and the time of the day to elicit different patterns of malware

communication [43]. For example, the malware binaries in [60] were collected during 2 years from different sources like Honeypots, spam traps, and anti-malware vendors.

Vulnerabilities are the entrances to a system by which threats are realized. If there are no vulnerabilities, then there is no risk of threats [37]. Honeypots are usually categorized as either *high-interaction* or *low-interaction* as explained below [24]:

1. High-interaction: Are systems with a real, non-emulated OS installed on them that can be exploited as a trap to be accessed and explored by attackers. Virtual machines or physical machines can be the compromised systems that can be reset after they are compromised [18].
2. Low-interaction: Are systems that only simulate parts of an operating system, such as certain network protocols. These systems are most commonly used to collect malwares by being "exploited" by other malware-infected systems. One example of a low-interaction Honeypot is *Nepenthes* [4]. *Nepenthes* is one of the most popular and commonly deployed low-interaction Honeypots on the Internet. The *Nepenthes* Honeypot has the ability to emulate Microsoft vulnerabilities that systems scanning can remotely exploit by the Internet. Nepenthes can be run on different operating systems, such as *Windows* via **Cygwin**, *Mac OS X*, *Linux*, and *BSD* [24].
3. Hybrid Systems: This type involves combining information from both as noted in some of the lessons learned by the authors of [3]. One example of a hybrid system is *SweetBait* [39]. This system is used in detecting 0 day exploits. The authors of previous approaches were criticized for having four problems: (1) false positives, (2) no checks for variations of a worm, just instances, (3) no capture of activity level of worm, and (4) only effective at known attacks.

Table 7.1 presents a comparison between Honeypots in terms of their focus, type, approach, and their references.

Table 7.1 Comparison of honeypots

Reference	High/Low	Focus	Approach
[18]	High	Malicious web pages	Logarithmic divide-and-conquer (LDAC)
[45]	High	Malicious web pages	Divide and conquer
[3]	High	SSH attacks	GNU Linux 2.6 with PAX patch
[39]	High	Fast worms	Argo
[33]	Low	General malware	Hardware FPGA
[4]	Low	General malware	Software emulation (Nepenthes)
[40]	Low	General malware	Software emulation (Dionaea)
[24]	Low	General attacks	Software emulation (mwcollectd)
[39]	Low	Fast worms	SweetSpot
[6, 41]	Low	General malware	Honeyd
[39]	Hybrid	Fast worms	SweetBait (Argo + SweetSpot)

The next section explains how the IDS deal with raw input data directly to collect some information that helps to detect intrusions.

7.4   Raw Input Data

Each intrusion detection system needs to deal with data as an input in order to analyze the status of the system, if it is under attack or not. This raw input data can be system calls, some layer information, a user profile, or network traffic. Collecting this data can help in constructing the normal behavior (baseline) of the legitimate user to enable IDS to differentiate between the legitimate user and potential intruder. In addition, these datasets can be considered as benchmarks that serve as a useful comparison point between competing IDSs. Some of them are used in order to optimize the search space of the searching algorithm or used for optimizing the storage space like the encoding used in [60] for the system calls. However, raw input data is often divided into datasets and data components as it is explained in the next sections.

7.4.1   Datasets

The incoming raw input data can be converted into an organized dataset by extracting some useful features for network intrusion detection purposes. A lot of efforts have been made in extracting features and building standard datasets for intrusion detection. Datasets can be customized by researchers depending on the attacks, or standardized benchmark datasets such as DARPA and KDD may be used. More details will be provided on these in the following subsections.

7.4.1.1   DARPA Dataset

Some IDS deal with standard datasets to train/test the system rather collecting the information and building a custom dataset. As mentioned previously, the benefit of these standardized datasets is to allow for easy comparison and evaluation of different IDS approaches. In this subsection, some research examples which use the DARPA datasets are given. For the work presented in [63], the selected dataset was the DARPA dataset. The DARPA dataset was constructed by *MIT's Lincoln Laboratory* for ID purposes. The first step of the experiment was taking a subset of DARPA dataset which consists of huge number of connections. Those connections represent either normal or malicious traffic, and fall in one of the following five classes: **Normal, Denial of Service (DoS)** attacks, **Remote to Local (R2L)** attacks, **User to Root (U2R)** attacks, and **Probing attacks (Probe)**. The second step involved dividing the entire dataset into two portions: a *training dataset* and a *testing*

Table 7.2 References that use training/testing datasets for intrusion detection

Reference	Training dataset	Testing dataset
[21, 46, 49, 63]	Yes	Yes
[43, 60]	Yes	No
[9]	No	Yes

dataset. The training dataset contained 494,020 connections and the testing dataset contained 311,029 connections [63]. However, not all the IDSs use a predefined testing dataset for testing purposes as can be seen in Table 7.2.

For the Enhanced Support Vector Machine (SVM) in [49], a DARPA IDS dataset was collected at the *MIT Lincoln Labs* which contained a huge variety of intrusions for IDS training and testing purposes. The TCP/IP network traffic was recorded in TCPdump format [61]. The data consisted of 3 weeks of *training* data and 2 weeks of *testing* data.

There are conflicting views on how useful the DARPA dataset is for training and evaluating IDSs. Thomas et al. [59] concluded it was useful with *Snort, Cisco IDS, PHAD* [27], and *ALAD*. However, they also noted that improvements to DARPA 1999 could be made to make the dataset more "real" and fairer to a wider variety of IDSs. Conversely, Brugger and Chow [57] claim that Snort does not perform well with the DARPA dataset. They reasoned that the DARPA set "includes only a limited number of attacks that detectable with a fixed signature." Brugger and Chow conclude that while any IDS should be able to perform well on the DARPA dataset, an IDS that does perform well is not necessarily a good IDS. They argue that further training and testing with other datasets is required to fully evaluate an IDS.

7.4.1.2 KDD Dataset

Lincoln Laboratory at *MIT* developed a dataset for intrusion detection purposes called KDD dataset in 1999. The *Knowledge Discovery and Data mining* (KDD) dataset contains network data extracted by TCPdump. The KDD training dataset consists of 4,900,000 single connection vectors where each single connection vector consists of 41 features. Same as with the DARPA dataset, the connections are labeled to one of the same five categories [48].

Some researchers do not use the whole KDD dataset for their experiments, rather they just extract a subset from the entire dataset. In [29], they used NSL-KDD dataset for the experiment which contains 22 different types of attacks as well as one type of normal data. The NSL-KDD dataset is an improved version of the KDD dataset with selected records from the KDD dataset and with the same number of features. For example, in [13], the authors used NSL-KDD for both binary and five class classification. However, some of the features are claimed to be not important in implementing an IDS. Therefore, the proposed model in [11] excluded some features from the NSL-KDD dataset in order to reduce model building time or training time of the classifier. Another issue with KDD-99 is that the dataset may contain some redundant records. Experimentally, in [55], the authors found that the KDD-99 dataset contains many redundant records which may bias IDS training towards frequently occurring records. The NSL-KDD set fixes this problem. Another example is presented in [64] which used a mobile agent to detect anomalies [29].

Similar to the DARPA dataset, the KDD dataset has also been criticized for being "too synthetic" [17]. Kayacik and Zincir-Heywood [17] showed that both DARPA and KDD have low dispersion compared to 2 gigabytes of real-world data collected from one day in December 2003 from a University of Dalhousie computer science server. Furthermore, they found that KDD is dissimilar to the real-world data collected. Both the similarity and the dispersion were computed using Euclidean distances between training centroids of the datasets.

7.4.1.3   ISCX Dataset

This dataset has been generated by the *Information Security Centre of Excellence (ISCX)* at **University of New Brunswick** in 2012 [50]. The dataset involves real traces analyzed to create profiles for agents that generate real traffic for HTTP, SMTP, SSH, IMAP, POP3, and FTP. The generated dataset contains different features including full packet payloads in addition to other relevant features such as total number of bytes sent or received. The full dataset has been captured in a period of 7 days and involved about more than one million network trace packets and 20 features, and every data example has been labeled as one of the two classes (normal or attack) [66]. The ISCX dataset has acquired the security community's attention and become a benchmark dataset for intrusion detection research purposes due to its realistic traffic, labeled connections, and its multiple attack scenarios [50]. The dataset has been designed to overcome the technical limitations of other intrusion detection datasets, and to prepare network traces by capturing contemporary legitimate and intrusive network behaviors and patterns [54].

However, the DARPA, KDD, and ISCX datasets are not the only datasets used by the network security community. We are mentioning only their names here for reference for space restrictions. They are the **LBNL** dataset [35] and **CAIDA** [1].

7.4.1.4   Custom Datasets

There has been an extensive effort by many researchers to construct a report or a dataset that contains the required information for detecting specific intrusion(s) on the network. The advantage behind building a custom dataset over using a benchmark (like KDD, DARPA, or ISCX) is that new attacks evolve every day. Benchmarks may not contain information about those attacks. Building a dedicated dataset may be the best way to detect a new type of intrusion. However, in order to compare two or more methods on such a novel dataset, all methods need to be implemented and applied to the dataset. This can be both time-consuming and

Table 7.3 Features used in custom datasets and their usage

Feature	Reference	Used for
System calls (MIST)	[60]	Detecting malwares
Unix commands	[21]	Detecting anomalies
Network flows	[43]	Detecting malwares
RSSI	[8]	Detecting anomalies in WSN
TCP connection features	[22]	Detecting DoS, U2R, R2L, and probe
KDD dataset records	[29]	Detecting DoS, U2R, R2L, and probe
User profile (CPU and memory usage, visited websites, number of windows, typing habits, application usage)	[38]	Detecting masquerading
Some KDD dataset features	[64]	Detecting anomalies
TCP/IP network traffic info using (TCPdump)	[46]	Detecting DoS and probe, IP sweep, ping of death, smurf

difficult if the authors of previous approaches do not share their software. These custom datasets or reports are built from multiple sources like *UNIX commands*, *system calls*, *network flows*, and others. Table 7.3 lists the features that are used in constructing custom datasets, their references and the purpose of using those features.

An instruction set report is an example of custom dataset which results from a behavior analysis technique which is used for detecting malicious software. In [60], obtaining a custom dataset begins with defining the instructions where each instruction describes a system call with its arguments using short numeric identifiers. That resulted in designing a whole instruction set for malware programs called *Malware Instruction Set* (MIST). From this point, each system call is encoded using MIST instructions to form the MIST report which will contain the encoding of every system call during run-time of a malware. Then, the resulting report is converted into a vector in order to check the similarity using geometric distances. This representation can be obtained directly during analysis of malware using a behavior monitoring tool or by converting existing behavior reports [60]. For further reading about custom datasets, readers can find these techniques: user masquerading dataset [21, 44, 68], tracking phoning home for malware capturing [43], and intrusion detection in Wireless Sensor Networks (WSN) [8, 47].

7.4.2   Data Components

The second part of the raw input data is the data components. These components can be used to extract some features that help to detect network intrusions. Utilizing one or more of those data components in collecting the input data for the IDS can be very beneficial and expressive since it provides new information about new attacks that might not be available in KDD, DARPA, or ISCX datasets. Some IDSs do not deal with standard datasets nor build custom datasets; instead they deal with the raw input data that come from the real world directly. The reason behind this trend is that the real world contains several types of raw input data that can help to detect an attack. Data components can be separated into network, system, and user components. In the following sections, these data components will be discussed in more detail.

7.4.2.1 Network Components

It is started with network components because this is the lowest level of attack into a system which has not been compromised locally. Network layers can give useful information in intrusion detection systems. One example is *Packet Header Anomaly Detection* (PHAD) [27]. This approach tried to learn normal ranges of values for fields within packet headers over time and then probabilistically indicated the likelihood of an anomaly. While many approaches consider only a single network layer at a time when training and deciding on anomalies, it is also possible to perform cross-layer anomaly detection. In [63], Wang et al. proposed a cross-layer detection system for wireless mesh networks which includes information from the **physical**, **link**, and **network** layers. Some of the extracted features include "channel assignment and channel switching frequency at physical layer, expected transmission time at MAC layer, and routing activities and data forwarding behavior at network layer." From the physical layer also, the *Received Signal Strength Indicator* (RSSI) value is used to detect masquerades, while in the MAC layer with the TDMA technique, it checks if an adversary follows the TDMA schedule, but with S-MAC it checks if the sender is supposed to be sleeping. Both techniques are used to detect masquerades [8].

Network traffic can be useful in extracting some information for the IDS since network traffic is a deep source of information for the ongoing traffic over the network. This deep information represents the incoming and outgoing packets with all their meaningful bits and bytes. In order to analyze network traffic data, some tool is needed to capture the real-world network activity. One of the available and commonly used tools for network analysis is **TCPdump**. **TCPdump** is a UNIX tool used to collect data (packets) from the network, decrypt the bits, and then show the output in a meaningful style. The collected data can be TCP/IP packets or other packets that can be specified from the TCPdump filters [37]. For space restrictions, we are only mentioning some other works that used network components such as defending against DoS SYN flooding attack [15], detecting black-hole attacks [32], collaborative module [56], and using TCPDump to record network traffic [46]. Table 7.4 gives a summary of the most important features used in custom datasets and their usage in addition to the references they are mentioned in.

Table 7.4 Features extracted from network components and their usage

Reference	Approach	Network layer(s)	Network components/features
[27]	PHAD	Network, transport	33 Fields of ethernet, IP, TCP, UDP, and ICMP headers
[63]	Cross-layer wireless mesh	Physical, data link, network, transport	Channel assignment and channel switching frequency at physical layer, expected transmission time at MAC layer, and routing activities and data forwarding behavior at network layer
[8]	Wireless sensor anomaly detection	Physical, medium access control, network, application	RSSI, RTT
[46]	Specification-based intrusion detection	Network	Various IP header fields
[14]	NIDS	Network	Source port, destination port, sequence number, acknowledgement number, URG flag, ACK flag, PSH flag, RST flag, SYN flag, rsv1, rsv2 window size, urgent flag
[14]	NIDS	Transport	Priority, don't fragment flag, more fragment flag, offset, identification number, TTL, protocol, source IP, destination IP
[14]	NIDS	Transport	Source port, destination port
[34]	NIDS	Transport	Port number (Server), port number (Client), data packets in the flow, pushed data packets (client), pushed data packets (server), minimum segment size, average segment size, initial window (client), initial window (server), total no. of RTT samples, median of bytes in IP samples, variance of bytes in IP packet, variance of bytes in Ethernet Packet, application class
[15]	PSO	Transport	SYN packets
[32]	DSR	Network	Data packets
[56]	IDS	Transport, network	Network packets
[2]	NIDS	Transport, network	Network packets

While it is possible to determine anomalous behavior through network traffic and layers, it is also possible to locally examine system calls. The next subsection explains how system calls can be utilized to detect anomalies.

7.4.2.2 System Calls

Anomalies and intrusions may also be detected by attempting to identify irregular usage patterns from users and programs. In [22], Lane and Brodley acquired the input data to build the user profile by collecting training data using the sequence actions of UNIX shell commands. A user profile consists of an order set of fixed-length groups of contiguous commands with their arguments. The pattern of typical commands for a particular user is learned allowing the system to determine when an illegitimate user may be accessing the system. This method has proved its success in detecting anomalies on networks [22].

A similar approach is taken in [63] by Warrander et al.. Instead of tracking user commands, however, system calls within a process or execution thread are tracked. A pattern is established for normal operation of a particular program. When the system calls deviate from this pattern it may be an indication that an intrusion or attack is occurring. Other approaches that used system calls in intrusion detection can be found in [60] that we could not provide more details about for space restrictions.

7.4.2.3 User Profile

To accurately determine the behavior of a user, behavioral IDSs collect data from different sources, which will detect intruders in the system. In this subsection, some of the characteristics of user profiles which are used by IDSs are discussed. In [38], the implemented system acquires data using some characteristic sources in order to create a complete user profile such as Applications running, Number of windows, Websites viewed, Application performance, and Keystroke analysis [38].

The running applications characteristic uses a list of current running processes to identify the processes currently running and to identify if a process has been run on that machine before [38].

The second characteristic, the number of windows is used to acquire data from the graphical user interface layer to identify the user's style of use [38].

The third characteristic is the "websites viewed characteristic" which is used to grab web history data from the browsers installed on the user's computer, by seeking the number of new sites visited by the user and/or the number of times a user has revisited a web page per hour.

The application performance characteristic is used to gain CPU and memory usage data for each application running on the computer to determine if applications are both being used and acting in the desired way.

Table 7.5 User profile characteristics and the references used them

Characteristic name	Reference(s) that used it
Application running	[38]
Websites viewed	[38]
Application performance (CPU and Memory)	[38]
Visited websites	[38]
Keystrokes	[7, 38]
Login frequency	[67]
Session time	[67]
Password failure	[67]
Resource operating frequency	[67]
File operation frequency	[67]
File records (read/write)	[67]

Finally, the fifth characteristic is the keystroke analysis characteristic, which is used to observe delays between typed keys to determine user behavior. This characteristic depends on calculating digraph delays between the user's typing the letters. This digraph is calculated for the previous 100 delays and all have been collected in the real-time of the system to build the user behavior [38].

For further reading about using user profile in collecting data components, readers can also read [67].

In Table 7.5, the previously discussed user profile characteristics are shown along with the references of some of the common approaches in this area.

Now by comparing the above-mentioned data components, several conclusions can be drawn: the network components can give deep insight into network traffic flows. These data components can help in analyzing the incoming traffic and building a signature database of intrusions (in case of misuse detection systems) and can be used in building a model of normal human behavior (in case of anomaly systems). The variety of the obtained information can be considered an advantage, and this advantage comes from the fact that the network component can be extracted from different layers as shown in Table 7.4. The second type of data component is the system calls which are used in building a user profile of normal operation of a particular program. This approach suffers from false positive errors. In addition, the database needs a periodic update since normal user actions sometimes change [22]. The third type is the user profile, which can be used to build an effective IDS sometimes, but still suffers from false positives as the second type.

It is also worth mentioning here that there are some IDS applications for academic and personal use such as **Snort** and **Bro**. **Snort** is an open-source software that is provided with the sniffing capability to sniff network packets and help in analyzing these packets using some rules [30]. **Bro** is also an open-source, powerful network analysis framework that contains number of various rules to identify anomaly behavior in the data [5].

Now, having explained the most well-known raw-data types for IDS, we need to consider how they will be pre-processed and why; this is the subject of the next section.

7.5 Pre-processing Phase

Generally, the intrusion detection algorithms deal with one or more of the raw input data types mentioned above. The values of data types can represent: *network traffic*, *Unix shell commands*, *user profiles*, *system-calls*, *custom dataset*, or *standard dataset* as can be seen in Fig. 7.1. These values need some preliminary processing in order to obtain more expressive information from them to simplify the classification phase and make it more accurate, efficient, and faster. Different algorithms use different pre-processing steps according to the intrusion detection algorithm, type of the intrusion being detected, and the type of dataset used. The rest of this section is broken down according to the types of pre-processing approaches applied in a variety of references.

7.5.1 Specification Method

In the specification-based method [46] which the researcher used to detect anomalies, phase 1 is divided into two steps: **Extended Finite State Automata (EFSA)** representation and the **specification development**.

The first step is the *EFSA representation*, which is used for network protocol modeling. Three states are used in this representation: *INIT*, *PCT RCVD*, and *DONE*. This representation includes adding each IP machine for each IP packet received. The representation is limited to capturing the packets coming from the internet rather than those going to the internet. The packet is given to every IP state machine, then each can make a transition based on this packet. Any state machine that reaches the *DONE* state is deleted from the list. Each transition has a label which represents the condition of moving from the source state to the destination state. For example, when the packet is directed to the gateway itself, then it takes the transition from *INIT* state to DONE state according to the transition label (to the gateway) between them which specifies if the packet is going to the gateway [46].

The second step in phase 1 of the *specification-based approach* is the specification development when only important details are captured in most protocols. In anomaly detection: net packets are critical sources of information because of the large size of data and raw data is difficult to analyze and differentiate from noise.

In the next stage, mapping packet sequence properties to properties of state-machine transitions is applied to convert the raw data into state-machine transitions. This stage involves partitioning the sequence of packets at the gateway into one of the three types of traces.

This partitioning has some advantages such as it structures the packets in an organized fashion and minimizes the storage space required by the possible properties of interest. In addition, it gives good clues to the properties that might be of interest. Two categories of properties are identified in this approach related to individual transitions. *Type 1* checks if a particular transition on the state machine is taken by a trace. *Type 2* monitors a certain variable or a packet field when a transition is traveled over by a trace.

7.5.2   Signature Generation

The pre-processing used in [43] is slightly different than the methods we have seen before. The pre-processing phase involved signature generation which uses the recorded information to infer typical and invariant contents in malware communication. This technique focuses on the contents of individual network traffic for signature generation. The signature is defined as a tuple (T, θ) where T represents a set of strings related with probabilities and θ is a threshold value. The i-th string of a signature is denoted as a token t_i and the corresponding probability is referred as its support $s_i \in (0,1)$. The signature is generated using two datasets: the malicious network traffic dataset collected during the repeated execution of a malware which is referred to as X^+, and the regular traffic dataset collected from the network which is referred to as X^-. Internally, the signature generation step consists of token extraction and signature assembly. The token extraction uses X^+ dataset tokens. Each substring included in at least d network flows with a minimum length l is chosen as a potential token for the signature. The signature then is assembled using the generated token, support values, and a threshold [43]. The signatures are stored in a tree structure, which allows for linear time retrieval. In some cases, this technique even allows for identification of malware in encrypted data streams. For instance, the "Storm" malware was found with 80% success despite encryption. In [36], the authors used a data mining tool on the normal log and attack log files (after some formatting) in order to generate the attack signature. They used the normal log to define the normal activities. The attack signature will be generated when a pattern does not appear on the normal log and but does appear on the attack log and will be saved in an attack database after confirming by the user.

Readers can also find good details in [10] about the signature generation procedure for intrusion detection purposes.

In addition, there are many other pre-processing techniques that we are not including them in this chapter for space restrictions such as Mobile agents [64], Host-based approaches [36, 67], Sub-Attribute utilization [12], Situation awareness-based approaches [20], Fuzzy logic-based approaches [48], MIST approaches [60], Clustering approaches [53, 65], and Feature selection approaches [23, 26, 42, 58].

Table 7.6 IDS pre-processing approaches and their advantages

Reference	Approach	Advantages
[46]	Specification method	Structures network packets in an organized fashion and minimizes the storage space required by the possible properties of interest
[43]	Signature generation	Efficient storage of patterns in network traffic. Success in identifying malware in encrypted network traffic
[31, 64]	Mobile agents	Improved robustness and real-time capabilities, efficient representation of data through encoding schemes
[67]	Host-based	Flexibility to process variety of local log files and pull out relevant information such as IPs, hostnames, and program names
[12]	Sub-attribute utilization	Efficient representation of binary, categorical, and continuous attributes
[20]	Situation awareness	Combines information from multiple sensors, reports frequencies and confidence levels
[48]	Fuzzy logic	Reduces the data down to subsets with fewer features
[60]	MIST	Efficient encoding allows for more automated analysis
[53, 65]	Clustering	Obtained better classification performance
[23, 26, 42, 58]	Feature selection	Decreases the computational cost of the ID process and does not compromise the performance of the detector

7.5.3 Comparison of Pre-processing Approaches

Normally, finishing the pre-processing phase will produce features needed later in the learning (if any) and in the classification phase. Different pre-processing approaches for IDS have been mentioned in the previous sections. Table 7.6 gives a brief summary and the advantages about each of these approaches. The extracted features from the pre-processing phase are explained in the next section.

With all the above-mentioned advantages in Table 7.6 of the pre-processing approaches, it is also important to know the limitations of these approaches. For the specification-based method, it was only used with received packets not the sent packets, while sent packets might contain significant information about a malware maintainer [43]. Although the signature generation proved its ability in detecting malwares even in the encrypted data in [43], the method will be obsolete if the signature database is not up to date. For the mobile agents approach, the main problem of the mobile agents is maintaining their security since the damage might hit the host from the agent side or vice versa if they are operated in an unsafe environment.

For the host-based approach used in [67], the log file analysis suffers from the growing of the log file size due to the large amount of traffic that needs to be pre-decoded and decoded before the analysis phase. The pre-processing approach used in [12] (which was sub-attribute utilization) did not consider applying feature selection to select the best subset of features after expanding each kind of attributes into multiple sub-attributes.

The Situation Awareness Based on Knowledge Discovery presented in [20] making the constraint requirement an attribute is better since dealing with an attribute is easier than dealing with a rule since the number of attributes can be reduced using attribute selection. The pre-processing phase of the model presented in [48] involved removing all the discrete attributes from the input dataset and just keeping the continuous ones. However, these discrete attributes may contain significant information and may help in obtaining better classification performance. Instead of removing the discrete attributes, the authors could have used attribute selection to select the best subset of attributes from both the discrete and the continuous ones to obtain better detection performance.

The encoding process used in [60] can be standardized to make standard sets for detecting malwares by adding all the possible system calls for every operating system. In the clustering approach presented in [53], the authors could not conduct a fair comparison between their approach (*K*-means clustering) and the supervised Naïve bayes classifier (without clustering) since they applied the two algorithms on different datasets. In [65], the pre-processing phase was able to achieve a successful separation between normal behavior and the other classes and led to obtaining low False Alarm Rate (FAR) and high detection rate.

7.6   Conclusion

In conclusion, in this chapter we have provided a new review of the input data and pre-processing techniques of the contemporary intrusion detection systems structured around a novel, component-oriented framework. The aim of the chapter is to provide the reader with a different and new review and taxonomy of the data and the pre-processing of any IDS. First, we started by explaining the real-world data of intrusions and their types. Then, the tools used for capturing the real-world data (Honeypots and their types) were also explained. Next, we explained all the raw input data that IDS deal with, including the standard and custom datasets. Lastly, we discussed the pre-processing phase of the raw input data. We listed and explained the extracted features from the pre-processing phase. The pattern analyzer (classifier) is beyond the scope of this chapter.

The value of this chapter lies not only on its treatment of the source papers discusses, but also in its novel style in presenting the information about IDSs to the reader. The chapter's concept is to make the reader flows with the stream of the data from the input until the internal processing. This manner gives researchers a comprehensive knowledge about ID and real-world data, honeypots used, and what has been done in this field until now in terms of input data preparation, and data pre-processing techniques.

In addition, this style helps the reader to find which feature can be used in detecting certain kind of intrusions and which papers have used that. Another benefit of this approach is that it can reveal which papers have used training and testing or testing data only. It is hoped that this chapter will have a significant impact on future research in the IDS area by providing readers new to this area which a "jumping-off point" into the source literature. Furthermore, the structure of the chapter should provide some perspective of how researchers can investigate specific aspects of IDS and what solutions have been previously explored within each aspect. In addition, the chapter conducted important comparisons and provided some critiques after each component of IDS supported by some tables to give the reader a better perspective about that particular component.

Intrusion detection will remain an interesting research topic for as long as there are intruders trying to gain illicit access to a network. The discipline represents a perpetual arms-race between those attempting to gain unauthorized control and those trying to prevent them. We hope that this chapter has provided an overview of this fascinating field and a starting point for future study.

References

1. Aghaei-Foroushani, V., & Zincir-Heywood, A. N. (2013). On evaluating ip traceback schemes: a practical perspective. In *2013 IEEE Security and privacy workshops (SPW)* (pp. 127–134). Piscataway, NJ: IEEE.
2. Al-Jarrah, O., & Arafat, A. (2015). Network intrusion detection system using neural network classification of attack behavior. *Journal of Advances in Information Technology, 6*(1), 291–295.
3. Alata, E., Nicomette, V., Kaâniche, M., Dacier, M., & Herrb, M. (2006). Lessons learned from the deployment of a high-interaction honeypot. In *Sixth European Dependable Computing Conference, 2006. EDCC '06* (pp. 39–46). doi:10.1109/EDCC.2006.17.
4. Baecher, P., Koetter, M., Dornseif, M., & Freiling, F. (2006). The nepenthes platform: An efficient approach to collect malware. In *Proceedings of the 9th International Symposium on Recent Advances in Intrusion Detection (RAID)* (pp. 165–184). Berlin: Springer.
5. Balkanli, E., & Zincir-Heywood, A. (2014). On the analysis of backscatter traffic. In *2014 IEEE 39th Conference on Local Computer Networks Workshops (LCN Workshops)* (pp. 671–678). doi:10.1109/LCNW.2014.6927719.
6. Baumann, R. (2005). Honeyd–a low involvement honeypot in action. Originally published as part of the GCIA (GIAC Certified Intrusion Analyst) practical (2003)
7. Bergadano, F., Gunetti, D., & Picardi, C. (2003). Identity verification through dynamic keystroke analysis. *Intelligent Data Analysis, 7*(5), 469–496. http://dl.acm.org/citation.cfm?id=1293861.1293866.
8. Bhuse, V., & Gupta, A. (2006). Anomaly intrusion detection in wireless sensor networks. *Journal of High Speed Networks, 15*(1), 33–51.
9. Casas, P., Mazel, J., & Owezarski, P. (2012). Unsupervised network intrusion detection systems: Detecting the unknown without knowledge. *Computer Communications, 35*(7), 772–783. http://dx.doi.org/10.1016/j.comcom.2012.01.016, http://www.sciencedirect.com/science/article/pii/S0140366412000266.
10. Chimedtseren, E., Iwai, K., Tanaka, H., & Kurokawa, T. (2014). Intrusion detection system using discrete Fourier transform. In *2014 Seventh IEEE Symposium on Computational Intelligence for Security and Defense Applications (CISDA)* (pp. 1–5). doi:10.1109/CISDA.2014.7035624.

11. Gaikwad, D., & Thool, R. C. (2015). Intrusion detection system using bagging ensemble method of machine learning. In *2015 International Conference on Computing Communication Control and Automation (ICCUBEA)* (pp. 291–295). Piscataway, NJ: IEEE.

12. Gong, Y., Mabu, S., Chen, C., Wang, Y., & Hirasawa, K. (2009). Intrusion detection system combining misuse detection and anomaly detection using genetic network programming. In *ICCAS-SICE, 2009* (pp. 3463–3467).

13. Ingre, B., & Yadav, A. (2015). Performance analysis of NSL-KDD dataset using ANN. In *2015 International Conference on Signal Processing and Communication Engineering Systems (SPACES)* (pp. 92–96). doi:10.1109/SPACES.2015.7058223.

14. Jadhav, A., Jadhav, A., Jadhav, P., & Kulkarni, P. (2013). A novel approach for the design of network intrusion detection system(NIDS). In *2013 International Conference on Sensor Network Security Technology and Privacy Communication System (SNS PCS)* (pp. 22–27). doi:10.1109/SNS-PCS.2013.6553828.

15. Jamali, S., & Shaker, V. (2014). Defense against {SYN} flooding attacks: A particle swarm optimization approach. *Computers and Electrical Engineering, 40*(6), 2013–2025. http://dx.doi.org/10.1016/j.compeleceng.2014.05.012, http://www.sciencedirect.com/science/article/pii/S0045790614001591.

16. Joo, D., Hong, T., & Han, I. (2003). The neural network models for IDS based on the asymmetric costs of false negative errors and false positive errors. *Expert Systems with Applications, 25*(1), 69–75.

17. Kayacik, H., & Zincir-Heywood, N. (2005). Analysis of three intrusion detection system benchmark datasets using machine learning algorithms. In P. Kantor, G. Muresan, F. Roberts, D. Zeng, F. Y. Wang, H. Chen, & R. Merkle (Eds.), *Intelligence and security informatics. Lecture notes in computer science* (Vol. 3495, pp. 362–367). Berlin/Heidelberg: Springer. doi:10.1007/11427995_29, http://dx.doi.org/10.1007/11427995_29.

18. Kim, H. G., Kim, D. J., Cho, S. J., Park, M., & Park, M. (2011). An efficient visitation algorithm to improve the detection speed of high-interaction client honeypots. In *Proceedings of the 2011 ACM Symposium on Research in Applied Computation* (pp. 266–271). New York: ACM. doi:10.1145/2103380.2103435, http://doi.acm.org/10.1145/2103380.2103435.

19. Kim, J., Bentley, P. J., Aickelin, U., Greensmith, J., Tedesco, G., & Twycross, J. (2007). Immune system approaches to intrusion detection–a review. *Natural Computing, 6*(4), 413–466.

20. Lan, F., Chunlei, W., & Guoqing, M. (2010). A framework for network security situation awareness based on knowledge discovery. In *2010 2nd International Conference on Computer Engineering and Technology (ICCET)* (Vol. 1, pp. 226–231). Piscataway, NJ: IEEE.

21. Lane, T. (2006). A decision-theoretic, semi-supervised model for intrusion detection. In *Machine learning and data mining for computer security* (pp. 157–177). London: Springer.

22. Lane, T., & Brodley, C. E. (1997). An application of machine learning to anomaly detection. In *Proceedings of the 20th National Information Systems Security Conference* (pp. 366–377).

23. Li, Y., Fang, B. X., Chen, Y., & Guo, L. (2006). A lightweight intrusion detection model based on feature selection and maximum entropy model. In *2006 International Conference on Communication Technology* (pp. 1–4). doi:10.1109/ICCT.2006.341771.

24. Ligh, M., Adair, S., Hartstein, B., & Richard, M. (2011). *Malware analyst's cookbook and DVD: Tools and techniques for fighting malicious code*. Hoboken: Wiley Publishing.

25. Lin, W. C., Ke, S. W., & Tsai, C. F. (2015). CANN: An intrusion detection system based on combining cluster centers and nearest neighbors. *Knowledge-Based Systems, 78*(0), 13–21. http://dx.doi.org/10.1016/j.knosys.2015.01.009, http://www.sciencedirect.com/science/article/pii/S0950705115000167.

26. Liu, H., & Yu, L. (2005). Toward integrating feature selection algorithms for classification and clustering. *IEEE Transactions on Knowledge and Data Engineering, 17*(4), 491–502. doi:10.1109/TKDE.2005.66.

27. Mahoney, M. V., & Chan, P. K. (2001). *Phad: Packet header anomaly detection for identifying hostile network traffic* (Tech. Rep. CS-2001-4), Florida Institute of Technology, Melbourne, FL, USA.

28. McGraw, G., & Morrisett, G. (2000). Attacking malicious code: A report to the infosec research council. *IEEE Software, 17*(5), 33–41.
29. MeeraGandhi, G., & Appavoo, K. (2010). Effective network intrusion detection using classifiers decision trees and decision rules. *International Journal of Advanced Networking and Applications, 2*(3), 686–692.
30. Mehta, V., Bahadur, P., Kapoor, M., Singh, P., & Rajpoot, S. (2015). Threat prediction using honeypot and machine learning. In *2015 International Conference on Futuristic Trends on Computational Analysis and Knowledge Management (ABLAZE)* (pp. 278–282). doi:10.1109/ABLAZE.2015.7155011.
31. Mo, Y., Ma, Y., & Xu, L. (2008). Design and implementation of intrusion detection based on mobile agents. In: *IEEE International Symposium on IT in Medicine and Education, 2008* (pp. 278–281). doi:10.1109/ITME.2008.4743870.
32. Mohanapriya, M., & Krishnamurthi, I. (2014). Modified DSR protocol for detection and removal of selective black hole attack in MANET. *Computers and Electrical Engineering, 40*(2), 530–538. http://dx.doi.org/10.1016/j.compeleceng.2013.06.001, http://www.sciencedirect.com/science/article/pii/S0045790613001596.
33. Muehlbach, S., & Koch, A. (2012). Malacoda: Towards high-level compilation of network security applications on reconfigurable hardware. In *Proceedings of the Eighth ACM/IEEE Symposium on Architectures for Networking and Communications Systems* (pp. 247–258). New York: ACM.
34. Muzammil, M., Qazi, S., & Ali, T. (2013). Comparative analysis of classification algorithms performance for statistical based intrusion detection system. In *2013 3rd International Conference on Computer, Control Communication (IC4)* (pp. 1–6). doi:10.1109/IC4.2013.6653738.
35. Nechaev, B., Allman, M., Paxson, V., & Gurtov, A. (2010). A preliminary analysis of TCP performance in an enterprise network. In *Proceedings of the 2010 Internet Network Management Conference on Research on Enterprise Networking, USENIX Association* (pp. 1–6).
36. Ng, J., Joshi, D., & Banik, S. (2015). Applying data mining techniques to intrusion detection. In *2015 12th International Conference on Information Technology – New Generations (ITNG)* (pp. 800–801). doi:10.1109/ITNG.2015.146.
37. Northcutt, S., & Novak, J. (2003). *Network intrusion detection*. Indianapolis: Sams Publishing.
38. Pannell, G., & Ashman, H. (2010). Anomaly detection over user profiles for intrusion detection. In *Proceedings of the 8th Australian Information Security Management Conference, School of Computer and Information Science, Edith Cowan University*, Perth, Western Australia (pp. 81–94)
39. Portokalidis, G., & Bos, H. (2007). Sweetbait: Zero-hour worm detection and containment using low-and high-interaction honeypots. *Computer Networks, 51*(5), 1256–1274.
40. Project, T. H. (2009). Dionaea. http://dionaea.carnivore.it. Accessed February 2013.
41. Provos N (2004) A virtual honeypot framework. In: Proceedings of the 13th Conference on USENIX Security Symposium - Volume 13, USENIX Association, Berkeley, CA, USA, SSYM'04, pp 1-14, http://dl.acm.org/citation.cfm?id=1251375.1251376.
42. Richharya, V., Rana, D. J., Jain, D. R., & Pandey, D. K. (2013). Design of trust model for efficient cyber attack detection on fuzzified large data using data mining techniques. *International Journal of Research in Computer and Communication Technology, 2*(3), 126–130.
43. Rieck, K., Schwenk, G., Limmer, T., Holz, T., & Laskov, P. (2010). Botzilla: Detecting the phoning home of malicious software. In *proceedings of the 2010 ACM Symposium on Applied Computing* (pp. 1978–1984). New York: ACM.
44. Schonlau, M., DuMouchel, W., Ju, W. H., Karr, A. F., Theus, M., & Vardi, Y. (2001). Computer intrusion: Detecting masquerades. *Statistical Science, 16*(1), 58–74.
45. Seifert, C., Welch, I., & Komisarczuk, P. (2008). Application of divide-and-conquer algorithm paradigm to improve the detection speed of high interaction client honeypots. In *Proceedings of the 2008 ACM Symposium on Applied Computing*, pp. 1426–1432. New York: ACM.
46. Sekar, R., Gupta, A., Frullo, J., Shanbhag, T., Tiwari, A., Yang, H., et al. (2002). Specification-based anomaly detection: A new approach for detecting network intrusions. In *Proceedings of the 9th ACM Conference on Computer and Communications Security* (pp. 265–274). New York: ACM.

47. Sen, J. (2010). Efficient routing anomaly detection in wireless mesh networks. In *2010 First International Conference on Integrated Intelligent Computing (ICIIC)* (pp. 302–307). doi:10.1109/ICIIC.2010.22.
48. Shanmugavadivu, R., & Nagarajan, N. (2011). Network intrusion detection system using fuzzy logic. *Indian Journal of Computer Science and Engineering (IJCSE), 2*(1), 101–111.
49. Sharma, V., & Nema, A. (2013). Innovative genetic approach for intrusion detection by using decision tree. In *2013 International Conference on Communication Systems and Network Technologies (CSNT)* (pp. 418–422). doi:10.1109/CSNT.2013.93.
50. Shiravi, A., Shiravi, H., Tavallaee, M., & Ghorbani, A. A. (2012). Toward developing a systematic approach to generate benchmark datasets for intrusion detection. *Computers and Security, 31*(3), 357–374. http://dx.doi.org/10.1016/j.cose.2011.12.012, http://www.sciencedirect.com/science/article/pii/S0167404811001672.
51. Shon, T., & Moon, J. (2007). A hybrid machine learning approach to network anomaly detection. *Information Sciences, 177*(18), 3799–3821.
52. Singh, S., & Silakari, S. (2009). A survey of cyber attack detection systems. *International Journal of Computer Science and Network Security (IJCSNS), 9*(5), 1–10.
53. Subramanian, U., & Ong, H. S. (2014). Analysis of the effect of clustering the training data in naive bayes classifier for anomaly network intrusion detection. *Journal of Advances in Computer Networks, 2*(1), 85–88.
54. Tan, Z., Jamdagni, A., He, X., Nanda, P., Liu, R. P., & Hu, J. (2015). Detection of denial-of-service attacks based on computer vision techniques. *IEEE Transactions on Computers, 64*(9), 2519–2533. doi:10.1109/TC.2014.2375218.
55. Tavallaee, M., Bagheri, E., Lu, W., & Ghorbani, A. A. (2009). A detailed analysis of the KDD CUP 99 data set. In *Proceedings of the Second IEEE Symposium on Computational Intelligence for Security and Defence Applications 2009* (pp. 53–58).
56. Teng, L., Teng, S., Tang, F., Zhu, H., Zhang, W., Liu, D., et al. (2014). A collaborative and adaptive intrusion detection based on SVMs and decision trees. In *2014 IEEE International Conference on Data Mining Workshop (ICDMW)* (pp. 898–905). doi:10.1109/ICDMW.2014.147.
57. Terry, S., & Chow, B. J. (2005). *An assessment of the DARPA IDS evaluation dataset using snort* (Tech. rep.), UC Davis Technical Report.
58. Thaseen, S., & Kumar, C. A. (2013). An analysis of supervised tree based classifiers for intrusion detection system. In *2013 International Conference on Pattern Recognition, Informatics and Mobile Engineering* (pp. 294–299). doi:10.1109/ICPRIME.2013.6496489.
59. Thomas, C., Sharma, V., & Balakrishnan, N. (2008). Usefulness of darpa dataset for intrusion detection system evaluation. In *SPIE Defense and Security Symposium, International Society for Optics and Photonics* (pp. 1–8)
60. Trinius, P., Holz, T., Willems, C., & Rieck, K. (2009). *A malware instruction set for behavior-based analysis* (Tech. Rep. TR-2009-07), University of Mannheim.
61. Van Jacobson, C. L., & McCanne, S. (1987). Tcpdump. http://www.tcpdump.org/tcpdump_man.html#index. Accessed January 7, 2014.
62. Wang, W., Guyet, T., Quiniou, R., Cordier, M. O., Masseglia, F., & Zhang, X. (2014). Autonomic intrusion detection: Adaptively detecting anomalies over unlabeled audit data streams in computer networks. *Knowledge-Based Systems, 70*(0), 103–117. http://dx.doi.org/10.1016/j.knosys.2014.06.018, http://www.sciencedirect.com/science/article/pii/S0950705114002391.
63. Warrender, C., Forrest, S., & Pearlmutter, B. (1999). Detecting intrusions using system calls: Alternative data models. In: *Proceedings of the 1999 IEEE Symposium on Security and Privacy, 1999* (pp. 133–145). doi:10.1109/SECPRI.1999.766910.
64. Xiaoqing, G., Hebin, G., & Luyi, C. (2010). Network intrusion detection method based on agent and SVM. In *2010 The 2nd IEEE International Conference on Information Management and Engineering (ICIME)* (pp. 399–402). Piscataway, NJ: IEEE.
65. Yanjun, Z., Jun, W. M., & Jing, W. (2013). Realization of intrusion detection system based on the improved data mining technology. In *2013 8th International Conference on Computer Science Education (ICCSE)* (pp. 982–987). doi:10.1109/ICCSE.2013.6554056.

66. Yassin, W., Udzir, N. I., Abdullah, A., Abdullah, M. T., Zulzalil, H., & Muda, Z. (2014). Signature-based anomaly intrusion detection using integrated data mining classifiers. In *2014 International Symposium on Biometrics and Security Technologies (ISBAST)* (pp. 232–237). doi:10.1109/ISBAST.2014.7013127.
67. Ying, L., Yan, Z., & Yang-Jia, O. (2010). The design and implementation of host-based intrusion detection system. In *2010 Third International Symposium on Intelligent Information Technology and Security Informatics (IITSI)* (pp. 595–598). doi:10.1109/IITSI.2010.127.
68. Zou, X., Pan, Y., & Dai, Y.-S. (2008). *Trust and security in collaborative computing*. Singapore: World Scientific.

Chapter 8
Security Protocols for Networks and Internet: A Global Vision

José María de Fuentes, Luis Hernandez-Encinas, and Arturo Ribagorda

8.1 Introduction

Communication networks have evolved significantly in the last years. Since the appearance of ARPANET in the 1970s, computer networks and the Internet are at the core of modern businesses.

This trend is becoming even more acute in recent years, when a plethora of resource-constrained devices are starting to connect. This so-called Internet of Things (IoT) opens the door to advanced, ubiquitous, and personalized services [13].

The increasing need for communication also raises concerns regarding the security of the information at stake. How to determine if a given data item has arrived correctly, that is, without any alteration? How to ensure that it comes from the authorized entity? Are the data protected from unauthorized parties? These questions refer to basic protections about integrity, origin authentication, and confidentiality of the transmitted data, respectively.

In order to offer these security properties, numerous protocols have been proposed so far. In this chapter, representative examples are described in a very general way. The purpose is not to give technical insights into every part of each protocol but to understand the foundations and its main security implications. The reader is pointed to the actual reference documents for further information. Moreover, some general practical remarks are highlighted for each family of protocols.

J.M. de Fuentes (✉) • A. Ribagorda
Computer Security Lab (COSEC), Carlos III University of Madrid, Avda. Universidad 30, 28911 Leganés, Spain
e-mail: jfuentes@inf.uc3m.es; arturo@inf.uc3m.es

L. Hernandez-Encinas
Institute of Physical and Information Technologies, Spanish National Research Council (CSIC), Serrano 144, 28006 Madrid, Spain
e-mail: luis@iec.csic.es

© Springer International Publishing AG 2018
K. Daimi (ed.), *Computer and Network Security Essentials*,
DOI 10.1007/978-3-319-58424-9_8

The remainder of this chapter is organized as follows. Section 8.2 focuses on authentication protocols, with emphasis on Kerberos. Section 8.3 describes protocols for secure communication among entities, focusing on SSL/TLS and IPSec. Afterward, Sect. 8.4 introduces SSH, the best representative for secure remote communication protocols. In order to cover wireless security, Sect. 8.5 describes WEP, WPA, and WPA2 protocols. Finally, Sect. 8.6 concludes the chapter.

8.2   Authentication Protocols

Networks are composed of communicating nodes. To enable their authentication, it is necessary to clarify how this process is performed at different levels. In the link layer (layer 2 within the Open Systems Interconnection or OSI model [16]), a pair of protocols are distinguished, namely the Password Authentication Protocol (PAP) defined in RFC 1334 [12] and the Challenge Handshake Authentication Protocol (CHAP) defined in RFC 1994 [20]. Both PAP and CHAP work over the Point-to-Point Protocol (PPP) which enables direct communication between nodes. Other relevant authentication and authorization protocol is Kerberos. It works at application level to facilitate mutual authentication between clients and servers.

This section introduces the essential aspects of PAP (Sect. 8.2.1), CHAP (Sect. 8.2.2), and Kerberos (Sect. 8.2.3). Some practical remarks about these protocols are shown in Sect. 8.2.4.

8.2.1   Password Authentication Protocol (PAP)

PAP is a simple authentication mechanism similar to the use of username and password. The node which wants to be authenticated sends its name and password to the authenticator which compares both values with stored ones and authenticates accordingly. PAP is vulnerable to third parties that intercept the communication and capture the password because it travels in plain text. It is also vulnerable against trial-and-error attacks. Thus, as this is far from being a robust authentication mechanism, the use of other more robust authentication mechanisms, such as CHAP, is recommended.

8.2.2   Challenge Handshake Authentication Protocol (CHAP)

CHAP verifies node's identity periodically, ensuring that the password remains valid and that the node has not been impersonated in some way. In this protocol, usernames and passwords are encrypted.

Once the authenticator and the node which wants to be authenticated (let us refer to it as *user*) know a common secret value, the authenticator sends a challenge to the user. The latter applies a hash over the challenge and the secret value previously shared. The result of this operation is sent to the authenticator which compares this value with the stored one. If both values are identical, the authentication is performed; otherwise, the process usually finishes. The authenticator periodically sends new challenges to the user. Note that challenges include an identifier which is incremented each time, avoiding the reuse of responses, called replay attack.

8.2.3 Kerberos Protocol

Kerberos was developed to facilitate centralized and robust authentication, being able to manage thousands of users, clients, and servers [14]. It was developed by Massachusetts Institute of Technology (MIT) in 1987. The first three versions were exclusively used in MIT, but the fourth one, v.4, was open to computer companies to be included in commercial authentication systems. Finally, version 5 was adopted in 1993 by the Internet Engineering Task Force (IETF) as an Internet standard, RFC 1510, updated in 2005 [15]. Since then, it has been updated several times; the last update was in 2016 [21].

The goal of Kerberos is to provide centralized authentication between clients (acting on behalf of users) and servers, and vice versa. Applying Kerberos terminology, clients and servers are called principals. Besides, clients and servers are usually grouped into different domains called *realms*.

Broadly speaking, Kerberos uses a Key Distribution Center (KDC) which acts as a Trusted Third Party (TTP). KDC is composed of an Authentication Server (AS) and a Ticket Granting Server (TGS). These components, though different, may be in the same system. Moreover, TGS can be unique or various of them can coexist, even if there is just one realm.

In general, depicted in Fig. 8.1, Kerberos consists of three components: a client (C) acting on behalf of a user, a server (S) whose services are accessed by the client, and the KDC. A client which wants to work with a server should be authenticated first by the KDC (steps 1–2, Fig. 8.1), providing the identification of the server. Then, the KDC provides the client credentials to be used in the authentication process with the server. These credentials are transmitted encrypted with a session key. Such a key is generated by the KDC and securely transmitted to the client and the server (steps 3–4, Fig. 8.1). Indeed, session keys are distributed through *tickets*. A ticket is a certificate (which contains data to be used in the authentication) issued by the KDC and encrypted with the server's master key. This ticket is processed by the server as a means to authenticate (and authorize) the requesting user (step 5, Fig. 8.1).

Fig. 8.1 Overview of
Kerberos

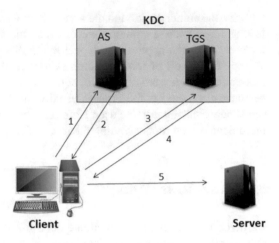

8.2.4   Practical Remarks

Authentication protocols are daily used for many purposes. For example, Single
Sign-On (SSO) architectures enable having a single entity in charge of authenticating the users. However, one critical remark is that the implementation of the
authentication protocol can introduce vulnerabilities that are not present in the
specification. For example, Microsoft Windows suffered from several Kerberos-
related issues[1] that were addressed in an update of August 2016. Thus, when
considering the use of a given authentication protocol, it is paramount to ensure
that software components are up to date.

8.3   Secure Communication Protocols

In this section, two well-known secure communication protocols are described.
In particular, Sect. 8.3.1 introduces SSL/TLS, whereas Sect. 8.3.2 describes IPSec.
Practical remarks of this family of protocols are given in Sect. 8.3.3.

8.3.1   Secure Sockets Layer (SSL)

Secure Sockets Layer (SSL) was originally developed by Netscape, being SSL 3.0
(in 1996) the first stable version [6].

---

[1] https://support.microsoft.com/en-us/kb/3178465, (access Dec. 2016).

Fig. 8.2 Overview of SSL.
The security provided by
each subprotocol is
highlighted. Key: SAu =
Server Authentication,
K = Agreement on key(s),
(CAu) = Client
Authentication (optional), DI
= Data Integrity, DC = Data
Confidentiality

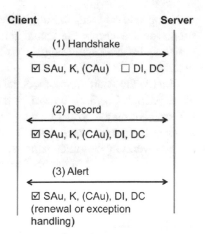

SSL provides the following three security services: (1) data confidentiality, (2) data integrity, and (3) server authentication. Optionally, client authentication can also be requested by the server.

SSL 3.0 was attacked in 2014 using a technique referred to as POODLE.[2] As a consequence, most browsers have discontinued the support of this mechanism (e.g., Microsoft's Internet Explorer 11[3]).

An alternative to SSL is called Transport Layer Security (TLS). TLS 1.0 appeared shortly after SSL 3.0 and was indeed significantly similar. However, its publication stated that it was not meant to interoperate (by default) with SSL [3]. TLS was also the target of a variant of the said POODLE attack. Indeed, TLS is still receiving attention and as of December, 2016, its version TLS 1.3 is still under development[4] and TLS 1.2 is the one that should be used [4].

Without entering into technical insights, both SSL and TLS share a common structure in what comes to their basis. Indeed, three big subprotocols can be identified even in the most modern version of TLS. They are called *Handshake*, *Record*, and *Alert* subprotocols (Fig. 8.2). Each one is described below.

In the Handshake subprotocol (step 1, Fig. 8.2), both parties agree on the set of protocols that are going to be used. Furthermore, the server is authenticated against the client, by means of a X.509 public key certificate. After this step (and upon successful authentication), both parties agree on a shared key for the encryption of the transmitted data. Remarkably, it must be noted that the set of cryptographic protocols are negotiated through a set of rounds in which the server proposes some protocols and the client determines whether they are suitable for its resources.

---

[2]http://cve.mitre.org/cgi-bin/cvename.cgi?name=CVE-2014-3566 (access December, 2016).

[3]https://blogs.microsoft.com/firehose/2015/04/15/april-update-for-internet-explorer-11-disables-ssl-3-0/#sm.0000x3es4m403dcm10bvx8k9qs1do (access December, 2016).

[4]https://tlswg.github.io/tls13-spec/ (access December, 2016).

Using this key and the algorithms defined in the previous phase, the Record subprotocol encrypts the actual data to be transmitted (step 2, Fig. 8.2). It also protects the message integrity using a Message Authentication Code (MAC) function.

Finally, the Alert subprotocol serves to notify when some abnormal issue takes place (step 3, Fig. 8.2). Indeed, it may serve to point out exceptions (from which the protocol may recover) or fatal, unrecoverable errors. An example of exception is when the server sends a certificate that is issued by an authority unknown to the receiver. On the other hand, fatal errors may happen, for example, when no agreement is reached in the handshake round.

8.3.2 IPSec

Internet Protocol Security (typically referred to as IPSec) is a technology for the protection of data authentication and encryption in a communication network [10]. One relevant aspect is that IPSec is not a protocol itself, but it is formed by a set of protocols, namely Internet Key Exchange (IKE), Authentication Header (AH), and Encapsulating Security Payload (ESP).

One critical remark is that IPSec operates at the network level, i.e., OSI level 3. This enables other applications and services belonging to upper layers to rely upon this technology. In the following, IKE, AH, and ESP are introduced.

8.3.2.1 IKE

Before two parties are able to exchange messages, it is necessary for them to agree on the set of protection mechanisms to be applied. This kind of agreement is called Security Association (SA) and is the rationale behind IKE [8]. In short, IKE enables setting up (and keep over time) SAs between two parties (Fig. 8.3).

IKE runs on top of the User Datagram Protocol (UDP) of the transport layer. As a practical remark, UDP does not offer any kind of reliable delivery. This means that every message may get lost without the sender noticing this issue. In order to cope with this issue, IKE is built in a challenge-response way which includes retransmission and acknowledgement mechanisms.

In an IKE run, two rounds are usually performed, namely IKE_SA_INIT and IKE_AUTH. The first one always takes place before any other round (step 1, Fig. 8.3). It enables agreeing on a shared key which is taken as a seed for two purposes—encrypting and authenticating exchanged data. IKE_SA_INIT is also applied to agree on the set of cryptographic algorithms that will be considered in the security association. This issue is also done in a challenge-response fashion, so the sender proposes a set of algorithms and the receiver either chooses one of them or returns an error if none is suitable.

Fig. 8.3 Outline of IKE. The security achieved after each IKE round is highlighted. Key: SAu = Server Authentication, K = Agreement on key(s), (CAu) = Client Authentication (optional), DI = Data Integrity, DC = Data Confidentiality

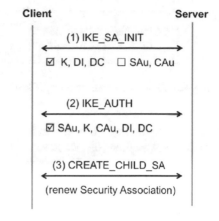

Fig. 8.4 Simplified header structure of AH

Next header	SA Identifier (SPI)	Sequence number	Integrity Check Value (ICV)

The SA itself is built in the IKE_AUTH round (step 2, Fig. 8.3). Using the key and the algorithms agreed in the previous round, both parties authenticate themselves and define the final issues of the SA. One important aspect is that the identities of the parties are encrypted using the shared key, thus ensuring the privacy of participants.

Every SA is meant to last for a given period of time. Indeed, the duration of a SA is agreed in this round. Once a SA expires, another SA comes into play. For this purpose, both parties may negotiate another SA using a CREATE_CHILD_SA round (step 3, Fig. 8.3). It must be noted that renewing SAs is beneficial from the security point of view—if a cryptographic key is used for a long time, it may become compromised in an easier way by an attacker.

8.3.2.2   Authentication Header (AH)

The main goal of the Authentication Header (AH) is to authenticate the packet content. Indeed, AH offers data integrity and sender authentication [22]. For this purpose, AH defines a header structure with four main fields (Fig. 8.4). First, an indication of the location of the next header. As AH is just one of the headers that can be included within an IPSec packet, this field points to the next in the packet to enable successful interpretation.

The next two fields are the identifier of the security association (referred to as Security Parameters Index, SPI) and the sequence number of the packet. Whereas the first field is critical to identify the security parameters to be applied, the second field enables the receiver to put packets in their correct order, no matter which

SA Identifier (SPI)	Sequence number	Initialization vector (if needed)	Payload (+padding)	Integrity Check Value (ICV) (optional)

Fig. 8.5 Simplified packet structure of ESP

packet arrives first. The last field is the Integrity Check Value (ICV), which is the central element of AH header. Indeed, ICV is the element that authenticates the packet information. For this purpose, a Message Authentication Code (MAC) value is calculated, using the keys that have been determined in the security association pointed out by SPI. Remarkably, ICV is calculated over all non-mutable (or predictable) fields of the IP packet and it is mandatory in AH (as opposed to ESP in which it is optional; see Sect. 8.3.2.3) [22].

8.3.2.3 Encapsulated Security Payload (ESP)

After setting up the security association, the Encapsulated Security Payload (ESP) protocol deals with the actual protection of data [9]. For this purpose, a novel packet structure is defined (Fig. 8.5).

The first two fields are the Security Parameters Index (SPI) and the sequence number, already explained for AH (recall Sect. 8.3.2.2). The core of the packet is formed by its payload, which may have variable size. In order to avoid any third party to learn the size of the actual payload, padding is introduced. The last part of the packet structure is given by the Integrity Check Value (ICV). The ICV is calculated over all previous fields, but only if it is defined as needed within the SA in force. Otherwise, the field is omitted. It may happen, for example, when the service that is making use of IPSec already takes care of integrity, so there is no need for IPSec to check this issue as well.

8.3.2.4 Practical Setting: Tunnel vs. Transport Modes

IPSec can be configured to protect different parts of the packet. In particular, two modes are defined, namely *tunnel* and *transport* modes [10]. In tunnel mode, the whole IP packet is enclosed within another (outer) IP packet. In this way, all its elements are protected, including the header. Hence, no external entity can learn the actual identity of both participants. This header protection is not applied in transport mode, which only protects the actual payload of the IP datagram.

8.3.3   Practical Remarks

Secure communication protocols usually rely on an agreement phase between participants. As it has been shown in this section, SSL/TLS includes a round to negotiate cryptographic algorithms, whereas IPSec relies upon the concept of Security Association. Thus, it must be noted that the effective security achieved depends on two factors. On the one hand, the correctness of the software implementing the protocol. Thus, a first practical remark is that updated and well-proven cryptographic components should be applied. In order to validate that a given component is error free, recent projects such as Google's Wycheproof[5] can be considered. On the other hand, the negotiation is usually carried out without human intervention. Thus, software components must be properly configured to avoid weak settings. For example, Google Chrome can be set up to avoid obsolete cryptography.[6]

8.4   Secure Remote Communication Protocols

With the spreading of communication networks, remote management has gained momentum. In order to connect to another machine, Secure SHell (SSH) protocol is the standard alternative. This section describes the main aspects of SSH, introducing its evolution (Sect. 8.4.1) and its structure (Sect. 8.4.2). Afterward, some practical remarks are presented in Sect. 8.4.3.

8.4.1   SSH Evolution

SSH was first proposed in 1995 as a means to enable remote login in other computers.[7] This version (called SSH-1) was intended to replace other existing alternatives such as Telnet or rlogin. As compared to these technologies, SSH-1 already provided data confidentiality and integrity protection, as well as authentication of the communicants. However, several weaknesses were found in SSH-1, such as the use of weak Cyclic Redundancy Check (CRC) for integrity preservation. The design of SSH-1, as a single, monolithic protocol, was also criticized as it was not beneficial for the sake of maintainability. In order to overcome these issues, in 2006 a new version, SSH-2, was standardized by IETF (RFC 4251) [25]. There are three major improvements that motivated this evolution [2]:

---

[5]https://github.com/google/wycheproof (access Dec. 2016).

[6]https://www.chromium.org/Home/chromium-security/education/tls (access Dec. 2016).

[7]https://www.ssh.com/ssh/ (access December, 2016).

- **Flexibility.** In SSH-2, encryption algorithms and integrity checking functions are negotiated separately, along with their respective keys. Moreover, passwords can be changed over time. SSH-2 is also formed by three subprotocols.
- **Security.** SSH-2 features strong integrity checking. Moreover, the client can now authenticate using several means in a single SSH session. Public key certificates are now allowed for this purpose. Regarding the session key, it is now negotiated using Diffie–Hellman key exchange [5].
- **Usability.** Several sessions can be run in parallel. Moreover, host authentication is independent from the IP address, which makes SSH suitable for environments such as proxy-based networks.

8.4.2 SSH Protocol Structure

According to RFC 4251 [25], SSH-2 is formed by three main components, namely Transport Layer Protocol (TLP), Authentication Protocol (AP), and Connection Protocol (CP) (see Fig. 8.6). Each one is presented below.

TLP is the lower layer protocol, which provides with security mechanisms for server authentication, data confidentiality, and integrity (step 1, Fig. 8.6). For network bandwidth reasons, it can also provide with data compression. As SSH is typically placed in the session layer[8] (layer 5 of OSI), it leverages the transport protocols of the lower OSI layer. In particular, RFC 4251 specifies that TLP could be run on top of TCP/IP, but it could be used on top of any reliable transport protocol.

On the other hand, AP offers client (i.e., user) authentication (step 2, Fig. 8.6). AP runs on top of TLP. For client authentication, three main mechanisms are

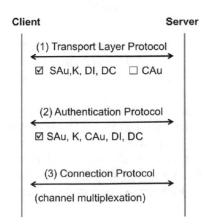

Fig. 8.6 Overview of SSH-2. The security provided by each subprotocol is highlighted. Key: SAu = Server Authentication, K = Agreement on key(s), CAu = Client Authentication, DI = Data Integrity, DC = Data Confidentiality

Client   **Server**

(1) Transport Layer Protocol

☑ SAu,K, DI, DC ☐ CAu

(2) Authentication Protocol

☑ SAu, K, CAu, DI, DC

(3) Connection Protocol

(channel multiplexation)

---

[8]https://www.sans.org/reading-room/whitepapers/protocols/understanding-security-osi-model-377.

allowed, namely public key authentication (using X.509 certificates), password, and host-based authentication. Only the first one is mandatorily supported for any implementation of SSH. The password-based method requires both parties to share a common secret (i.e., the password) in advance. Host-based is suitable for those sites that rely upon the host that the user is connecting from and the username within that host. As stated in RFC 4252, this form is optional and could not be suitable for high-sensitivity environments [24].

Last but not least, CP runs on top of AP and it is meant to enable channel multiplexation (step 3, Fig. 8.6). Thus, several SSH sessions can run simultaneously over a single connection. These sessions may serve to execute remote commands or to run x11-related software, that is, software programs that require graphical user interface.

8.4.3 Practical Remarks

SSH is used not only for remote communications but also for other purposes such as file transfer (Secure Copy Protocol, SCP). Thus, it is important to spread these remarks to all protocols that are based on SSH.

SSH has to be configured in the server, determining which are the considered cryptographic protocols. For example, in Ubuntu Linux systems these settings are located into the */etc/sshd/sshdconfig* file.[9] In this regard, one important aspect is to define which cryptographic algorithms are applied, avoiding weak (or vulnerable) ones. In the said file, directives *Ciphers*, *MACs*, and *KexAlgorithms* determine which encryption, MAC, and key exchange methods are allowed, respectively.

Moreover, as SSH is typically implemented through libraries or specialized software modules, it is essential to keep up to date on existing vulnerabilities. Indeed, as of December 2016 more than 360 vulnerabilities[10] can be found within the Common Vulnerabilities and Exposures (CVE) database, with some relation to SSH. It must be noted that some vulnerabilities are highly critical, even allowing unauthorized access to systems (e.g., vulnerability[11] CVE-2016-6474).

8.5 Secure Wireless Communication Protocols

Since the appearance of wireless networks, connectivity has become almost ubiquitous in developed countries and modern societies. However, security in these networks cannot be taken for granted. Thus, security protocols have been proposed

---

[9]https://help.ubuntu.com/community/SSH/OpenSSH/Configuring.

[10]http://cve.mitre.org/cgi-bin/cvekey.cgi?keyword=SSH (access Dec. 2016).

[11]https://web.nvd.nist.gov/view/vuln/detail?vulnId=CVE-2016-6474 (access Dec. 2016).

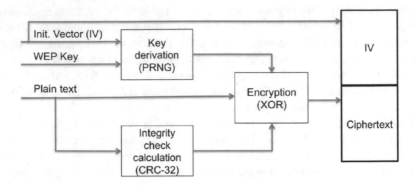

Fig. 8.7 Overview of WEP

since several decades. This section introduces the main examples of wireless security, namely WEP (Sect. 8.5.1), WPA, and WPA2 (Sect. 8.5.2). Practical considerations are introduced in Sect. 8.5.3.

8.5.1 Wired Equivalent Privacy (WEP)

WEP was included in 1997 as part of the wireless connectivity standard IEEE 802.11 [7]. This technology remained in the standard until 2004, when a revision made WEP to be superseded by WPA and WPA2 (explained below).

WEP offers data confidentiality. For this purpose, data is encrypted using algorithm RC4 [19] (Fig. 8.7). This algorithm is a stream cipher, which means that data is encrypted in a continuous manner, as opposed to block ciphers in which data is encrypted in a block-by-block basis. For this purpose, it is necessary that the encryption key comes in the form of pseudo-random sequence, which in the case of WEP is produced by a Pseudo-Random Number Generator (PRNG). Thus, PRNG is seeded with part of the WEP key, called Initialization Vector (IV).

WEP also provides data integrity. This is achieved by applying a Cyclic Redundancy Check (CRC) algorithm. In particular, CRC-32 is applied [11].

The short key length (initially, 64 bits), the lack of key renewal, as well as the election of cryptographic algorithms were the source of vulnerabilities in WEP. Furthermore, it must be noted that no explicit authentication is carried out from the access point. This facilitates launching attacks impersonating these nodes.

8.5.2   Wireless Protected Access (WPA and WPA2)

In order to overcome the limitations of WEP, in 2003 a novel protection mechanism called Wireless Protected Access (WPA) was developed [7]. The idea was to develop a novel technique that could be run directly on existing hardware.

WPA introduced Temporal Key Integrity Protocol (TKIP), a technique to improve key usage for encryption purposes [18]. In particular, TKIP enables mixing up the initialization vector with the root key, and using the result as input for the PRNG. Thanks to this action, the information available for the attacker was significantly reduced. Furthermore, TKIP features a sequence control mechanism, which is useful to counter replay attacks (i.e., attacks by repeating packets already sent).

Apart from a better key usage, WPA featured the use of an additional message authentication technique, called Michael. Thanks to Michael, should the access point receive incorrect integrity values within a period, a new session key would be applied for encryption. This is very beneficial to prevent external attackers gaining access to the network.

Despite these benefits, WPA relied upon the same cryptographic algorithms as WEP. Thus, although the attack chances were reduced, vulnerabilities were discovered as well. To address these issues, WPA2 was developed in 2004.

As opposed to WPA, WPA2 makes use of a different set of algorithms. In particular, AES-CCMP is applied. This algorithm comes from a particular instantiation of AES encryption algorithm. Remarkably, it also offers data integrity protection. Nowadays, WPA2 is resilient against the attacks that were feasible for its predecessors WEP and WPA [1].

8.5.3   Practical Remarks

Protocols for secure wireless communication can be configured in terms of the involved cryptographic algorithms. Specifically, for hardware-constrained devices it is important to carefully choose these algorithms, since there is a technical trade-off between security and performance. Thus, Potlapally et al. [17] have studied the impact of cryptographic algorithms for constrained devices. Although the study is focused on SSL, the implications are also valid for wireless protocols.

Apart from performance, cryptographic robustness is also relevant. As such, Tews and Beck [23] reported several practical attacks against WEP and WPA, along with some countermeasures. Remarkably, remediation usually involves tuning some parameters. Thus, as a practical recommendation, default settings should be revised by users to achieve the desired security level.

8.6 Conclusion

Computer networks and the Internet have greatly evolved in the last years. As a consequence, they are an integral part of any modern information technology system. In order to address their underlying security issues, a plethora of techniques have been proposed in the last decades.

In this chapter, an overview of network security-related protocols has been presented. They are focused on different areas, such as user authentication, secure communications, remote login, and wireless networks. For each protocol, a historical overview has been presented and the main features have been pointed out. The vast majority of technical issues have been left out of the discussion so that the reader gets the big picture of network security. Thus, Table 8.1 summarizes the main discussed aspects for each protocol.

Despite the amount of protocols described, many others have been intentionally left out of the scope of this chapter for space restrictions. Remarkably, other authentication technologies such as Radius or lower-level authentication protocols such as L2TP have not been addressed. However, we believe that the current overview is representative enough to show the recent evolution of these technologies.

Acknowledgements This work was supported by the MINECO grant TIN2013-46469-R (SPINY: Security and Privacy in the Internet of You), by the CAM grant S2013/ICE-3095 (CIBERDINE: Cybersecurity, Data, and Risks), which is co-funded by European Funds (FEDER), and by the MINECO grant TIN2016-79095-C2-2-R (SMOG-DEV—Security mechanisms for fog computing: advanced security for devices). Authors would like to thank the anonymous reviewers for their useful comments.

References

1. Adnan, A. H., Abdirazak, M., Sadi, A. S., Anam, T., Khan, S. Z., Rahman, M. M., et al. (2015). A comparative study of WLAN security protocols: WPA, WPA2. In *2015 International Conference on Advances in Electrical Engineering (ICAEE)* (pp. 165–169). Piscataway, NJ: IEEE.
2. Barrett, D., Silverman, R., & Byrnes, R. (2005). *SSH, the secure shell: The definitive guide* (2nd ed.). Sebastopol: O'Reilly.
3. Dierks, T., & Allen, C. (1999). *The TLS Protocol Version 1.0*. RFC 2246 (Proposed Standard). http://www.ietf.org/rfc/rfc2246.txt. Obsoleted by RFC 4346, updated by RFCs 3546, 5746, 6176, 7465, 7507, 7919.
4. Dierks, T., & Rescorla, E. (2008). *The Transport Layer Security (TLS) Protocol Version 1.2*. RFC 5246 (Proposed Standard). http://www.ietf.org/rfc/rfc5246.txt. Updated by RFCs 5746, 5878, 6176, 7465, 7507, 7568, 7627, 7685, 7905, 7919.
5. Diffie, W., & Hellman, M. (2006). New directions in cryptography. *IEEE Transactions on Information Theory, 22*(6), 644–654.
6. Freier, A., Karlton, P., & Kocher, P. (2011). *The Secure Sockets Layer (SSL) Protocol Version 3.0*. RFC 6101 (Historic). http://www.ietf.org/rfc/rfc6101.txt.

Table 8.1 Summary of considered protocols

	Authentication			Secure networks			Remote connection		Wireless networks		
	PAP	CHAP	Kerberos	IPSec	SSL (v. 3.0)	TLS (1.2)	SSH-1	SSH-2	WEP	WPA	WPA2
Data confidentiality	–	Yes	Yes	Yes (ESP)	Several options. Agreed between client and server		Yes	Several symmetric algorithms (e.g., AES, 3DES, etc.)	RC4 (40 bits)	TKIP (per packet)	AES-CCMP
Data integrity	–	–	Yes	Yes (AH, ESP)	Several options. Agreed between client and server		Yes		CRC-32 (24 bits IV)	Michael	Improved MIC
Client authentication	Yes	Yes	Yes	Yes (AH, ESP)	Optional		Yes	Public key/Password/Hostbased	Yes		
Server/Access point authentication	–	Optional	Yes	Yes (IKE)	X.509 public key certificates		Yes	Public key/Certificate	Optional	Optional (EAP)[a]	Optional (WPA2-PSK)
OSI Layer	2 (Data link)		5 (Session)[b]/7 (Application)[c]	4 (Transport)	4 (Transport) and upwards		5 (Session)		2 (Data link)		

(continued)

Table 8.1 (continued)

	Authentication			Secure networks			Remote connection		Wireless networks		
	PAP	CHAP	Kerberos	IPSec	SSL (v. 3.0)	TLS (1.2)	SSH-1	SSH-2	WEP	WPA	WPA2
Date proposed	1992		1993	2005[d]	1996[e]	2008	1995	2006	1997	2003	2004
Date superseded/Obsoleted/Declared as insecure	2013[f]		None	None	2015[g]	None	2006	None	2003	2004	None

[a] https://www.sans.org/reading-room/whitepapers/wireless/evolution-wireless-security-80211-networks-wep-wpa-80211-standards-1109 (access Dec. 2016)

[b] https://www.sans.org/reading-room/whitepapers/protocols/understanding-security-osi-model-377 (access Feb. 2017)

[c] http://www.networksorcery.com/enp/default.htm (access Feb. 2017)

[d] https://tools.ietf.org/html/rfc4301 (access Feb. 2017)

[e] https://web.archive.org/web/19970614020952/http://home.netscape.com/newsref/std/SSL.html (access Feb. 2017)

[f] https://datatracker.ietf.org/doc/rfc1994/ (access Feb. 2017)

[g] https://tools.ietf.org/html/rfc7568 (access Feb. 2017)

7. Group, W. W. L. W. (2012). *802.11-2012 – IEEE Standard for Information technology–Telecommunications and information exchange between systems Local and metropolitan area networks–Specific requirements Part 11: Wireless LAN Medium Access Control (MAC) and Physical Layer (PHY) Specifications.*

8. Kaufman, C., Hoffman, P., Nir, Y., Eronen, P., & Kivinen, T. (2014). *Internet Key Exchange Protocol Version 2 (IKEv2).* RFC 7296 (Internet Standard). http://www.ietf.org/rfc/rfc7296.txt. Updated by RFCs 7427, 7670.

9. Kent, S. (2005). *IP Encapsulating Security Payload (ESP).* RFC 4303 (Proposed Standard). http://www.ietf.org/rfc/rfc4303.txt.

10. Kent, S., & Seo, K. (2005). *Security Architecture for the Internet Protocol.* RFC 4301 (Proposed Standard). http://www.ietf.org/rfc/rfc4301.txt. Updated by RFCs 6040, 7619.

11. Koopman, P. (2002). 32-bit cyclic redundancy codes for internet applications. In *Proceedings International Conference on Dependable Systems and Networks* (pp. 459–468).

12. Lloyd, B., & Simpson, W. (1992). *PPP Authentication Protocols.* RFC 1334 (Proposed Standard). http://www.ietf.org/rfc/rfc1334.txt. Obsoleted by RFC 1994.

13. Mattern, F., & Floerkemeier, C. (2010). Chap. From the internet of computers to the internet of things. *From active data management to event-based systems and more* (pp. 242–259). Berlin/Heidelberg: Springer. http://dl.acm.org/citation.cfm?id=1985625.1985645.

14. Neuman, B.C., & Ts'o, T. (1994). Kerberos: An authentication service for computer networks. *IEEE Communications Magazine, 32*(9), 33–38. doi:10.1109/35.312841.

15. Neuman, C., Yu, T., Hartman, S., & Raeburn, K. (2005). *The Kerberos Network Authentication Service (V5).* RFC 4120 (Proposed Standard). http://www.ietf.org/rfc/rfc4120.txt. Updated by RFCs 4537, 5021, 5896, 6111, 6112, 6113, 6649, 6806, 7751.

16. ISO, I. (1994). IEC 7498-1: 1994 information technology-open systems interconnection-basic reference model: The basic model. ISO standard ISO/IEC, 7498-1.

17. Potlapally, N. R., Ravi, S., Raghunathan, A., & Jha, N. K. (2006). A study of the energy consumption characteristics of cryptographic algorithms and security protocols. *IEEE Transactions on Mobile Computing, 5*(2), 128–143.

18. Potter, B. (2003). Wireless security's future. *IEEE Security and Privacy, 1*(4), 68–72.

19. Rivest, R. L., & Schuldt, J. C. (2014). Spritz-a spongy rc4-like stream cipher and hash function. In *Proceedings of the Charles River Crypto Day,* Palo Alto, CA, USA (Vol. 24).

20. Simpson, W. (1996). *PPP Challenge Handshake Authentication Protocol (CHAP).* RFC 1994 (Draft Standard). http://www.ietf.org/rfc/rfc1994.txt. Updated by RFC 2484.

21. Sorce, S., & Yu, T. (2016). *Kerberos Authorization Data Container Authenticated by Multiple Message Authentication Codes (MACs).* RFC 7751 (Proposed Standard).

22. Stallings, W. (2002). *Cryptography and network security: Principles and practice.* Edinburgh: Pearson Education.

23. Tews, E., & Beck, M. (2009). Practical attacks against WEP and WPA. In *Proceedings of the Second ACM Conference on Wireless Network Security* (pp. 79–86). New York: ACM.

24. Ylonen, T., & Lonvick, C. (2006). *The Secure Shell (SSH) Authentication Protocol.* RFC 4252 (Proposed Standard). http://www.ietf.org/rfc/rfc4252.txt.

25. Ylonen, T., & Lonvick, C. (2006). *The Secure Shell (SSH) Protocol Architecture.* RFC 4251 (Proposed Standard). http://www.ietf.org/rfc/rfc4251.txt.

Chapter 9
Differentiating Security from Privacy in Internet of Things: A Survey of Selected Threats and Controls

A. Al-Gburi, A. Al-Hasnawi, and L. Lilien

9.1 Introduction

This chapter studies security and privacy issues that are major concerns for the Internet of Things due to the nature of this emerging technology.

9.1.1 Internet of Things

The Internet of Things (IoT) can be defined as "a new paradigm that links the objects of the real world with the virtual world, thus enabling anytime, anyplace connectivity for anything and not only for anyone. It refers to the world where physical objects and beings, as well as virtual data and environments, all interact with each other in the same space and time" [1]. It aims at enabling efficient connectivity, communication, and information exchange among large collections

A. Al-Gburi
Department of Computer Science, Western Michigan University, Kalamazoo, MI, USA

On leave from Department of Computer Science, Al-Mustansiriyah University, Baghdad, Iraq
e-mail: ahmedaliahmed.algburi@wmich.edu

A. Al-Hasnawi (✉)
Department of Computer Science, Western Michigan University, Kalamazoo, MI, USA

On leave from Department of Electrical Engineering, Al-Furat Al-Awsat Technical University, Najaf, Iraq
e-mail: abduljaleelmoh.alhasnawi@wmich.edu

L. Lilien
Department of Computer Science, Western Michigan University, Kalamazoo, MI, USA
e-mail: leszek.lilien@wmich.edu

© Springer International Publishing AG 2018
K. Daimi (ed.), *Computer and Network Security Essentials*,
DOI 10.1007/978-3-319-58424-9_9

153

of heterogeneous things and beings—to cooperatively provide useful services. However, IoT brings new security and privacy issues that need to be addressed to provide appropriate controls (the term "controls" is used in the area of Computer Security and Privacy to mean "solutions" [2]).

9.1.2 Definitions of Security and Privacy

A textbook definition of *computer security* considers it "the protection of the items you value, called the assets, of a computer or a computer system," and a textbook definition of *information privacy* means by it "the right to control who knows certain aspects about you, your communication, and your activities" [2].

In the context of this chapter, "security" and "privacy" are used as synonyms of "computer security" and "information privacy."

We decided to use simpler definitions of security and privacy, boiling down to their most essential characteristics. Our guide was the famous Cooley's classic definition of personal immunity as "a right of complete immunity: to be let alone" [3]. This phrase was soon adapted for definition of privacy. Being provided by a lawyer, it includes physical aspects of privacy—critical in the real world but not essential in the virtual world; as will be clear from our definitions of security and privacy in the next paragraph, we see these aspects more as security characteristics than privacy characteristics.

In this vein, we propose to define *security* as the right not to have one's activities adversely affected via tampering with one's objects, and *privacy* as the right to have information about oneself left alone. We use these definitions to divide "security and privacy" issues into pure security issues, or pure privacy issues, or intertwined security and privacy issues.

For example, contemplate an attack on data integrity that affects security but not privacy. Suppose that all Jane's identity data in her personal profile are completely distorted. This means that integrity of the profile, and hence security of the profile (since integrity is one of the components of the classic *C-I-A security triad* [2]) is violated. This distortion does not negatively affect Jane's privacy. As a matter of fact, it improves it—since nobody looking at the data after the attack can link de-identified data do Jane.

As another example, look now at an attack on data anonymity that violates privacy of the subject described by these data but does not affect system security. Suppose that an *insider* is authorized to view company's sensitive data linking Michael to some de-identified data. If—in a criminal act—he copies and sells the sensitive data, Michael's privacy is violated. However, system security is not affected since the insider did not affect system activities.

As a final example, consider an attack that affects both security and privacy. Suppose that an attack on Michael's data anonymity is executed not by an insider but by an *outsider* (i.e., an entity without any authorization to access any system data). The outsider breaks into the system, views and steals sensitive data linking Michael to some de-identified data. In this case, system security must be successfully compromised (which obviously involves system's activities adversely affected via tampering with system's objects) before Michael's privacy is violated.

9.1.3   Differentiating Security from Privacy

The common approaches to security and privacy do not clearly differentiate or separate security threats and controls from privacy threats and controls. Often the technical aspects of privacy—the ones that must be investigated by computer scientists—are viewed as "inextricably linked to computer security" [2]. We hypothesize that explicitly differentiating security from privacy wherever possible will provide a proper focus for considering threats and facilitate the search for controls. Hence, we separate the two, presenting as security issues or privacy issues those for which security or privacy, respectively, is the only or at least the primary concern. Our differentiation is not dichotomic, since security issues may include privacy aspects and vice versa.

9.1.4   Chapter Contributions and Organization

This chapter makes two main contributions. First, it differentiates security from privacy. Second, it introduces classification of selected security threats and controls and selected privacy threats and controls based on the framework provided by the IoT reference model (discussed below).

The chapter is organized as follows. Section 9.2 presents the four-layer IoT reference model used by us, emphasizing the most common components and main functions of each layer. The proposed classifications of selected threats and controls for IoT security and privacy are described in Sects. 9.3 and 9.4, respectively. Section 9.5 concludes the chapter.

9.2   IoT Reference Model

The IoT is expected to interconnect billions of heterogeneous devices or objects via the Internet. To facilitate IoT analysis and design, IoT reference models were developed. In this chapter, we use a known four-layer IoT reference model [4] shown in Fig. 9.1. The layers are briefly discussed below in the order from the bottom up.

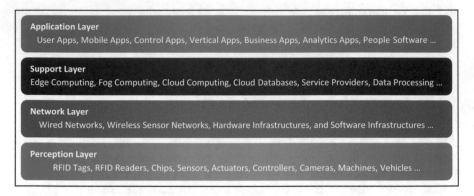

Fig. 9.1 IoT reference model

9.2.1 Perception Layer

The IoT requires a comprehensive perception of the environment, which means that the perception layer must be able to obtain information about objects anytime and anywhere. Thus, the *perception layer* (a.k.a. the *device layer*) must contain a large collection of heterogeneous physical sensing devices and controllers, which are the "things" in the IoT. These are typically *thin clients* or *lightweight devices*, with low processing and storage capabilities. They are capable of generating data, collecting data, responding to queries, communicating data about themselves and their environment, and being controlled over the net. Typically, data collected in this layer are about location, orientation, object motion, temperature, humidity, air pollution, etc. Data are then passed to the network layer through channels assumed to be secure.

9.2.2 Network Layer

The *network layer* (a.k.a. the *transmission layer*) includes all hardware and software entities for communication networks. Examples of network layer entities are routers, switches, gateways, servers, protocols, APIs, etc. The primary function of this layer is to assure reliable and timely data transmission. It manages communication between things, between devices and the network, across the network, as well as between network and data processing entities. It includes implementations of various communication protocols (e.g., MQTT, XMPP, AMQP, and DDS) [5].

9.2.3 Support Layer

The *support layer* (a.k.a. the *middleware layer*) includes all middleware tech-nologies that implement IoT services, and integrate services and applications. Middleware technologies include cloud computing and opportunistic resource utilization networks (a.k.a. oppnets) [6, 7]; IoT services include cloud services and helper IoT services in oppnets (such as sensing traffic flow); and integrated services and applications include smart home, smart city, and smart vehicles.

This layer supports many types of data processing such as data aggregation, data accumulation, and data abstraction. The primary functions of this layer include reformatting network packets into relational database tables, and transforming event-based computing into query-based computing; both are needed to make data usable to the IoT application layer.

Trust management—which supports security, privacy, cooperation, etc.—is another function of the support layer. Trust management provides analysis of trustworthiness of IoT entities based on their past behavior in the system using either the first-hand evidence of their behavior, or their second-hand reputation. A trustworthy IoT system can significantly reduce unwanted attackers' activities [8].

9.2.4 Application Layer

The *application layer* provides global management for applications based on information processed in the support layer. It is considered as the hub of services requested by end users. For instance, the application layer can provide diagrams with temperature and air humidity information to the customers. The primary function of this layer is providing high-quality *smart* services to meet customers' needs. Examples of the supported applications are smart health, smart home, smart city, smart farming, smart environment, intelligent transportation, etc.

9.3 Using IoT Reference Model for a Classification of Security Threats and Controls

This section proposes a classification of security threats and controls shown in Fig. 9.2. Due to space limitations, for each layer discussed below we list only a few security threats and a few security controls. These are, in our opinion, the most dangerous threats and the most capable controls. In a few cases, a threat or control indicated by us as a major one for one layer is also indicated as a major one for another layer. For example, Denial of Service (DoS) is listed as a major threat for both perception and application layers.

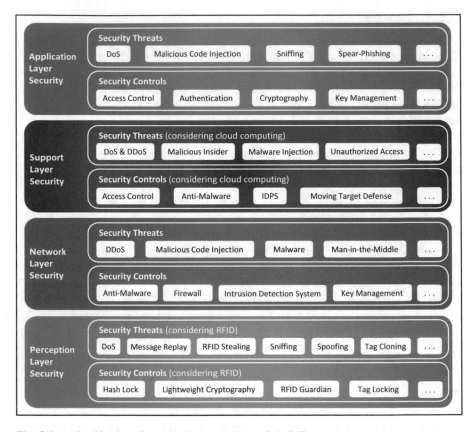

Fig. 9.2 A classification of security threats and controls in IoT

It should be noted that a threat or control chosen as a major one for one layer might also be a minor threat or control for another layer but—as a minor one—it is not enumerated by us for the other layer. For example, a firewall is listed as a (major) control for the network layer but not for the application layer—because an application-oriented firewall is considered by us a less promising control for the application layer than the controls listed by us for this layer.

9.3.1 Perception Layer Security

Diverse sensor technologies are utilized in the perception layer to collect data of interest for IoT services and applications. Radio-frequency identification (RFID) is one of the most common technologies used here. Due to space limitations, we limit our discussion for this layer to selected threats and controls for RFID technology.

9.3.1.1 Security Threats in Perception Layer

In IoT, RFID technology is used in the perception layer mainly in the form of RFID tags. The following are among the most dangerous security threats for the use of RFID tags.

Denial of Service (DoS) is aimed at disrupting communications between RFID tags and readers. A DoS attack can use multiple tags or specially designed tags to overwhelm an RFID reader's capacity with requests [9]. This renders the system inoperative, since the reader becomes unable to distinguish different tags, and the legitimate tags are unable to communicate with the reader.

Message replay occurs when an RFID message is recorded (e.g., via eavesdropping), and "replayed" later [10]. The attacker might replay an encrypted message that he is unable to decrypt. Message replay may confuse the receiving device if it does not check for redundant messages.

RFID stealing [11] is the physical security threat of stealing an RFID tag for a period of time without the prover (e.g., an owner) being able to notice this fact. Countermeasures require not only software but also hardware controls.

Sniffing happens when an attacker intercepts communication between an RFID tag and a reader [4]. The attacker can then falsify the message and redirect it to parties other than the original addressee—these activities do *not* occur in message replay.

Spoofing occurs when an adversary impersonates a valid RFID tag to gain its privileges when communicating with an RFID reader [10]. Adversaries impersonate RFID tags using emulating devices with capabilities exceeding those of RFID tags.

In *tag cloning*, tag data from an original and legitimate RFID tag are first stolen and transferred onto a cloned RFID tag owned by an adversary [9]. The data on the cloned tag are then altered to suit the needs of the attacker. Finally, the cloned tag is inserted into an RFID system to realize the planned threat.

9.3.1.2 Security Controls for Perception Layer

Many security controls for the RFID technology used in the perception layer are proposed in the literature to overcome the security threats. The following are a selected set of these controls.

The *hash lock* encryption scheme is designed to fit on the tags that have only little memory available [9]. Each of the hash-enabled tags operates in a locked and unlocked stage and has a small amount of its memory reserved for a temporary hash-encrypted ID. In the locked stage, the broadcast data is no longer readable to eavesdroppers, since they are unable to decrypt these data locked by the hash-encrypted ID.

Lightweight cryptography provides cryptographic algorithms and protocols tailored for implementation in resource-constrained environments, with thin clients or lightweight devices [12]. Since the RFID technology works in such IoT environments, lightweight cryptography is essential.

An *RFID Guardian* is a battery-powered device that looks for, records, and displays all RFID tags that it is able to detect in its vicinity via scanning [9]. It also manages RFID keys, authenticates nearby RFID readers, and blocks attempted accesses to the user's RFID tags from unauthorized RFID readers.

The *tag locking* mechanism allows locking RFID tag data other than the tag ID, preventing their transmission [9]. Once the *lock* mode (protected by a PIN number) is entered, the RFID tag replies with its ID but does not send any other data (note that this still enables tracking a person or entity by correlating the locations where the ID was read). After unlocking (with the PIN number), data transmission by the tag is reactivated.

9.3.2 Network Layer Security

Network layer must provide secure hardware and software infrastructures to enable reliable transmission of sensing data. This section discusses selected threats and controls for the networking layer.

9.3.2.1 Security Threats in Network Layer

We indicate here some of the most dangerous threats to the security of communicating IoT hosts.

A *Distributed Denial-of-Service (DDoS)* threat is a potential for a coordinated attack on the availability of network services of a single target system by multiple compromised computing systems [13], each executing a DoS scenario. It results in denying legitimate users access to the resources they expect to use.

Malicious code injection is a process of injecting malicious code into a legitimate network node by an adversary attempting to control the node or the network (which might result even in an overall shutdown of the network) [14].

Malware is any malicious code/software specifically designed by attackers to damage data, hosts, or networks [15]. Malware includes Trojan horses, worms, spam, and spyware. It has the ability to spread rapidly across an IoT network.

In a *Man-in-the-Middle (MITM)* threat, the malicious entity takes over the control of communications between two hosts, becoming an illegitimate and covert "mediator" [14]. The MITM attacker violates at least confidentiality of the intercepted messages, but can also violate the integrity of the intercepted messages (since these messages can be modified in arbitrary ways, including injection of false or confusing information).

9.3.2.2 Security Controls for Network Layer

The following are some of the most promising security controls for the IoT network layer.

Anti-malware is a specific type of software designed to detect, prevent, and eliminate malware from network devices [15]. Typically, anti-malware scans files looking for known types of malware that match the ones predefined in the dictionary. Once the malware is detected, anti-malware can alert the system or users, disable malware or neutralize malware to protect the network.

A *firewall* protects an IoT network by blocking unauthorized access identified by a set of rules [15]. Typically, a firewall allows choosing a level of security suitable for specific IoT services and applications.

An *Intrusion Detection System (IDS)* provides continuous monitoring and logging of the behavior of a network for suspicious activity [14]. Logs facilitate identification of malicious nodes and detecting intrusions.

Key management is required to support secure communication channels among Internet hosts and sensor nodes in IoT [16]. They allow network devices to negotiate appropriate security credentials (e.g., certificates, signatures, or secret keys) for secure communications, which protect information flow. Key management includes secret key creation, distribution, storage, updating, and destruction—all of them in a secure manner.

9.3.3 Support Layer Security

The support layer consists of the IoT services including typical cloud computing services: Infrastructure as a Service (IaaS), Platform as a Service (PaaS), and Software as a Service (SaaS) [17]. The IoT services process data collected by the perception layer, and store these data in an organized (e.g., indexed) manner, making them available for different IoT applications. Due to space limitations, we limit our discussion to threats and controls in cloud computing, which is a prime example of technology used at the support layer.

9.3.3.1 Security Threats in Support Layer

The following are among the most dangerous security threats that can target cloud computing used in the IoT support layer.

A *Denial of Service (DoS)* and *Distributed DoS (DDoS)* were already discussed in Sects. 9.3.1.1 and 9.3.2.1—for perception layer and network layer, respectively.

A *malicious insider* is among the most difficult to detect and the riskiest in cloud computing [17]. It occurs when an *authorized* cloud user steals cloud data, tampers with cloud data, or harms cloud operation in another way. A prime example is an attack by a disgruntled employee on the employer's cloud.

An attacker performing *malware injection* attempts to inject a malicious service or virtual machine into a cloud system and make it appear as a valid service or virtual machine running in this cloud system [18]. A malicious VM instance in an IoT cloud enables the attacker to execute malicious data or cloud system modifications, affecting the overall cloud functionality.

An *unauthorized access* usually targets confidentiality of IoT data stored in the cloud or software that manages access to these data (e.g., an SQL database) [17]. The consequence of a successful attack includes compromising the attacked system.

9.3.3.2   Security Controls for Support Layer

The following are some of the most promising security controls for cloud computing used in the IoT support layer.

Access control manages how resources of a system can be accessed by other parties based on a set of access policies [19]. In an IoT cloud system, assuring a fine-grained access control to the resources including data, software, and hardware increases the security level of the system. Access control in an IoT cloud must support scalability, context-awareness, and flexibility.

An *anti-malware* threat is analogous to the one discussed for the network layer.

An *Intrusion Detection and Prevention System (IDPS)* is a system that not only is an intrusion detection system (IDS) but also has capabilities to prevent some of the incidents [20]. Placing IDPS at the edge of the IoT network, as close as possible to the sensing devices, significantly decreases the possibility of intrusion attacks on cloud data.

The *Moving Target Defense (MTD)* relies on changing the set of virtual machines (VMs) used for executing an application in a cloud [21]. Suppose that an application runs on the set S1 of VMs provided by a cloud. An attacker can compromise the application by penetrating one or more of the VMs from the set S1. The attacker needs at least time interval T to do so. Before time T expires, the application is moved by the cloud to another set, S2, of VMs. Any gains in attacking the VMs from S1 are now useless.

9.3.4   Application Layer Security

This section outlines selected security threats that target the IoT application layer, and discusses the most promising security controls for this layer.

9.3.4.1 Security Threats in Application Layer

The following are among the most dangerous security threats for the IoT application layer.

Two of these threats, namely *DoS* and *sniffing* were discussed in Sect. 9.3.1.1, and another, namely *malicious code injection* was discussed in Sect. 9.3.2.1.

Spear-phishing is a social-engineering attack in which an attacker uses spoofed emails (e.g., associated with a trusted brand) to trick IoT end users into providing him with unauthorized access to their own information or installing malware on their own devices [22]. To gain credibility, attackers use specific knowledge about individuals and their organizations.

9.3.4.2 Security Controls for Application Layer

The following are some of the most promising security controls for the IoT application layer, which can protect most applications.

Controls for *access control* and *key management* were already discussed above in Sects. 9.3.2.2 and 9.3.3.2—for network layer and support layer, respectively.

Authentication of sensor nodes and end users is essential for secure communication in IoT [15]. An IoT system needs to identify its users, services, and devices as a prerequisite for enabling authentication based on specific credentials. Authentication should include a lightweight version that can be handled even by simple IoT applications running on thin devices.

Cryptography is the most pervasive mechanism used for security and should be used to protect IoT transactions [15]. For thin IoT devices, lightweight cryptographic algorithms and protocols must be used.

9.4 Using IoT Reference Model for a Classification of Privacy Threats and Controls

This section proposes a classification of privacy threats and controls shown in Fig. 9.3. As we did for the IoT security section, due to space limitations also for this IoT privacy section we list only a few threats and controls for each layer. These are, in our opinion, the most dangerous privacy threats and the most capable privacy controls.

As explained before, a given threat or control might be relevant for more than one layer. In a few cases, a threat or control indicated by us as a major one for one layer is also indicated as a major one for another layer. It should be noted that a threat or control chosen as a major one for one layer might also be a minor threat or control for another layer but—as a minor one—it is not enumerated by us for the other layer.

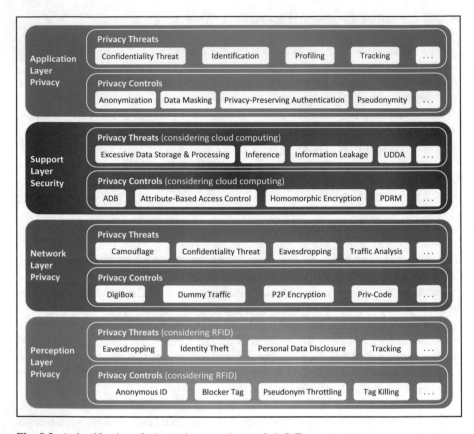

Fig. 9.3 A classification of privacy threats and controls in IoT

9.4.1 Perception Layer Privacy

As before, due to space limitations, we limit our discussion for the perception layer to threats and controls for the RFID technology, one of the most common technologies used here.

9.4.1.1 Privacy Threats in Perception Layer

The following are among the most dangerous privacy threats for the use of RFID in the IoT perception layer.

For RFID, *eavesdropping* is an unauthorized listening to the wireless communication between RFID tags and readers in order to extract confidential information [9].

Identity theft is a form of RFID cracking in which an adversary uses someone else's personal identity information to obtain access to privileged system areas with sensitive information [23].

Personal data disclosure is an illegitimate disclosure of personal data by an adversary (either an outsider or an insider) [24].

A *tracking* threat is an illegitimate or unwanted disclosure of a location of a person or an object [9]. In RFID technology, each RFID tag contains an ID code that enables readers to uniquely identify the tag. Once a specific tag or a set of tags is associated with a particular person or object, tracking becomes possible.

9.4.1.2   Privacy Controls for Perception Layer

The following are some of the most promising privacy controls for the RFID technology used in the IoT perception layer.

An *anonymous ID* scheme uses an encrypted tag ID instead of the clear text tag ID to prevent adversaries from identifying a tag (and its user) by the tag ID [9]. Asymmetric key encryption can be used to encrypt and decrypt tag's ID as well as RFID messages.

A *blocker tag* is a specific kind of RFID tag used to block communications between RFID tags and RFID readers [9]. A blocker tag can respond with a fake message to the RFID reader's request. It can in this way prevent an unauthorized RFID reader from obtaining, e.g., the actual tag ID. Consequently, it prevents tracking the tag owner by adversaries.

Pseudonym throttling is a pseudonym authentication scheme using a circular list of pseudonyms stored in a tag [9]. When a tag receives the next query from an RFID reader, it responds with the next pseudonym from the list. The corresponding tag reader checks the received pseudonym with its stored list of pseudonyms to find the actual tag ID associated with that pseudonym. Once the ID is found, the tag is authenticated.

Tag killing enables the destruction of sensitive tag information when it is no longer needed; thus information becomes unavailable for any disclosure [9]. Killing a tag requires providing a PIN code.

9.4.2   Network Layer Privacy

After data leave their source and are transmitted over an IoT network layer, they are subject to many privacy threats, some of them discussed in this section. Controls for reducing or overcoming these privacy threats are also discussed here.

9.4.2.1 Privacy Threats in Network Layer

The following are among the most dangerous privacy threats for the IoT network layer.

Camouflage occurs when a malicious node hides itself and masquerades as a normal node in an IoT network, which can result in compromising the network [25]. A camouflaged node may start advertising false routing information to other nodes in the network in order to attract traffic from them. Intercepted packets can be manipulated (which includes discarding them), or analyzed for extracting private information.

A *confidentiality* threat, if materialized, results in unauthorized sensitive data disclosures and disseminations [26]. It includes exposing sensitive routing data when a routing entity allows these data to be exposed to an outside entity, e.g., due to misconfiguration. It also includes exposing routing data to an insider in an uncontrolled manner, e.g., exceeding the need-to-know boundaries.

Eavesdropping (not only for RFID, as was discussed for the perception layer) is a threat that an unauthorized entity listens to information sent between IoT nodes [25]. One of the results of materialization of this threat is *communication tracking*, which reveals identities of communicating IoT nodes.

Traffic analysis is the process of intercepting and examining data flow in an IoT network in order to extract sensitive information, including network routing and flow patterns [27]. In particular, malicious nodes can identify the most active nodes, and make them targets of subsequent privacy attacks (assuming that the most active nodes are also most important or most knowledgeable).

9.4.2.2 Privacy Controls for Network Layer

Some of the most promising controls for the privacy of IoT hosts communicating at the IoT network layer are as follows.

The *DigiBox self-protecting container* technology provides data containers designed to protect private data during transmission through network nodes [27]. DigiBox is cryptographically protected to secure private data transmission. It also controls enforcement of data rules to ensure data confidentiality.

Dummy traffic (a.k.a. *link padding*) prevents traffic analysis by disseminating dummy data packets (with no useful content) or injecting deliberate artificial delays into data flow [28]. Nodes producing and transmitting dummy traffic appear more active than they really are, which can contribute to increasing unobservability and hiding traffic patterns from adversaries (including hiding the most active nodes).

In *Point-to-Point Encryption (P2PE)*, the encryption process is initiated by a designated and independently validated device. Encrypted data is subsequently sent as ciphertext (unreadable for intermediate devices) to be decrypted by another designated and independently validated device [29].

Priv-Code prevents an attacker from identifying traffic senders or receivers by making all nodes in the network transmit with the same traffic pattern [30]. It is based on network coding for data communication. It uses concurrent unicast sessions in multi-hop wireless networks. Only packets from the same session are encoded together.

9.4.3 Support Layer Privacy

This section discusses privacy in IoT services from the data perspective, especially from the point of view of private or sensitive data that are processed or stored by all kinds of middleware in the IoT support layer. We limit our discussion to threats and controls in cloud computing, which is a prime example of technology used by the support layer.

9.4.3.1 Privacy Threats in Support Layer

The following are among the most dangerous privacy threats for the use of cloud computing in the IoT support layer.

The *excessive data storage and processing* is the threat of storing sensitive data longer than needed or processing these data in an uncontrolled manner that exceeds the (stated or implied) purpose of collecting these data [31]. This increases the window of vulnerability for the system, which can be exploited by an attacker.

An *inference attack* exploits secondary sources of data (also off-line sources, provided by, e.g., social engineering) to infer meaningful information about a hidden action violating the privacy of the actor [32].

Information leakage can be caused by various cloud services or other IoT services [33]. Typically, a service user must provide some personal data (including authentication data) to the service provider, in a way that places these data under the service provider's control. Inadequate privacy protection by the service provider (e.g., unprotected index files) may result in information leakage.

An *Unauthorized Direct Data Access (UDDA)* is a threat of storing gathered sensitive data in the cloud on devices vulnerable to unauthorized access [34].

9.4.3.2 Privacy Controls for Support Layer

The following are some of the most promising privacy controls for the IoT support layer.

Active Data Bundles (ADBs) are software constructs encapsulating sensitive data (to be protected) with metadata (including privacy policies) and a virtual machine (serving as a policy enforcement engine) [26]. They can be used to protect sensitive

IoT data from unauthorized dissemination and disclosure. They can also reduce exceeding the need-to-know boundaries by authorized users (insiders).

Attribute-Based Access Control (ABAC) grants access to sensitive IoT data based on attributes of the party that requests access to these data [35]. This approach combines user, device, and other attributes (e.g., user id, IP address, MAC address, location) to make a context-aware decision at run-time.

Homomorphic encryption allows processing of encrypted data by services that do not know what the data are about [36]. It is a very powerful concept for preventing information disclosure. Unfortunately, it is very limited in practice due to the fact that servers working in the homomorphic encryption mode are limited to a small set of simple operations on data (including addition and multiplication operations).

Personal Digital Rights Management (PDRM) is a self-protecting data approach enabling individuals to protect their privacy rights for their sensitive data [37]. PDRM attaches to data a software construct named a *detector*, generated using an Artificial Immune System (AIS) technique. The detector assesses the use of the data to which it is attached, and denies data access if it recognizes an unusual access pattern.

9.4.4 Application Layer Privacy

This section outlines selected privacy threats that target the IoT application layer, and discusses relevant controls for this layer.

9.4.4.1 Privacy Threats in Application Layer

The diversity of IoT applications serving end users leads to many privacy vulnerabilities. The following are among the most dangerous privacy threats for the IoT application layer.

Confidentiality and *tracking* threats were discussed in Sects. 9.4.2.1 and 9.4.1.1—for network layer and perception layer, respectively.

The *identification* threat means associating an identity with a particular context including action, location, etc. [38]. Many technologies used in the IoT application layer enable identification; examples include surveillance cameras, fingerprinting, speech recognition, etc.

Profiling means inferring and recording information on personal interests and habits of individuals [38]. This includes obtaining psychological and behavioral characteristics that can be correlated with other data to identify a particular person or group of people. Profiling, often perceived as a privacy violation, can be performed by *data marketplaces* and sold to advertisers and business wishing to target customers.

9.4.4.2   Privacy Controls for Application Layer

The following are some of the most promising privacy controls for the IoT application layer, which can protect most applications.

Anonymization disassociates all identifiers from user's data, making the user anonymous [39]. Anonymization is a form of *de-identification* used when there is no need for re-identification (since, if properly and honestly executed, anonymized data cannot be re-identified by any party other than the entity that anonymized them).

Data masking provides a copy of the original data in which sensitive data items are changed or made invisible while maintaining the "statistical" data integrity [40]. It used to protect an original dataset (e.g. individuals' records or parts of it) by transforming it into the corresponding masked dataset. This is a useful technique for exchanging sensitive data among IoT applications.

Privacy-preserving authentication scheme integrates two cryptographic primitives, namely a blind signature and a hash chain, in order to hide the association between user's real identity and the user's authorization credential [41]. It provides mutual authentication between a user and a service interacting with the IoT application layer, without revealing user's real identity.

Pseudonymization replaces user's real identity with a pseudonym [39]. It is an efficient mechanism preventing identification and tracking in IoT applications. Pseudonymization is a form of de-identification (different than de-identification used for anonymization) for applications that might require re-identification. In contrast to a user of anonymized data, a user of pseudonymized data can re-identify them easily (when provided with the pseudonym-to-identifier mapping by the party that pseudonymized the data).

9.5   Concluding Remarks

9.5.1   Lessons Learned

In the beginning, we hypothesized that differentiating security from privacy will be beneficial for research. We performed this short survey following this hypothesis. After completing it, we are convinced that differentiating security from privacy leads to a clearer and better organized classification of threats and controls.

The evaluation of research papers for this survey resulted in classifying threats and controls into categories based on a four-layer IoT reference model [4].

9.5.2 Conclusions

The security and privacy issues are major concerns for the Internet of Things due to the nature of this emerging technology, which utilizes huge numbers of sensing devices to collect data about people and their environments. In contrast to the common approach, we differentiated security threats and controls (solutions) from privacy threats and controls—by defining *security* as the right not to have one's activities adversely affected via tampering with one's objects, and *privacy* as the right to have information about oneself left alone. Of course, many security issues (or privacy issues) include some privacy aspects (or security aspects, respectively). However, a clear indication of the primary concern (as either security or privacy) provided a useful focus for considering threats, and facilitated identifying controls.

This chapter is organized upon the framework provided by the four-layer IoT reference model [4]. Our main contribution lies in differentiating security from privacy in IoT as well as proposing a classification of security threats and controls, and a separate classification of privacy threats and controls. Both classifications were presented within the framework of the IoT reference model. In other words, for each of the four layers of the IoT reference model we provided an overview of what we perceive as the most dangerous threats and the most capable controls. In most cases security or privacy threats and security or privacy controls are unique to each layer, although in some cases identical threats might endanger two or more layers and same controls might help two or more layers.

References

1. Sundmaeker, H., Guillemin, P., Friess, P., & Woelfflé, S. (2010). *Vision and challenges for realising the Internet of Things*. Cluster of European Research Projects on the Internet of Things, European Commission (CERP-IoT). doi: 10.2759/26127
2. Pfleeger, C. P., Pfleeger, S. L., & Margulies, J. (2015). *Security in computing* (5th ed.). Englewood Cliffs, NJ: Prentice Hall.
3. Cooley, T. M. (1879). *Treatise on the law of torts or the wrongs which arise independent of contract*. Chicago: Callaghan.
4. Yang, G., Xu, J., Chen, W., Qi, Z. H., & Wang, H. Y. (2010). Security characteristic and technology in the Internet of Things. *Journal of Nanjing University of Posts and Telecommunications, 30*(4), 20–29.
5. Al-Fuqaha, A., Guizani, M., Mohammadi, M., Aledhari, M., & Ayyash, M. (2015). Internet of Things: A survey on enabling technologies, protocols, and applications. *IEEE Communications Surveys & Tutorials, 17*(4), 2347–2376.
6. Lilien, L., Kamal, Z., Bhuse, V., & Gupta, A. (2006). Opportunistic networks: the concept and research challenges in privacy and security. *Proceedings of International Workshop on Research Challenges in Security and Privacy for Mobile and Wireless Networks*, Miami, FL, pp. 134–147.
7. Lilien, L., Gupta, A., Kamal, Z., & Yang, Z. (2010). Opportunistic resource utilization networks—a new paradigm for specialized ad hoc networks [Special Issue: Wireless Ad Hoc, Sensor and Mesh Networks, Elsevier]. *Computers and Electrical Engineering, 36*(2), 328–340.

8. Yan, Z., Zhang, P., & Vasilakos, A. V. (2014). A survey on trust management for Internet of Things. *Journal of Network and Computer Applications, 42*, 120–134.
9. Spruit, M., & Wester, W. (2013). *RFID security and privacy: Threats and countermeasures.* Utrecht: Department of Information and Computing Sciences, Utrecht University.
10. Mitrokotsa, A., Rieback, M. R., & Tanenbaum, A. S. (2010). Classification of RFID attacks. *Journal of Information Systems Frontiers, 12*(5), 491–505.
11. De Fuentes, J. M., Peris-Lopez, P., Tapiador, J. E., & Pastrana, S. (2015). Probabilistic yoking proofs for large scale IoT systems. *Ad Hoc Networks, 32*, 43–52.
12. Katagi, M., & Moriai, S. (2011). *Lightweight cryptography for the Internet of Things* (Technical Report). Tokyo: Sony Corporation. Online: http://www.iab.org/wp-content/IAB-uploads/2011/03/Kaftan.pdf
13. Specht, S. M., & Lee, R. B. (2004). Distributed denial of service: taxonomies of attacks, tools, and countermeasures. *Proceedings of ISCA International Conference on Parallel and Distributed Computing Systems (PDCS)*, San Francisco, CA, pp. 543–550.
14. Farooq, M. U., Waseem, M., Khairi, A., & Mazhar, S. (2015). A critical analysis on the security concerns of Internet of Things (IoT). *International Journal of Computer Applications, 111*(7), 1–6.
15. Mahmood, Z. (2016). *Connectivity frameworks for smart devices.* Cham: Springer International Publishing.
16. Roman, R., Alcaraz, C., Lopez, J., & Sklavos, N. (2011). Key management systems for sensor networks in the context of the Internet of Things. *Computers & Electrical Engineering, 37*(2), 147–159.
17. Alani, M. M. (2016). *Elements of cloud computing security: A survey of key practicalities. Springer Briefs in Computer Science.* Berlin: Springer International Publishing.
18. Zunnurhain, K., & Vrbsky, S. V. (2010). Security attacks and solutions in clouds. *Proceedings of the 1st International Conference on Cloud Computing*, Tuscaloosa, AL, pp. 145–156.
19. Anggorojati, B. (2015). *Access control in IoT/M2M-cloud platform.* Ph.D. dissertation, The Faculty of Engineering and Science, Aalborg University, Aalborg, Denmark.
20. Patel, A., Taghavi, M., Bakhtiyari, K., & Júnior, J. C. (2013). An intrusion detection and prevention system in cloud computing: A systematic review. *Journal of Network and Computer Applications, 36*(1), 25–41.
21. Ahmed, N. (2016). *Designing, implementation and experiments for moving target defense.* Ph.D. dissertation, Department of Computer Science, Purdue University, West Lafayette, IN.
22. Hong, J. (2012). The state of phishing attacks. *Communications of the ACM, 55*(1), 74–81.
23. Muir, B. (2009). Radio frequency identification: privacy & security issues (slides). *Slide Share.* Online: http://www.slideshare.net/bsmuir/rfid-privacy-security-issues-31614795
24. Thompson, D. R., Chaudhry, N., & Thompson, C. W. (2006). RFID security threat model. In *Proceedings of Conference on Applied Research in Information Technology*, Conway, AR.
25. Virmani, D., Soni, A., Chandel, S., & Hemrajani, M. (2014). *Routing attacks in wireless sensor networks: A survey.* arXiv preprint arXiv:1407.3987.
26. Ben Othmane, L., & Lilien, L. (2009). Protecting privacy in sensitive data dissemination with active bundles. In *Proceedings of Seventh Annual Conference on Privacy, Security and Trust (PST)* (pp. 202–213). Saint John, NB.
27. Sibert, O., Bernstein, D., & Van Wie, D. (1995). The DigiBox: A self-protecting container for information commerce. *Proceedings of First USENIX Workshop on Electronic Commerce*, New York, NY, pp. 15–15.
28. Berthold, O., & Langos, H. (2002). Dummy traffic against long term intersection attacks. In *Proceedings of International Workshop on Privacy Enhancing Technologies* (pp. 110–128). Berlin: Springer.
29. PCI Security Standards Council. (2010). Initial roadmap: point-to-point encryption technology and PCI DSS compliance. *Emerging Technology Whitepaper.* Online: https://www.pcisecuritystandards.org/documents/pci_ptp_encryption.pdf

30. Wan, Z., Xing, K., & Liu, Y. (2012). Priv-Code: Preserving privacy against traffic analysis through network coding for multi-hop wireless networks. *Proceedings of 31st Annual IEEE International Conference on Computer Communications (INFOCOM)*, Orlando, FL, pp. 73–81.
31. Pearson, S. (2009). Taking account of privacy when designing cloud computing services. *Proceedings of the ICSE Workshop on Software Engineering Challenges for Cloud Computing*, Vancouver, BC, pp. 44–52.
32. Waterson, D. (2015). IoT inference attacks from a whole lotta talkin' going on. *Thoughts on Information Security*. Online: https://dwaterson.com/2015/08/26/iot-inference-attacks-from-a-whole-lotta-talkin-going-on/
33. Squicciarini, A., Sundareswaran, S., & Lin, D. (2010). Preventing information leakage from indexing in the cloud. *Proceedings of 3rd IEEE International Conference on Cloud Computing*, Miami, FL, pp. 188–195.
34. Nasim, R. (2012). Security threats analysis in Bluetooth-enabled mobile devices. *International Journal of Network Security & its Applications, 4*(3), 41–56.
35. Monir, S. (2017). *A Lightweight attribute-based access control system for IoT*. Ph.D. dissertation, University of Saskatchewan, Saskatoon, SK.
36. Tebaa, M., & Hajji, S. E. (2014). Secure cloud computing through homomorphic encryption. *International Journal of Advancements in Computing Technology (IJACT), 5*(16), 29–38.
37. Tchao, A., Di Marzo, G., & Morin, J. H. (2017). Personal DRM (PDRM)—A self-protecting content approach. In F. Hartung et al. (Eds.), *Digital rights management: Technology, standards and applications*. New York: CRC Press, Taylor & Francis Group.
38. Ziegeldorf, H., Morchon, G., & Wehrle, K. (2014). Privacy in the Internet of Things: Threats and challenges. *Security and Communication Networks, 7*(12), 2728–2742.
39. Pfitzmann, A., & Hansen, M. (2010). *A terminology for talking about privacy by data minimization: Anonymity, unlinkability, undetectability, unobservability, pseudonymity, and identity management (Version v0.34)*. Online: https://dud.inf.tu-dresden.de/literatur/Anon_Terminology_v0.34.pdf
40. Duncan, G., & Stokes, L. (2009). Data masking for disclosure limitation. *Wiley Interdisciplinary Reviews: Computational Statistics, 1*(1), 83–92.
41. Ren, K., Lou, W., Kim, K., & Deng, R. (2006). A novel privacy preserving authentication and access control scheme for pervasive computing environments. *IEEE Transactions on Vehicular Technology, 55*(4), 1373–1384.

Chapter 10
Reliable Transmission Protocol for Underwater Acoustic Networks

Xiujuan Du, Meiju Li, and Keqin Li

10.1 Challenges of UANs

Recently, Underwater Acoustic Networks (UANs) research has attracted significant attention due to the potential for applying UANs in environmental monitoring, resource investigation, disaster prevention, and so on [1–10]. UANs use acoustic communication, but the acoustic channel is characterized by high bit errors (on the order of magnitude of 10^{-3}–10^{-7}), long propagation delay (at a magnitude of a few seconds), and narrow bandwidth (only scores of kbps). The result is that the terrestrial-based communication protocols are either inapplicable or inefficient for UANs. Compared with conventional modems, the acoustic modems used in UANs consume more energy. However, the nodes are battery-powered and it is considerably more difficult to recharge or replace nodes in harsh underwater environments. Furthermore, underwater nodes are usually deployed sparsely, move passively with water currents or other underwater activity, and some nodes will fail due to energy depletion or hardware faults; therefore the network topology of UANs usually changes dynamically, which causes significant challenges in designing protocols for UANs.

X. Du (✉)
School of Computer Science, Qinghai Normal University, Xining, 810008, Qinghai, China

Key Laboratory of the Internet of Things of Qinghai Province, Xining, 810008, Qinghai, China
e-mail: dxj@qhnu.edu.cn

M. Li
School of Computer Science, Qinghai Normal University, Xining, 810008, Qinghai, China
e-mail: 1143828260@qq.com

K. Li
Department of Computer Science, State University of New York, New Paltz, NY, 12561, USA
e-mail: lik@newpaltz.edu

© Springer International Publishing AG 2018
K. Daimi (ed.), *Computer and Network Security Essentials*,
DOI 10.1007/978-3-319-58424-9_10

Applications of UANs in areas such as business, scientific research, and military are usually sensitive: outsiders are not allowed to access the sensitive information, and anonymous secure communication is broadly applied. However, thus far, to the best of our knowledge, there are few papers concerning secure communications protocols for UANs [11–14]. The nature of opening and sharing of underwater acoustic channel makes communications inherently vulnerable to eavesdropping and interference. Because of the highly dynamic nature of UANs, as well as their lack of centralized management and control, designing secure routing protocols that support anonymity and location privacy is a large challenge.

In UANs with dynamic topology and impaired channel, network efficiency following the traditional five-layered architecture was obtained by cross-layer designs, which cause numerous complicated issues that are difficult to overcome. The chapter introduces a three-layer protocol architecture for UANs, which includes application layer, network-transport layer, and physical layer and is named Micro-ANP. Based on the three-layer Micro-ANP architecture, the chapter provides a handshake-free Media Access Control (MAC) protocol for UANs, and achieves reliable hop-by-hop transmissions.

The remainder of the chapter is organized as follows. Section 10.2 presents the Micro-ANP architecture. Section 10.3 reviews the research on reliable transmission mechanism so far. Section 10.4 details the handshake-free reliable transmission protocol for UANs based on Micro-ANP architecture and RLT code. Section 10.5 makes a conclusion and has a discussion about new trends of UANs research.

10.2   Micro-ANP Architecture

The majority of research on UANs has focused primarily on routing or MAC protocols, and few studies have investigated protocol architecture for UANs. The energy, computation, and storage resources of UANs are seriously constrained; consequently, the protocol stack running on UANs nodes should not be complicated. However, most research on UANs so far has followed the traditional five-layered architecture in network design, and in tough condition such as dynamic topology, seriously impaired channel, and scarce resources, network efficiency was obtained by cross-layer designs, which cause numerous complicated issues that are difficult to overcome. UANs need a simple and efficient protocol architecture. Du et al. provided a three-layered Micro-ANP architecture for UANs, which is composed of an application layer, a network-transport layer, and a physical layer as well as an integrated management platform, as shown in Fig. 10.1 [15].

The network-transport layer in Micro-ANP is primarily responsible for reliable hop-by-hop transmission, routing, and channel access control. In Micro-ANP, broadcasting, Level-Based Adaptive Geo-Routing (LB-AGR), and a secure anonymous routing are the three major routing protocols that are applicable to dynamic underwater topology [7, 16]. A secure anonymous routing protocol can achieve anonymous communication between intermediate nodes as well as two-way

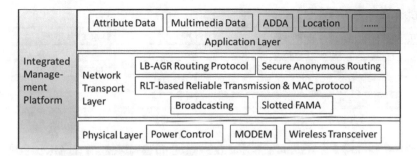

Fig. 10.1 Micro-ANP architecture

Table 10.1 Head fields of micro-ANP

Bits: 8	8	8	2	6	1	1	24	8
Level of sender	Sender ID	Receiver ID	Type 00: Data 01: Ack 10: Control	Frame sequence number	Immediately ack 1: yes 0: no	If block 1: Yes 0: No	IDs of original packets	Block ID
Bits:6	1	2	1	48	4	8	Variable	16
Block size	Direction 0: down 1: up	Sink ID	(Source|destination) 0: position 1: node ID	(Source|destination) position or ID Full "1" for broadcast	Application priority (application type)	Load length	Data	FCS

authentication between source and destination nodes without any real-time online Public Key Generator (PKG), thus decreases the network delay while improving network scalability. In Micro-ANP, slotted Floor Acquisition Multiple Access (slottedFAMA) and a RLT Code-based Handshake-Free (RCHF) reliable MAC protocol are the two-channel access control mechanism [9, 17].

Micro-ANP is a three-layered architecture that allows intermediate nodes to perform Application Dependent Data Aggregation (ADDA) at the application layer. Without requiring a cross-layer design, Micro-ANP can make efficient use of scarce resources. Moreover, Micro-ANP eliminates inapplicable layers and excessive repeated fields such as address, ID, length, Frame Check Sequence (FCS), and so on, thus reducing superfluous overhead and energy consumption. The head fields of the network-transport layer are listed in Table 10.1.

The application priority field is used to distinguish between different applications as shown in Table 10.2. This is because different applications have different priorities and require different Quality of Service (QoS) and their messages are transmitted using different routing decisions. Other fields in Table 10.2 will be explained in the respective protocol overview of the network-transport layer.

Table 10.2 Application priority

Priority	Upper protocol	Priority	Upper protocol
0	Attribute data	4	Video
1	Integrated management	5	Emergency alarm
2	Image	6	
3	Audio	7	

From Table 10.1, we can see that the common head-length of Micro-ANP is less than 20 bytes. In comparison, the total head-length of well-known five-layer models is more than 50 bytes. Therefore, Micro-ANP protocol greatly improves data transmission efficiency.

10.3 Overview of Reliable Transmission Mechanism

Considering the challenges for UANs, the existing solutions of terrestrial Radio Frequency (RF) networks cannot be applied directly to UANs, regardless of the MAC mechanism used, the reliability of data transmission, or the routing protocol. Sustained research work over the last decade has introduced new and efficient techniques for sensing and monitoring marine environments; several issues still remain unexplored. The inapplicability of conventional reliable transport mechanisms in UANs is analyzed as follows:

1. The high bit error rates of acoustic channels lead to high probability of packet erasure and a low probability of success in hop-by-hop transfers. Therefore, traditional end-to-end reliable transport mechanisms may incur too many re-transmissions and experience too many collisions, thus reducing channel utilization.
2. The low propagation speed of acoustic signals leads to long end-to-end delays, which causes issues when controlling transmissions between two end-nodes in a timely manner.
3. The Automatic Repeat Request (ARQ) mechanism re-transmits lost packets, but it requires an ACK (acknowledgement) for packets received successfully. It is well known that the channel utilization of the simple stop-and-wait ARQ protocol is very low in UANs due to long propagation delays and low bit rates. In addition, acoustic modems adopt half-duplex communication, which limits the choices for efficient pipelined ARQ protocols. Even worse, if the ACKs are lost, the successfully received packets will be re-transmitted by the sender, further increasing the bandwidth and energy consumed.

Some reliable transport protocols resort to Forward-Error-Correcting (FEC) to overcome the inherent problems with ACKs. FEC adopts erasure codes and redundancy bits. The payload bits of FEC are fixed prior to transmission. Before transmitting, the sender encodes a set of n original packets into a set of N ($N \geq n$) encoded packets. Let $m = N - n$, and m redundant packets are generated. To reconstruct the n original packets, the receiver must receive a certain number (larger than n) of encoded packets. The stretch factor is defined as N/n, which is a constant that depends on the erasure probability. However, the error probability of UANs channels is dynamic; overestimated error probability will incur additional overhead and underestimated error probability will lead to transmission failure.

Reed and Solomon proposed the Reed–Solomon code based on some practical erasure codes [18]. Reed–Solomon code is efficient for small n and m values. However, the encoding and decoding algorithms require field operations, resulting in a high computation overhead that is unsuitable for UANs due to the nodes' limited computational capabilities. Luby et al. studied a practical Tornado code which involves only XOR operations [19]. In addition, the encoding and decoding algorithms are faster than those used for Reed–Solomon code. However, the Tornado code uses a multi-layer bipartite graph to encode and decode packets, resulting in a high computation and communication overhead for UANs. Xie et al. presented a Segmented Data Reliable Transfer (SDRT) protocol [20]. SDRT adopts Simple Variant of Tornado (SVT) code to improve the encoding/decoding efficiency. Nevertheless, after pumping the packets within a window into the channel quickly, the sender sends the packets outside the window at a very slow rate until it receives a positive feedback from the receiver, which reduces channel utilization. Mo et al. investigated a multi-hop coordinated protocol for UANs based on the GF(256) random-linear-code to guarantee reliability and efficiency [21]. However, the encoding vectors are generated randomly; consequently, the probability of successfully recovering K data packets from K encoded packets could not be guaranteed. Moreover, the decoding complexity was higher than other sparse codes. Furthermore, the multi-hop coordination mechanism requires time synchronization and is restricted to a string topology in which there is a single sender and a single receiver.

Digital fountain codes are sparse codes on bipartite graphs that have high performance [21, 22, 23]. They are rate-less, i.e., the amount of redundancy is not fixed prior to transmission and can be determined on the fly as the error recovery algorithm evolves. These codes are known to be asymptotically near-optimal for every erasure channel, and they allow for lightweight encoder and decoder implementations. Luby proposed the LT code, in which the decoder is capable of recovering the original symbols at a high probability from any set of output symbols whose size is close to the originals [24]. However, the LT code was designed for large numbers of data packets, which is not typically the case in UANs—especially for mobile networks where the transmission time between two nodes is very limited because of node mobility. Furthermore, the degree distribution used in LT code results in a large number of nodes in the graph, causing a large overhead for each packet.

10.4   Reliable Transmission Protocol for UANs

In this section, based on digital fountain code, a Recursive LT (RLT) code with a small degree distribution is proposed along with a reliable and handshake-free MAC protocol called as RCHF MAC protocol.

10.4.1   RLT Code

The coding scheme can greatly impact system performance. In this section, we present a Recursive LT (RLT) code, which achieves fast encoding and decoding. Given that packet loss is independent, we use a bipartite graph $G = (V, E)$ with two levels to represent the RLT code, where E is the set of edges and V is the set of nodes in the graph. $V = D \bigcup C$, where D is the set of input packets and C is the set of encoded packets. The edges connect the nodes in D and C.

1. Encoding

Consider a set of k input (original) packets, each having a length of l bits. The RLT encoder takes k input packets and can generate a potentially infinite sequence of encoded packets. Each encoded packet is computed independently of the others. More precisely, given k input packets $\{x_1, x_2, \cdots, x_k\}$ and a suitable probability distribution $\Omega(d)$, a sequence of encoded packets $\{y_1, y_2, \cdots, y_j, \cdots, y_n\}, n \geq k$, are generated as shown in Fig. 10.2. The parameter d is the degree of the encoded packets—the number of input packets used to generate the encoded packets and $d \in \{1, 2, \cdots, k\}$ (e.g., the degree of packet y_2 is 2 while the degree of packet y_8 is 3 in Fig. 10.2).

To restore all the k original packets at the receiver, the number of encoded packets received successfully is subject to be greater than k. Let $n = (k + \xi)/(1 - P_p)$; here, P_p is the erasure probability of an underwater acoustic channel (i.e., the PER), and $\xi(\xi > 0)$ corresponds to the expected number of redundant encoded packets received. The ξ redundant packets are used to decrease the probability that the

Fig. 10.2 Encoding graph of RLT code

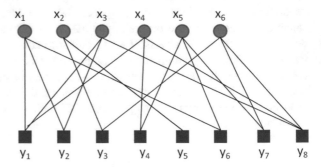

receiver fails to restore the original k input packets in only one transmission phase. The sequence of encoded packets is $y_1, y_2, \cdots, y_j, \cdots, y_n \in C$. The RLT encoding procedure is as follows:

a. From D, the set of input packets, successively XOR the k packets to generate one encoded packet with degree k, then duplicate the packet to obtain $\lceil 1/(1 - P_p) \rceil$ copies.

b. From set D, select $\lceil m/(1 - P_p) \rceil$ distinct packets randomly to constitute a seed set S_1, and generate $\lceil m/(1 - P_p) \rceil$ encoded packets with degree one. Here, m is the expected number of encoded packets received successfully with degree one. In reality, we can set $1 \leq m \leq \max(\lfloor k/4 \rfloor, 1)$.

c. Let $S_2 = D - S_1$. From the set S_2, uniformly select $\lceil k/(2(1 - P_p)) \rceil$ input packets at random, and perform the XOR operation, randomly selecting one packet in the set S_1 to generate $\lceil k/(2(1 - P_p)) \rceil$ encoded packets with degree two.

d. Let $S_3 = D - S_1 - S_2$. If S_3 is not null, select $\lceil k/(6(1 - P_p)) \rceil$ input packets at random from set S_3; otherwise, from set D, perform the XOR operation using one packet from S_2 and another from S_1 to generate $\lceil k/(6(1 - P_p)) \rceil$ encoded packets with degree three.

e. Let $S_4 = D - S_1 - S_2 - S_3$. If S_4 is not null, randomly select $\lceil (\xi + k/3 - m - 1)/(1 - P_p) \rceil$ input packets from set S_4; otherwise, from set D, perform the XOR operation using three packets from S_1, S_2, and S_3, respectively, to generate $\lceil (\xi + k/3 - m - 1)/(1 - P_p) \rceil$ encoded packets with degree four.

2. Decoding

When an encoded packet is transmitted over an erasure channel, it is either received successfully or lost. The RLT decoder tries to recover the original input packets from the set of encoded packets received successfully. The decoding process of RLT is as follows:

a. Find an encoded packet y_j which is connected to only one input packet x_i. If the receiving node fails to find any such encoded packet, stop decoding.

b. Set $x_i = y_j$.

c. Set $y_m = y_m \oplus x_i$ for each encoded packet which is connected to x_i, denoted by y_m. Here, $\oplus$ indicates the XOR operation.

d. Remove all the edges connected to x_i.

e. Go to Step a.

3. Degree distribution.

The limited delivery time between two nodes caused by node mobility leads to the constraint that digital fountain codes must work with small k values in UANs communications. In RLT, to reconstruct the input packets, the degree distribution of the received encoded packets should have the following properties:

a. The received encoded packets should connect all the input packets.

b. The process of encoding and decoding should not involve too many XOR operations.

c. At least one encoded packet with degree one should be successfully received by the receiver.

Given the high bit error, P_b, which is on the order of magnitude of 10^{-3}–10^{-7}, the PER, P_p, is given by Eq. (10.1):

$$p_p = 1 - (1 - p_b)^l, \tag{10.1}$$

where l is the packet size. As discussed earlier, in Micro-ANP architecture, the optimal packet size is greater than 100 bytes, and P_p is non-negligible in Eq. (10.1). Considering the k input packets, to address the properties of degree distribution discussed above, the degree distribution of the encoded packets in the sending nodes is given by Eq. (10.2):

$$\Omega(d) = \begin{cases} \frac{m}{\xi+k}, & d = 1; \\ \frac{k}{d(d-1)(\xi+k)}, & d = 2, 3; \\ \frac{\xi+(1/3)k-(m+1)}{\xi+k}, & d = 4; \\ \frac{1}{\xi+k}, & d = k; \end{cases} \tag{10.2}$$

where $\sum_d \Omega(d) = 1$.

Lemma 1 *The average degree of encoded packets $\lambda \approx 3.7$.*

Proof From the degree distribution given by Eq. (10.2), we obtain:

$$\lambda = E(d) = \sum_{d=1}^{4} (d \times \Omega(d))$$

$$= \frac{1 \times m}{\xi + k} + \frac{2 \times k}{2 \times 1 \times (\xi + k)} + \frac{3 \times k}{3 \times 2 \times (\xi + k)}$$

$$+ \frac{4 \times (\xi + 1/3k - (m + 1))}{\xi + k} + \frac{k}{\xi + k}$$

$$= 3\frac{2}{3} + \frac{\frac{\xi}{3} - 3m - 4}{\xi + k}.$$

Usually, $|(\xi/3) - 3m - 4| \ll |\xi + k|$, so $\lambda \approx 3\frac{2}{3} \approx 3.7$.

Given the block size k, from Lemma 1, we can derive the decoding complexity of RLT is about 3.7 which is irrelevant to the number of input packets. A comparison of the encoding/decoding complexity of various codes is shown in Table 10.3.

In this section, based on the digital fountain code, we propose a Recursive LT (RLT) code with small degree distribution, and introduce the erasure probability of channel P_p into the RLT code for the first time to improve the decoding probability at the receiving node. RLT is applicable to dynamic UANs with limited transmission time between two nodes; it reduces the overhead of encoding and decoding and substantially improves the efficiency of decoding process.

Table 10.3 Decoding complexity comparison

Code	Encoding/decoding complexity
GF (256) in [21]	$O(k^3)$
LT	$k\ln_e^k$
SDRT in [20]	$k \cdot \ln(1/\varepsilon)$
RS	$k(N-k)\log_2^N$
RLT	3.7

10.4.2 RCHF: RLT Code-Based Handshake-Free Reliable Transmission Protocol

After solving the problems of degree distribution, encoding and decoding of RLT in advance, a reliable RLT-based media access control protocol should be presented that nodes can use to communicate in real time. Wireless transceivers usually work in half-duplex mode: a sending node equipped with a single channel is unable to receive packets while it is transmitting; therefore, the RCHF solution is supposed to avoid interference caused by transmitting to a node in a sending state. So far, in MAC solutions of wireless multi-hop packet networks, an RTS/CTS handshake is used to dynamically determine whether the intended receiver is ready to receive a frame. For underwater sensors, the rate at which data bits can be generated is approximately 1–5 bps and the optimal packet-load for UANs is about 100 bytes. In contrast, the length of an RTS frame is a few dozen bytes. Therefore, RTS/CTS frames are not particularly small compared with data frames; consequently, the benefits from using RTS/CTS handshake are unremarkable. Moreover, considering the characteristics of acoustic communication (i.e., low bandwidth, long propagation delay, etc.), RTS/CTS handshake decreases channel utilization and network throughput dramatically while prolonging end-to-end delay. Therefore, coupled closely with the RLT code, we propose a RCHF protocol which is a state-based handshake-free reliable MAC solution for UANs.

10.4.2.1 Reliable Transmission Mechanism

In the RCHF MAC solution, a source node first groups input packets into blocks of size k (i.e., there are k input packets in a block). Then, the source node encodes the k packets, and sends the encoded packets to the next hop. When k is equal to 50, the minimum time interval for transmitting a block between two neighbor nodes is approximately 60 s, which is in compliance with the requirements of the limited transmission time between two neighbor nodes in dynamic UANs. By setting the block size k appropriately, RCHF can control the transmission time, allowing the receiver to be able to receive sufficient encoded packets to reconstruct the original block even when the nodes are moving. Application data are transferred from a source node to a sink node block by block and each block is forwarded via RLT coding hop-by-hop.

In the RCHF protocol, a node sending packets is considered to be in the transmission phase. To facilitate receiving an ACK for transmitted packets, avoid conflicts between transmitting and receiving, and compromise between transmission efficiency and fairness, two transmission constraints are defined as follows:

1. The maximum number of data frames allowed to be transmitted in one transmission phase is N_{max}.
2. The minimum time interval between two tandem transmission phases of the same node is T_a. The node waiting for T_a expiration is considered to be in a send-avoidance phase. At present, underwater acoustic modems are half-duplex, the delay for state transition between sending and receiving usually ranges from hundreds of milliseconds to several seconds, which is close to the magnitude of the maximum round-trip time (RTT) [18]. Therefore, to facilitate the receiver to switch to the sending state to transmit the ACK, we set $T_a = 2 \times RTT$.

After transmitting N ($N \le N_{max}$) encoded packets, the sender switches to the receiving state and waits for the receiver's ACK. To have a high probability of being able to reconstruct the original k input packets at the receiver, the number of encoded packets received successfully is supposed to be larger than k, denoted as $k + \xi$. Considering the high packet error rate, P_p, we set $N = (k + \xi)/(1 - P_p)$. The parameter ξ, ($\xi > 0$) is fixed and corresponds to the expected number of redundant encoded packets the receiver will receive. The ξ redundant packets are used to decrease the probability that the receiver fails to restore the original k input packets in the transmission phase, and the factor $1/(1 - P_p)$ is used to compensate for channel errors.

The ACK frame includes the number of frames received at the receiver as well as the indices of unrecovered input packets. The number of frames received successfully can be used to update the packet error rate P_p on the fly. If the receiver can reconstruct the whole block, it sends back an ACK with "null" in the index field.

Given k_1 input packets unrecovered after the previous transmission phase, the sender encodes and transmits N_1 encoded packets with the degree distribution given by Eq. (10.2) in which k is replaced by k_1. $N_1 = (k_1 + \xi)/(1 - P_p)$. Then the sender collects the feedback from the receiver again. This process repeats until the sender receives an ACK with "null" in the index field.

10.4.2.2 State-Based Handshake-Free Media Access Control

After network initialization, each node maintains one dynamic neighbor table that includes a state field containing the real-time state of neighbor nodes as shown in Table 10.4. Here, state "0" indicates that the neighbor node is in sending state, state "1" indicates that the neighbor node is receiving frames from other nodes, "2" denotes an unknown state, and "3" means the neighbor node is in the send-avoidance phase.

The format of frames in our protocol is shown in Table 10.5. The level field contains the forwarder's level, the frame sequence number is used to identify the

Table 10.4 The state table of neighbor nodes

Value	State
0	Sending state
1	Receiving frame from other nodes
2	Unknown state
3	Transmission–avoidance

Table 10.5 The format of data frame

Bits: 8	8	8	2	6	1	1
Level of sender	Sender ID	Receiver ID	Type 00: Data 01: Ack 10: Control	Frame sequence number	Immediately ack 1: yes 0: no	If block 1: Yes 0: No
Bits:24	8	6	–	8		Variable
IDs of original packets	Block ID	Block size	–	Load length		Data

frame in one frame-sequence during one transmission phase, the original packet ID field is used to indicate the IDs of packets that are XORed, and the immediate ACK field is used to inform the receiver whether to return an ACK immediately, where "1" means "yes" and "0" means "no." The first nine bytes are used by the RCHF MAC protocol to realize reliable transmission hop-by-hop; the fields are updated hop-by-hop. The fields from the tenth to the sixteenth bytes are used by the LB-AGR routing protocol and are omitted here for simplicity.

When a node has packets to send, it searches the neighbor table for the state field of the intended receiver. If the state is "0" or "1," it will delay delivery until the state is greater than one; otherwise, the node becomes a sender, switches into the transmission phase, and starts to deliver frames. The pseudocode for sending packets is omitted.

10.4.3   Simulation Result of RCHF

In this section, we evaluate the performance of the RCHF protocol by simulation experiments. All simulations are performed using Network Simulator 2 (NS2) with an underwater sensor network simulation package extension (Aqua-Sim). Our simulation scenario is similar to reality; 100 nodes are distributed randomly in an area of 7000 m × 7000 m × 2000 m. The simulation parameters are listed in Table 10.6.

The protocol is evaluated in terms of average end-to-end delay, end-to-end delivery ratio, energy consumption, and throughput. We define the delivery ratio and throughput of the RCHF protocol as follows:

Table 10.6 Simulation parameters

Parameter	Value
Block size k	50
Packet length l	160 bytes
Bandwidth	10 kbps
Routing protocol	Static
Traffic	CBR
Transmission range	1500 m
MAC protocol	802.11

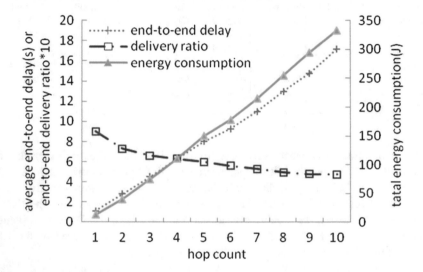

Fig. 10.3 Performance vs. hop COUNT

1. The end-to-end delivery ratio is defined by Eq. (10.3):

$$\text{end-to-end delivery ratio} = \frac{\#\text{of packets received successfully at sink}}{\#\text{of packets generated at sources}} \qquad (10.3)$$

2. The throughput is defined as the number of bits delivered to the sink node per second (bps)

As shown in Fig. 10.3, the end-to-end delivery ratio of the RCHF protocol is close to "1" when the hop count is "1" and decreases slightly as the hop count increases, which is considered good performance for UANs from a delivery ratio aspect. Figure 10.3 also shows that the end-to-end delay and total energy consumption rise with the hop count which is understandable. Note that the real value of the end-to-end delivery ratio is the value of the ordinate axis divided by 10.

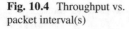

Fig. 10.4 Throughput vs.
packet interval(s)

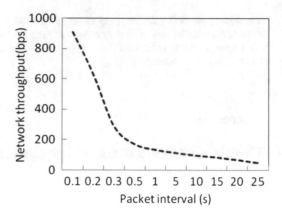

As shown in Fig. 10.4, the network throughput of RCHF decreases as the interval time between two successive packets generated by the source node increases. This occurs because as the interval time increases, fewer packets are generated, which reduces the network load.

10.5   Conclusion

In this chapter, a three-layer Micro-ANP protocol architecture for UANs is introduced. Further, a kind of digital fountain code which is called as RLT is presented. RLT is characterized by small degree distribution and recursive encoding, so RLT reduces the complexity of encoding and decoding. Based on the Micro-ANP architecture and RLT code, a handshake-free reliable transmission mechanism-RCHF is presented. In RCHF protocol, frames are forwarded according to the state of the receiver which can avoid the sending–receiving collisions and overhearing collisions. Simulations show that RCHF protocol can provide higher delivery ratio, throughput, and lower end-to-end delay.

As a new trend, how to combine the specific underwater application scenarios, transform the negative factors of UANs into favorable factors is an interesting research. For example, the mobility of nodes brings about extra routing overhead, and reduces end-to-end performance. However, the mobility of Autonomous Underwater Vehicle (AUV) and the policy of cache-carry-forward help to improve the data forwarding rate.

Meanwhile, under the precondition of less resource consumption, guaranteed channel utilization and network throughput, combining the technologies of channel coding, cognitive underwater acoustic communication, data compression, and post-quantum public key cryptography, studying on secure and reliable data transmission is another future work.

Acknowledgments This work is supported by the National Natural Science Foundation Projects of China (61162003), Key laboratory of IoT of Qinghai Province (2017-Z-Y21), Qinghai Office of Science and Technology (2015-ZJ-904), Hebei Engineering Technology Research Center for IOT Data acquisition & Processing.

References

1. Zhou, Z., Peng, Z., Cui, J. H., & Jiang, Z. (2010). Handling triple hidden terminal problems for multi-channel MAC in long-delay underwater sensor networks. In *Proceedings of international conference on computer communications (INFOCOM)* (pp. 1–21). San Diego, USA: IEEE Computer Society.
2. Pompili, D., & Akyildiz, I. F. (2010). A multimedia cross-layer protocol for underwater acoustic sensor networks. *IEEE Transaction on Wireless Communications, 9*(9), 2924–2933.
3. Pompili, D., Melodia, T., & Akyildiz, I. F. (2010). Distributed routing algorithms for underwater acoustic sensor networks. *IEEE Transaction on Wireless Communications, 9*(9), 2934–2944.
4. Huang, C. J., Wang, Y. W., & Liao, H. H. (2011). A power-efficient routing protocol for underwater wireless sensor networks. *Applied Soft Computing, 11*(2), 2348–2355.
5. Zhou, Z., & Cui, J. H. (2008). Energy efficient multi-path communication for time-critical applications in underwater sensor networks. In *Proceedings of the 9th ACM international symposium on mobile ad hoc networking and computing*, Hong Kong, China (pp. 1–31). New York, USA: ACM.
6. Hao, K., Jin, Z., Shen, H., & Wang, Y. (2015). An efficient and reliable geographic routing protocol based on partial network coding for underwater sensor networks. *Sensors, 15*, 12720–12735.
7. Du, X., Huang, K., & Lan, S. (2014). LB-AGR: Level-based adaptive geo-routing for underwater sensor networks. *The Journal of China Universities of Posts and Telecommunications, 21*(1), 54–59.
8. Du, X., Peng, C., Liu, X., & Liu, Y. (2015). Hierarchical code assignment algorithm and state-based CDMA protocol for UWSN. *China Communications, 12*(3), 50–61.
9. Du, X., Li, K., Liu, X.Su, Y. (2016 RLT code based handshake-free reliable MAC protocol for under-water sensor networks. *Journal of Sensors.* doi:10.1155/2016/3184642
10. Du, X., Liu, X., & Su, Y. (2016). Underwater acoustic networks testbed for ecological monitoring of Qinghai Lake. In *Proceedings of oceans16 Shanghai* (pp. 1–10).
11. Dong, Y., & Liu, P. (2010). Security consideration of underwater acoustic networks. In *Proceedings of International Congress on Acoustics, ICA*.
12. Cong, Y., Yang, G., Wei, Z., & Zhou, W. (2010). Security in underwater sensor network. In *Proceedings of international conference on communication and mobile computing* (pp. 162–168).
13. Dini, G., & Lo Duca, A. (2011). A cryptographic suite for underwater cooperative applications. In *Proceedings of IEEE symposium on computers & communications* (pp. 870–875).
14. Peng C., Du X., Li K., & Li M.. (2016 An ultra lightweight encryption scheme in underwater acoustic networks. *Journal of Sensors.* doi:10.1155/2016/8763528
15. Du, X.2014 Micro-ANP protocol architecture for UWSN. China Patent ZL201210053141.0.
16. Molins, M., & Stojanovic, M. (2006). Slotted FAMA: A MAC protocol for underwater acoustic networks. In *Proceedings of IEEE OCEANS'06* (pp. 16–22), Singapore.
17. Reed, I., & Solomon, G. (1960). Polynomial Codes over certain finite fields. *Journal of the Society for Industrial and Applied Mathematics, 8*(2), 300–304.
18. Luby, M., Mitzenmacher, M., Shokrollahi, A., & Spielman, D. (1997). Practical loss-resilient codes. In *ACM STOC* (pp. 150–159).

19. Xie, P., Zhou, Z., Peng, Z., Cui, J., & Shi, Z. (2010). SDRT: A reliable data transport protocol for underwater sensor networks. *Ad Hoc Networks, 8*(7), 708–722.
20. Mo, H., Peng, Z., Zhou, Z., Zuba, M., Jiang, Z., & Cui, J. (2013). Coding based multi-hop coordinated reliable data transfer for underwater acoustic networks: Design, implementation and tests. In *Proceedings of Globecom 2013, wireless network symposium* (pp. 5066–5071).
21. MacKay, D. J. C. (2005). Fountain codes. In *Proceedings of IEEE communications* (pp. 1062–1068).
22. Shokrollahi, A. (2006). Raptor codes. *IEEE Transactions on Information Theory, 52*(6), 2551–2567.
23. Luby, M. (2002). LT codes. In *Proceedings of the 43rd annual IEEE symposium on foundations of computer science* (pp. 271–280).
24. Xie, P., Cui, J.-H., & Lao, L. (2006). VBF: Vector-based forwarding protocol for underwater sensor networks. In *Proceedings of IFIP networking*.

Chapter 11
Using Sports Plays to Configure Honeypots Environments to form a Virtual Security Shield

Tyrone S. Toland, Sebastian Kollmannsperger, J. Bernard Brewton, and William B. Craft

11.1 Introduction

Society has become increasingly dependent on sharing electronic information. That is, companies can provide access to customer information, share marketing information, advertise job openings, and so on. As organizations provide e-business access via the cloud (i.e., Amazon [6]), threats to information security and privacy become a challenge.

While not new, securely exchanging information has always been of concern. For example, pre-computer days used cipher to encrypt information (e.g., Caesar Cipher) [14]. With the emergence of the personal computer (now mobile devices) and the Internet, stronger measures are needed to enforce security. This is of course compounded by the globalization of the World's economy. That is, one can live in one country and purchase goods and services from another country, which increases the threat to sensitive information. Organizations must be mindful of both information and infrastructure (i.e., corporate network) security concepts. Organizations use several defense mechanisms, e.g., firewalls [2, 5], encryption tools [9, 24], access control systems [3, 12], intrusion detection systems (IDS) [23, 25], to secure the corporate infrastructure. However, malicious attackers still succeed in unauthorized access to networks and ultimately sensitive data.

For example, in 2007 The TJX Companies, Inc. (TJX) was hacked in which 45 million credit card information was reported as being stolen in [4], but in [18] it was reported as actually being as high 90 million cards. In 2013, Target reported that

T.S. Toland (✉) • S. Kollmannsperger • W.B. Craft
University of South Carolina Upstate, 800 University Way, Spartanburg, SC 29303, USA
e-mail: ttoland@uscupstate.edu; kollmans@email.uscupstate.edu; craftwb@email.uscupstate.edu

J.B. Brewton
City of Spartanburg, 145 W. Broad Street, Spartanburg, SC 29306, USA
e-mail: brewtonb@bellsouth.net

© Springer International Publishing AG 2018   189
K. Daimi (ed.), *Computer and Network Security Essentials*,
DOI 10.1007/978-3-319-58424-9_11

their computer system was hacked in which approximately 40 million credit and debit card information "may have been impacted" [18]; Hardekopf [7] reported that credit card and debit information had been stolen from Target. In each of these examples as well as other examples in [7], a malicious user had penetrated the security measures to gain unauthorized access to an organization's computer system.

Prevention, detection, and response are three information security tasks considered in security management[14]. Prevention is the attempt to protect resources from danger and harm. Preparations of mechanisms to protect information technology (IT) should be accomplished as efficiently and effectively as possible. The goal is to ultimately make it as hard as possible for intruders and hackers to access resources. Common prevention tools are firewalls, password protections, encryption tools, and digital signatures.

When prevention is not effective, detection becomes an important process. The goal of detection is to find out (1) if the system was compromised and (2) what is the source of the attack. Detection acts as a passive tool to monitor. An IDS system is an example of a passive security tool.

Once an intruder has been detected, one needs to respond to an unauthorized access. Every action in a system gets recorded and stored by one of the detection tools. That is, an intruder generates evidences of the unauthorized access. Analyzing this evidence can reveal the following about the unauthorized access of the infrastructure: (1) how the attacker penetrated the security measures, (2) what was accessed, and (3) what was "possibly" manipulated. This information can be used to take steps to react to the unauthorized access.

This chapter extends our preliminary work in [15]. In particular, we discuss how American Football (football) play formations can be used to configure honeypot environments to form a Virtual Security Shield (VSS). Honeypots are fake computer systems that masquerade as a real computer system with real sensitive information, i.e., the honey [17, 19]. The goal of a honeypot is to attract (via the honey) malicious users to access the sensitive information. The information generated from accessing the honeypot can be used to strengthen security measures to prevent future unauthorized access. VSS uses various football play formations to simulate moving a honeypot to different locations in the network to generate valuable information about a malicious attacker. This information can be analyzed and used to defend future unauthorized access to an infrastructure. In addition to football, other sports can also be used by our novel approach in configuring honeypot environments.

This chapter is organized as follows. In Sect. 11.2, we discuss honeypot environments. Section 11.3 gives an overview of American football. In Sect. 11.4 we show how to configure honeypot environments using football. Section 11.5 gives a proof of concept implementation and results. Section 11.6 discusses related work. Section 11.7 concludes the chapter.

11.2   Honeypot Overview

11.2.1   Honeypots

Compared to other approaches in information security, honeypots are a more aggressive and an active form of defense against malicious attacks [19]. Honeypots are defined in several ways. That is, honeypot can be defined as a computer system whose value lies in being probed, attacked, or compromised [17]. This chapter builds on the definition of a honeypot as an IT resource with the goal to attract potential malicious attackers. That is, any access of the honeypots is examined and recorded to be used to deter similar attacks from occurring in the future. Contrary to other components of an IT system, it is desired that the honeypot gets attacked and probed. Since honeypots are masquerading as sensitive resource, they do not provide any functionality for an organization. Therefore, if a malicious user accesses a honeypot, then this access can be seen as unauthorized intrusion [19]. Honeypots can be categorized as either a production honeypot or a research honeypot as follows [17]:

- **Production Honeypot**: These kind of honeypots are used in a production environment. Their main purpose is to gather information for a specific organization about intrusions. They add value to an organizations information security.
- **Research Honeypot**: These honeypots are used primarily in a research environment to gather information about potential attackers. They do not add value to a specific organization. Information from Research Honeypots can be used to learn about techniques and resources from attackers which can help to prepare the production system for attacks.

11.2.2   Honeypots Benefits

Honeypots are flexible tools that contribute to three security tasks in the following manner [17]:

- **Prevention**: Honeypots can help to prevent attacks because of deception and deterrence. Deception means in the sense that potential attackers may waste time and resources on honeypots. Without knowing, attackers interact with a computer system that imitates a valuable resource (i.e., a honeypot). During this interaction, organizations may have the chance to react to the attack. After all, it may even be possible to stop attacks in which sensitive information is leaked. Preventive measures may contribute to deterrence. That is, honeypots can scare off attackers because of the warning measures associated with some security measures, e.g., your activities are being logged. When attackers know that an organization uses honeypots, they may not be willing to try to attack. As we can see, honeypots contribute to the prevention of attacks in a certain degree. Nonetheless, traditional prevention tools like firewalls are more efficient.

- **Detection**: Honeypots have the biggest impact in detection. For many organizations, detection is a difficult topic. Mairh et al. [17] identify three challenges when it comes to detection: false positives, false negatives, and data aggregation. False positives are mistakenly reported alerts. This happens, when the system interprets normal network traffic as an attack. The opposite false negatives are missed alerts for attacks that the system does not notice. Finally, data aggregation is the struggle to collect the data and transform it into valuable information. Common IDSs struggle in these three aspects. IDSs act like a watchdog over a company's IT infrastructure. They monitor the traffic and identify whether an access is authorized or not. Therefore, IDSs generate a lot of data, resulting in an overload of information. Honeypots, however, help us to eliminate these negative aspects. Because every interaction with a honeypot can be seen as unauthorized, honeypots only register these interactions. The problem with data aggregation and false positives can be eliminated. False negatives can still occur, i.e., if an intrusion does not affect the honeypot, but this risk can be mitigated by placing the honeypot in an attracting position. Consequently, honeypots help us to detect intrusions more effectively.
- **Response**: After an intrusion is detected, response is the next step to take. Honeypots help us to identify evidence via log files. That is, the user can analyze log files that are generated by honeypots to find out how the attacker gained access to the system. With the information collected by a honeypot, we can construct countermeasures to prevent similar attacks from occurring in the future.

It should be noted that the goal of a honeypot is not to prevent attacks, but to attract, detect, and monitor malicious attacks, so honeypots should be combined with other security tools (e.g., firewalls, encryption, password protection). Figure 11.1 illustrates a honeypot integrated within an IT infrastructure with existing computer systems (e.g., mail server, web server). The *Credit Card Info* honeypot should prove to be an inviting (i.e., honey) target for a malicious user to attack.

This chapter uses service (e.g., mail server, web server, etc.) and honey interchangeably.

11.3   American Football Overview

We now provide a brief overview of American Football (football). In football there are two teams of 11 players. Each team takes turns defending their goal. That is, the defending team wants to prevent the opposing team from taking the football into their end zone to score (e.g., touchdown, field goal, touch back). The teams are divided into offense and defense. The team that has the football is the offense and the other team is the defense.

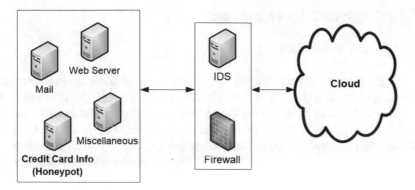

Fig. 11.1 Information technology infrastructure with honeypot

Fig. 11.2 Offense and defense formation

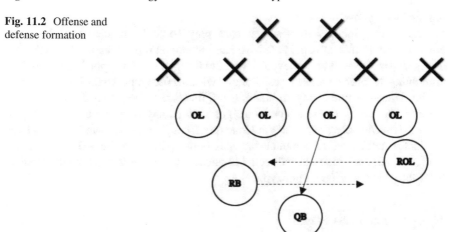

Offensive Line Formation

Although in real football there are eleven players per team, we will only consider seven players. Our offensive formation consists of five players that form the offensive line (OL + ROL). The offensive line has the task of keeping the ball away from the defending team. Behind the offensive line we have the Quarterback (QB) and Running Backs (RB). The job of the QB is to control the play. The RB, on the other hand, tries to outrun the defense. Figure 11.2 shows the offense represented as an **OL**.

Defensive Line Formation

The defensive formation consists of five defensive linemen and two Linebackers (LB). The defensive linemen try to attack either the QB or the ball carrier. The LBs are there to provide additional support for the defense. Sometimes the LBs also try to sack (e.g., attack) the opposing QB. Ultimately, the goal of the defense is to get the ball and stop the attack that advances the offense into their end zone. Figure 11.2 shows the defense represented as an **X**.

11.3.1   Offense Play Formations

11.3.1.1   The Screen Play

The screen pass is a popular play in football that is used when the defensive team is constantly rushing (i.e., all of the defensive team's players) the offensive side's QB [22]. The screen play is designed to take advantage of the fact that most of the defense's team is within close range of the football. The play begins with the QB signaling that he will make a long range pass, hoping to trick the defense into aggressively pressuring him. While this is happening, an offensive RB will take advantage of this by slipping past the advancing defensive line and positioning himself behind them relatively close to the QB. Once this has happened, the QB will then pass the ball to this RB, who should be in a good position to receive the ball and gain yards.

   The basic principles of a screen pass play could be applied to a honeypot environment. In this case, the defensive team would be represented as the malicious attackers, the QB will be Honeypot 1, and the RB will be Honeypot 2. The attackers attempting to probe the services (honey) within Honeypot 1 can be represented as the defensive line trying to attack the QB. Once it has been determined that the attackers are trying to attack Honeypot 1, its honey could be deactivated (i.e., rendered unavailable); The honey at Honeypot 2 can then be activated (i.e., rendered available), which would simulate the ball being passed to the RB. That is, the malicious attacker has been redirected to another honeypot to see how the attacker behaves against a different honeypot.

11.3.1.2   The Draw Play

The draw play is a running play in football which aims to misdirect the defensive line. In this play, the offense counters the defense's blocking positions, while the QB telegraphs that he is going to pass. From this position, the QB has two options. He can either hand the ball off to a nearby RB, or he can choose to run the ball himself past the advancing defense. "The idea behind a draw play is to attack aggressive, pass-rushing defenses by 'drawing' the defensive linemen across the line of scrimmage towards the passer while the linebackers and defensive backs commit to positioning themselves downfield in anticipation of a pass" [16].

   The mechanics of the draw play can be applied to a honeypot environment as follows. The QB can be represented as a honeypot while the RB can be represented as another honeypot on the network. Initially, the QB honeypot would activate its honey to draw the attention of the attackers. Similar to the draw play in football, the RB can either receive the pass from the QB (i.e., deactivate the honey at Honeypot 1, while activating the honey at Honeypot 2), or the QB can try to run the ball by himself, i.e., the honey at Honeypot 1 continues to be active, while the honey at Honeypot 2 continues to be deactivated. This can benefit a honeypot system by allowing for misdirection of malicious traffic if a honeypot is in danger of being corrupted or damaged by the attackers.

11.3.1.3   The Counter Run Play

The counter run play is a popular play in football in which the offense will try to deceive the defense into believing they will move the ball in a different direction from the direction that is initially telegraphed [21]. This is generally signaled by a RB initially moving in the opposite direction of the final receiving direction. Once the defense begins to commit to the misdirection on the offensive side's part, the RB will then receive the ball via a handoff from the QB in the opposite direction. The goal of the play is to exploit the holes in the defensive team's coverage due to the defensive team committing to the wrong side. The counter run play is generally used when the defense is very aggressive in their pursuit of the ball.

This play can be represented via a honeypot environment as well. That is, the malicious attackers would be represented by the defensive side. Two honeypots (e.g., one and two) would represent the two RBs that initiate and complete the misdirection, respectively. To simulate the play, one can picture the attackers trying to send packets to Honeypot 1 at the beginning of the play to access the active honey. Once Honeypot 1 receives the malicious traffic (simulating the defense's commit to one direction), Honeypot 1 can deactivated its honey. Then, Honeypot 2 will activate its honey, simulating the misdirection of the defensive team, i.e., the misdirection of the attackers from Honeypot 1.

11.3.1.4   Double Reverse Flea Flicker Play

The double reverse flea flicker is one of the many different football plays. It involves three players, the QB, the RB, and one player of the OL, called the right offensive lineman (ROL). For the purpose of this play, the ROL starts in a different position. Figure 11.3 shows the starting position. The dashed lines show the running paths of the players. The continuous line shows the path of the ball. So in the first move, the ball travels from the center of the offensive line to the QB. The ROL and RB run their paths. In Fig. 11.3 we can see the subsequent moves. When the RB crosses the QB, the ball travels from the QB to the RB (1). The next move happens, when the RB crosses the ROL. The ball travels from the RB to the ROL (2). The final move happens, when the ROL crosses the QB. The ball goes from the ROL to the QB (3). During the play, the QB does not move. However, the RB and the ROL cross and switch their sides. The ball travels from the center to the QB to the RB to the ROL and back to the QB. The goal of that play is to distract the defenders and create room for the QB to pass the ball. The defender cannot identify the location of the ball to tackle the correct player. The following shows how a honeypot environment can be configured using a football formation.

Fig. 11.3 Double reverse flea
flicker formation

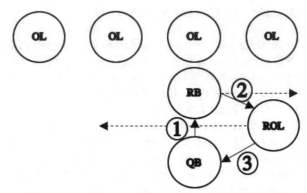

11.4 Honeypot Virtual Security Shield

Configuring the honeypot environment using a football play formation forms what
we call a *Virtual Security Shield* (VSS). VSS deflects the attacker from accessing
the honey at one honeypot to the honey at a subsequent honeypot. We recognize
that information about the malicious attacker will be generated when the initial
honeypot (i.e., Honeypot 1) is accessed; however, accessing subsequent honeypots
provides additional opportunities to gather information about malicious activities.
An additional benefit of VSS is that it generates different honeypot environment
configurations without physically adding or deleting a honeypot within the network.
That is, these virtual honeypot configurations are generated when subsequent
honeypots are accessed.

Our approach will simulate football play formation in a honeypot environment
by using a boolean switch to manage the availability of the honey. If the switch is
set to true at a honeypot, then the honey is active, i.e., available to attract malicious
attackers; otherwise, if the switch at the honeypot is set to false, then the honey
is not active. We acknowledge by definition in [17, 19] that when the honey is
deactivated, then the computer system is no longer a honeypot; however, activating
and deactivating the honey simulates physically moving a honeypot to a different
location within the network to form a different honeypot environment configuration.
We now present some formal definitions.

Definition 1 (Honeypot) Let $honeypot = [\{service_1, service_2, \ldots, service_n\},$
$Active]$, where $service_i (1 \leq i \leq n)$ are data items and *Active* is boolean switch.
When *Active* is set to *true*, then the $service_i$ (i.e., honey) is available to an attacker;
otherwise, when *Active* is set to *false*, *service* is unavailable to an attacker.

Definition 2 (Virtual Security Shield) Let $hp = \{hp_1, hp_2, \ldots, hp_n\}$ be a set of
honeypots as defined in Definition 1. The location of the active honeypot $hp_i \in$
$hp (1 \leq i \leq n)$ is determined by football play formation. The set hp is called a
Virtual Security Shield.

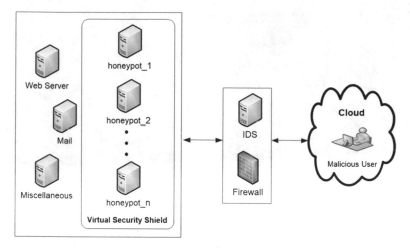

Fig. 11.4 Honeypot virtual defense shield configuration

Passing the football (i.e., the honey) can now be simulated by setting the active switch to true in one of honeypot, while setting the active switch to false in other honeypot(s). As the malicious attackers attempt to locate the honeypot that has the honey, each of the honeypots will collect valuable information about by the attacker. So, what we propose in this chapter goes above and beyond the conventional honeypot environments in [17, 19]. Our approach is to simulate moving the honey between honeypots with the goal to gather additional information as the malicious user accesses an active honeypot.

Figure 11.4 shows how VSS can be incorporated into a honeypot environment.

11.4.1 Virtual Security Shield Example

We now show how to build a VSS using the football play in Sect. 11.3. We configure the honeypot environment using the Flea Flicker play from Sect. 11.3.1.4. See Fig. 11.5.

In Fig. 11.5, $honeypot_1$ acts like the RB, $honeypot_2$ acts like the QB, and $honeypot_3$ acts like the ROL. Initially, the honey at $honeypot_2$ (i.e., QB) is available, i.e., $honeypot_2 : Active = true$; so, the attackers will attempt to access $honeypot_2$. When this happens, the honey at $honeypot_1$ becomes active. So, $honeypot_1$ becomes active (i.e., $honeypot_1.Active = true$) and $honeypot_2$ becomes inactive (i.e., $honeypot_2.Active = false$). Once $honeypot_2$ has been accessed by the attacker, then $honeypot_3$ will need to be activated. So, $honeypot_3$ is set to active (i.e., $honeypot_3.Active = true$), while $honeypot_1$ is set to inactive (i.e., $honeypot_1.Active = false$). Finally, $honeypot_2$ again becomes active (i.e., $honeypot_2.Active = true$), while $honeypot_3$ becomes inactive (i.e.,

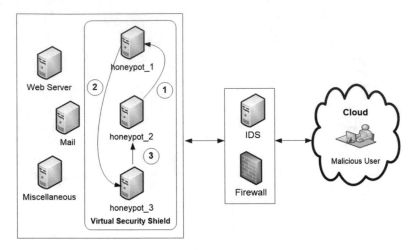

Fig. 11.5 Football formation mapped into honeypot environment

honeypot₃.Active = *false*). Again, by setting the active switches to true or false simulates the ball being passed, i.e., the honey being available or unavailable at the respective honeypot machine.

11.5   Implementation and Experiment

An experiment was conducted using a framework we developed in Java 8 [11].

11.5.1   Implementation

To show a proof of concept, we developed the following three programs:

- HoneypotServer (HPTS) is a program that simulates the honeypot. The program uses a Boolean variable (e.g., activeData) to simulate access to the honey. If activeData is true, then the access to honey is available via HPTS; otherwise, if activeData is false, then the honey is currently not available via access to this machine.
- HoneypotManager (HPTM) is a program that sends a message to either activate or deactivate access to the honey on HPTS. When the access to the honey has been deactivated on one HPTS (i.e., hp_i), then the honey on a different HPTS (i.e., hp_j) is activated.
- HoneypotAttacker (HPTA) is a program that simulates the attacker. This program attempts to access honey on a HPTS. The attacker sends an access message request (i.e., a malicious attack message) to the HPTS. If HPTS has access

capability to the honey (i.e., activeData is true), then an active message is generated that contains: the value A (i.e., access to the honey is active), HPTA IP address, the attack message arrival time on HPTS, and the attack message departure time from HPTS. Otherwise, an inactive message is generated that contains: the value N (i.e., access to the honey is not active), HPTA IP address, the attack message arrival time on HPTS, and the attack message departure time from HPTS.

11.5.2 Experiment

We ran our experiment in a test networking lab. To simulate the example in Sect. 11.4, we ran HPTS on three computers (i.e., HP1, HP2, and HP3). We ran HPTA on a separate computer to simulate the attacks. We ran HPTM on a separate computer. HPTM transmitted a sequence of activate and deactivate messages to HP1, HP2, and HP3. For our experiments, HPTS only listens on port 9001. All machines in Fig. 11.6 were running Windows 7.

HP2 is initially activated, while HP1 and HP3 are deactivated. The attacker can now search for the active honeypot using HPTA. To accomplish this, the attacker

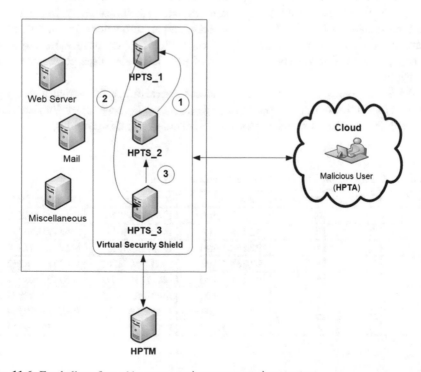

Fig. 11.6 Football configured honeypot environment experiment setup

successively tries to connect to the honeypots. Once the attacker finds the active honeypot (i.e., activeData is true), the manager deactivates that honeypot (i.e., activeData is set to false) and then activates the next honeypot in the play sequence. Then, the attacker searches for the next active honeypot and the process continues per Fig. 11.5.

Table 11.1 shows the result from this experiment. The attacker does follow the sequence of the play in Fig. 11.5 when accessing active data items. As we proposed, we could gather information from the malicious user in Msg 2 at HP2, in Msg 4 at HP1, in Msg 7 at HP3, and in Msg 8 again at HP2. That is, we can gather information from a malicious user at a given machine at a specified time.

We note that our implementation may appear to retrieve a limited amount of information (i.e., msg#, honeypot #, etc.); however, the Java implementation allows for additional security information to be captured by modifying the test programs as needed. This should proved to be a useful and beneficial exercise for a either an Internet Security course or a Networking Programming course.

11.5.3 Discussion

The experiment shows that our approach is feasible. Our approach provides a guaranteed time interval for which we can evaluate malicious activity. In particular, we can evaluate malicious activity when accessing an active honeypot and/or when searching for an active honeypot. Based on Table 11.1, we have extracted a set of active honeypot access times and a set of time intervals to search for an active honeypot.

We define $T_{\text{FoundHoneypot}}$ as a set of access arrival times for which a message arrives at an active honeypot. We define $T_{\text{SearchingForHoneypot}}$ as a set of time intervals in which the attacker is searching for the active honeypot; this provides,

Table 11.1 Experimental results with time in milliseconds

Msg#	HP#	Active	IP address	ArrivalTime	DepartureTime
1	1	N	192.168.1.100	1463775948262	1463775948262
2	2	A	192.168.1.100	1463775950737	1463775950737
3	3	N	192.168.1.100	1463775957258	1463775957258
4	1	A	192.168.1.100	1463775958262	1463775958262
5	2	N	192.168.1.100	1463775966977	1463775966977
6	1	N	192.168.1.100	1463775967575	1463775967575
7	3	A	192.168.1.100	1463775970534	1463775970534
8	2	A	192.168.1.100	1463775979379	1463775979379

- $T_{\text{FoundHoneypot}} = \{\text{Msg2.ArrivalTime, Msg4.ArrivalTime, Msg7.ArrivalTime,}$
  $\text{Msg8.ArrivalTime}\}$
- $T_{\text{SearchingForHoneypot}} = \{[\text{Msg1.ArrivalTime, Msg2.ArrivalTime}],$
  $[\text{Msg3.ArrivalTime, Msg4.ArrivalTime}], [\text{Msg5.ArrivalTime, Msg7.}$
  $ArrivalTime], [\,]\}$

We defined sets of times which potentially provide more information than conventional honeypot solutions.

11.6 Related Work

Honeypot has been well studied. For an overview of honeypots, the interested reader is referred to [17, 19, 20].

Bait and Switch [1] is a honeypot solution which aims to utilize honeypots in a way that is different from the typical way. In most cases, honeypots play a passive role in terms of network intrusion in that they do not prevent any intrusions from occurring. Honeypots simply monitor the networks for possible malicious activity. Bait and Switch change this approach by creating a system which detects hostile traffic and instantly attempts to redirect this traffic to a honeypot which mirrors the real system; hence, the attacker would be accessing the honeypot instead of a real system [1]. Bait and Switch as a honeypot solution has an advantage over most other solutions in that it actually redirects malicious traffic away from the real data. This allows any network system that utilizes this for honeypots to become more secure by a considerable margin. The VSS in this chapter redirects malicious users by way of activating and deactivating the switch on the respective honeypot. Unlike Bait and Switch, we do not instantly redirect the attacker. VSS allows the attacker time to access the honeypot to allow for valuable information to be gathered by HoneypotManager. Then, the attacker is redirected with the attempt to gain additional information.

"Honeyd [simulates] . . . small daemon that creates virtual hosts on a network" [8]. The virtual hosts that Honeyd creates are configurable. They can accept instructions to make them appear as though they are running different operating systems. For example, on a network, there could be a virtual host that appears to be running Linux, and another virtual host that appears to be running Windows. When used in a honeypot system, Honeyd's main goal is to hide the host which contains the real honey, amongst a crowd of virtual hosts which can emulate the real systems in order to (1) waste attackers' time and (2) protect the real data. Honeyd has the advantage of being able to use the daemons to create multiple virtual hosts, which makes it difficult for attackers to distinguish which of the hosts it connects to is the one containing the valuable data.

KFSensor is a honeypot IDS that is a Windows based software package [10]. "KFSensor acts as a honeypot, designed to attract and detect hackers and worms by simulating vulnerable system services and trojans" [13]. The software emulates real services (e.g., mail server, web server, etc.). This allows one to reveal what the intentions are of an attacker, e.g., gain unauthorize access to email data. KFSensor also has the ability to send alerts via email, filter out events, and examine events using the administration console. Results from any incoming data are able to be represented graphically via reports and graphs. Another feature of the software is the ability to filter malicious attack results by attack type, time period as well as other parameters.

Honeyd [8] and KFSensor [10] can be used to configure VSS. That is, one can simulate the various services running on real machine (e.g., mail server). Then, sequence the honeypots via an activate switch to have these services be active or non-active to simulate various football play formations.

11.7 Conclusion

We have shown how an American Football formation can be used to configure a honeypot environment to gather information about cyber-attacks. In particular, we have shown how we can develop a Virtual Security Shield using a honeypots configured as a football play. We have also provided a proof of concept experiment to show that our approach is feasible. Our novel approach can be used to gather valuable information about single and ultimately coordinated attacks using well-established American Football play formations.

Future research involves implementing additional play formation using the test framework presented in this chapter. We would like to investigate the effects of coordinated tasks using our approach. We further propose that we can use plays from other sports to configure a honeypot environment to construct a Virtual Security Shield.

Acknowledgements The authors would like to thank Dr. Frank Li, Dr. Jerome Lewis, and Dr. Bernard Omolo for their support in using the Networking Lab in the Division of Mathematics and Computer Science. The authors would also like to thank the reviewers for their invaluable feedback.

References

1. Bait and Switch Honeypot. (2016). Retrieved June 2016, http://baitnswitch.sourceforge.net/.
2. Basile, C., & Lioy, A. (2015). Analysis of application-layer filtering policies with application to http. *IEEE/ACM Transactions on Networking, 23*(1), 28–41. doi:10.1109/TNET.2013.2293625. http://dx.doi.org/10.1109/TNET.2013.2293625.
3. Bobba, R., Fatemieh, O., Khan, F., Khan, A., Gunter, C. A., Khurana, H., et al. (2010). Attribute-based messaging: Access control and confidentiality. *ACM Transactions on Information and System Security, 13*(4), 31:1–31:35. doi:10.1145/1880022.1880025. http://doi.acm.org/10.1145/1880022.1880025.
4. Brand, M. 'Marketplace' Report: TJX Data Breach: NPR. http://www.npr.org/templates/story/story.php?storyId=9209541.

5. Chen, H., Chowdhury, O., Li, N., Khern-am nuai, W., Chari, S., Molloy, I., et al. (2016). Tri-modularization of firewall policies. In *Proceedings of the 21st ACM on Symposium on Access Control Models and Technologies, SACMAT '16* (pp. 37–48). New York, NY: ACM. doi:10.1145/2914642.2914646. http://doi.acm.org/10.1145/2914642.2914646.

6. Free Cloud Services. AWS Free Tier. //aws.amazon.com/free/.

7. Hardekopf, B. (2014). The big data breaches of 2014.   http://www.forbes.com/sites/moneybuilder/2015/01/13/the-big-data-breaches-of-2014/.

8. Honeyd. (2016). Retrieved June 2016, http://www.citi.umich.edu/u/provos/honeyd/.

9. Hunter, D., Parry, J., Radke, K., & Fidge, C. (2017). Authenticated encryption for time-sensitive critical infrastructure. In *Proceedings of the Australasian Computer Science Week Multiconference, ACSW '17* (pp. 19:1–19:10). New York, NY: ACM. doi:10.1145/3014812.3014832. http://doi.acm.org/10.1145/3014812.3014832.

10. Intrusion Detection, Honeypots and Incident Handling Resources. (2016).   Retrieved June 2016, http://www.honeypots.net/honeypots/products.

11. Java. (2016). Retrieved June, 2016 https://www.java.com/en/.

12. Kechar, M., & Bahloul, S. N. (2015). An access control system architecture for xml data warehouse using xacml.   In *Proceedings of the International Conference on Intelligent Information Processing, Security and Advanced Communication, IPAC '15* (pp. 15:1–15:6). New York, NY: ACM.   doi:10.1145/2816839.2816897.   http://doi.acm.org/10.1145/2816839.2816897.

13. KFSensor. (2016). Retrieved June 2016, http://www.keyfocus.net/kfsensor/.

14. Kim, D., & Solomon, M. G. (2018). *Fundamentals of information systems security* (3rd ed.). Burlington, MA: Jones and Bartlett.

15. Kollmannsperger, S., & Toland, T. (2016). Using football formations in a honeypot environment. In *Proceedings of The 2016 International Conference on Security and Management, SAM'16* (pp. 299–303). Athens: CSREA Press.

16. Learn and talk about Draw play, American football plays. (2016).   Retrieved June 2016, http://www.digplanet.com/wiki/Draw_play/.

17. Mairh, A., Barik, D., Verma, K., & Jena, D. (2011). Honeypot in network security: A survey. In *Proceedings of the 2011 International Conference on Communication, Computing & Security, ICCCS '11* (pp. 600–605). New York, NY: ACM. doi:10.1145/1947940.1948065. http://doi.acm.org/10.1145/1947940.1948065.

18. Memmott, M. Breach At Target Stores May Affect 40 Million Card Accounts: The Two-Way: NPR. http://www.npr.org/sections/thetwo-way/2013/12/19/255415230/breach-at-target-stores-may-affect-40-million-card-accounts.

19. Mokube, I., & Adams, M. (2007). Honeypots: Concepts, approaches, and challenges.   In *Proceedings of the 45th Annual Southeast Regional Conference, ACM-SE 45* (pp. 321–326). New York, NY: ACM. doi:10.1145/1233341.1233399. http://doi.acm.org/10.1145/1233341.1233399.

20. Pisarčík, P., & Sokol, P. (2014). Framework for distributed virtual honeynets. In *Proceedings of the 7th International Conference on Security of Information and Networks, SIN '14* (pp. 324:324–324:329). New York, NY: ACM. doi:10.1145/2659651.2659685. http://doi.acm.org/10.1145/2659651.2659685.

21. Running Plays in Football. (2016).   Retrieved June 2016, http://www.dummies.com/sports/football/offense/running-plays-in-football/.

22. Screen Pass. (2016). Retrieved June 2016, http://nflbreakdowns.com/beginner-series-screen-pass/.

23. Soleimani, M., Asl, E. K., Doroud, M., Damanafshan, M., Behzadi, A., & Abbaspour, M. (2007). Raas: A reliable analyzer and archiver for snort intrusion detection system.   In *Proceedings of the 2007 ACM Symposium on Applied Computing, SAC '07* (pp. 259–263). New York, NY: ACM. doi:10.1145/1244002.1244067. http://doi.acm.org/10.1145/1244002.1244067.

24. Tarle, B. S., & Prajapati, G. L. (2011). On the information security using fibonacci series. In *Proceedings of the International Conference & Workshop on Emerging Trends in Technology, ICWET '11* (pp. 791–797). New York, NY: ACM. doi:10.1145/1980022.1980195. http://doi.acm.org/10.1145/1980022.1980195.
25. Wang, X., Kordas, A., Hu, L., Gaedke, M., & Smith, D. (2013). Administrative evaluation of intrusion detection system. In *Proceedings of the 2nd Annual Conference on Research in Information Technology, RIIT '13* (pp. 47–52). New York, NY: ACM. doi:10.1145/2512209.2512216. http://doi.acm.org/10.1145/2512209.2512216.

Part III
Cryptographic Technologies

Chapter 12
Security Threats and Solutions for Two-Dimensional Barcodes: A Comparative Study

Riccardo Focardi, Flaminia L. Luccio, and Heider A.M. Wahsheh

12.1 Introduction

One-dimensional (1D) barcodes store data in special patterns of vertical spaced lines, while two-dimensional (2D) barcodes store data in special patterns of vertical and horizontal squares and thus have a higher capacity. 2D barcodes are widely used and can be placed on any surface or location to store textual description, Uniform Resource Locators (URLs), contact information, and specific parameters such as coordinates for maps [19, 24].

There are many types of barcodes available; the most widely used are the Quick Response (QR) codes [13]; very common are also the Data Matrix [10], PDF417 [12], and Aztec Codes [11]. Table 12.1 presents a comparison between different standards [27] and shows how 2D barcode types differ both in the storage capacity and in the practical applications. The QR code is the most popular one, is commonly used in Japan (as it can encode Kanji characters), and has the largest data storage capacity (in version 40). It is used in different applications such as advertising, digital government and public services, physical access control, and mobile payments [19]. All the other listed barcodes have been developed in the USA: the Aztec code has a good storage capacity, and it is widely used in patient-safety applications. The Data Matrix barcode is commonly used for item marking and can be printed in a small area size, but it has less data capacity. Finally, the PDF417 barcode has a small data capacity and is commonly used in logistic and governmental applications.

R. Focardi • F.L. Luccio (✉) • H.A.M. Wahsheh
Department of Environmental Sciences, Informatics and Statistics (DAIS), Ca' Foscari University of Venice, via Torino 155, 30172 Venezia, Italy
e-mail: focardi@unive.it; luccio@unive.it; heider.wahsheh@unive.it

© Springer International Publishing AG 2018
K. Daimi (ed.), *Computer and Network Security Essentials*,
DOI 10.1007/978-3-319-58424-9_12

Table 12.1 Different 2D barcodes standards

	QR code	Data matrix	Aztec code	PDF417
Max capacity (numeric)	7089	3116	3832	2710
Max capacity (alphanum)	4296	2335	3067	1850
Country	Japan	US	US	US
Notes	Most popular 2D barcode. Used for advertising, government and public services, physical access control, and mobile payments	Used for marking small containers	Used for patient identification wristbands and medicines	Used in logistics and in governmental applications

The increasing use of 2D barcodes has attracted the attention of cyber attackers trying to break users' privacy by accessing personal information, or to directly compromise users' smartphones and any other connecting device. Thus, understanding possible attacks to barcodes and studying protection techniques is a very challenging and important issue. In 2011, the Kaspersky Lab detected the first dangerous attack to a QR code, which consisted of encoding a malicious URL inside the barcode, and using phishing and malware propagation to get the users' personal information from the connecting devices [15]. This attack is based on the lack of content authentication and could be mounted, in principle, on all of the most commonly used barcodes.

Previous studies discussed different attacks to 2D barcodes and proposed various solutions to protect them. This paper aims at summarizing all the existing attacks to the barcodes and at presenting the available techniques to protect them. All the existing protecting method weaknesses will be highlighted, compared, and evaluated based on their security level and the adopted cryptographic mechanisms. In fact, although many of the available barcode security systems offer cryptographic solution, they do not always adhere to the latest recommendations and might be still vulnerable due, e.g., to the adoption of deprecated cryptographic hash functions and to the usage of short keys. In some cases, cryptographic solutions do not even provide enough detail to evaluate their effective security. We finally revise potential weaknesses and suggest remedies based on the recommendations from the European Union Agency for Network and Information Security (ENISA) [4].

The remainder of the paper is organized as follows: Sect. 12.2 presents an overview of attack scenarios for 2D barcodes; Sect. 12.3 revises secure systems in which 2D barcodes are used as a fundamental component; Sect. 12.4 explores security-enhanced barcodes and readers; Sect. 12.5 summarizes and compares the different studies and discusses limitations and possible improvements; the last section presents concluding remarks and future work.

12.2 Attack Scenarios for 2D Barcodes

Barcodes are used in various scenarios for different purposes. A typical application is to encode a URL that links to a related Web page containing detailed information about a product or service. When the barcode is scanned, the link is usually shown to the user who can decide whether to open it or not in the browser. Barcodes are also used for physical access control, identification, and logistics. In these cases, they contain data that are given as input to back-end applications, which interpret them and act consequently.

In general, barcodes are just a way to provide input to users or applications, and, since they do not offer any standard way to guarantee content authentication, the input they provide is in fact *untrusted*. Potential security risks regard the encoding of malicious URLs that look similar to the honest ones and the encoding of data that trigger vulnerabilities in the back-end applications. Moreover, the barcode reader application may become a point of attack since, independently of the use case, the barcode content passes through it and might trigger vulnerabilities directly on the user device.

In the following, we discuss different attack scenarios for 2D barcodes such as phishing, malware propagation, barcode tampering, SQL and command injection, cross-site scripting (XSS), and reader applications attacks.

Phishing In a barcode phishing attack, the attacker tries to get sensitive information such as the login details and the credit card number of a user, by encoding a malicious Web address inside the barcode that redirects the user to a fake Web page (usually a login Web page). This fake page appears very similar to the legitimate one; thus, unintentionally the victim accesses the page and provides the login details to the attacker [18, 19]. The study of [28] presents an analysis of QR code phishing, which authors call QRishing. The authors conducted two main experiments, the first one aiming at measuring the proportion of users that scan a QR code and decide to visit the associated Web page, and the second one aiming at understanding the user interaction with QR codes. The results are that the majority (85%) of the users visited the associated Web page and that the main motivation for scanning QR codes is curiosity or just fun.

Malware Propagation In [16] it is discussed how QR codes can be used by attackers to redirect users to malicious sites that silently install a malware by exploiting vulnerable applications on the device. This is typically done through an exploit kit that fingerprints the device and selects the *appropriate* exploit and malware. The experiments used crawlers and were run on 14.7 million Web pages over a ten-month period. The crawlers extracted 94,770 QR codes from these Web pages that mainly included marketing products or services. The results showed that 145 out of 94,770 QR codes had a malicious behavior. They contained attractive words such as free download and personal/business websites. The authors also found that 94 out of 145 QR codes redirected the users to intermediate sites containing malware that could cause damage to the users' mobile devices.

Barcode Tampering and Counterfeiting Since 2D barcodes are typically used in advertisement and e-commerce to indicate detailed information about the products or to perform the purchase process, an attacker can benefit from the companies' reputation by pasting fake 2D barcodes on the real posters. These fake 2D barcodes might advertise false product information or false special offers in which, in fact, the adversary will sell another product to the victims [18]. Interestingly, the study of [2] demonstrates that it is possible to generate 2D barcodes that adhere to multiple standards and that might be decoded, non-deterministically, in multiple ways. One way to achieve this "barcode-in-barcode" is to embed one barcode into another one, so that the decoded content will depend on which of the two is detected by the reader. The authors show how to embed a QR code, an Aztec Code, and a Data Matrix inside a QR code barcode. The error correction feature of QR codes allows for reconstructing the missing part, so that the hosting barcode is not compromised by embedding another one inside it. The experiments demonstrate that the decoded content depends on the smartphone and reader application used to scan the barcode. This is interesting because it opens the way to stealthy barcode-based attacks that only affect a small number of devices and are thus harder to detect.

SQL and Command Injections The studies of [18, 19] discuss scenarios in which the attacker can encode SQL statements in the barcode in order to attack a database system. The study of [18] refers to automated systems that use the information encoded in the barcodes to access a relational database. If the string in the barcode is appended to the query without a proper input sanitization, the attacker may encode inside the barcodes the SQL commands together with the normal information. For example, this could be done by adding a semicolon ; followed by SQL statements such as `drop table <tablename>`, causing the destruction of a database table. Similarly the attacker might retrieve or modify sensitive information stored in the database. Both papers also describe possible scenarios in which the content of the barcode is used as a command-line parameter. In this case, it might be possible to directly inject commands and take control of the server host. For example, in [19] the authors mention how Samsung phones may be attacked by embedding malicious Man–Machine Interface (MMI) instructions, normally used to change phone settings, into a barcode. Once the barcode is scanned, it triggers the execution of these malicious instructions that, e.g., erase all phone data. These attacks happen when developers assume that the information in barcodes cannot be manipulated by attacks and consider it as a trusted input.

Cross-Site Scripting Attacks (XSS) Mobile apps are often based on Web technology and this may allow malicious JavaScript code to be injected into trusted HTML pages and executed in the app. The simplest case is when the attacker includes JavaScript code into input forms so that, if the server does not sanitize the form data and data are eventually rendered in a page (e.g., as in a blog post), the script would appear and run in the context of a trusted page accessed by the user. This attack is called Cross-Site Scripting (XSS) and can be mounted also using barcodes [14]. The study of [14] discusses risks in HTML5-based mobile applications, in which new forms of XSS attacks using several, unexpected channels are discussed. For example, authors discuss how the Calendar provider in Android might become a dangerous internal channel in which the attacker inserts malicious JavaScript code that is executed when a vulnerable HTML5-based application displays a Calendar event. The authors show a very interesting example of XSS attack for a barcode reader application. The application reads the QR code and then displays its content to the user. However, this is done by putting the content of the barcode in a HTML5 page that is then displayed to the user. This, of course, triggers the attack by executing whatever script is included in the barcode.

Reader Applications Attacks During the installation process, many of the 2D barcode reader applications ask for full permissions to access user's smartphone resources such as the device location, the contact list, and the photos. If a reader application has a vulnerability that can be triggered by a suitable crafted barcode, this might allow the attacker to get access to private user's data [17].

Table 12.2 summarizes the above attack scenarios for 2D barcodes, classifies the attacks into standard and novel, and summarizes the role of the barcode in the attack. In particular, attack novelty indicates to which extent the attack is a novel one, specific for barcodes, or just a variation of a standard attack. The role of the barcode indicates if the barcode is used to redirect to a malicious website or if, instead, it contains the attack payload.

Table 12.2 Summary of the attacking scenarios to 2D barcodes

Attack scenario	Attack novelty		Role of barcode	
	Standard	Novel	Redirect	Payload
Phishing	✓		✓	
Malware propagation	✓		✓	
Barcode tampering and counterfeiting		✓	✓	✓
SQL and command injections	✓			✓
XSS	✓			✓
Reader applications		✓		✓

12.3   Secure Systems Based on 2D Barcodes

In this section, we present some studies that do not focus on how to directly protect a 2D barcode, but on how the barcode can be used as a component of a bigger security system that aims, e.g., at protecting physical documents or operations such as bank transactions. Barcodes may directly enhance security by adding sensitive information into printed documents [29] or may simply provide a human-usable way to implement security protocols, as in the case of [5, 25]. Below, we describe these systems in more detail.

Quick Response—Transaction Authentication Numbers (QR-TAN) [25] is a transaction authentication technique based on QR codes. More precisely, QR-TAN is a challenge-response protocol based on a shared secret and uses QR codes for the transmission of information between the user's computer and the mobile device, which is assumed to be trusted. The protocol works as follows: transaction data and a nonce (the challenge) from the server are encoded in a QR code which is displayed on the screen of the untrusted computer. The user can use her trusted mobile device to scan it and check that the transaction data are correct. If the user approves, the device secret will be used to authenticate the transaction data together with the nonce through the generation of an HMAC. The user is required to manually enter the first characters of the (alphanumeric version of the) HMAC into her computer that will send it to the server for the final verification. Since the device secret is shared with the server, the server can recompute the HMAC and check that it is consistent with the fragment inserted by the user.

In [5], a mobile payment system that is pervasively based on Data Matrix barcodes is presented. Barcodes include product information and merchant URL, so that when a client wants to buy some product, she can scan the barcode and connect to the merchant website. At this point, the client can issue a purchase request which is also encoded as a barcode; the merchant server generates another barcode for the purchase invoice and sends it back to the client; finally, the client sends a barcode payment request to the payment server. All transactions are encoded as barcodes that are digitally signed using Elliptic Curve Digital Signature Algorithm (ECDSA) in order to guarantee authentication. Authors describe application scenarios for mobile purchasing and payment, but no evaluation of the proposed system is provided.

CryptoPaper [29] is a system that allows to include secure QR codes in printed documents containing sensitive information. The QR code stores both the encrypted sensitive information and the meta-information which is used for the decryption process. In order to read the QR code, the scanner needs an authorized access to the key which is stored in a cloud database. If the access is granted, the scanner automatically gets the key (through QR code meta-information) and produces the plaintext. Authentication is achieved through a digital signature and confidentiality through AES encryption. Cryptographic keys are stored in the cloud databases. The system allows to include sensitive information in printed documents and to regulate access through a cloud server. In this way, it is possible to dynamically grant or remove access and, at the same time, the cloud server does not have access to sensitive information.

12.4 Security Enhanced Barcodes and Readers

We now overview technological solutions and research proposals aiming at improving the security of applications using 2D barcodes. We first revise solutions and studies that extend barcodes through security mechanisms and cryptographic techniques (cf. Sect. 12.4.1). Then, we describe solutions and research work aiming at preventing attacks directly in the reader applications (cf. Sect. 12.4.2).

12.4.1 Security Enhanced Barcodes

Technology and Applications Secret-function-equipped QR Code (SQRC) is a type of QR code which can store additional private information, only accessible through a special reader with the correct cryptographic key. One of the features of SQRC is that, when accessed through a standard reader, it is indistinguishable from a normal QR code. There is no publicly available description of SQRC, and the official website states that SQRC can only be read by "scanners with the same password (cryptography key) as the one set when the SQRC is generated." However, in a note it is reported that "this function does not provide any security guarantee" [3], which sounds a bit contradictory. However, because of lack of documentation, we cannot evaluate the security of SQRC.

2D Technology Group (2DTG) commercializes a product named Data Matrix Protection/Security Suite (DMPS) [1], based on a patented Barcode Authentication technology [30]. DMPS protects against barcode counterfeiting and data tampering through a symmetric-key based "signature" algorithm.[1] The motivation for adopting this proprietary technology is to overcome the excessive computational load of standard asymmetric key signature schemes. However, as far as we know, there is no security analysis/proof of the patented technology.

Research Work In [9], a tamper detection system for QR codes based on digital signature is proposed: a digital signature of the barcode content is embedded into the error correcting area using a stenographic technique. Authors have implemented a prototype and performed experiments finding that the technique could not scale well to QR code version 12. However, they do not give insights about this limitation. Using the stenographic technique, they are able to embed just 324 bits of information in the error correcting area. The embedding of actual signatures is left as a future work.

In [22], the author foresees a scenario in which attackers might spam the Internet of Things (IoT) by flooding the physical space with fake or tampered barcodes pointing to unrelated pages, with the specific purpose of increasing the traffic

---

[1]The use of word "signature" for a symmetric-key based algorithm is quite unusual since any entity knowing the symmetric key might provide a valid "signature."

towards those pages. Independently of the plausibility of the above scenario, the underlying problem is barcode counterfeiting and, more generally, phishing. The proposed solution is to use ECDSA in order to provide authentication and integrity guarantees of a scanned barcode: the content of the barcode will be trusted only if it contains a valid signature from a recognized content creator. Experimental results on different key lengths and hash functions for ECDSA show a reasonable time/space overhead.

In [20], a group of students from MIT have performed interesting experiments about enhancing QR codes with cryptography and digital signature. They have also pointed out potential vulnerabilities of two QR code applications: ZXing [6] and SPayD [7].

12.4.2   Security Enhanced Barcode Readers

Technology and Applications The Norton Snap QR code reader is an Android mobile application which automatically reads the QR code and checks the content to establish the safety of any URL embedded inside the QR code [26]. The features of the Norton Snap QR code include identification of safe websites that are loaded immediately; blocking of malicious, phishing sites, preventing them from being loaded in the browser; and expansion of full website address so that users know the final URL before they click it. Norton Snap QR code protects users from phishing, automatic download of malware, and form of frauds where the user is redirected to malicious websites. It does not prevent command/SQL injection, XSS, and attacks on the reader application.

Secure QR and barcode reader is an Android mobile application capable of scanning several barcodes [23]. It improves smartphone security by following a simple principle: when installed, it does not ask for permission to access personal information such as user location, contact numbers, and photos. This mitigates the consequences of attacks that might leak personal information.

Research Work The study of [31] investigates the security features of existing QR code scanners for preventing phishing and malware propagation. Authors considered 31 QR code scanner applications. The results showed that 23 out of 31 have a user confirmation feature that gives the user the choice to continue/discontinue visiting the URL; however, users typically click on the displayed URL without thinking about the possible consequences. Only two QR code readers out of 31 have security warning features, but authors show that the detection rate is unsatisfactory with too many false negatives. For this reason, authors developed a new scanner, named SafeQR, based on two existing Web services: the Google Safe Browsing API [8] and the Phishtank API [21]. Google Safe Browsing API tests the websites under Google blacklists of phishing and malware URLs, while Phishtank API provides a phishing checking service based on users' feedbacks about possible phishing websites. The experiments showed that SafeQR performs a better phishing and malware detection and has a more effective warning user interface when compared with available QR code readers.

12.5   Summary and Comparison

In this section, we summarize, compare, and, to some extent, evaluate the various solutions, applications, and research proposals discussed in previous sections. In the tables we are going to present, we follow the order in which works have been presented and we refer to each of them through their proper names, when available, or using concise but descriptive ones. We always include the appropriate citation and a reference to the section in which we have described the work.

Summary of the Relevant Features   In Table 12.3, we summarize various relevant features of the works: the supported barcodes and whether or not the proposed solution is required to be online; mitigation and prevention of the attacks discussed in Sect. 12.2, grouped by the barcode role (cf. Table 12.2); and the provided security properties. Notice that we have grouped authenticity and integrity since solutions that provide one of the two properties also provide the other one.

From Table 12.3, we observe that the situation is quite variegate. In particular, some proposals and applications only work if the smartphone/reader is online. This is an important requirement that needs to be taken into account when adopting one of those solutions. Note that the proposals for enhanced barcodes might use an Internet connection to download missing certificates or to deal with key revocation; however, since this does not require continuous connection we did not mark them as online.

Table 12.3 Summary of the relevant features of solutions, applications, and research proposals

		Barcode			Attack prevention		Security properties	
Paper/application	Ref.	QR code	Data matrix	Online	Redirect	Payload	Auth. and integrity	Confidentiality
QR-TAN [25]	Sect. 12.3	✓		✓	N/A	N/A	✓[a]	✓[a]
Payment sys. [5]	Sect. 12.3		✓	✓	N/A	N/A	✓[a]	✓[a]
CryptoPaper [29]	Sect. 12.3	✓		✓	N/A	N/A	✓[a]	✓[a]
SQRC [3]	Sect. 12.4.1	✓						✓
DMPS [1]	Sect. 12.4.1		✓		✓	✓	✓	✓
Enhanced barcode [9]	Sect. 12.4.1	✓			✓	✓	✓	
Enhanced barcode [22]	Sect. 12.4.1	✓			✓	✓	✓	
Enhanced barcode [20]	Sect. 12.4.1	✓			✓	✓	✓	✓
Norton snap [26]	Sect. 12.4.2	✓		✓	✓[b]			
QR and BC reader [23]	Sect. 12.4.2	✓				✓[b]		
Enhanced reader [31]	Sect. 12.4.2	✓		✓	✓[b]			

[a]Properties guaranteed by the system which, in turn, is based on barcodes
[b]Attacks are only mitigated by checking the safety of URLs or by limiting access to resources

Table 12.4 Cryptographic mechanisms and experimental results

Research paper	Ref.	ECC	ECDSA	RSA	AES	HMAC	Key length	Signature hash	# Tested	Delay (ms)
QR-TAN [25]	Sect. 12.3			✓		✓	N/A	N/A	N/A	N/A
Payment sys. [5]	Sect. 12.3	✓	✓[a]		✓		256, 128	SHA-2 256	N/A	N/A
CryptoPaper[b] [29]	Sect. 12.3				✓[c]		N/A	N/A	5/test	N/A
Enh. barcode [9]	Sect. 12.4.1			✓			N/A	N/A	N/A	N/A
Enh. barcode [22]	Sect. 12.4.1		✓[a]				224	SHA-2 224	50	3210
							256	SHA-2 224	50	3290
							384	SHA-2 384	50	7300
							521	SHA-2 512	50	9000
Enh. barcode [20]	Sect. 12.4.1			✓	✓		N/A, 128	N/A	N/A	N/A

[a]ECDSA should only be used for legacy applications [4]
[b]The proposed system also used asymmetric cryptography but does not provide details
[c]Uses Electronic Codebook (ECB) mode for confidentiality which is insecure [4]

The systems proposed in [5, 25, 29] do not aim at securing barcodes in general, so attack prevention does not apply here (written N/A). They, however, give forms for authentication, integrity, and confidentiality at the system level (see note a in the table). Techniques to enhance barcodes in order to provide authentication and integrity (cf. Sect. 12.4.1) can prevent all the attack scenarios discussed in Sect. 12.2, since the attacker cannot counterfeit or modify barcodes. For these solutions, a tick on Authentication and Integrity implies the two ticks on Attack Prevention. Finally, enhanced barcode readers can only mitigate attacks since, for example, they cannot provide a comprehensive detection of any phishing or malware propagation URL (see note b in the table).

Cryptographic Mechanisms and ENISA Recommendations Table 12.4 reports the cryptographic algorithms, key lengths, hashes used for digital signatures, and the performed experimental results, when available. We analyze the results along the European Union Agency for Network and Information Security (ENISA) recommendations about cryptographic algorithms, key size, and parameters [4]. In particular, we observe that ECDSA has weak provable security guarantees and should only be used for legacy applications (cf. [4, Sect. 4.8]). Solutions adopting RSA do not report the key length, but it should be noticed that a length of at least 3072 bits is recommended (cf. [4, Sect. 3.5]), which would imply a big space overhead on the barcode. Following ENISA suggestions, and considering space limitations of barcodes, we suggest to try experiments using ECKDSA, a variant of ECDSA with strong provable security guarantees. CryptoPaper uses Electronic Codebook (ECB) mode to encrypt sensitive data bigger than the cipher block, which is considered insecure. Other block cipher modes should be used instead (cf. [4, Sect. 4.1]).

12.6   Conclusion and Future Work

In recent years, the barcode use has spread in most marketing companies around the world. The main aim of these barcodes is to store the information and to let the customers of the products that contain them to easily read them using smartphones or other scanning devices. There are several types of barcodes with different data capacity storage, and this study is dedicated to 2D barcodes. We have discussed many different works, and we have presented several potential attacking scenarios such as phishing, malware propagation, barcode tampering and counterfeiting, SQL and command injections, XSS, and reader application attacks.

We have summarized the available research studies and applications that developed and proposed several techniques to protect 2D barcodes. We have found that some of them lack of important detailed information such as key lengths, encryption algorithms, and hash functions. However, other studies provided these details. We have compared the methods, highlighted the limitations and weaknesses of their mechanisms, and, to some extent, evaluated their security level. Among other things, our report shows that protecting 2D barcodes against several security threat scenarios using standard state of the art cryptographic techniques is still an open issue.

As a future work, we plan to investigate new comprehensive solutions for all possible attack scenarios and for different barcodes types, and test them using various cryptographic mechanisms and security parameters, in order to determine the optimal security/feasibility trade-off.

References

1. 2D Technology Group Inc. (2016). Barcode security suite. http://www.2dtg.com/node/74.
2. Dabrowski, A., Krombholz, K., Ullrich, J., & Weippl, E. (2014). QR inception: Barcode-in-barcode attacks. In *Proceedings of the 4th ACM CCS Workshop on Security and Privacy in Smartphones and Mobile Devices (SPSM'14)*, November 7, Scottsdale, Arizona, USA (pp. 3–10).
3. Denso Wave Inc. (2017). SQRC® Secret-function-equipped QR Code. https://www.denso-wave.com/en/adcd/product/software/sqrc/sqrc.html.
4. European Union Agency for Network and Information Security (ENISA) (2014). Algorithms, key size and parameters report 2014. https://www.enisa.europa.eu/publications/algorithms-key-size-and-parameters-report-2014.

5. Gao, J., Kulkarni, V., Ranavat, H., Chang, L., & Mei, H. (2009). A 2D barcode-based mobile payment system. In *Third International Conference on Multimedia and Ubiquitous Engineering (MUE'09)*, Qingdao, China, June 4–6 (pp. 320–329)
6. GitHub. Official ZXing "Zebra Crossing" project home (website). https://github.com/zxing/zxing/.
7. GitHub. Short Payment Descriptor project home (website). https://github.com/spayd/spayd-java.
8. Google. Google Safe Browsing API (website). https://developers.google.com/safe-browsing/.
9. Ishihara, T., & Niimi, M. (2014). Compatible 2D-code Having tamper detection system with QR-code. In *Proceedings of the Tenth International Conference on Intelligent Information Hiding and Multimedia Signal Processing (IIHMSP'14)*, Kitakyushu, Japan, August 27–29 (pp. 493–496). Piscataway, NJ: IEEE.
10. ISO/IEC Standard (2006). ISO/IEC 16022:2006, Information technology – Automatic identification and data capture techniques – Data Matrix Bar code Symbology Specification.
11. ISO/IEC Standard (2008). ISO/IEC 16022:2008, Information technology – Automatic identification and data capture techniques – Aztec Bar code Symbology Specification.
12. ISO/IEC Standard (2015). ISO/IEC 15438:2015, Information technology – Automatic identification and data capture techniques – PDF417 Bar code Symbology Specification.
13. ISO/IEC Standard (2015). ISO/IEC 18004:2015, Information technology – Automatic identification and data capture techniques – QR code 2005 Bar code Symbology Specification.
14. Jin, X., Hu, X., Ying, K., Du, W., Yin, H., & Peri, G. (2014). Code injection attacks on HTML5-based mobile for apps: characterization, detection and mitigation. In *Proceedings of the 2014 ACM SIGSAC Conference on Computer and Communications Security (CCS'14)* (pp. 66–77).
15. Kaspersky Lab (2011). Malicious QR Codes: Attack Methods & Techniques Infographic. http://usa.kaspersky.com/about-us/press-center/press-blog/2011/malicious-qr-codes-attack-methods-techniques-infographic.
16. Kharraz, A., Kirda, E., Robertson, W., Balzarotti, D., & Francillon, A. (2014). Optical delusions: A study of malicious QR codes in the wild. In *44th Annual IEEE/IFIP International Conference on Dependable Systems and Networks (DSN'14)*, 23–26 June, Atlanta, GA, USA (pp. 192–203)
17. Kieseberg, P., Leithner, M., Mulazzani, M., Munroe, L., Schrittwieser, S., Sinha, M., & Weippl, E. (2010). QR code security. In *Proceedings of the 8th International Conference on Advances in Mobile Computing and Multimedia (MoMM'10)*, Paris, France, November 8–10 (pp. 430–435)
18. Kieseberg, P., Schrittwieser, S., Leithner, M., Mulazzani, M., Weippl, E., Munroe, L., & Sinha, M. (2012). Malicious pixels using QR codes as attack vector. In *Trustworthy ubiquitous computing*. Atlantis Ambient and Pervasive Intelligence (Vol. 6, pp. 21–38).
19. Krombholz, K., Fruhwirt, P., Kieseberg, P., Kapsalis, I., Huber, M., & Weippl, E. (2014). QR code security: A survey of attacks and challenges for usable security. In *Proceedings of the Second International Conference on Human Aspects of Information Security, Privacy, and Trust (HAS'14)*, 8533 (pp. 79–90).
20. Peng, K., Sanabria, H., Wu, D., & Zhu, C. (2014). Security overview of QR codes. MIT Student Project: https://courses.csail.mit.edu/6.857/2014/files/12-peng-sanabria-wu-zhu-qr-codes.pdf.
21. Phishtank: Phishtank API (website). https://www.phishtank.com/.
22. Razzak, F. (2012). Spamming the Internet of Things: A possibility and its probable solution. In *Proceeding of the 9th International Conference on Mobile Web Information Systems (MobiWIS'12)*, Niagara Falls, Canada, August 27–29 (pp. 658–665).
23. Red Dodo. (2014). QR & barcode reader (secure). http://reddodo.com/qr-barcode-scanner.php.
24. Soon, T. J. (2008). QR code. *Synthesis Journal*, 59–78. https://foxdesignsstudio.com/uploads/pdf/Three_QR_Code.pdf.

25. Starnberger, G., Froihofer, L., & Goschka, K. (2009). QR-TAN: Secure mobile transaction authentication. In *International Conference On Availability, Reliability and Security (Ares '09), Fukuoka, Japan*, March 16th–19th (pp. 16–19).
26. Symantec Corporation. (2015). Norton snap QR code reader. https://support.norton.com/sp/en/us/home/current/solutions/v64690996_EndUserProfile_en_us.
27. Tec-it. (2015). Overview: 2D Barcode Symbologies. http://www.tec-it.com/en/support/knowbase/barcode-overview/2dbarcodes/Default.aspx.
28. Vidas, T., Owusu, E., Wang, S., Zeng, C., Cranor, L., & Christin, N. (2013). QRishing: The susceptibility of smartphone users to QR code phishing attacks. In *17th International Conference on Financial Cryptology and Data Security (FC'13), Okinawa, Japan*, April 1, LNCS, 7862 (pp. 52–69). Berlin: Springer.
29. Wang, P., Yu, X., Chen, S., Duggisetty, P., Guo, S., & Wolf, T. (2015). CryptoPaper: Digital information security for physical documents. In *Proceedings of the 30th Annual ACM Symposium on Applied Computing (SAC'15), Salamanca, Spain*, April 13–17 (pp. 2157–2164).
30. Yakshtes, V., & Shishkin, A. (2012). Mathematical method of 2-D barcode authentication and protection for embedded processing. https://www.google.com/patents/US8297510.
31. Yao, H., & Shin, D. (2013). Towards preventing QR code based for detecting QR code based attacks on android phone using security warnings. In *Proceedings of the 8th ACM SIGSAC Symposium on Information, Computer and Communications Security (ASIA CCS'13), Hangzhou, China*, May 8–10 (pp. 341–346)

Chapter 13
Searching Encrypted Data on the Cloud

Khaled A. Al-Utaibi and El-Sayed M. El-Alfy

13.1 Introduction

Today, efficient data storage, management, and processing are fundamental requirements not only for business and governmental organizations but also for individual
users. Management of large amounts of heterogeneous data types such as emails,
images, videos, documents, and financial transactions requires expensive resources
in terms of hardware, software, and professional staff. As cloud computing is
becoming popular with numerous advantages (e.g., lower cost, enhanced services,
improved data sharing, better reliability and availability, etc.), more and more users
continue to shift to store their data on cloud servers maintained by professional
and specialized companies to provide high-quality data storage services. By doing
so, data owners can be relieved from the overhead of data storage, administration,
and maintenance. However, since cloud servers are untrusted by data owners, a
mechanism is needed to protect the privacy of the data against unauthorized access
and information leakage.

A common practice is to encrypt sensitive data before outsourcing. This introduces another challenging task when an authorized user wants to retrieve some
information from the outsourced data. Since the server cannot search encrypted
data directly, one obvious solution is to download all documents, then decrypt, and
search on the user's local machine. Obviously, this solution is inefficient as it is
time and bandwidth consuming. Alternatively, the cloud server should implement

K.A. Al-Utaibi
College of Computer Sciences and Engineering, University of Hail, Hail, Saudi Arabia
e-mail: alutaibi@uoh.edu.sa

E.-S.M. El-Alfy (✉)
Information and Computer Science Department, College of Computer Sciences and Engineering,
King Fahd University of Petroleum and Minerals, Dhahran 31261, Saudi Arabia
e-mail: alfy@kfupm.edu.sa

© Springer International Publishing AG 2018   221
K. Daimi (ed.), *Computer and Network Security Essentials*,
DOI 10.1007/978-3-319-58424-9_13

a technique to search encrypted documents using keyword(s) specified by the user and then return only relevant documents. The requirement of this approach is that the server should not be able to obtain any information about the keywords or the content of the document.

The rest of this chapter is organized as follows. Section 13.2 defines the problem and describes a framework. Section 13.3 presents a taxonomy of various searchable encryption techniques for different application scenarios. A discussion on searchable symmetric techniques is given in Sects. 13.4 and 13.5 for single-keyword search and multi-keyword search, respectively. Subsequently, public-key searchable encryption is reviewed in Sect. 13.6. Section 13.7 describes an interesting area of research that allows an approximate search. Finally, a summary is presented in Sect. 13.8.

13.2   Problem Definition and Framework

Search over encrypted data in cloud computing involves interaction among three main entities (as shown in Fig. 13.1):

1. **The Server** S: This is a remote storage system that provides outsourcing services and is considered as an untrusted party by the two other entities.
2. **The Data Owner/Producer** X: This is the entity that generates the data and sends an encrypted version of it to S for storage.
3. **The Data User/Consumer** Y: This entity sends queries to S to retrieve specific data stored by X. It can be the same as X or a different entity depending on the problem setting.

Fig. 13.1 Entities of searchable encryption

Searching outsourced encrypted data can be defined as follows. Assume $\mathcal{X}$ has a set of data records (e.g., text documents, emails, relational data, etc.) $D = \{d_1, d_2, \ldots, d_n\}$ stored on the server $\mathcal{S}$. Each element $d_i \in D$ can be referenced by a set of keywords as in the case of text documents or by a set of fields as in the case of relational databases. Let such set be $W_i = \{w_{i,1}, w_{i,2}, \ldots, w_{i,k}\}$. To retrieve a certain data element based on a set of keywords/fields, a data user $\mathcal{Y}$ sends a query of encrypted keywords/fields, $q = \{x_1, x_2, \ldots, x_m\}$ to $\mathcal{S}$ who responds with the set of documents/records satisfying the keyword(s)/field(s) in the query:

$$\text{Ans}(q) = \{d_i \in D | \forall x_j \in q, x_j \in W_i\} \tag{13.1}$$

To allow the server to search over encrypted data, the data owner $\mathcal{X}$ encrypts the data using what is called a *trapdoor*. Whenever $\mathcal{Y}$ wants to retrieve certain data based on some keyword/field (w), he generates a trapdoor (T_w) corresponding to w and sends it to $\mathcal{S}$. Upon receiving, T_w, $\mathcal{S}$ searches the encrypted data for T_w, and if there is a match, it returns the corresponding encrypted data to $\mathcal{Y}$ (without knowing the actual keyword/field or the data content).

The general framework of searchable encryption consists of four functions [11]:

1. *Setup*$(1^\lambda) \rightarrow K$: Takes a security parameter λ as an input and outputs a set of secret keys K to be used by the scheme.
2. *Encrypt*$(K, D) \rightarrow (I, C)$: Takes the user's data D and the set of keys K as inputs and outputs encrypted data C and keywords index I.
3. *Trapdoor*$(K, w) \rightarrow T_w$: Generates trapdoor T_w for given keyword w and keys K.
4. *Search*$(T_w, I) \rightarrow \{0, 1\}$: Takes T_w and the encrypted keywords index I as inputs and returns 1 if the search succeeds or 0 otherwise.

13.3   Taxonomy of Searchable Encryption Techniques

In recent years, several encryption techniques have been developed to allow secure search over encrypted data. These techniques can be classified based on different application scenarios as shown in Fig. 13.2. These categories are not exclusive, i.e., a particular scenario can belong to more than one category.

Searchable encryption can be classified based on their key-encryption scheme into two categories: *symmetric-key* search and *public-key* search. The first category involves two parties only: a data owner $\mathcal{X}$ and an untrusted server $\mathcal{S}$. $\mathcal{X}$ encrypts his documents using his own private key prior to uploading them on $\mathcal{S}$. Unlike classical symmetric-key encryption techniques, the encryption algorithm used here is designed to allow secure search on encrypted data using a trapdoor. The untrusted server $\mathcal{S}$ stores the encrypted documents without knowing their cleartext content since the secret key is known only to the data owner. Whenever $\mathcal{X}$ wants to search for documents containing a keyword, w, he generates a trapdoor (T_w) for this

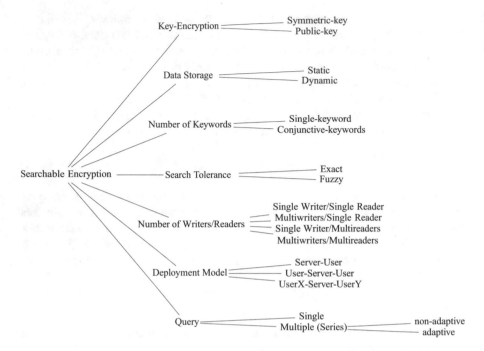

Fig. 13.2 Taxonomy of searchable encryption techniques

keyword and sends it to S to search the encrypted documents for a match of T_w and return the corresponding ones. In the second category, encrypted data is produced by a third party $\mathcal{X}$ using the public key of the data consumer $\mathcal{Y}$. When $\mathcal{Y}$ wants to retrieve a certain document based on a keyword w, he uses his private key to generate a trapdoor T_w and sends it to S.

Searchable encryption can be classified with regard to data storage scheme into two main categories: *static storage* and *dynamic storage*. In the first category, the uploaded files are not changed, and hence, the keyword index does not need to be updated. On the other hand, dynamic-storage schemes support index update functionality to allow the user to update his uploaded data.

Searching over encrypted data can be conducted using a single keyword or multiple keywords. Based on this search scheme, searchable encryption techniques can be classified into two categories: *single-keyword* searchable encryption techniques and *multi-keywords (conjunctive)* searchable encryption techniques.

In addition to the classification criteria discussed above, searchable encryption techniques can be classified based on the search tolerance. Some searchable encryption techniques return only documents/records that exactly match the given keyword(s). Other techniques can return documents/records matching the given keyword(s) with some *tolerance*.

In a survey by Bösch et al. [4], searchable encryption techniques are classified according to the number of involved readers (data owners) and writers (data consumers) into four models: *single writer/single reader*, *multiwriters/single reader* model, *single writer/multireaders* model, and *multiwriters/multireaders* model.

Another survey by Han et al. [11] classified searchable encryption based on the deployment model into three categories: *server-user* model, *user-server-user* model, and *userX-server-userY* model. In the first model, data is stored and owned by the server (i.e., $\mathcal{X}$ and $\mathcal{S}$ are the same). The user can search on encrypted data to avoid leaking private search information to the server. In the second model, the user performs a secure search on the encrypted data owned and uploaded by himself (i.e., $\mathcal{X}$ and $\mathcal{Y}$ are the same). In the third model, the encrypted data is owned and uploaded by user $\mathcal{X}$ who authorizes user $\mathcal{Y}$ to perform a secure search on the data (here, $\mathcal{X}$, $\mathcal{Y}$, and $\mathcal{S}$ are different).

Based on the query type, the search may consist of a *single* word or a *stream* of words. In the latter case, the query words may be independent of the results for previous words, which is termed as *non-adaptive*; otherwise, it is called *adaptive*.

13.4 Single-Keyword Search

Research on searchable encryption started with single-keyword search techniques. Clearly, this type of search is too restrictive for real-world applications and cannot retrieve very relevant documents. However, it provides a basis for more advanced search techniques. In this section, we discuss three approaches under this category.

13.4.1 Sequential Scan

This technique was proposed by Song et al. [14] using symmetric encryption. Its basic idea is to encrypt individual keywords of the plain text with a sequence of pseudo-random bits with a special structure. This structure allows searching the encrypted data for a certain keyword without revealing the keyword itself. The detailed scheme is described by the algorithm shown in Fig. 13.3.

Data Encryption The *encryption* function operates on a document D containing a sequence of l words $w_1, w_2, \ldots,$ and w_l. Each word, w_i, has a fixed length of n bits. Typically each word corresponds to an English language word where extra padding is added to make all words equal in length. The encryption function encrypts each keyword w_i using a two-layered encryption construct as shown in Fig. 13.4a. In the first layer, w_i is encrypted using a deterministic encryption function E_{k_e}. The resulting encrypted keyword X_i is split into two parts: (1) a left part L_i of size $(n-m)$ bits and (2) a right part R_i of size m bits. In the second layer, X_i is XORed with a special hash code Y_i. This hash code is computed by applying a hash function F_{k_i} on a pseudo-random number s_i. The hash key k_i is computed based on the left part of X_i using another hash function f_{k_f}. The pseudo-random number s_i is generated using a pseudo-random number generator G_{k_g}.

- Let $D = \{w_1, w_2, \ldots, w_l\}$ be a document with l keywords.
- Assume that each keyword $w_i \in D$ has a fixed length of n bits.
- $Setup(1^\lambda)$:

    (1) Generate a set of secret keys $K = \{k_e, k_f, k_g\} \in \{0,1\}^\lambda$.
    (2) Choose a pseudo-random number generator $G : \{0,1\}^\lambda \to \{0,1\}^{n-m}$ for some positive integers $n > m$.
    (3) Choose a keyed-hash-function $f : \{0,1\}^\lambda \times \{0,1\}^{n-m} \to \{0,1\}^\lambda$.
    (4) Choose a keyed-hash-function $F : \{0,1\}^\lambda \times \{0,1\}^{n-m} \to \{0,1\}^m$.
    (5) Choose a deterministic encryption function $E : \{0,1\}^\lambda \times \{0,1\}^n \to \{0,1\}^n$.

- $Encrypt(K, D)$:

    (1) Generate a sequence of pseudo-random numbers: $S = \{s_1, s_2, \ldots, s_l\}$ using G_{k_g}.
    (2) For each keyword $w_i \in D$:
        a. Encrypt w_i and split the result into two parts: $X_i = E_{k_e}(w_i) = \langle L_i, R_i \rangle$.
        b. Compute the hash-key: $k_i = f_{k_f}(L_i)$.
        c. Compute the hash-code: $Y_i = \langle s_i, F_{k_i}(s_i) \rangle$.
        d. Output the cipher-text: $C_i = X_i \oplus Y_i$.

- $Trapdoor(K, w)$:

    (1) Encrypt the keyword w and split the result into two parts: $X_w = E_{k_e}(w) = \langle L_w, R_w \rangle$.
    (2) Compute the hash-key: $k_w = f_{k_f}(L_w)$.
    (3) Output the trapdoor: $T_w = \langle X_w, k_w \rangle$.

- $Search(T_w)$

    (1) For each cipher-text C_j:
        a. Compute the hash-code: $Y_j = C_j \oplus T_w = \langle s_j, s_j' \rangle$.
        b. Check if $s_j' = F_{k_w}(s_j)$, then return 1; otherwise return 0.

Fig. 13.3 Algorithm description of the sequential scan scheme

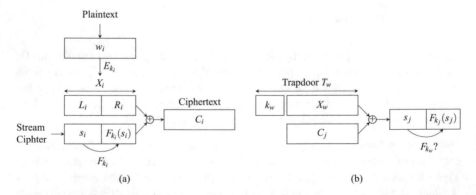

Fig. 13.4 The encryption and search functions of the sequential scan scheme. (**a**) Keyword encryption. (**b**) Keyword search

Trapdoor Generation The *trapdoor* function generates a trapdoor T_w to allow the server to search the encrypted data for a keyword w. The trapdoor consists of a tuple of two elements: (1) an encrypted keyword X_w and (2) a hash key k_w. Both X_w and k_w are generated based on the keyword w in the same manner used to generate the encrypted keyword X_i and the secret key k_i in the *encryption* function.

Secure Data Search The *search* function scans the encrypted keywords, $C = \{C_1, C_2, \ldots, C_l\}$ trying to match the hash code embedded within each encrypted keyword with the hash code corresponding to the keyword w. For each encrypted keyword, C_j, the function XORs C_j with X_w to extract the embedded pseudo-random number s_j and its corresponding hash code $F_{k_j}(s_j)$ as shown in Fig. 13.4b. Then, it uses the hash key k_w to check if the extracted hash code matches that of the keyword w. That is, whether $F_{k_w}(s_j)$ equals $F_{k_j}(s_j)$ or not.

The sequential scan method is not efficient when the data size is large. Song et al. [14] suggested to use an encrypted index of stored documents to speed up the search operation. However, this modification requires index update whenever the data owner modifies his documents.

13.4.2 Secure Indexes

Goh [9] defined secure indexes and proposed a single-keyword symmetric-key searchable encryption technique. In this technique, a secure index based on Bloom filter [1] is used. A Bloom filter (BF) is defined as a data structure that uses an array of m bits (called the filter) to represent a set $S = \{s_1, s_2, \ldots, s_n\}$ of n elements. Initially, all bits of the array are set to 0. The filter represents S by applying r independent hash functions $h_1, h_2, \ldots, h_r$ on the individual elements of S, where $r << m$. These hash functions are defined as $h_i : \{0, 1\}^* \rightarrow [1, m]$ for $i \in [1, r]$. For each element $s \in S$, all the bits at the positions $h_1(s), h_2(s), \ldots, h_r(s)$ are set to 1. Figure 13.5a shows an example of constructing a 16-bit BF for a set S of two elements s_1 and s_2 using four hash functions. Note that a bit can be set to 1 multiple times by individual elements of S. To determine if a given element s' belongs to S, we check the bits of the Bloom filter at positions $h_1(s'), h_2(s'), \ldots, h_r(s')$. If all these bits are set to 1, then we consider that $s' \in S$. On the other hand, if any checked bit is found to be 0, then s' is definitely not a member of S. An example of testing the membership of two elements s_1 and s_3 is shown in Fig. 13.5b. This check, however, may result in a false positive but no false negative, because some bits may have been set by some elements other than s'.

Basically, the secure indexes method (shown in Fig. 13.6) uses a data structure (Bloom filter in this case) that allows an untrusted party to test in constant time, $O(1)$, if the index contains a certain keyword w using a trapdoor T_w. The secure index reveals no information about its content without a valid trapdoor which is generated by the data owner using a secret key.

Index Construction The *build-index* function constructs a BF-based index for a document D comprising a unique identifier D_{id} and a list of t keywords w_1, $w_2, \ldots,$ and w_t. The document identifier is used to make sure that all Bloom filters look different, even for documents with the same keyword set. Thus, this technique avoids leaking document similarity to untrusted servers. The build-index function processes each keyword w_i twice using the pseudo-random function f

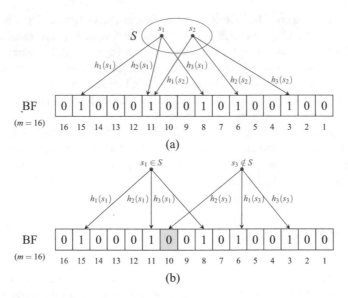

Fig. 13.5 Example of constructing and testing the membership of a Bloom filter. (**a**) BF construction for the set $S = \{s_1, s_2\}$. (**b**) Membership testing for $s_1 \in S$ and $s_3 \notin S$

- Let $D = \{w_1, w_2, \ldots, w_t\}$ be a document of t keywords.
- Let $D_{id} \in \{0,1\}^n$ be a unique identifier for the document D.
- $Setup(1^s)$:

  (1) Generate a set of r secret keys $K = \{k_1, k_2, \ldots, k_r\} \in \{0,1\}^{sr}$.
  (2) Choose a pseudo-random number function $f : \{0,1\}^n \times \{0,1\}^s \rightarrow \{0,1\}^s$.

- $BuildIndex(K, D, D_{id})$:

  (1) For each keyword $w_i \in D$:
     a. Compute the trapdoor $X_i = \{x_1 = f(w_i, k_1), x_2 = f(w_i, k_2), \ldots, x_r = f(w_i, k_r)\} \in \{0,1\}^{sr}$.
     b. Compute the codeword $Y_i = \{y_1 = f(D_{id}, x_1), y_2 = f(D_{id}, x_2), \ldots, y_r = f(D_{id}, x_r)\} \in \{0,1\}^{sr}$.
     c. Insert the codeword Y_i into document D_{id}'s Bloom filter BF.

- $Trapdoor(K, w)$:

  (1) Output the trapdoor: $T_w = \{x_1 = f(w, k_1), x_2 = f(w, k_2), \ldots, x_r = f(w, k_r)\}$.

- $Search(T_w = \{x_1, x_2, \ldots, x_r\}, \mathcal{I}_{D_{id}} = \{D_{id}, BF\})$:

  (1) Compute the codeword $Y_w = \{y_1 = f(D_{id}, x_1), y_2 = f(D_{id}, x_2), \ldots, y_r = f(D_{id}, x_r)\} \in \{0,1\}^{sr}$.
  (2) Test if BF contains 1's in all bit-positions denoted by $(y_1, y_2, \ldots, y_r)$.
  (3) Output 1 if the test is positive; otherwise, output 0.

Fig. 13.6 Algorithm description of the secure indexes scheme

before inserting it into the Bloom filter. It computes a trapdoor X_i by applying the function f on w_i using r secret keys $(k_1, k_2, \ldots, k_r)$. Then, it computes a codeword Y_i by applying f on the document identifier D_{id} using the r elements of the trapdoor X_i.

Trapdoor Generation The *trapdoor* function generates a trapdoor T_w by applying the pseudo-random function f on w using the r secret keys $(k_1, k_2, \ldots, k_r)$.

Secure Data Search The *search* function takes as input a trapdoor T_w for the keyword to be searched and an index $\mathcal{I}_{D_{id}}$ comprising the identifier of the document to be searched and its corresponding Bloom filter *BF*. The function starts by computing a codeword Y_w by applying the pseudo-random function f on the elements of the trapdoor. Then, it tests whether the keyword w belongs to the document identified by D_{id}. This test is done by checking if the *BF* contains 1's in all locations denoted by the elements of Y_w.

The secure indexes method addresses some limitations of the sequential scan method. Unlike sequential scan method which is restricted to fixed-size word, the secure indexes method can handle variable-length keywords. The search time of the secure indexes method is linear in the number of documents, whereas the search time of sequential scan is proportional to the number of keywords.

There are two problems in using Bloom filters in searchable encryption techniques. The first problem is the possibility of false positives. The probability of false positive can be reduced by choosing appropriate parameter settings. The second problem is that the Bloom filters leak the number of keywords in each document as the number of 1's in the *BF* is directly proportional to the number of unique keywords in each document. To avoid this problem, we can add random keywords to the *BF* to make the number of 1's in the *BF* almost equal for different documents.

13.4.3 Inverted Indexes

Curtmola et al. [6, 7] revised the previous security definitions and introduced two standard adversarial models for searchable symmetric encryption (SSE). The first is a non-adaptive model which only considers adversaries who make search queries without seeing the outcome of previous searches. The other model is an adaptive model which considers adversaries who can make search queries as a function of the outcome of previous searches. The proposed schemes are based on inverted indexes and considered to be secure under the two adversarial models. Figure 13.7 shows an example of forward indexes and their corresponding inverted indexes. This scheme uses a look-up table constructed using the FKS hashing technique [8] which is an efficient data structure that requires $O(n)$ storage and $O(1)$ look-up time. This makes it the first-ever sublinear searchable symmetric scheme. Despite its efficiency in searching encrypted data, the inverted indexes scheme can only support a single-keyword search and a limited notion of document updates.

13.5 Multi-Keyword Search

A common limitation to the previous schemes is that they only allow searching encrypted data using a single keyword. Golle et al. [10] proposed the first scheme that allows conjunctive keyword queries on encrypted data. Their model assumes

keyword	document ids
w_1	$id(d_2), id(d_4)$
w_2	$id(d_3), id(d_4)$
w_3	$id(d_1), id(d_3), id(d_4)$
w_4	$id(d_2)$
w_5	$id(d_1), id(d_2), id(d_3), id(d_4)$
w_6	$id(d_2), id(d_4)$
w_7	$id(d_1), id(d_3)$
w_8	$id(d_1), id(d_2)$

document id	keywords
$id(d_1)$	w_3, w_5, w_7, w_8
$id(d_2)$	$w_1, w_4, w_4, w_5, w_6, w_8$
$id(d_3)$	w_2, w_3, w_5, w_7
$id(d_4)$	$w_1, w_2, w_3, w_3, w_5, w_6$

(a)   (b)

Fig. 13.7 Example of (a) forward and (b) inverted indexes

that there are n documents each associated with m keyword fields. For example, one might consider the following keyword fields for emails: "From," "To," "Date," and "Subject." The details of secure conjunctive keyword search scheme are given by the algorithm shown in Fig. 13.8.

Data Encryption A document d_i is encrypted as a set of elements from the group G having the form g^r. The integer r is computed as a multiplication of two values $V_{i,j}$ and a_i (except for the first element, where $r = a_i$). The number a_i is a uniform random variable, while $V_{i,j}$ is a pseudo-random value generated by the function f based on jth keyword of the document d_i.

Trapdoor Generation The *trapdoor* function takes a set of keyword field indices $J = \{j_1, \ldots, j_t\}$, and the set of keywords $W = \{w_{j_1}, \ldots, w_{j_t}\}$ associated with these fields, and generates a trapdoor T_W consisting of four elements. The first element is a set of hash codes Q generated by hashing n elements of the group G of the form $g^{a_i s}$, where a_i is the random value associated with document d_i and s is a uniform random value. The second element of T_W is a single value C computed by adding s to the sum of the pseudo-random values generated by the function f based on the keywords in W. The third element of T_W is the set of keyword field indices J. The forth element is the random value s.

Secure Data Search Given a trapdoor T_W, the *search* function searches the encrypted documents as follows. For each encrypted document X_i, the function computes a code Y_i as follows:

$$Y_i = (X_i[1])^C . (X_i[j_1])^{-1} \ldots (X_i[j_t])^{-1}$$
$$= g^{a_i C} . g^{-a_i V_{i,j_1}} \ldots g^{-a_i V_{i,j_t}}$$
$$= g^{a_i C} . g^{-a_i (\sum_{x=1}^{t} (V_{i,j_x}))}$$

- Let $D = \{d_1, \ldots, d_n\}$ be a collection of n documents.
- Let $d_i = \{w_{i,1}, \ldots, w_{i,m}\}$ denote the ith document with m keyword fields, where $w_{i,j}$ is the keyword of document d_i in the jth keyword field.
- Let $J = \{j_1, \ldots j_t\}$ be a set of keyword field indices, where $1 \leq t \leq m$.
- $Setup(1^\lambda)$:

    (1) Choose a group G of order q in which DDH is hard, and let g be a generator of G.
    (2) Choose a keyed-hash function $f : \{0,1\}^\lambda \times \{0,1\}^* \rightarrow \mathbb{Z}_q^*$.
    (3) Choose a hash function h.
    (4) Generate a secret key $k \in \{0,1\}^\lambda$.
    (5) Generate a set of random values $A = \{a_1, \ldots, a_m\}$, where a_i is chosen uniformly at random from $\mathbb{Z}_q^*$.
    (6) Output the set of parameters $\rho = \{G, g, f, h, A\}$.

- $Encrypt(\rho, k, d_i)$:

    (1) Compute $V_{i,j} = f_k(w_{i,j})$ for $j = 1, \ldots, m$.
    (2) Output the encrypted document as $X_i = \langle g^{a_i}, g^{a_i V_{i,1}}, g^{a_i V_{i,2}}, \ldots, g^{a_i V_{i,m}} \rangle$.

- $Trapdoor(\rho, k, J = \{j_1, \ldots, j_t\}, W = \{w_{j_1}, \ldots, w_{j_t}\})$:

    (1) Let s be a value chosen uniformly at random from $\mathbb{Z}_q^*$..
    (2) Compute $Q = \{h(g^{a_1 s}), h(g^{a_2 s}), \ldots, h(g^{a_n s})\}$.
    (3) Compute $C = s + (\Sigma_{x=1}^t f_k(w_{j_x}))$.
    (4) Output $T_W = \langle Q, C, J, s \rangle$

- $Search(\rho, T_W)$:

    (1) For every encrypted document X_i:
        a. Compute $Y_i = g^{a_i C} \cdot g^{-a_i(\Sigma_{x=1}^t (V_{i,j_x}))}$.
        b. Let $S_i = Q[i] = h(g^{a_i s})$.
        c. If $h(Y_i) = S_i$, then return true; otherwise return false.

Fig. 13.8 Algorithm description of the secure conjunctive keyword scheme

$$= g^{a_i(s + (\Sigma_{x=1}^t f_k(w_{j_x})))} \cdot g^{-a_i(\Sigma_{x=1}^t (V_{i,j_x}))}$$

$$= g^{a_i s} \cdot g^{a_i(\Sigma_{x=1}^t (V_{w_{j_x}}))} \cdot g^{-a_i(\Sigma_{x=1}^t (V_{i,j_x}))}$$

If X_i contains the keywords in W at the keyword fields specified by indices in J, then Y_i will reduce to $g^{a_i s}$. Thus, we can check if there is a match by comparing $h(Y_i)$ with the ith element of Q which is $h(g^{a_i s})$.

13.6   Searchable Public-Key Encryption

This type of searchable encryption schemes varies from the previous schemes in the sense that the producer of the data (i.e., data owner) and the data consumer are different. As an example, consider a scenario where an untrusted email gateway stores incoming emails from multiple users to a particular recipient. For example, If Bob wants to send an email containing sensitive information to Alice, he would encrypt the email using Alice's public key. On the other side, if Alice would like to search encrypted emails in her mailbox using some keyword(s), then she will generate a trapdoor of the required keyword(s) and send it to the untrusted mail gateway. The untrusted mail server will use the trapdoor to search the encrypted emails and retrieve the corresponding ones for Alice.

Searchable public-key encryption was first introduced by Boneh et al. [2]. They referred to their scheme as *"Public-key Encryption with Keyword Search* (PEKS)." This approach works as follows: the data producer encrypts the message/document using the consumer's public key and appends the encrypted message/document with a set of additional codewords corresponding to a set of unique keywords from the message/document itself. The codewords are generated by a PEKS algorithm using the data consumer public key in a way that allows the untrusted server to perform a secure search using a trapdoor. To send a message/document M with keywords $w_1, w_2, \ldots,$ and w_m, the data producer encrypts the message and its keywords using the following formula:

$$E_{k_{\text{pub}}}(M) \| \text{PEKS}(k_{\text{pub}}, w_1) \| \text{PEKS}(k_{\text{pub}}, w_2) \| \ldots \| \text{PEKS}(k_{\text{pub}}, w_m)$$

The data consumer can give the untrusted server a trapdoor T_w to search for messages containing the keyword w. The untrusted server uses the given trapdoor to search within the codewords appended with each message sent by the data consumer and retrieves all messages containing the required keyword w learning nothing about the keyword itself or the content of the codewords.

The PEKS algorithm is based on *admissible bilinear maps*. A bilinear map is a function $e : G_a \times G_b \to G_c$, where G_a, G_b, and G_c are cyclic groups of order p, such that for all $g \in G_a$, $h \in G_b$, and $x, y \in [1, p]$, $e(g^x, h^y) = e(g, h)^{xy}$. A bilinear map e is said to be admissible, if $e(g, h)$ generates G_c, and e is efficiently computable. The PEKS algorithm uses two groups G_1 and G_2 of prime order p and an admissible bilinear map $e : G_1 \times G_1 \to G_2$ between them. The details of secure conjunctive keyword search scheme are given by the algorithm shown in Fig. 13.9.

Data Encryption The *PEKS* function encrypts the given keyword w using the public key $\langle g, g^\alpha \rangle$ of the data consumer. It applies a two-level encryption using two hash functions (H_1, H_2) and a bilinear map e. In the first level, it generates an element $t = e(H_1(w), g^{r\alpha}) \in G_2$. Since $H_1(w) \in G_1$, we can rewrite t in terms of the group generator as $t = e(g^\beta, g^{r\alpha})$ for some integer $\beta \in \mathbb{Z}_p^*$. Knowing that e is a bilinear map, we can further simplify the result as $t = e(g, g)^{r\alpha\beta}$. In the second level, the *PEKS* function generates the code $c = \langle g^r, H_2(e(g, g)^{r\alpha\beta}) \rangle$. The *Encrypt* function encrypts a given message M_i as follows. First, it encrypts the message M_i using an encryption function $E_{k_{\text{pub}}}$. Then, it encrypts every keyword w_j in M_i using the function *PEKS* and appends them to the encrypted message.

Tapdoor Generation Given a keyword w, the *Trapdoor* function generates $T_w = (H_1(w))^\alpha$. Since $(H_1(w))^\alpha \in G_1$, T_w can be rewritten in terms of the group generator as $T_w = g^{\alpha\beta}$.

Secure Data Search The *Search* function scans the encrypted keywords of the stored messages. For every encrypted keyword, $c_j = \langle g^r, H_2(e(g, g)^{r\alpha\beta_j}) \rangle$, the function computes $H_2(e(T_w = g^{\alpha\beta}, g^r))$ which equals $H_2(e(g, g)^{r\alpha\beta})$. Then, the function checks whether $H_2(e(g, g)^{r\alpha\beta})$ equals $H_2(e(g, g)^{r\alpha\beta_j})$ or not.

- Let $M = \{M_1, \ldots, M_n\}$ be a set of messages generated by user-X and consumed by user-Y.
- Let $M_i = \{w_1, \ldots, w_m\}$ be the set of keywords in message $M_i \in M$.
- $Setup(1^\lambda)$:

  (1) Choose a large prime number p based on the security parameter λ.
  (2) Choose two groups G_1 and G_2 of order p.
  (3) Choose two hash functions $H_1 : \{0,1\}^* \rightarrow G_1$ and $H_2 : G_2 \rightarrow \{0,1\}^{\log p}$.
  (4) Choose a random number $\alpha \in \mathbb{Z}_p^*$ and a generator g of G_1.
  (5) Generate a public-key $k_{pub} = \langle g, g^\alpha \rangle$.
  (6) Generate a private key $k_{priv} = \alpha$.
  (7) Output the set of keys $K = \{k_{pub}, k_{priv}\}$.

- $PEKS(k_{pub}, w)$:

  (1) Compute $t = e(H_1(w), g^{r\alpha}) \in G_2$ for a random number $r \in \mathbb{Z}_p^*$.
  (2) Output $c = \langle g^r, H_2(t) \rangle$

- $Encrypt(k_{pub}, M_i)$:

  (1) For each keyword $w_j \in M_i$: compute $c_j = PEKS(k_{pub}, w_j)$.
  (2) Output $M_i' = \langle E_{k_{pub}}(M_i), c_1, c_2 \ldots, c_m \rangle$.

- $Trapdoor(k_{priv}, w)$:

  (1) Output $T_w = (H_1(w))^\alpha \in G_1$.

- $Search(k_{pub}, T_w)$:

  (1) For every encrypted message M_i', and every encrypted keyword $c_j \in M_i'$:
    a. Let $c_j = \langle x, y \rangle$.
    b. If $H_2(e(T_w, x)) = y$, then return 1; otherwise return 0.

Fig. 13.9 Algorithm description of the public-key encryption with keyword search (PEKS) scheme

A major problem with this scheme is that the trapdoor of a keyword is never refreshed. This makes the scheme subject to offline keyword-guessing attacks.

Some changes of the basic approach have been proposed such as public-key encryption with conjunctive field keyword search, which allows searching with conjunctive keywords [13]. Another extension allows conjunctive, subset, and range queries [3]. In [16], the authors studied the search problem of conjunctive subset search function and proposed a more efficient mechanism known as Public-Key Encryption with Conjunctive-Subset Keywords Search (PECSK) scheme. A more recent approach based on prime-order bilinear groups is proposed in [15] which can simultaneously support conjunction and disjunction keyword search.

13.7 Fuzzy Keyword Search

The previous schemes allow users to search encrypted data using exact keywords. Fuzzy keyword searchable encryption schemes, on the other hand, allow users to perform more flexible and error tolerating search.

13.7.1 Error-Tolerant Searchable Encryption

Bringer et al. [5] proposed the first construction both for error-tolerant searchable encryption and for a biometric identification protocol over encrypted data. They used locality-sensitive hashing (LSH) and Bloom filters with storage (BFS) to permit searching encrypted data with an approximation of some keyword. The basic idea of this scheme is to use LSH to enable error-tolerant keyword search. If two keywords w and w' are similar, then the LSH will output the same hash values for these two keywords, allowing error-tolerant queries.

13.7.2 Wildcard-Based Fuzzy Keyword Search

Li et al. [12] proposed a fuzzy keyword searchable scheme based on wildcards. The details of this scheme are given by the algorithm shown in Fig. 13.10.

Data Encryption The *encryption* function encrypts a set of keywords $W = \{w_1, \ldots, w_n\}$ by computing fuzzy keyword sets $S_{w_i,d} = \{S'_{w_i,0}, S'_{w_i,1}, \ldots, S'_{w_i,d}\}$ with

---

- Let $W = \{w_1, \ldots, w_n\}$ be the set of distinct keywords.
- Let ID_{w_i} be the set of file identifiers containing the keyword w_i.
- $Setup(1^s)$:

    (1) Generate a secret key $k_s \in \{0,1\}^s$.
    (2) Choose a hash function $H : \{0,1\}^* \rightarrow \{0,1\}^*$.
    (3) Choose a deterministic encryption function $E : \{0,1\}^s \times \{0,1\}^* \rightarrow \{0,1\}^*$.

- $Encrypt(k_s, W)$:

    (1) For each keyword $w_i \in W$:
        a. Compute fuzzy keyword sets $S_{w_i,d} = \{S'_{w_i,0}, S'_{w_i,1}, \ldots, S'_{w_i,d}\}$, where d is a predefined edit distance.
        b. Set $T_{w'_i,d} = \varphi$.
        c. For each $w'_i \in S_{w_i,d}$: Compute $T_{w'_i,d} = T_{w'_i,d} \cup H(k_s || w'_i)$.
        d. Generate the index $X_i = \langle T_{w'_i,d}, E_{ks}(\text{ID}_{w_i} || w_i) \rangle$.
        e. Add X_i to the index table $\mathcal{I}$.

- $Trapdoor(k_s, w, k)$:

    (1) Compute fuzzy keyword sets $S_{w,k} = \{S'_{w,0}, S'_{w,1}, \ldots, S'_{w,k}\}$ for the k edit distance.
    (2) Set $T_{w',k} = \varphi$.
    (3) For each $w' \in S_{w,k}$: Compute $T_{w',k} = T_{w',k} \cup H(k_s || w')$.
    (4) Output $T_{w',k}$.

- $Search(\mathcal{I}, T_{w',k})$:

    (1) For each index X_i in $\mathcal{I}$:
        a. Compute $S_i = T_{w',k} \cap T_{w'_i,d}$.
        b. If $S_i \neq \varphi$, then return 1; otherwise return 0.

Fig. 13.10 Algorithm description of the wildcard-based fuzzy keyword search scheme

edit distance d for each keyword w_i. The edit distance $ed(w, w')$ between two words w and w' is the number of operations (e.g., character substitution, deletion, and insertion) required to transform one word to the other. For example, the wildcard-based fuzzy keyword set for the keyword "CAT" with edit distance 1 is given by $S_{CAT,1}=\{CAT, *CAT, *AT, C*AT, C*T, CA*T, CA*, CAT*\}$. Next, the function computes a trapdoor set $T_{w'_i,d}$ for each fuzzy keyword $w'_i \in S_{w_i,d}$ using a hash function H and a secret key k_s shared between data owner and authorized users. Finally, it generates an index $X_i = \langle T_{w'_i,d}, E_{k_s}(ID_{w_i}||w_i)\rangle$ and adds it to the index table. The index table and encrypted data files are outsourced to the cloud server for storage.

Trapdoor Generation To search for a keyword w with edit distance k, an authorized user computes the trapdoor set $T_{w',k}$ for each fuzzy keyword w' in the fuzzy keyword sets $S_{w,k}$ and sends it to the server.

Secure Data Search Upon receiving the trapdoor set, the server compares it with the trapdoors in the index table and returns all possible encrypted file identifiers matching the trapdoor set (i.e., $\{E_{k_s}(ID_{w_i}||w_i)|T_{w',k} \cap T_{w'_i,d} \neq \phi\}$).

13.8 Conclusion

The field of secure search over encrypted data is gaining more importance with the massive increase in storing confidential data in external untrusted servers. Several approaches have been proposed in the literature which vary in many aspects including complexity, security, and search capabilities. In this chapter, we reviewed the basic concepts and constructs for searching encrypted data over cloud computing. The principal idea of these techniques is to encrypt the data in a way that allows an untrusted server to perform a keyword search using a trapdoor without revealing any information about the keyword or the content of the encrypted data. We presented a taxonomy of searchable encryption schemes based on various application scenarios. Then, we presented detailed descriptions and explanations of fundamental schemes from the main classes of searchable encryption which inspired many subsequent approaches and advances in the field.

Acknowledgements The authors would like to thank both King Fahd University of Petroleum and Minerals and Hail University for the support during this work.

References

1. Bloom, B. H. (1970). Space/time trade-offs in hash coding with allowable errors. *ACM Communications, 13*, 422–426.

2. Boneh, D., Di Crescenzo, G., Ostrovsky, R., & Persiano, G. (2004). Public key encryption with keyword search. In *EUROCRYPT 2004*. Lecture notes in computer science (Vol. 3027, pp. 506–522).
3. Boneh, D., & Waters, B. (2007). Conjunctive, subset, and range queries on encrypted data. In *Theory of Cryptography Conference* (pp. 535–554).
4. Bösch, C., Hartel, P., Jonker, W., & Peter, A. (2014). A survey of provably secure searchable encryption. *ACM Computing Surveys, 47*(2), 18:1–18:51.
5. Bringer, J., Chabanne, H., & Kindarji, B. (2009). Error-tolerant searchable encryption. In *Proceedings of IEEE International Conference on Communications (ICC'09)* (pp. 768–773).
6. Curtmola, R., Garay, J., Kamara, S., & Ostrovsky, R. (2006). Searchable symmetric encryption: Improved definitions and efficient constructions. In *Proceedings of 13th ACM Conference on Computer and Communications Security, CCS '06* (pp. 79–88).
7. Curtmola, R., Garay, J., Kamara, S., & Ostrovsky, R. (2011). Searchable symmetric encryption: Improved definitions and efficient constructions. *Journal of Computer Security, 19*(5), 895–934.
8. Fredman, M. L., Komlós, J., & Szemerédi, E. (1984). Storing a sparse table with 0(1) worst case access time. *Journal of the ACM, 31*(3), 538–544.
9. Goh, E. J. (2003). Secure indexes. Cryptology ePrint Archive, Report 2003/216.
10. Golle, P., Staddon, J., & Waters, B. (2004) Secure conjunctive keyword search over encrypted data. In *Applied cryptography and network security*. Lecture notes in computer science (Vol. 3089, pp. 31–45). Berlin: Springer.
11. Han, F., Qin, J., & Hu, J. (2016). Secure searches in the cloud: A survey. *Future Generation Computer Systems, 62*, 66–75.
12. Li, J., Wang, Q., Wang, C., Cao, N., Ren, K., & Lou, W. (2010). Fuzzy keyword search over encrypted data in cloud computing. In *Proceedings of 29th Conference on Information Communications (INFOCOM'10)* (pp. 441–445).
13. Park, D. J., Kim, K., & Lee, P. J. (2004). Public key encryption with conjunctive field keyword search. In *International Workshop on Information Security Applications* (pp. 73–86).
14. Song, D., Wanger, D., & Perrig, A. (2000). Practical techniques for searches on encrypted data. In *IEEE Symposium on Security and Privacy* (pp. 44–55).
15. Xiao, S., Ge, A., Zhang, J., Ma, C., & Wang, X. (2016). Asymmetric searchable encryption from inner product encryption. In *International Conference on P2P, Parallel, Grid, Cloud and Internet Computing* (pp. 123–132).
16. Zhang, B., & Zhang, F. (2011). An efficient public key encryption with conjunctive-subset keywords search. *Journal of Network and Computer Applications, 34*(1), 262–267.

Chapter 14
A Strong Single Sign-on User Authentication Scheme Using Mobile Token Without Verifier Table for Cloud Based Services

Sumitra Binu, Mohammed Misbahuddin, and Pethuru Raj

14.1 Introduction

Cloud computing is "gracefully losing control while maintaining accountability even if the operational responsibility falls upon one or more third parties" [1]. Cloud computing is an evolving computing model that concentrates on delivering computing resources over the Internet, in a shared manner that allows on-demand scalability, self-service, and typically a pay-for-usage pricing model. Cloud computing offers individuals and companies' affordable storage, professional maintenance, and adjustable space without much investment in new infrastructure, training, or software licensing. The elasticity and scalability of resources, combined with "pay-as-you-go" resource usage, have heralded the rise of cloud computing. In a survey that was conducted by Rackspace, and in 2016, it was reported that 95% of the organizations are using cloud services, and of that 89% have shifted their operations on to public cloud [2]. Rackspace reports that this pay-as-you-go service saves around 58% of cost [3].

By 2016 more than 50% of global 1000 companies are projected to store their sensitive data in public clouds [4]. Anticipating this switch over, many large technology companies such as Amazon and Google have built huge data centers to offer cloud computing services with self-service interface so that cloud users can use on-demand resources with location independence.

S. Binu (✉)
Christ University, Bangalore, India
e-mail: Sumitra.binu@christuniversity.in

M. Misbahuddin
C-DAC, Electronic City, Bangalore, India

P. Raj
IBM India Pvt. Ltd., Bangalore, India

© Springer International Publishing AG 2018
K. Daimi (ed.), *Computer and Network Security Essentials*,
DOI 10.1007/978-3-319-58424-9_14

Though the characteristics and attractions of cloud are well understood from a business perspective, the security state of cloud computing still lacks clarity. Although there is a growth in cloud computing which implies that many businesses have adopted this computing model, there are several security issues which are preventing organizations from migrating their sensitive data onto cloud.

Though the self-service interface provided by cloud enables users to access the resources without human interaction with the service provider, the indirect control of the physical infrastructure introduces many vulnerabilities unknown in a non-cloud environment. The cloud model for delivering computing and processing power has raised many security concerns such as data security, identity and access management, key management, and virtual machine security which could limit the use of cloud computing. Armbrust et al. in their report [5] have mentioned hearing multiple times "My sensitive corporate data will never be in the cloud." Security, interoperability, and portability have been identified by NIST, as the major barriers for a whole-hearted adoption of cloud [6]. In addition, a survey was conducted in 2009 [7] by International Data Corporation (IDC), a market research and analysis firm, to identify the most disturbing cloud issues. The results of the survey which was attended by masses belonging to varied levels from IT executives to top CEOs highlight that security is the major concern as it ranked first with 87.5% of the votes, 12.9 % more than the study of the previous year [8].

14.1.1 Cryptography and Security

The introduction of computers as data processing equipment has contributed a lot towards efficient storage and processing of data. This automated processing of data also raised the need to have automated tools for protecting the sensitive information stored in the computers. Another major change that raised serious concerns about security was the introduction of communication networks and distributed systems which demands security of data in transit as well.

Information security plays a crucial role in safe guarding the resources and services that are available online. As per the NIST standard, Confidentiality, Integrity, Availability, Authenticity, Accountability and Non-Repudiation, should be the security objectives of any Information systems [9]. To achieve these security objectives, cryptographic mechanisms are employed. Application of cryptographic mechanisms such as encryption and hashing enables us to store sensitive information or transmit it across insecure networks without worrying about breach of confidentiality, authenticity, integrity, and so on. There are three classes of approved cryptographic algorithms based on the number of keys that are used in conjunction with the algorithm. This includes hash functions, symmetric key algorithms, and public key algorithms [10].

14.2 Authentication

Security of any network depends on the attainment of two simple objectives: (1) ensuring that unauthorized persons are denied access to resources and (2) ascertaining that authorized users can access the resources they need. These objectives can be attained in many ways, and one among them is to assign access permissions to resources which specify the category of users who can or cannot access a resource. Nevertheless, permission to access a resource can be granted only after verifying the claimed identity of the individual attempting to access a resource, and that's where authentication has a role to play. Authentication is the act or process of verifying the identity [11], claimed by an individual or an object prior to disclosing any sensitive information. Authentication process in turn allows authorized users and services to access sensitive resources in a secure manner, while denying access to unauthorized users, thereby supporting confidentiality and access control. Consequently, in most applications where security has top priority, it is necessary to attain authenticity which is an indispensable element of a typical security model.

Furthermore, in [7, 12] authors have pointed out that identity and access management issues in cloud require immediate attention of cloud service providers (CSPs) to accelerate the adoption of cloud. A survey by Fujitsu Research Institute [13] reveals that 88% of prospective customers are worried about unauthorized access to their data in the cloud.

To provide secure access to sensitive data, CSPs need to ensure that only valid users are accessing the resources and services hosted in the cloud and to make this possible, they need to adopt strong user authentication mechanisms.

14.2.1 User Authentication Mechanisms

Authentication of remote users is a matter of importance, as they are more susceptible to security risks when compared to onsite users. Single Sign-on (SSO) authentication feature allows users to use a single credential (password, smart card, biometrics, etc.) and prove their authenticity to multiple servers in a network without repeatedly submitting the credentials. This relieves the user of the pain of remembering multiple passwords as well as the going through the authentication process multiple times to access multiple resources.

14.2.1.1 Authentication Types

There are different means for providing the authentication credentials to the verification system. The simplest and the commonly used remote user authentication mechanism is password authentication, though it is not the most secure. In general, authentication can be classified into single factor, two factor, and multi factor based on the type and number of factors used [15]. An authentication factor is an independent category of credential that uniquely identifies an entity, and it is a secret that is known to, possessed by, or inherent to the owner. For instance, password is a secret known only to the owner and hence can be considered as an authentication factor, whereas user-ID is public information and hence fails to qualify as an authentication factor. When network resources include highly sensitive data, authentication mechanisms that offer more protection such as two-factor and multi-factor authentication mechanisms based on smart cards, crypto-tokens, biometric authentication, etc. are preferred.

Most widely used authentication factor includes knowledge factor or something the user knows such as password and PIN, possession factor or something the user has (USB token, smart card, crypto-token, mobile-token), inherence factor or something the user is (physiological or behavioral biometrics), and location factor or somewhere the user is (geographic location at the time of login). Combination of any two of these factors offers higher level of security strength than single-factor authentication and is referred to as two-factor authentication.

14.2.2 Authentication Challenges in Cloud

Wide array of cloud services and ever growing number of cloud service providers are beneficial to users from the perspective of scalability, ease of maintenance, elasticity, etc. Permission to access the secure resources hosted by cloud service providers is granted to the user only after successful user authentication which is the process of verifying the identity claimed by an individual (Meyer [10]). However, users accessing cloud services/resources from multiple cloud service providers need to address many authentication challenges (Granneman [14]; Misbahuddin [15]).

- Customers are requested by the cloud service providers to store their account related information in the cloud. This information can be accessed by the service providers, and customers are worried about unauthorized access to their stored information and service providers misusing the information.
- Majority of the cloud service providers use weak authentication mechanisms to authenticate users. Password based authentication is the most commonly used mechanism as it is simple, cheap, and easy to deploy. However, human beings tend to choose simple and easy to remember passwords often leading to data breaches.
- Password based authentication requires the cloud service providers to store the password information of the user. Owing to security reasons, this is stored either in the hashed form or salted hashed form in the server's database. However, if an attacker manages to gain access to the server's database, then he can retrieve this stored information and launch an offline dictionary attack.
- A user who uses different cloud services will need to store his/her password information or authentication credential with every service provider. Many a

times, a user uses the same password for different services and if a hacker manages to get hold of the password of an account of a legitimate user, then he can use the same password to login into another account of the victim. This redundancy of information is a concern to both the customers and the service providers.

- When a user maintains different accounts to access different cloud services, then he will have to undergo multiple authentication processes. While authenticating to each service provider, he needs to exchange his authentication credentials. This redundant exchange of information can be exploited by an attacker to create a security loop hole in the system.

- The SLAs of cloud service providers contain information pertaining to the mechanisms followed by the service providers to ensure the security of the information stored in the cloud. However, from a user's perspective, verifying whether the rules are being enforced properly or not is a very laborious process. This makes it difficult for the user to monitor the security of stored information.

14.2.3   Authentication Attacks in Cloud

This section discusses various attacks that are launched by exploiting the loopholes in the authentication process.

14.2.3.1   Guessing Attack

Easy to remember passwords chosen by users make them susceptible to guessing attack [16]. Based on some password related information obtained by the adversary, he guesses a password and tries to verify the correctness by logging in multiple times until he succeeds. Probability of guessing correctly is high in an offline scenario, as there is no restriction on the number of login attempts. However, the system places a restriction on the number of attempts in an online scenario, which makes guessing difficult.

14.2.3.2   Malicious Insider Attack

Insider attack is performed by a system administrator or an employee of the service provider who has access to the secret information of the user. From a convenience perspective, users tend to use one password to access multiple applications such as e-mail and online banking. If an insider with privileges of an administrator can access the password information of a registered user maintained by the authentication server, they can use it to impersonate the user or leak out the information to others.

14.2.3.3 Replay Attack

A replay attack involves sniffing of an authentication message exchanged between two honest communication partners and resending the same after some point in time [16, 17]. This replayed message contains an authentication token that was previously exchanged, and hence the solution to handle replay attack is to ensure that there is some content that change every time, the message is transmitted.

14.2.3.4 Stolen Verifier Attack

Many authentication servers store user passwords in a hashed form or a hash of the salt value combined with the user password in the database [15]. To launch a stolen verifier attack, the adversary steals this verification table and attempts to guess the password using an offline guessing attack. The attacker compares entries in the verification table with the message digest of entries in a dictionary of passwords or he compares the entries with a rainbow table. The adversary will arrive at the right password when there is a match.

14.2.3.5 Impersonating Attack

The attacker pretends to be an authorized entity or a valid server and lures a valid user to share his/her credentials which in turn is used to impersonate the user. A verifier impersonation attack involves an adversary who assumes to be a valid verifier and lures the client to reveal the authentication keys or information [18], which can be used by the adversary to falsely authenticate to the verifier. A phishing attack, which also comes under this category, is launched by making the users to believe that a valid server is communicating with them, by displaying a web page that resembles a valid server page [19].

14.2.3.6 Denial-of-Service Attack

There are three different ways in which a denial-of-service attack can be launched [15]. (1) Assume that an administrator who has access to the user database stored in the server modifies the secret information used to authenticate the user. In that case, a legitimate user, who attempts to login with his valid credentials, will be denied from accessing the resources he is authorized to access. (2) Another possible scenario where a denial-of-service attack can happen is when the crypto-token is possessed by the attacker. In this case, the attacker attempts to login to the account of the owner of the token using a random password and he will be denied access. Now if the attacker modifies the current password with his own password, then the legitimate user will be denied access in his future login attempts. (3) Generating login requests without password verification at the client side can also lead to denial-of-service attack, since the resources of the server will ultimately get blocked in verifying the received login requests. To resist denial-of-service attack in the first case, the authentication protocols without verification table at the server were designed. In the second and third case, denial-of-service can be resisted by verifying the password before permitting password update and before generating a login request.

14.2.3.7 Crypto-Token Lost Attack

If an adversary gains possession of the token of a legitimate user, then he can attempt various nefarious activities such as offline password guessing, impersonate the user, and gain unauthorized access to resources by logging into his account, modify the stored contents of the token [15], etc. The authentication protocol and the contents of the token should be designed such that it is infeasible for the attacker to derive secret information from the token or launch the discussed attacks.

Cost-effectiveness, ease of implementation, and simplicity make password based authentication (PBA) the most preferred authentication mechanism. For password based authentication, when attempting to logon to access a resource, the user is required to submit the user name/user-ID and password corresponding to an account, as the authentication factor. The authentication server maintains a database of user accounts holding the credentials of authorized users, and the submitted password is verified against the entries in the database.

In the case of password based authentication systems the security of the entire system depends entirely on a secret password. However, human beings tend to choose simple and easy to remember and hence easy to guess passwords making them vulnerable to several attacks. To resist passwords from being guessed, users are recommended to secure their accounts with high entropy passwords [20].

It is typically difficult for a human being to guess a password without having some information about the owner of the password and if the value of the password is something representative of the user. However, there are software programs called "password crackers" which can be used by human beings to launch a "Brute force" attack on password systems [21]. This means that an application program tries out each word in a pre-computed dictionary of terms until the correct combination of characters breaks the password [22]. To prevent such attacks, it is advisable to choose strong passwords having alphabets, numbers, and symbols, and passwords should have high entropy with a minimum length of eight characters.

Nevertheless, PBA requires the authentication system to store passwords and user name in a database against which the password information submitted by the user can be verified. If the server is not provided with strong security, the stored passwords can be retrieved/modified by an adversary [23] who manages to gain unauthorized access to the database. To address this issue, many authentication systems rather than storing plain text password store the password in its hashed form [24]. This involves using a hash function, which takes the plain text password as input and produces a unique non-reversible digest as output [17]. If the database is breached the attacker will be able to read only the hash of the passwords and not the original password. However, storing password hashes are not an ultimate solution to ensure the security of stored passwords, since the attackers can use a rainbow table which is a pre-computed table for cracking password hashes [25].

Passwords are also prone to shoulder surfing attack [19] and sniffing attack [26], which mostly happens when you attempt to log into various websites using passwords while in a public place such as Internet cafes, CCD, libraries, and air terminals.

Over the years, many enhanced two-factor authentication schemes have been proposed to overcome the limitations of password based schemes. Taking into consideration the storage and computational capabilities of smart cards, several password based authentication schemes with smart cards have been proposed [27, 28]. Most of the proposed schemes assume that the smart card is tamper resistant and recent research results have revealed that the secret information stored in the smart card could be extracted by some means such as monitoring the power consumption [29] and analyzing the leaked information [30]. Therefore, such schemes based on the tamper resistance assumption of smart cards are prone to attacks such as impersonation attack and password guessing attack once an adversary has obtained the secret information stored in a smart card. Biometric authentication mechanisms are also quite popular and biometric identifiers are difficult to forge. Biometrics is unique to the individual and non-transferable, but biometric authentication mechanisms have the drawback of being costly as they need additional hardware to read and process the stored data. Hence, there is an immediate requirement to design strong authentication mechanisms that maintains a good level of usability.

The rest of the chapter is organized as follows. Section 14.3 reviews the related work. Section 14.4 discusses the authentication architecture and protocol, and Sect. 14.5 analyzes the security of the proposed scheme. Section 14.6 includes the efficiency analysis of the proposed scheme, Sect. 14.7 discusses the formal analysis of the protocol using Scyther tool and Sect. 14.8 concludes the work done.

14.3   Related Work

This section discusses a few user authentication schemes proposed for cloud environment.

Hao et al. [30] in 2011 proposed a time-bound ticket based mutual authentication scheme for cloud computing. In their scheme the authors follow an authentication model like that of Kerberos where in a user, to access the services from an application server should first authenticate himself to and get tickets from a ticket granting server. Authors claimed that their scheme provides mutual authentication and is secure against lost smart card attack, offline password guessing attack, lost ticket attack, masquerade attack, and replay attack.

Jaidhar [31] in his work mentions that among the security issues of cloud computing, authentication is considered as one among the most important issues. He proved that the scheme [30] is susceptible to denial-of-service (DoS) attack, and the password change phase requires the involvement of the server. Author proposed an improved mutual authentication scheme which inherited the security features of Hao et al.'s scheme and was resistant to DoS attack.

Choudhary et al. proposed user authentication framework for cloud environment [32] that provides two-step verification using password, smart card, and out of band (OOB) authentication token. The scheme which uses light-weight XOR and hash operations applies a two-step verification to authenticate a user, and verification is done using password, smart card, and out-of-band authentication in which a one-time key is send as SMS via HTTP/SMS gateway. Authors claimed that their scheme provides identity management, mutual authentication, session key agreement, etc. and is resistant to various attacks.

Rui Jiang [33] in 2013 proved that their scheme is prone to masquerade user attack, the OOB attack, and has a flaw in the password change phase. They proposed a modified scheme that addresses the security issues of [32], but uses time stamps which can lead to time synchronization problems in a distributed cloud environment especially when client and server are from two different time zones. Also, the protocol requires the server to store a variant of the user password, which can result in a stolen verifier attack.

Sanjeet et al. [34] proposed a user authentication scheme which uses symmetric keys to exchange communication between user and server in which case key distribution may be a challenge. The protocol uses a one-time token which is sent to the registered users e-mail ID. In this scenario, the authentication process will require logging into two accounts which may cause user inconvenience.

Though there are various two-factor authentication schemes proposed using hash functions for cloud environment, every scheme is found to have some limitation in terms of desirable security features. Moreover, none of the schemes provide perfect security and is thus susceptible to various attacks. The proposed two-factor authentication schemes using smart cards for cloud environment require the user to directly register at the service providing servers who will then issue the smart card which serves as an authentication factor. Hence in a scenario where the user needs to access multiple cloud services, the user should undergo multiple registration processes, maintain multiple accounts, and remember multiple identities. These limitations of the currently available two-factor authentication schemes, viz. susceptibility to attacks, need for multiple registration, maintaining multiple accounts, and carrying different authentication tokens (smart cards/crypto-tokens) when accessing the services of multiple service providers, etc. are being addressed by the proposed research work. The research proposes a hash function based, two-factor authentication scheme using mobile tokens. The proposed scheme uses nonce values to resist replay attacks and does not require the server to maintain a verifier table (Table 14.1).

Table 14.1 Comparison of proposed scheme with similar authentication schemes

Scheme	Authentication factors	Provides Single Sign-on	Requires verification table	Password change requires server intervention
Hao et al. [30]	Password, smart card	No	No	No
Jaidhar [31]	Password, smart card	No	Yes	Yes
Choudhary et al. [32]	Password, smart card, OTP	No	No	No
Rui Jiang [33]	Password, smart card	No	Yes	Yes
Proposed scheme	Password, mobile token	Yes	No	No

14.4 Proposed Scheme

Direct authentication by cloud service providers is not always a viable solution in scenarios where users need to access different services simultaneously, in the same session without requiring to login for every service. Services provided via a web portal or services provided by service providers that are functioning in a collaborative environment can be accessed simultaneously by a user. For example, logged-in users of research analyst site Gartner are allowed access to research produced by research analyst site Forrester. Similarly, users may access e-mail service by Gmail, CRM services by Sales Force, and storage services provided by Dropbox, simultaneously. In a scenario, where users are directly authenticated by individual service providers, users must go through multiple authentication processes to access these services. This requires redundant storage of information, repetitive exchange of credentials, and repeated execution of authentication protocol.

14.4.1 Brokered Authentication Scheme

Brokered authentication effectively solves the problem of direct authentication by having an authentication broker who does the authentication on behalf of the rest of the service providers. By doing so, the service providers are relieved from the task of identifying and authenticating users, and the users are provided with Single Sign-on functionality, where in they are required to authenticate only once during a session. However, in many cases users need to use services from different domains. These services belonging to different providers need to have interoperability to accept the tokens issued by the central authentication broker/identity provider (IdP).

For brokered authentication, the proposed protocols require a security token service (STS) whose functionality is executed by an identity provider. The authentication server of the IdP authenticates the user by executing the two-factor authentication protocol and generates a SAML token, which is signed by the IdP and sent to the service provider (relying party). The IdP also provides a Single Sign-on (SSO) functionality using Security Assertion Markup Language (SAML) tokens. The SAML protocol is an open standard for exchanging security information between hosted SAML enabled applications [35]. SAML enables a user who has established and verified his identity in one domain to access services hosted in another domain.

14.4.1.1   Proposed Brokered Authentication Architecture

The proposed architecture for a cloud environment includes four participants, viz. a registration server (RS), an authentication server (AS), service providers (SPs), and users. The RS and AS are in the same trusted domain and together they provide the functionality of the identity provider (IdP). The users and SPs comprising the proposed architecture need to register with the registration server of the IdP. When a SP registers with the IdP, he submits his identity information and the details of the services provided. The CSPs and IdP work in a trust based environment.

In this two-factor authentication scheme, user's password and a registered crypto-token/mobile-token serve as the authentication factors. When a user wants to get the service of a CSP, he is re-directed to the IdP by the SP if he is not a registered user. In such a scenario, the user needs to do a single registration at IdP as illustrated in Fig. 14.1, by providing the user-ID and password. On successful registration, IdP provides the user with a secret file containing the security parameters, which is downloaded into user's smart phone to serve as a mobile token. The server IDs of all the participating service providing servers and the details of their services are also communicated to the user via an e-mail. The login and authentication phase of the proposed scheme runs on the IdP and the service providers redirect the users requesting their services to the IdP for authentication. A user who wants to access the services of a SP tries to login to the provider's web page by submitting the login request. The user is re-directed to IdP and authentication module within the IdP executes the proposed protocol. The second authentication factor of the proposed protocols contains only a few hashed values generated from user's ID, password, and the secret key of the server. The proposed protocols do not require the support of PKI.

The protocols do not require the server to maintain a password verification table. The registration and authentication process flow is illustrated in Fig. 14.1.

In a Single Sign-on platform, if users are authenticated at one service, they do not have to re-enter their credentials to log on to access another service [36]. Most of the existing Single Sign-on (SSO) solutions typically rely on browser cookies for maintaining state and exchanging identity information. Cookie poisoning is an

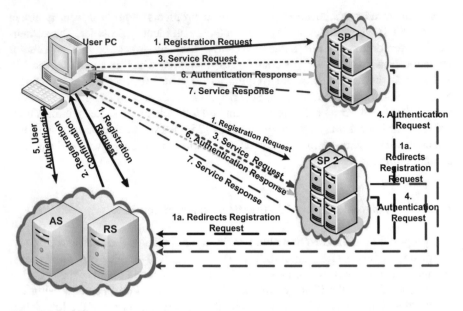

Fig. 14.1 Registration and authentication process

authentication attack, which involves the modification of cookies of an authorized user to gain unauthorized access to resources. Hence cookies are not a reliable mechanism for sending authentication information. Browser cookies are not transferrable across DNS domains, and hence the browser cookies, created from one security domain, for security reasons (same origin policy), can't be read from another one [37]. Therefore, to solve cross domain SSO, proprietary mechanisms to pass the authentication data between security domains have been used. This solution which works fine for a single enterprise becomes impractical when different organizations using different mechanisms collaborate. The proposed brokered authentication protocol uses SAML to exchange authentication information, and the information is contained in an encrypted SAML token. To maintain information about the sessions of authenticated users, SAML protocol uses session cookies which contains only information such as session ID of the user and the domain information of the IdP.

The kind of authentication broker required by the discussed brokered authentication scheme is a security token service (STS) that issues SAML tokens. In the proposed brokered authentication scheme, IdP is representative of the STS who does the role of the authentication authority and provides SSO functionality using the SAML tokens. Here both IdP who authenticates the user and issues the SAML assertion and the service providers who accepts the SAML assertions from IdP should be enabled with SAML. The IdP carries out the two-factor authentication protocol exchange with the user who is re-directed to the IdP by the service provider for the authentication process. If the authentication is successful, IdP generates a signed SAML token and the user is redirected to the service provider. The service

Table 14.2 Notations used in the protocol

IdP, SP, Sj	Identity provider, service provider in the cloud, jth SP
U_i, ID_i, PW_i	ith User, unique identification of U_i, password of U_i
S	Secret key of IdP
R	Random number generated by crypto-token
$h(.), \oplus, \|$	One-way hash function, XOR operation, concatenation operation
N_i, N_j	Nonce values of U_i and RS, respectively

provider verifies the SAML token and ascertains the origin and the content of the response, before providing the requested service.

14.4.1.2 A Strong Single Sign-on User Authentication Protocol Using Mobile Token for Cloud Based Services

Phases of the Proposed Protocol

The proposed protocol consists of three phases, viz. registration, login, and authentication, and the password change phase. The notations used are listed in Table 14.2.

Registration Phase

During the registration process, user submits his credentials to RS of IdP. RS generates a set of security parameters using the submitted credentials and his key values. RS stores the security parameters within a secret file which is downloaded and stored in a secure location within the user's mobile phone. The secret file is encrypted using the password (PBE) of the user which ensures that only a valid user will be able to store the token into his mobile phone and use the same to avail secure access to the cloud services.

The registration process illustrated in Fig. 14.2 can be explained as follows:

R1: To proceed with the registration process, U_i needs to download the mobile app into his smart phone. IdP prompts U_i to submit her identity ID_i and PW_i. The user submits his $h(ID_i)$ and $h(PW_i)$.

R2: IdP checks whether $h(ID_i)$ already exists in its user table. If so U_i is prompted to select a new ID_i.

R3: IdP creates a file containing the authentication parameters K_i, M_i, J_i, $h(.)$ and the file is encrypted using password of U_i and a salt value. The values of K_i, M_i, J_i are generated by performing hash and XOR operations on ID_i, PW_i, S.
$V_i = h(h(ID_i) \| h(S))$, $K_i = V_i \oplus h(h(ID_i) \| h(PW_i))$, $M_i = h(h(ID_i) \oplus h(S)) \oplus J_i$, $J_i = h(V_i)$

R4: IdP generates a QR code embedding service provider Name, Salt and the URL for downloading the secret file.

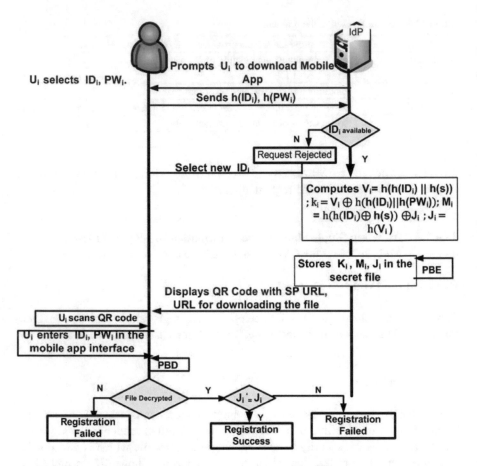

Fig. 14.2 Registration phase of brokered authentication using mobile token

R5: The QR code will be displayed on the service provider's page and the user will be prompted to download the mobile application.

R6: The mobile app invokes the scanning application, and the user can scan the code. The user will get user name, salt, the service provider URL, and the link to download the secret file.

R7: The user will be prompted to enter his ID_i, PW_i. The app attempts to decrypt the file using password given as input by the user and the salt value attached to the file. If the decryption is successful, the secret file contents will be accessed.

R8: When the user touches the register button in the mobile app, mobile app calculates $V_i' = K_i \oplus h(h(ID_i') \parallel h(PW_i'))$ and $J_i' = h(V_i')$. J_i' is compared with J_i stored in the mobile token and if equal, the registration process is considered successful and the app sends $E_i = J_i \oplus h(ID_i)$. The server computes $E_i' = h(h(h(ID_i) \parallel h(S))) \oplus h(ID_i)$ and compares with received E_i. If equal the account is created, and the user will get the "Registration Successful" message.

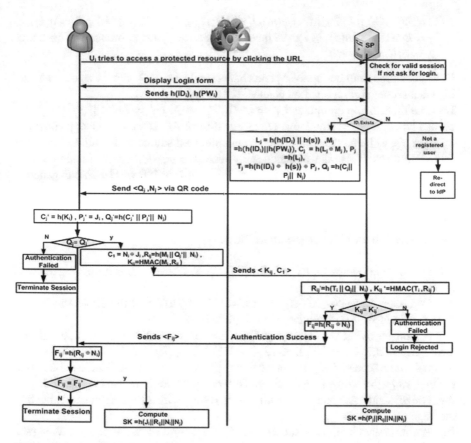

Fig. 14.3 Login phase

R9: The file will be stored in a safe location within the user's phone in the form of a mobile token. IdP stores the user-ID, i.e., $h(ID_i)$ and other profile information in its user table.

Login and Authentication Phase

In this phase, the user attempts to access a protected resource of a service provider (SP). If there is no existing session between the browser and the SP, then SP generates a login session and authenticates the user by executing the authentication phase, as illustrated in Fig. 14.3. The user uses his password and the parameters stored within the mobile token, to authenticate himself to the SP. The procedure can be explained as follows:

L1: Authentication server (AS) displays the login page and prompts the user to enter user's identity (ID_i) and password (PW_i). The values are sent over the communication channel as $h(ID_i)$ and $h(PW_i)$). AS calculates:

$L_j = h(h(\text{ID}_i) \| h(S))$, $M_j = h(h(\text{ID}_i) \| h(\text{PW}_i))$, $C_j = h(L_j \oplus M_j)$, $P_j = h(L_j)$, $T_j = h(h(\text{ID}_i) \oplus h(S)) \oplus P_j$. AS generates a nonce N_j and computes the challenge $Q_j = h(C_j \| P_j \| N_j)$

L2: The random number nonce N_j and challenge Q_j, i.e., $<Q_j, N_j>$ is send to the user U_i, via a secure communication channel (QR code)

L3: The mobile app computes $C_j' = h(K_i)$, $P_j' = J_i$, $Q_j' = h(C_j' \| P_j' \| N_j)$ and checks whether $Q_j' = $ Challenge Q_j, received from AS. If so, mobile app considers the message as being received from an authenticated source and continues with the following steps. This step is included to avoid the possibility of phishing attack, since only the server which holds the shared key $h(S)$ of IdP will be able to generate this message.

Authentication and Key Agreement Phase

A1: Mobile app computes $R_{ij} = h(M_i \| Q_j' \| N_i)$, where N_i is a nonce generated by U_i.

A2: Computes $C_1 = N_i \oplus J_i$, $K_{ij} = \text{HMAC}(M_i, R_{ij})$ to AS of the SP. U_i sends $<K_{ij}, C_1>$ to AS via Wi-Fi or cellular network (GSM).

A3: AS on receiving the message $<K_{ij}, C_1>$ computes $N_i' = C_1 \oplus h(h(h(\text{ID}_i) \| h(S)))$, $R_{ij}' = h(T_j \| Q_j \| N_i')$, and $K_{ij}' = \text{HMAC}(T_j, R_{ij}')$. AS checks whether K_{ij}' is equal to the received K_{ij}. If equal SP considers the user as authenticated and that the integrity of message is maintained. Otherwise the login request is rejected.

A4: The SP sends a response $F_{ij} = h(R_{ij} \oplus N_i)$ along with a successful authentication message.

A5: If the authentication is successful, then SP notifies the user's browser of a successful login. The user on receiving F_{ij} computes $F_{ij}' = h(R_{ij} \oplus N_i)$ and verifies the authenticity of the server.

A6: Both U_i and SP compute the session key as $\text{SK}_{us} = h(J_i \| R_{ij} \| N_i \| N_j)$ and $\text{SK}_{su} = h(P_j \| R_{ij}' \| N_i \| N_j)$, respectively.

Password Change Phase

The password change phase, as shown in Fig. 14.4, is invoked when the user wishes to change his password without the intervention of the IdP or the SP and is carried out as follows:

P1: User enters his identity (ID_a) and password (PW_a) and executes the "Password Change" request. The mobile app computes $V_i' = K_i \oplus h(h(\text{ID}_i) \| h(\text{PW}_i))$ and checks if $h(V_i')$ it is equal to stored J_i. If equal, the mobile app prompts the user to enter the new password "PW_{inew}." Otherwise the password change request is rejected.

P2: The app calculates $K_{\text{inew}} = K_i \oplus h(h(\text{ID}_i) \| h(\text{PW}_i)) \oplus h(h(\text{ID}_{\text{inew}}) \| h(\text{PW}_{\text{inew}}))$. Then the app computes $M_{i\,\text{new}} = h(\text{PW}_{\text{inew}}) \oplus M_i \oplus h(\text{PW}_i)$. The app replaces the existing values in the file with the new values.

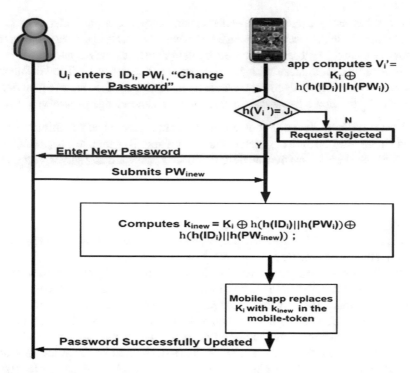

Fig. 14.4 Password change phase

14.5   Security Analysis of Proposed Protocol

Security analysis is carried out to analyze the resistance of the protocol to various attacks. The proposed protocol is secure against the following attacks:

A. *Security against Dictionary Attack*: The aim of this attack is to find out the password of the user. Assume that the adversary manages to get the secret file containing $<K_i, M_i, J_i, h(.)>$. If he tries to get the password from K_i, then he should know V_i, which is not stored in the secret file. Otherwise, to obtain V_i, he should be knowing the secret key of the server. In the case of android phones, the secret file is stored in a private location accessible only to the mobile app within the phones memory. Hence even the owner of the file will not be able to access its contents which rules out the possibility of a valid user getting $h(S)$ using his own password and then trying to guess another user's password by stealing his secret file.

B. *Security against Replay Attack*: The attack launched by replaying the same message received from the sender or the receiver at a later point of time is known as replay attack. Many authentication schemes use time stamps to resist replay attacks which may lead to time synchronization problems if the clocks at the sender and the receiver are not synchronized. The proposed authentication scheme uses nonce values N_i and N_j generated by the user and server, respectively. These nonce values which are generated independently are unique to a particular session and are included in the messages exchanged between the user and the server. Hence an adversary cannot get access to a system by using previous messages.

Assume that the attacker replays a message $<Q_j^*, N_j^*>$ to U_i. Then the attacker can receive an acknowledgement message from U_i as $<K_{ij}, C_1>$. Now to generate the response F_{ij} he should know R_{ij} which requires the knowledge of password and S.

C. *Server Spoofing Attack*: For an adversary to masquerade as a legitimate service provider, he must be able to generate the session key. To generate $SK_{su} = h(P_j || R_{ij} || N_i || N_j)$ he should have the knowledge of user's password and server's secret "S."

D. *Insider and Stolen Verifier Attack*: Insider attack is launched by an administrator who deliberately leaks secret information resulting in security flaws of the authentication scheme. In the proposed scheme both during registration and login phase, the $h(PW_i)$ is send to the server. Deriving the password from $h(PW_i)$ within a specific time interval is very difficult. The proposed scheme does not maintain any verifier table and hence it is secure against stolen verifier attack.

E. *Two-Factor Security*: In a scenario where both the user's mobile token and his password are stolen, then there is no way to prevent the attacker from masquerading as the user. Hence the security of the proposed two-factor authentication scheme can be guaranteed when either the mobile token or the password is stolen but not both. This security property is referred to as two-factor security. In the discussed scheme the secret parameters $<K_i, M_i, J_i, h(.)>$ of the mobile token are difficult to be derived if the attacker has obtained the user's password alone and not the mobile token. Now if the attacker also intercepts the challenge $Q_j = h(C_j || P_j || N_j)$, it is a laborious process to extract M_j (which contains PW_i) from C_j and P_j due to the irreversible property of one-way hash functions.

Again if the attacker intercepts the response $<K_{ij}, C_1>$ from the user, it is infeasible to derive $h(S)$ or $h(PW_i)$ from HMAC (M_i, R_{ij}) as they are based on symmetric encryption and hash functions.

On the other hand, if the attacker manages to get the mobile token and extracts the values $<K_i, M_i, J_i, h(.)>$ using power analysis attacks suggested by Messergers et al., he still cannot obtain PW_i directly from any of these stored values.

F. *Known-Key Security*: The known key security means that even if the session key of any of the previous sessions is compromised, the attacker should not be able to derive the session key of any of the future sessions. In the proposed scheme, the session key is calculated using p_j and R_{ij} as $SK_{su} = h(P_j || R_{ij} || N_i || N_j)$ which require the knowledge of password and server's secret key, which is not known to the adversary. The irreversible property of hash functions ensure that P_j and R_{ij} cannot be derived from the past session keys, which makes it difficult for the attacker to derive the future keys. Also the session key calculation involves nonce values generated randomly and independently by both the user and the server. Hence even the valid user and the server will not able to predict the future session keys.

G. *Forward Key Secrecy*: The forward key secrecy property requires that a compromise of the master key of the system should not help the adversary to calculate the previously established session keys. In the proposed scheme, even if the master key of the IdP is compromised, the adversary cannot compute any of the previous session keys without knowing the password PW_i of the user.

H. *Mutual Authentication*: When the user receives the challenge Q_j from the server, it is verified as $Q_j{}' = h(C_j{}' \,||\, P_j{}' \,||\, N_j)$, $C_j{}'$ and $|\,P_j{}'$ are calculated using parameters in the secret file. A response to this challenge is generated by using M_i, which is extracted from the mobile token and is not there in the challenge received from server. The server calculates $T_j = M_i$ using the user's password and its own secret key. A successful verification proves the authenticity of the user. Again the response send from the server, F_{ij} is verified by the user. Thus the proposed achieves the requirement of mutual authentication which is required in a multi-server environment.

14.6 Efficiency Analysis of Proposed Protocol

This section analyzes the efficiency of the proposed mobile token based scheme in terms of the computational and the communication cost. It is assumed that nonce values are 128 bits long and the output of hash function (SHA-2) is 256 bits long. Let T_h, T_x, and T_c denote the time complexity for hashing, XOR, and concatenation, respectively. In the protocol, the parameters stored in the secret file are K_i, M_i, J_i and the memory (E1) needed in the mobile is 768 (3*256) bits. Communication cost of authentication (E2) includes the capacity of transmitting parameters $(h(ID_i), h(PW_i), k_{ij}, N_i, Q_j, N_j, F_{ij})$ which makes E2 equal to 1536 (256 + 256 + 256 + 128 + 256 + 128 + 256) bits. The computation cost of user registration (E3) is the total time of all operations executed in this phase by the user and IdP and is equal to $12T_h + 7T_x$. The computation cost of the user (E4) and the server (E5) is the total time of all operations executed by the mobile app and the service provider during login and authentication. During authentication, the mobile app performs 7 hash functions, 1 XOR, and 6 concatenation making E4 equal to $7T_h + 1T_x + 6T_c$. Similarly, E5 is $10T_h + 4T_x + 8T_c$.

14.7 Formal Analysis Using Scyther Tool

A. *Scyther Tool*

Automatic tools are preferred in protocol analysis and among the various available tools, Scyther is used for the verification of the proposed protocol. Scyther was developed by Cas Cremers in 2007 for automatic verification of security protocols [38].

B. *Modeling Proposed Protocol in SPDL Language*

A security protocol specification describes the communication parties, the protocol events to be executed, the order of the events, and initial knowledge required for communication parties. The description of a protocol and the claims are written in Security Protocol Description Language (SPDL), and the proposed protocol can be written in SPDL as follows:

```
//Login & Authentication Phase of Brokered Authentication
   Protocol Using Mobile-Token
const  exp: Function;const hash: Function; hashfunction h;
   const XOR: Function;
const HMAC:Function; secret SK:Function;const Fresh:Function;

protocol brokeredauthMobileProtocol-login(I,R){
role I {
const IDi,PWi,S,mg1,mg2;var Nj:Nonce; var SK: SessionKey;
fresh Ni :Nonce;
send_1(I,R,h(IDi), h(PWi));
recv_2(R,I, h(h(XOR(h(h(IDi), h(S)), h(h(IDi), h(PWi))),
   h(h(h(IDi), h(S))), Nj)), Nj);//Qj
send_3(I,R, HMAC(XOR(h(XOR(h(IDi), h(S))), h(h(h(IDi), h(S)) ))),
   h(XOR(h(PWi), h(h(h(IDi), h(S)) )),
   h(h(XOR(h(h(IDi), h(S)), h(h(IDi), h(PWi)),
   h(h(h(IDi), h(S))), Nj)), Ni)), Ni); //Kij
recv_4(R,I, h(XOR(h(XOR(h(XOR(h(IDi), h(S))), h(h(h(IDi),
   h(S)) )), h(h(XOR(h(h(IDi), h(S)), h(h(IDi), h(PWi)),
   h(h(h(IDi), h(S))), Nj)), Ni), Ni)))); //Fij
macro Ji = h(h(IDi), h(S));
macro Mi= XOR(h(XOR(h(IDi), h(S))), h(h(h(IDi), h(S)) ));
macro SK = h(Ji,Mi,Ni,Nj);
secret SK:Function;
recv_6(R,I,{mg1}SK(R));
send_7(I,R,{mg2}SK(I));
claim_i1(I, Secret, h(IDi)); // IDi
claim_i2(I, Secret, h(PWi)); // PWi
claim_i3(I,Secret,h(h(XOR(h(h(IDi), h(S)), h(h(IDi), h(PWi))),
   h(h(h(IDi), h(S))), Nj))); //Qj'
claim_i4(I, Secret, Nj); // Nj
claim_i3(I,Secret,HMAC( (XOR(h(XOR(h(IDi), h(S))), h(h(h(IDi),
   h(S)) ))), h(XOR(h(PWi), h(h(h(IDi), h(S)))), h(h(XOR(h(h(IDi),
   h(S)), h(h(IDi), h(PWi)), h(h(h(IDi), h(S))), Nj)),
   Ni)))); //Kij
claim_i4(I,Secret,h(h(h(IDi), h(S)))); //Ji
claim_i5(I,Secret, h(XOR(h(XOR(h(XOR(h(IDi), h(S))), h(h(h(IDi),
   h(S)) )), h(h(XOR(h(h(IDi), h(S)), h(h(IDi), h(PWi)),
   h(h(h(IDi), h(S))), Nj)), Ni), Ni))));//Fij
claim_i6(I,Secret,h(S)); claim_i7(I,Secret,h(Ni));
   claim_i8(I,Niagree);claim_i9(I,Nisynch);
claim_i10(I, Alive);claim_i11(I,Weakagree);claim_113(I,Secret,SK);
   claim_i12(I,Empty,(Fresh,SK));
claim_i14(I,Commit,R,Ni,Nj);claim_i15(I,SKR,SK);
}
```

```
role R {
const IDi,PWi,S; const SK: Function;fresh Nj:Nonce;var Ni:Nonce;
   fresh SK:SessionKey;
const mg1,mg2;
macro Lj = h(h(IDi), h(S)); macro Tj = XOR(h(PWi), h(h(h(IDi),
   h(S)) ));macro SK = h(Lj,Tj,Ni,Nj);
secret SK:Function;
recv_1(I,R,h(IDi), h(PWi));
send_2(R,I, h(h(XOR(h(h(IDi), h(S)), h(h(IDi), h(PWi))),
   h(h(h(IDi), h(S))), Nj)), Nj);//Qj
recv_3(I,R, HMAC(XOR(h(XOR(h(IDi), h(S))), h(h(h(IDi), h(S)) ))),
   h(XOR(h(PWi), h(h(h(IDi), h(S)) )), h(h(XOR(h(h(IDi),
   h(S)), h(h(IDi), h(PWi)), h(h(h(IDi), h(S))),
   Nj)), Ni)), Ni); //Kij
send_4(R,I, h(XOR(h(XOR(h(XOR(h(IDi), h(S))), h(h(h(IDi),
   h(S)) )), h(h(XOR(h(h(IDi), h(S)), h(h(IDi), h(PWi)),
   h(h(h(IDi), h(S))), Nj)), Ni), Ni)))); //Fij
/*Testing the sending of messages encrypted using the generated
   session key*/
send_6(R,I,{mg1}SK(R));
recv_7(I,R,{mg2}SK(I));
claim_r1(R, Secret,h(IDi)); // IDi claim_r2(R, Secret,
   h(PWi)); // PWi
claim_r3(R,Secret,h(h(XOR(h(h(IDi), h(S)), h(h(IDi), h(PWi))),
   h(h(h(IDi), h(S))), Nj))); //Qj
claim_r4(R, Secret, Nj); // Nj
claim_r5(R,Secret,HMAC( (XOR(h(PWi), h(h(h(IDi), h(S)) ))),
   h(XOR(h(PWi), h(h(h(IDi), h(S)) )), h(h(XOR(h(h(IDi), h(S)),
   h(h(IDi), h(PWi)), h(h(h(IDi), h(S))), Nj)), Ni)))); //Kij'
claim_r6(R,Secret,h(h(h(IDi), h(S))))); //Pj
claim_r7(R,Secret, h(XOR(h(XOR(h(XOR(h(IDi), h(S))), h(h(h(IDi),
   h(S)) )), h(h(XOR(h(h(IDi), h(S)), h(h(IDi), h(PWi)),
   h(h(h(IDi), h(S))), Nj)), Ni), Ni))));//Fij
claim_r8(R,Secret,h(S)); claim_r9(R,Secret,h(Ni));
   claim_r10(R,Alive);
claim_r11(R,Niagree); claim_r12(R,Nisynch);claim_r11(R,Weakagree);
   claim_r13(R,Secret,SK);
claim_r12(R,Empty,(Fresh,SK));claim_r14(R,Running,I,Ni,Nj);
   claim_r15(R,SKR,SK);
}
}
   const Eve: Agent; untrusted Eve;compromised SK(Eve);
```

C. *Scyther Analysis Results and Interpretation*

To analyze the protocol, we assume the existence of an adversary in the communication network. The adversary's capabilities are as defined by Dolev–Yao Network threat model [20], and it is assumed that the network is completely or partially under the control of the adversary. The complete results of the analysis of the proposed protocol are shown in Fig. 14.5. The output of the verification process is described per the following Scyther attributes.

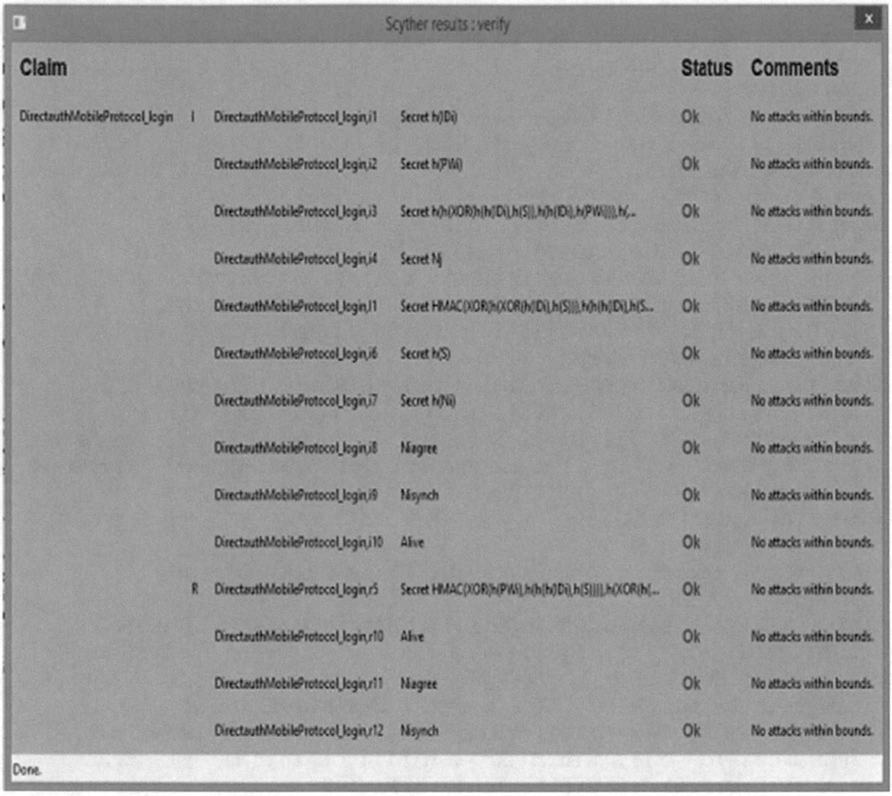

Claim				Status	Comments
		Scyther results : verify			
DirectauthMobileProtocol_login	I	DirectauthMobileProtocol_login_i1	Secret h)Di)	Ok	No attacks within bounds.
		DirectauthMobileProtocol_login_i2	Secret h(PWi)	Ok	No attacks within bounds.
		DirectauthMobileProtocol_login_i3	Secret h(h(XOR)h(h(Di),h(S)),h(h(Di),h(PWi))),h(...	Ok	No attacks within bounds.
		DirectauthMobileProtocol_login_i4	Secret Nj	Ok	No attacks within bounds.
		DirectauthMobileProtocol_login_I1	Secret HMAC(XOR)h(XOR(h(Di),h(S)),h(h(h(Di),h(S...	Ok	No attacks within bounds.
		DirectauthMobileProtocol_login_i6	Secret h(S)	Ok	No attacks within bounds.
		DirectauthMobileProtocol_login_i7	Secret h(Ni)	Ok	No attacks within bounds.
		DirectauthMobileProtocol_login_i8	Niagree	Ok	No attacks within bounds.
		DirectauthMobileProtocol_login_i9	Nisynch	Ok	No attacks within bounds.
		DirectauthMobileProtocol_login_i10	Alive	Ok	No attacks within bounds.
	R	DirectauthMobileProtocol_login_r5	Secret HMAC(XOR)h(PWi),h(h(h)Di),h(S)))),h(XOR(h(...	Ok	No attacks within bounds.
		DirectauthMobileProtocol_login_r10	Alive	Ok	No attacks within bounds.
		DirectauthMobileProtocol_login_r11	Niagree	Ok	No attacks within bounds.
		DirectauthMobileProtocol_login_r12	Nisynch	Ok	No attacks within bounds.

Done.

Fig. 14.5 Scyther analysis

Secrecy: The first claim is that the protocol ensures the confidentiality of the user's credentials. After analyzing, it is obvious from the results that the user's credentials are not revealed to the adversary when communicated over an untrusted network. As shown in Fig. 14.5, the authentication parameters $\{N_i, N_j, F_{ij}, S, PW_i, ID_i, Q_j\}$ retain the confidentiality during 10 protocol runs.

Non-Injective Agreement (NiAgree): Niagree claim made claims that sender and the receiver agree upon the values of variables exchanged and the analysis results justify the correctness of this claim.

Synchronization: Ni-Synch or non-injective synchronization property requires that the corresponding send and receive events (1) are executed by the runs indicated by the cast function, (2) happened in the correct order, and (3) have the same contents. The proposed protocol satisfies this claim as indicated by the result of Scyther analysis.

14.8 Conclusion

This chapter elaborated an authentication scheme that can be adopted by service providers who would prefer a strong two-factor authentication mechanism to authenticate users of its services. The proposed scheme can be adapted by those service providers who work in a collaborative environment and by service providers who offer their services via a web portal. To provide a seamless authentication experience to users who access different services during the same session, these service providers prefer Single Sign-on functionality. Hence in the proposed scheme for brokered authentication, users are authenticated by a third-party identity provider, who does the authentication of the users re-directed to it by the service providers. Security Assertion Markup Language (SAML) protocol which is used to exchange authentication related information about users between the identity provider and service providers is required to provide Single Sign-on functionality.

In the proposed brokered authentication protocol, the authentication broker stores only the profile information of users (e.g., user-ID, first name, last name, e-mail address, etc.). The authentication broker does not maintain the password information or the user keys. Also, the proposed brokered authentication protocols use two-factor authentication which requires the user to provide both his password and the parameters stored within the crypto-token/mobile-token to prove his identity to the authentication server.

The proposed authentication protocols do not require the server to maintain a verifier table. The chapter also includes security analysis of proposed protocols to validate the resistance to various common attacks on authentication protocols. In addition to security analysis, efficiency analysis is done to compare the computational efficiency of the proposed protocol with similar two-factor authentication schemes for cloud. Formal verification is done using Scyther which verifies the validity of security claims made with respect to the protocol.

References

1. CSA. (2009). *Security guidance for critical areas of focus in Cloud Computing V2.1*, Prepared by the Cloud Security Alliance.
2. Weins, K. (2017). *Cloud computing trends: State of the cloud survey* [Online], Available: http://www.rightscale.com/blog/cloud-industry-insights/cloud-computing-trends-2017-state-cloud-survey
3. Smith, D. M., Natis, Y. V., Petri, G., Bittman, T. J., Knipp, E., Malinverno, P., et al. (2011). *Predicts 2012: Cloud computing is becoming a reality* (Technical report, as G00226103). Gartner.
4. Armbrust, M., Fox, A., Griffith, R., Joseph, A. D., Katz, R., Konwinski, A., et al. (2009). *Above the clouds: A Berkeley view of cloud computing* (Technical report UCB/EECS-2009-28). Electrical Engineering and Computer Sciences, University of California.
5. NIST. (2012). *NIST cloud computing program* [Online], Available: http://www.nist.gov/itl/cloud/

 6. Gens, F. (2009). *New IDC IT cloud services survey: Top benefits and challenges, IDC Exchange* [Online]. Available: http://blogs.idc.com/ie/?p=730
 7. Gens, F. (2008). *IT cloud services user survey, pt.2: Top benefits and challenges, IDC* [Online]. Available: http://blogs.idc.com/ie/?p=210
 8. Mell, P., & Grance, T. (2011). *The NIST definition of cloud computing* (NIST special publication 800-145) [Online]. Available: http://csrc.nist.gov/publications/nistpubs/800-145/SP800-145.pdf
 9. Barker, E., Barker, W., Burr, W., Polk, W., & Smid, M. (2012). *NIST special publication 800-57, Recommendation for key management-part 1: General (revision 3)* [Online]. Available: http://csrc.nist.gov/publications/nistpubs/800-57/sp800-57_part1_rev3_general.pdf
10. Meyer, R. (2007). *Secure authentication on the Internet*, SANS Institute Infosec Reading Room [Online]. Available: https://www.sans.org/reading-room/whitepapers/securecode/secure-authentication-internet-2084
11. Ponemon, L. (2009). *Security of cloud computing users* (Ponemon Institute Research Report May 2010). Challenges, IDC Exchange, http://www.ca.com/files/industryresearch/security-cloud-computing-users_235659.pdf
12. Fujitsu. (2010). *Personal data in the cloud: A global survey of consumer attitudes* (Technical report). Fujitsu research Institute.
13. Liang, C. (2011). *The five major authentication issues in the current cloud computing environment* [Online]. Available: https://chenliangblog.wordpress.com/tag/e-commerce/
14. Granneman, J. (2012, August). *Password-based authentication: A weak link in cloud authentication* [Online]. Available: http://searchcloudsecurity.techtarget.com/tip/Password-based-authentication-A-weak-link-in-cloud-authentication
15. Misbahuddin, M. (2010). *Secure image based multi-factor authentication (SIMFA): A novel approach for web based services*. PhD thesis, Jawaharlal Nehru Technological University [Online]. Available: http://shodhganga.inflibnet.ac.in/handle/10603/3473
16. Stallings, W. (2011). Cryptography and network security, principles and practices (5th ed.). Upper Saddle River, NJ: Pearson Publications.
17. NIST. (2006, April). *Verifier impersonation attack, electronic authentication guideline* (NIST special publication 800-63, Version 1.0.2).
18. Raza, M., Iqbal, M., Sharif, M., & Haider, W. (2012). A survey of password attacks and comparative analysis on methods for secure authentication. *World Applied Sciences Journal, 19*(4), 439–444.
19. Cristofaro, C. E., Hongle, D., Freudiger, J. F., & Norcie, G. (2014). A comparative study of two factor authentication. In *Proceedings on the workshop on usable security USEC'14*, San Diego, CA, USA.
20. Password Cracking. *Wikipedia* [Online]. Available: https://en.wikipedia.org/wiki/Password_cracking
21. Dictionary Attack. *Wikipedia* [Online]. Available: https://en.wikipedia.org/wiki/Dictionary_attack
22. Lee, C., Lin, T., & Chang, R. (2011). A secure dynamic ID based remote user authentication scheme for multi-server environment using smart cards. *Expert Systems with Applications, 38*, 13863–13870.
23. Misbahuddin, M., Aijaz, A. M., & Shastri, M. H. (2006). A simple and efficient solution to remote user authentication using smart cards. In *Proceedings of IEEE innovations in information technology conference (IIT 06)*, Dubai.
24. Rainbow Table. *Wikipedia* [Online]. Available: https://en.wikipedia.org/wiki/Rainbow_table
25. Kulshrestha, A, & Dubey, S. K. (2014). A literature review on sniffing attacks in computer networks. *International Journal of Advanced Engineering Research and Science, 1*(2), 32–37.
26. Ku, W. C., & Chen, S. M. (2004). Weaknesses and improvements of an efficient password based remote user authentication scheme using smart cards. *IEEE Transactions Consumer Electronics, 50*(1), 204–207.
27. Chen, Y. C., & Yeh, L. Y. (2005). An efficient nonce-based authentication scheme with key agreement. *Applied Mathematics and Computation, 169*(2), 982–994.

28. Kocher, P., Jaffe, J., & Jun, B. (2010). Differential power analysis. In M. Wiener (Ed.) *CRYPTO 1999. LNCS: Vol. 1666* (pp. 388–397). Heidelberg: Springer.
29. Messerges, T. S., dabbish, E. A., & Sloan, R. H. (2002). Examining smart card security under the threat of power analysis attacks. *IEEE Transactions on Computers, 51*(5), 541–552.
30. Hao, Z., Zhong, S., & Yu, N. (2011). A time-bound ticket based mutual authentication scheme for cloud computing. *International Journal of Computers, Communications & Control, 6*(2), 227–235.
31. Jaidhar, C. D. (2013). Enhance mutual authentication scheme for cloud architecture. In: *Proceeding 3rd IEEE International advanced computing conference (IACC).*
32. Choudhary, A. J., Kumar, P., Sain, M., Lim, H., & Lee, H. J. (2011). A strong user authentication framework for cloud computing. In *IEEE Asia Pacific services computing conference.*
33. Jiang, R. (2013). Advanced secure user authentication framework for cloud computing. *International Journal of Smart Sensing and Intelligent Systems, 6*(4), 1700–1724.
34. Sanjeet, K. N., Subashish, M., & Bansidhar, M. (2012). An improved mutual authentication framework for cloud computing. *IJCA, 52*(5), 36–41.
35. OASIS. (2005, February). *Security Assertion Mark Up Language (SAML) 2.0 Technical overview, working draft 03.* Available: https://www.oasis-open.org/committees/download.php/27819/sstc-saml-tect-overview-2.0-cd-02.pdf
36. Hillenbrand, M., Gotze, J., Muller, J., & Muller, P. (2005). A single sign-on framework for web-services-based distributed applications. In *Proceedings of 8th international conference on telecommunications, ConTEL 2005* (pp. 273–279).
37. Trosch, J. (2008). *Identity federation with SAML 2.0* [Online]. Available http://security.hsr.ch/theses/DA_2008_IdentityFederation_with_SAML_20.pdf
38. Cremers, C., & Casimier, J. F. (2006). *Scyther - Semantics and verification of security protocols.* PhD thesis [Online]. Available: http://alexandria.tue.nl/extra2/200612074.pdf

Chapter 15
Review of the Main Security Threats and Challenges in Free-Access Public Cloud Storage Servers

Alejandro Sanchez-Gomez, Jesus Diaz, Luis Hernandez-Encinas, and David Arroyo

15.1 Introduction

Data outsourcing is one of the most conspicuous characteristics of the ongoing paradigm shift in the current technological context. Certainly, we are witnessing a transition from configurations where the data owners determined and deployed the means to store their information assets, to scenarios where those owners trust in third parties to handle such tasks. This being said, from a general point of view it encompasses a series of considerations regarding the manner trust is articulated.

This chapter is focused on pinpointing the underlying security assumptions in the adoption of free-access public cloud storage, but also on highlighting the means to enhance the protection of outsourced data. To achieve such goals, in Sect. 15.2 we discuss ten security challenges in cloud storage. The main concern of each of these challenges is summarised, and we provide a set of recommendations to tackle them. These recommendations are intended to identify cryptographic procedures and software solutions that can help both SMEs and end users to implement usable and low-cost security solutions upon free-access cloud storage. Certainly, the proper combination of standard cryptographic measures and the functionality provided by

---

A. Sanchez-Gomez • D. Arroyo (✉)
Departamento de Ingeniería Informática, Escuela Politécnica Superior, Universidad Autónoma de Madrid, Madrid, Spain
e-mail: asgsanchezgomez@gmail.com; david.arroyo@uam.es

J. Diaz
BEEVA, Madrid, Spain
e-mail: jesus.diaz@beeva.com

L. Hernandez-Encinas
Institute of Physical and Information Technologies (ITEFI), Spanish National Research Council (CSIC), Madrid, Spain
e-mail: luis@iec.csic.es

© Springer International Publishing AG 2018
K. Daimi (ed.), *Computer and Network Security Essentials*,
DOI 10.1007/978-3-319-58424-9_15

free cloud storages can lead to appealing *serverless* solutions. This being the case, our effort is on configuring a guideline to develop client-side software for the secure and privacy-respectful adoption of cloud storage services. This study is very helpful to discuss and define the main requirements of client-side software products to access cloud services. Furthermore, it configures a checklist to take into account in the risk analysis of the related cloud services, which would also be useful for Service Providers (SPs) in the design of their platforms. Finally, we conclude this chapter in Sect. 15.3.

15.2   Main Threats and Challenges in Cloud Storage

Next, we describe the main challenges in cloud storage derived from the paradigm shift outlined in the Introduction, also sketching possible solutions.

15.2.1   Challenge 1: Authentication

The protection of information assets demands the convenient concretion of access control methodologies, i.e. of authentication and authorisation procedures. In short, the vulnerabilities in the binomial authentication-authorisation can be a consequence of inner attacks, bad access control policies or use of software with some vulnerability.

In the case of cloud services, authentication is usually performed by checking users' credentials in a safe way. However, Cloud Providers (CPs) often use basic authentication mechanisms, in which the user has to provide her credentials to her provider. Therefore, this user has to trust the fact that providers do not have access to her credentials and they also manage these credentials in a secure way. Nevertheless, there are some relevant examples of a neglecting management of user's credentials by CPs.[1]

In addition, it is not difficult to find cases where authentication or authorisation can be circumvented due to the failures in the implementation of some security protocol. On this point, it is relevant to consider the problems regarding the implementation of the OAuth 2.0 protocol [36], which has been exploited to mount a Man-In-The-Cloud (MITC) attack [30].

---

[1] http://tinyurl.com/hkypel5, http://tinyurl.com/hdqoum3, http://tinyurl.com/kryf254 Accessed 2016-12-27.

15.2.1.1 Solutions

In order to face the problem of credentials leakage, we should use CPs that employ advanced password-based authentication techniques, like the Password-based Authenticated Key Exchange (PAKE) [1] or Secure Remote Password (SRP) [62] protocols, which effectively conveys a zero-knowledge password proof from the user to the server and thus hampers eavesdropper or man in the middle attacks.

Furthermore, this verification process of user's identity could be better by adding Two-Step Verification (2SV) mechanisms, which is present in several free cloud storage services.[2] In Google Drive, the user could use Google Authenticator as a two-factor verification mechanism.[3] Dropbox has also an optional 2SV (via text message or Time-Based One-Time Password apps). Box also provides the use of 2SV. In this case, the process requires a user to present two authenticating pieces of evidence when they log in: something they know (their Box password) and something they have (a code that is sent to their mobile device).[4] Moreover, all OneDrive users can protect the login via One-Time Code app or a text message.[5]

On the other hand, in some cloud storage services, sign-up is based on users' Personally Identifiable Information (PII). Consequently, the provider can monitor the storage record of its users having access to a complete behavioural profile of its users. Therefore, anonymous authentication mechanisms are mandatory if the user really needs to get an adequate privacy in the cloud [40]. A way to achieve this type of authentication is through non-conventional digital signatures, like group signatures, which allow members of a group of signers to issue signatures on behalf of the group that can be verified without telling which specific member issued it [6].

Finally, we would want to point out some possible solutions to the MITC attack. One possible strategy could be, firstly, to identify the compromise of a file synchronisation account, using a Cloud Access Security Broker solution that monitors access and usage of cloud services by the users, and secondly, to identify the abuse of the internal data resource, deploying controls such as Dynamic Authorisation Management[6] around the data resources identifying abnormal and abusive access to the data [30]. However, we consider that the best way to solve this security issue is being very carefully in the OAuth implementation, using along with it a second authentication factor in order to have a high security level. To achieve this goal, CPs should be aware of the current potential weaknesses of this protocol and follow the well-known recommendations made by the cryptographic community about how to implement OAuth 2.0 in a secure way [23].

---

[2] http://tinyurl.com/jdonyg7. Accessed 2016-12-27.
[3] https://support.google.com/accounts/answer/1066447. Accessed 2016-12-27.
[4] http://tinyurl.com/jpqtndx. Accessed 2016-12-27.
[5] http://tinyurl.com/p2s8dlw. Accessed 2016-12-27.
[6] https://tinyurl.com/gr86xxu. Accessed 2016-12-27.

15.2.1.2 Limitations

In this and other scenarios, the deployment of security protocols implies a risk that has to be taken into account. As much effort is used in the authentication process, we can always find a poor implementation of the cryptographic functionality used (e.g. the Imperva attack on OAuth 2.0 [30], Heartbleed in the case of TLS,[7] Ghost for Linux servers,[8] etc.).

Regarding 2SV mechanisms, it has been shown they are not completely secure [18]. Besides, if we are using a service from our mobile phone which includes 2SV sending a text message to this phone, the authentication process would be almost the same than if we only have to provide our username and password straightaway to this service [18]. Furthermore, it is necessary to consider the recent security concerns about the Signalling System No 7 (SS7) and the NIST advice against adopting Short Message Service (SMS) as out-of-band verifier [27].

Another key element in authentication is usability. The adequate combination of usability goals and security concerns is not easy to achieve, and the design of any procedure to improve usability should be thoroughly analysed before deploying the corresponding authentication process. This study, for example, should have been applied to confirm that security is not degraded by the usability goals of the so-called Single Sign-On (SSO) mechanism. SSO is a process that allows a user to enter a username and password to access multiple applications. It enables authentication through a token, and thus, it increases the usability of the whole system. Nevertheless, if these tokens are not adequately managed, SSO could be vulnerable against replay attacks [37].

15.2.2 Challenge 2: Information Encryption

The trust model and the risk model are very different when the CP is in charge of data encryption. If so, users are implicitly trusting CP, and this assumption can entail a threat to security and/or privacy [57]. With the current free cloud storage solutions, users have to completely trust their CPs [53]. Indeed, an important question concerns who controls the keys used for encryption of data at rest. If they are under the control of the CPs, the user is implicitly trusting that CPs manage them securely and use it only for legitimate purposes. Furthermore, providers could be also compromised because of hardware failures [48], software flaws [39], or misbehaviour of system administrators [33]. Once one of these happens, a tremendous amount of data might get corrupted or lost [64].

---

[7]http://tinyurl.com/lhjr7zf. Accessed 2016-12-27.
[8]http://tinyurl.com/h3fbqdx. Accessed 2016-12-27.

15.2.2.1   Solutions

Solutions for protecting privacy require encrypting data before releasing them to the CPs, and this measure has gained a great relevance since Snowden's leaks [51]. Client-side encryption could be properly articulated to obtain privacy and data integrity protection.

Nonetheless, encryption can bring several difficulties in scenarios where querying data is necessary. Then, we could use fragmentation instead of encryption and maintain confidential the associations among data values. This technique protects sensitive associations by splitting the concerned pieces of information and storing them in separate un-linkable fragments [53]; also, in this set-up CPs could detect redundant file blocks and perform deduplication.

15.2.2.2   Limitations

Client-side encryption implies that the user is in charge of keys generation and management,[9] which could suppose a high workload for her, along with a security risk (e.g. the user could loss a cryptographic key and thus she could not access to her data). Although it is possible to reduce such a burden by adopting password-based tools for the automatic generation and management of cryptographic keys, these solutions represent a security risk as they place all the protection around a single component. This Single Point of Failure (SPOF) can be avoided through distributed authentication via two or more servers. Hence, the user only has to know her password, and the compromise of one server exposes neither any private data nor any password [24].

Apart from the key management problem, data encryption thwarts traditional deduplication mechanisms and querying by keywords in the cloud [53]. However, the new paradigm called Secure Computation (SC) solves this problem, since it is essentially based on processing encrypted data [3]. Indeed, Multi-Party Computation and homomorphic encryption have been discarded in the past, since they are computationally expensive. Nonetheless, currently there exist efficient implementations (e.g. Dyadic[10] and Sharemind[11]).

Finally, sharing encrypted documents over the cloud requires a mechanism to share the encryption key of each encrypted file among the members of the group. A first approach to solve this problem could be to make up a data packet based on the concept of the digital envelope container.[12] However, this incurs in high management costs for the owner of the data and the group, requiring granting/revoking access as users join/leave the group, as well as delays for

---

[9]http://tinyurl.com/hy6pyqr. Accessed 2016-12-27.

[10]www.dyadicsec.com. Accessed 2016-06-04.

[11]http://sharemind-sdk.github.io/.

[12]http://tinyurl.com/z78dssy. Accessed 2016-12-27.

acquiring the keys. Here, it is possible to generate and store the secret keys in a centralised cloud service. This implies that the user trusts completely the CP, which becomes a SPOF [59].

A third approach includes key management through a trusted client-side authority (the manager) different from the CP [59]: users are segmented into populations called groups, each of which has read and write access to a different data partition. Access to each data partition is managed (through keys generation and users' authentication) by the manager.

A fourth approach is based on the combination of proxy signatures, a Tree-Based Group Diffie–Hellman (TGDH) key exchange protocol and proxy re-encryption [63]. This solution is privacy-respectful and supports better the updating of encryption keys, because it transfers most of the computational complexity and communication overhead to cloud servers.

15.2.3   Challenge 3: Inappropriate Modifications of Assets

Both location and data's owner are blurry and not clearly identified, which makes the old perimeters obsolete as a means to secure IT assets. In fact, when the user uploads her assets to the cloud, she cannot assure that they are not going to be modified by unauthorised third parties (not even when client-side encryption is carried out).

Additionally, for security considerations, previous public auditing schemes for shared cloud data hid the identities of the group members. However, the unconstrained identity anonymity enables a group member to maliciously modify shared data without being identified. This is a threat for coherent data availability, and, consequently, identity traceability should also be retained in data sharing [64].

15.2.3.1   Solutions

The first solution is to perform a hash function in each file to upload to her CP. These hashes can be verified after files are downloaded, but the problem is that the user needs to store all of them. Alternatively, we can consider retrievability proofs by encrypting the information and inserting data blocks (the sentinels) in random points [32]. The provider only sees a file with random bits, and she is not able to distinguish between data and sentinels. Integrity verification is performed by asking for a random selection of the sentinels; as the provider does not know where the sentinels are, any change in data can be detected. However, this solution supposes an overload for the user.

At this point, we should distinguish between content integrity (achieved with hashes, CRC, sentinels, etc.) and both content and source integrity, which can be carried out with digital signatures. However, in modern IT networks conventional digital signatures do not offer the whole of set of features that may be required. Depending on the specific needs, different schemes may be required, such as group signatures, multi- and aggregate signatures or blind signatures [6]. This being the case, group signatures could be used as anonymous authentication methods in the cloud. On the other hand, multi- and aggregate signatures could be used when a file is digitally signed and shared between several users in a cloud storage service. Finally, identity-based signatures could eliminate the need of distributing public keys in the cloud, allowing the verification of digital signatures just from the identity that the signer claims to own.

Moreover, the audit of shared cloud data demands the traceability of users' operations without eroding privacy. Therefore, it is required an efficient privacy-respectful public auditing solution that guarantees the identity traceability for group members. In [64], identity traceability is achieved by two lists that record members' activity on each data block. Besides, the scheme also provides data privacy during the authenticator generation by utilising a blind signature technique.

15.2.3.2   Limitations

Digital signatures are often performed using the asymmetric cryptographic algorithm RSA. The main disadvantage of this algorithm is the time required for its key generation, although it usually has to be done just once. Nevertheless, we could improve this time using elliptic curves [28], which provide the same security as RSA but with smaller key size. In fact, RSA key generation is significantly slower than ECC key generation for RSA key of sizes 1024 bits and greater [50]. In addition, the cost of key generation can be considered as a crucial factor in the choice of public key systems to use digital signatures, especially for smaller devices with less computational resources [31]. Nonetheless, in some cases it is not necessary to generate RSA keys for each use. For this situation, we would remark that the problem mentioned before is not so dramatic, since RSA is comparable to ECC for digital signature creation in terms of time, and it is faster than ECC for digital signature verification.[13]

Furthermore, we could incorporate Merkle trees [8], such that we will reduce the number of digital signatures to handle. Regarding non-conventional digital signatures, the problem comes for the scarcity of standard and thoroughly verified software libraries [6].

---

[13]http://tinyurl.com/3uc96d. Accessed 2016-12-27.

15.2.4   Challenge 4: Availability

Users typically place large amounts of data in cloud servers, and CPs need to ensure availability even when data are not accessed during long periods of time. Despite redundancy techniques (software and hardware) or error correction codes, from a general point of view, users cannot be sure whether their files are vulnerable or not against hard drive crashes [12].

15.2.4.1   Solutions

Recent studies propose the utilisation of RAID-like techniques over multiple cloud storage services [65]. Some cloud storage managers have been proposed to handle cloud storage in multiple CPs.[14] Moreover, in [65] it is proposed a solution for mobile devices that unifies storage from multiple CPs into a centralised storage pool.

15.2.4.2   Limitations

The above solutions are private, so the user would have to pay fees to use this services and to trust third-party companies to manage the data securely and privately.

15.2.5   Challenge 5: Data Location

When the CP guarantees that the data are stored within specific geographic area, users might not have the assurance about this fact. On this point, it had a great relevance to the decision of the European Court of Justice made on October 2015: the annulment of the EU-US data sharing agreement named Safe Harbour.[15] This revocation prevented the automatic transfer of data of European citizens to the United States of America (USA). However, since February 2016 there is a new framework that protects the fundamental rights of anyone in the EU whose personal data is transferred to the United States [21]. These agreements prove the huge legal complexity of the ubiquity in cloud storage systems.

---

[14]http://tinyurl.com/jtvq2o4. Accessed 2016-12-27.

[15]https://tinyurl.com/pnax3go. Accessed 2016-01-08.

15.2.5.1   Solutions

Cloud service latency is a hint to infer the geographic location of data as stored by CP. Nevertheless, these measures have to be carried out with high degree of precision since the information moves really quickly in electronic communications. An example of location proof is the use of distance bounding protocols [13]. These protocols always imply a timing phase in which the verificator sends a "challenge" and the provider responds. The provider is first authenticated and then required to respond within a time limit which depends on the distance between provider and user.

15.2.5.2   Limitations

Network, processing and disk access delays undermine the precision of the above methods. In fact, those methods are really dependent on the CP. For instance, Dropbox has to decrypt the information before sending it (as the encryption/decryption is made in the server), whereas this is not necessary in Mega (which performs this operation in the client side).

15.2.6   Challenge 6: Data Deduplication

The simple idea behind deduplication is to store each piece of data only once. Therefore, if a user wants to upload already existing data, the CP will add the user in the list of owners of that data. Deduplication is able to save space and costs, so that many CPs are adopting it. However, the adoption of deduplication is not a straightforward decision for a CP. The CP must decide between file or block level deduplication, server-side vs. client side deduplication, and single user vs. cross-user deduplication [56]. Some of these possibilities determine side channels that pose privacy matters [43]. Again we can opt for protecting privacy through encryption, but data encryption prevents straightforward deduplication techniques. Moreover, privacy against *curious* CPs cannot be ensured for unencrypted data [41].

15.2.6.1   Solutions

Several solutions have been proposed to mitigate privacy concerns with deduplication [43]. One of these solutions is Convergent Encryption (CE), which consists of using the hash of the data as encryption key. Consequently, two equal files will generate the same encrypted file, which allows both deduplication and encryption [35].

An alternative solution is ClouDedup [41], a tool that proposes the inclusion of an intermediate server between users and CPs. Users send to the intermediate server their blocks encrypted with CE, along with their keys. Afterwards, the intermediate server deduplicates blocks from all users and executes a second non-convergent encryption. As a result, the intermediate server sends to the CP only blocks that are not duplicated.

Other privacy-respectful deduplication techniques are more focused on providing benefits to the clients. This is the case of ClearBox [4], which endows cloud users with a means to verify the real storage space that their (encrypted) data is occupying in the cloud; this allows them to check whether they qualify for benefits such as price reductions.

15.2.6.2 Limitations

CE allows dictionary attacks. For instance, client-side CE allows malicious users to know if some particular information already exists in the cloud. As a possible solution, one can classify data segments into popular and unpopular chunks of data [42]. Popular pieces of information (those which demand deduplication) would be encrypted using CE; unpopular ones are encrypted using symmetric encryption. This achieves a high level of confidentiality and privacy protection, since the unpopular data (more likely associated to PII) is protected with semantically secure encryption. Nevertheless, the user has to decide on the popularity of data fragments. The usability of the previous scheme can be further improved by handling blocks popularity according to a previously established set of metadata and the use of an additional server [41].

Another option is proposed in [25], where the authors claim to impede content guessing attacks by implementing CE at block level under a security parameter.

15.2.7   Challenge 7: Version Control of Encrypted Data

Version Control Systems (VCSs) are a must as part of the recuperative controls of any architecture for securing information assets. But its deployment is not straightforward for encrypted data. Indeed, the recommended operating modes for symmetric block ciphers prevent VCSs to know which data segment was modified, as modifications in earlier blocks affect the subsequent ones. Consequently, one copy is needed per each version of a file, introducing too much overhead.

15.2.7.1 Solutions

VCSs can be built by splitting the files in data objects [46]. If some modifications are produced in a big file, the user would only upload the objects that have been modified. This enables a more efficient backup mechanism, as an object appearing in multiple backups is stored in the cloud only once and it is indexed by a proper hashing-based pointer.

Two tools for version control with encrypted data are SparkleShare[16] and Git-crypt.[17] Both solutions enable using a non-fully trusted Git server [54].

15.2.7.2 Limitations

The main shortcomings of the previous version control systems are given by usability concerns. In the first scheme, the user must decide how to split files and classify the resulting pieces. In addition, the cryptographic keys of each data object must be properly stored and managed [46]. This burden is also undermining the usability of SparkleShare and Git-crypt [29]. Furthermore, SparkleShare does not provide any key change mechanism; the encryption password is saved on each SparkleShare client as plain text and can be read by everyone; and the filename is not encrypted at all, so an attacker can monitor which files are stored. On the other hand, in Git-crypt the filename is not encrypted, and the password can be found in plain text on the client's computer. As a matter of fact, the management of metadata in this VCS can imply some security risks [58]. In addition, key exchange is not an easy process, and it is still an issue to be solved in coherence with usability criteria [54].

15.2.8 Challenge 8: Assured Deletion of Data

It is not desirable to keep data backups permanently, since sensitive information can be exposed in the future due to a security breach or a poor management made by CPs [46]. Moreover, CPs can use multiple copies of data over cloud infrastructure to have a great tolerance against failures. As providers do not publish their replication policies, users do not know how many copies of their data are in the cloud, or where those copies are stored. Hence, it is not clear how providers can delete all the copies when users ask for removing their data.

---

[16]www.sparkleshare.org. Accessed 2016-08-15.

[17]https://github.com/AGWA/git-crypt. Accessed 2016-08-15.

15.2.8.1   Solutions

Client-side cryptographic protection is the easiest procedure to get assured deletion
of data: if a user destroy the secret keys needed to decrypt data, then these data is
not accessible anymore [46].

15.2.8.2   Limitations

As stated above, client-side encryption resorts to a key management system that is
totally independent of the cloud system. Therefore, the same limitations explained
in Sect. 15.2.2.2 can also apply for this section. On the other hand, note that client-
side encryption only protects users' data against CPs, but not against changes in
the users groups in which is performed a key distribution protocol. In this specific
case, if a user is revoked from a group, she could still access all previous versions
of shared files that belongs to this group if she has previously stored them locally.

15.2.9   Challenge 9: API's Validation

The current technological scenario is highly determined by the adoption of third-
party software products, and it has been coined as the era of containers (*container
age*). Most of these third-party solutions are given by APIs that have not always
been properly validated in terms of security[18,19], enclosing additional security risks.

15.2.9.1   Solutions

The new trust model should be handled by means of a Secure Development
Lifecycle (SDL) [47]. SDLs are the key step in the evolution of software security,
and they should be guided by automatic tools to analyse security properties through
formal methods [7]. These tools are intended to help not only the design stage
but also to face security and privacy threats arising in the production phase.
The dynamic and adaptive nature of the SDL has been underlined in different
methodologies to deploy systems according to the security-by-design principle [17].
In order to sustain such a methodology, there exist tools as Maude-NPA[20]and the

---

[18]http://tinyurl.com/pljob9s. Accessed 2016-12-27.

[19]http://tinyurl.com/jxp7jp8 Accessed 2016-12-27.

[20]http://maude.cs.uiuc.edu/tools/Maude-NPA. Accessed 2016-10-09.

framework STRIDE [55], and companies as Cryptosense,[21] which is a recent start-up which is looking to commercialise techniques for analysing the security of APIs [38].

15.2.9.2 Limitations

The human factor also takes part in the analysis of protocols procedure, so a bad formalisation can make that some errors are not detected. Then, as in any engineering process, the solution presented does not provide the 100% of success, but it helps avoid a great quantity of failures.

Finally, assuming that the CPs are validated by the cryptographic community, the solutions proposed for this challenge will be used to validate the new developments made by cloud developers. Nevertheless, the cloud users should verify that their cloud providers follow these good practices.

15.2.10   Challenge 10: Usable Security Solutions

Any change required to improve security should not erode users' acceptance of the so-modified cloud services [49, 60]. In other words, secure cloud services solutions should be also easy to use. Otherwise, these services will not be adopted.

15.2.10.1   Solutions

The so-called security-by-design and privacy-by-design paradigms [15] are hot topics in cryptographic engineering, and their fulfilment calls for software developing methodologies that integrate standards,[22] security reference architectures [22], and well-known technologies.

Furthermore, technologies usability has to be tested. In [34], Human–Computer Interaction (HCI) methods and practices are provided for each phase of the Software Development Life Cycle (SDLC), in order to create usable secure and privacy-respectful systems. Moreover, there are recent efforts on formalising and automatically assessing the role of human factors in security protocols [44, 45].

---

[21]https://cryptosense.com/. Accessed 2016-06-04.
[22]http://tinyurl.com/2642d8. Accessed 2016-12-27.

15.2.10.2 Limitations

Firstly, a key limitation to consider is that security and privacy are rarely users' main goal, and they would like privacy and security systems and controls to be as transparent as possible [34]. On the other hand, users want to be in control of the situation and understand what is happening. As a consequence, these two factors make harder to develop usable security applications.

Finally, in order to be successful in the development of usable security solutions, it is needed to have a multidisciplinary team including experts from the fields of IT, information science, usability engineering, cognitive sciences and human factors. This could suppose a main limitation, since no all companies have enough resources in order to form this kind of heterogeneous work group.

15.3 Conclusions

Nowadays, cloud storage is a relevant topic due to the increase in the number of users who place their assets onto the cloud. However, these users often do not trust about where their data are going to be stored and who is going to have access to these data. It is for this reason that many users feel the obligation of applying security measures in order to have a total control over their data. More concretely, the user could find authentication, integrity, availability, confidentiality and privacy problems. In the specific case of enterprises, these imply key considerations that should be included in any cloud service agreement.

To address these concerns, this chapter comprises the most relevant security problems that users of free cloud storage services can find. For each identified security challenge, we have outlined some solutions and limitations (see Table 15.1). In addition, we have discussed standards of information security that can be combined with the functionality of free cloud storage services according to a serverless architecture.

Finally, we have to take into account that the world of security information evolves really quickly every day. Therefore, we have to be aware with the new potential security problems which could affect the security and privacy of the user in cloud storage environments. In this evolving scenario, our work is intended to help cloud users to evaluate the cloud service agreements according to recommendations as the new ISO/IEC 19086-1 standard.

Acknowledgements This work was supported by Comunidad de Madrid (Spain) under the project S2013/ICE-3095-CM (CIBERDINE).

Table 15.1 Summary of security challenges and solutions in public cloud storage

Challenge	Solutions	Limitations	References
Authentication	PAKE; SRP; 2SV; anonymous authentication mechanisms; CASB; DAM; SSO	Poor implementation of the cryptographic functionalities; difficult combination of usability goals and security concerns	[1, 2, 6, 23, 26, 36, 52, 62]
Information encryption	Client-side encryption; data fragmentation	Key management: MPC systems; loss of functionalities: homomorphic encryption as solution; sharing of encrypted files: digital envelope container as solution	[11, 51, 53]
Inappropriate modifications of assets	Hash functions; retrievability proofs; digital signatures; identity traceability mechanisms	Performance	[5, 32, 64]
Availability	Multi-cloud environments	Trust in third parties in multi-cloud solutions	[10, 65]
Data location	Distance bounding protocols	Solutions dependent on the storage service and the Internet communication	[13, 21]
Data deduplication	Convergent encryption (CE) + popularity of data segments; ClouDedup approach	CE suffers of several weaknesses: confirmation of a file attack, learn the remaining information attack	[9, 41, 61]
Version control of encrypted data	Split the files uploaded to the cloud in data objects; SparkleShare; Git-crypt	High management costs for the user	[29, 46, 54, 58]
Assured deletion of data	Client-side encryption	Key management; data encryption does not protect against changes in the users groups	[46]
API's validation	Use of a secure development lifecycle (SDL) and automatic tools for software validation	The human factor	[16, 19, 38, 44, 45, 55]
Usable security solutions	Security-by-design and privacy-by-design paradigms; usability tests	Security and privacy are rarely the user's main goal but these solutions must be user friendly	[7, 14, 17, 20, 38, 55]

References

1. Abdalla, M., Fouque, P. A., & Pointcheval, D. (2005). Password-based authenticated key exchange in the three-party setting. In *Public key cryptography-PKC 2005* (pp. 65–84). Berlin: Springer.
2. Alphr. How secure are Dropbox, Microsoft OneDrive, Google Drive and Apple iCloud? [Online]. Available from: http://www.alphr.com/dropbox/1000326/how-secure-are-dropbox-microsoft-onedrive-google-drive-and-apple-icloud. Accessed December 29, 2015.
3. Archer, D. W., Bogdanov, D., Pinkas, B., & Pullonen, P. (2015). *Maturity and performance of programmable secure computation*. Technical Report, IACR Cryptology ePrint Archive.
4. Armknecht, F., Bohli, J. M., Karame, G. O., & Youssef, F. (2015). Transparent data deduplication in the cloud. In *Proceedings of the 22nd ACM SIGSAC Conference on Computer and Communications Security* (pp. 886–900). New York: ACM.
5. Arroyo, D., Diaz, J., & Gayoso, V. (2015). *On the difficult tradeoff between security and privacy: Challenges for the management of digital identities* (pp. 455–462). Cham: Springer International Publishing.
6. Arroyo, D., Diaz, J., & Rodriguez, F. B. (2015). Non-conventional digital signatures and their implementations - a review. In *CISIS'15* (pp. 425–435). Berlin: Springer
7. Bansal, C., Bhargavan, K., Delignat-Lavaud, A., & Maffeis, S. (2014). Discovering concrete attacks on website authorization by formal analysis. *Journal of Computer Security, 22*(4), 601–657.
8. Becker, G. (2008). *Merkle signature schemes, Merkle trees and their cryptanalysis*. Ruhr-Universität Bochum.
9. Bellare, M., Keelveedhi, S., & Ristenpart, T. (2013). Message-locked encryption and secure deduplication. In *Annual International Conference on the Theory and Applications of Cryptographic Techniques* (pp. 296–312). New York: Springer.
10. Best Backups. 7 cloud storage managers for multiple cloud storage services - Best backups.com [Online]. Available from: http://www.bestbackups.com/blog/4429/7-cloud-storage-managers-for-multiple-cloud-storage-services Accessed March 26, 2016.
11. Bogdanov, D., Laur, S., & Willemson, J. (2008). Sharemind: A framework for fast privacy-preserving computations. In *Computer Security-ESORICS 2008* (pp. 192–206). New York: Springer.
12. Bowers, K. D., van Dijk, M., Juels, A., Oprea, A., & Rivest, R. L. (2011). How to tell if your cloud files are vulnerable to drive crashes. In *Proceedings of the 18th ACM Conference on Computer and Communications Security* (pp. 501–514). New York: ACM.
13. Boyd, C. (2013). Cryptography in the cloud: Advances and challenges. *Journal of Information and Communication Convergence Engineering 11*(1), 17–23.
14. Butler, B. Researchers steal secret RSA encryption keys in Amazon's cloud [Online]. Available from: http://www.networkworld.com/article/2989757/cloud-security/researchers-steal-secret-rsa-encryption-keys-in-amazon-s-cloud.html. Accessed November 22, 2015.
15. Cavoukian, A., & Dixon, M. (2013). *Privacy and security by design: An enterprise architecture approach*. Ontario: Information and Privacy Commissioner.
16. Cryptosense. Cryptosense automated analysis for cryptographic systems [Online]. Available from: https://cryptosense.com. Accessed November 22, 2015.
17. Diaz, J., Arroyo, D., & Rodriguez, F. B. (2014). A formal methodology for integral security design and verification of network protocols. *Journal of Systems and Software, 89*, 87–98.
18. Dmitrienko, A., Liebchen, C., Rossow, C., & Sadeghi, A. R. (2014). Security analysis of mobile two-factor authentication schemes. *Intel® Technology Journal, 18*(4), 138–161.
19. Escobar, S., Meadows, C., & Meseguer, J. (2009). Maude-NPA: Cryptographic protocol analysis modulo equational properties. In *Foundations of security analysis and design V* (pp. 1–50). Berlin: Springer.

20. Escobar, S., Meadows, C., & Meseguer, J. (2012). The Maude-NRL protocol analyzer (Maude-NPA) [Online]. Available from: http://maude.cs.uiuc.edu/tools/Maude-NPA. Accessed October 9, 2016.
21. European Commission. European Commission launches EU-U.S. Privacy shield: stronger protection for transatlantic data flows [Online]. Available from: http://tinyurl.com/jeg3doq. Accessed September 12, 2016.
22. Fernandez, E. B., Monge, R., & Hashizume, K. (2015). Building a security reference architecture for cloud systems. *Requirements Engineering, 21*, 1–25.
23. Fett, D., Küsters, R., & Schmitz, G. (2016). A comprehensive formal security analysis of OAuth 2.0 (pp. 1–75). http://arxiv.org/abs/1601.01229.
24. Ford, W., & Kaliski, B. S., Jr. (2000). Server-assisted generation of a strong secret from a password. In *IEEE 9th International Workshops on Enabling Technologies: Infrastructure for Collaborative Enterprises, 2000. (WET ICE 2000). Proceedings* (pp. 176–180).
25. González-Manzano, L., & Orfila, A. (2015). An efficient confidentiality-preserving proof of ownership for deduplication. *Journal of Network and Computer Applications, 50*, 49–59.
26. Gordon, W. Two-factor authentication: The big list of everywhere you should enable it right now [Online]. Available from: http://www.lifehacker.com.au/2012/09/two-factor-authentication-the-big-list-of-everywhere-you-should-enable-it-right-now. Accessed December 31, 2015.
27. Grassi, P. A., Fenton, J. L., Newton, E. M., Perlner, R. A., Regenscheid, A. R., Burr, W. E., Richer, J. P., Lefkovitz, N. B., Choong, J. M. D. Y. Y., Mary, K. K. G., & Theofanos, F. (2016). *Digital authentication guideline; authentication and lifecycle management.* Technical Report Draft NIST SP 800-63B, National Institute of Standards and Technology.
28. Hankerson, D., Menezes, A. J., & Vanstone, S. (2004). *Guide to elliptic curve cryptography.* New York, NY: Springer.
29. Happe, A. Git with transparent encryption [Online]. Available from: https://snikt.net/blog/2013/07/04/git-with-transparent-encryption Accessed August 16, 2016.
30. Imperva. Man in the cloud attacks [Online] http://tinyurl.com/qf7n6s8. Accessed December 27, 2016.
31. Jansma, N., & Arrendondo, B. (2004). *Performance comparison of elliptic curve and RSA digital signatures.* Technical Report, University of Michigan College of Engineering (pp. 1–20).
32. Juels, A., & Kaliski, B. S., Jr. (2007). PORs: Proofs of retrievability for large files. In *Proceedings of the 14th ACM Conference on Computer and Communications Security* (pp. 584–597).
33. Kandias, M., Virvilis, N., & Gritzalis, D. (2011). The insider threat in cloud computing. In *International Workshop on Critical Information Infrastructures Security* (pp. 93–103). New York: Springer.
34. Karat, C. M., Brodie, C., & Karat, J. (2005). Usability design and evaluation for privacy and security solutions. In L. F. Cranor & S. Garfinkel (Eds.), *Security and usability* (pp. 47–74). O'Reilly Media, Inc.
35. Li, J., Chen, X., Xhafa, F., & Barolli, L. (2014). Secure deduplication storage systems with keyword search. In *Proceedings of 2014 IEEE 28th International Conference on Advanced Information Networking and Applications (AINA'14)* (pp. 971–977).
36. Li, W., & Mitchell, C. J. (2014). Security issues in OAuth 2.0 SSO implementations. In *Information Security - 17th International Conference, ISC 2014, Proceedings*, Hong Kong, China, October 12–14, 2014 (pp. 529–541).
37. Mainka, C., Mladenov, V., Feldmann, F., Krautwald, J., & Schwenk, J. (2014). Your software at my service: Security analysis of SaaS single sign-on solutions in the cloud. In *Proceedings of the 6th Edition of the ACM Workshop on Cloud Computing Security* (pp. 93–104). New York: ACM.
38. Meadows, C. (2015). Emerging issues and trends in formal methods in cryptographic protocol analysis: Twelve years later. In *Logic, rewriting, and concurrency* (pp. 475–492). New York: Springer.
39. Pasquier, T., Singh, J., Bacon, J., & Eyers, D. (2016). Information flow audit for PaaS clouds. In *International Conference on Cloud Engineering (IC2E)*. New York: IEEE.

40. Pulls, T., & Slamanig, D. (2015). On the feasibility of (practical) commercial anonymous cloud storage. *Transactions on Data Privacy, 8*(2), 89–111.
41. Puzio, P., Molva, R., Onen, M., & Loureiro, S. (2013). ClouDedup: secure deduplication with encrypted data for cloud storage. In *Proceedings of 2013 IEEE 5th International Conference on Cloud Computing Technology and Science (CloudCom'13)* (pp. 363–370).
42. Puzio, P., Molva, R. Önen, M., & Loureiro, S. (2016). PerfectDedup: Secure data deduplication. In J. Garcia-Alfaro, G. Navarro-Arribas, A. Aldini, F. Martinelli, N. Suri (Eds.), *Data Privacy Management, and Security Assurance: 10th International Workshop, DPM 2015, and 4th International Workshop QASA 2015, Vienna, Austria, September 21–22, 2015* (pp. 150–166). Cham: Springer International Publishing. doi:10.1007/978-3-319-29883-2_10, ISBN:978-3-319-29883-2, http://dx.doi.org/10.1007/978-3-319-29883-2_10.
43. Rabotka, V., & Mannan, M. (2016). An evaluation of recent secure deduplication proposals. *Journal of Information Security and Applications, 27*, 3–18.
44. Radke, K., Boyd, C., Nieto, J. G., & Bartlett, H. (2014). CHURNs: Freshness assurance for humans. *The Computer Journal, 58*, 2404–2425. p. bxu073.
45. Radke, K., Boyd, C., Nieto, J. G., & Brereton, M. (2011). Ceremony analysis: Strengths and weaknesses. In *Future challenges in security and privacy for academia and industry* (pp. 104–115). Berlin: Springer.
46. Rahumed, A., Chen, H. C. H., Tang, Y., Lee, P. P. C., & Lui, J. C. S. (2011). A secure cloud backup system with assured deletion and version control. In *Proceedings of the International Conference on Parallel Processing Workshops* (pp. 160–167).
47. Ransome, J., & Misra, A. (2013). *Core software security: Security at the source*. Boca Raton: CRC Press.
48. Razavi, K., Gras, B., Bosman, E., Preneel, B., Giuffrida, C., & Bos, H. (2016). Flip Feng Shui: hammering a needle in the software stack. In *Proceedings of the 25th USENIX Security Symposium*.
49. Renaud, K., Volkamer, M., & Renkema-Padmos, A. (2014). Why doesn't Jane protect her privacy? In *Privacy enhancing technologies* (pp. 244–262). New York: Springer.
50. Rifà-Pous, H., & Herrera-Joancomartí, J. (2011). Computational and energy costs of cryptographic algorithms on handheld devices. *Future Internet, 3*(1), 31–48.
51. Rusbridger, A. (2013). The Snowden leaks and the public.
52. Ruvalcaba, C., & Langin, C. (2009). Four attacks on OAuth - How to secure your OAuth implementation. *System, 1*, 19. https://www.sans.org/reading-room/whitepapers/application/attacks-oauth-secure-oauth-implementation-33644.
53. Samarati, P., & di Vimercati, S. (2016). Cloud security: Issues and concerns. In *Encyclopedia on cloud computing*. New York: Wiley.
54. Shirey, R. G., Hopkinson, K. M., Stewart, K. E., Hodson, D. D., & Borghetti, B. J. (2015). Analysis of implementations to secure Git for use as an encrypted distributed version control system. In *2015 48th Hawaii International Conference on System Sciences (HICSS)* (pp. 5310–5319). New York: IEEE.
55. Shostack, A. (2014). *Threat modeling: Designing for security*. New York: Wiley.
56. Srinivasan, S. (2014). Security, trust, and regulatory aspects of cloud computing in business environments. In *IGI Global*.
57. Strandburg, K. (2014). Monitoring, datafication and consent: Legal approaches to privacy in a big data context. In J. Lane, V. Stodden, S. Bender, & H. Nissenbaum (Eds.), *Privacy, big data, and the public good: Frameworks for engagement*. Cambridge: Cambridge University Press.
58. Torres-Arias, S., Ammula, A. K., Curtmola, R., & Cappos, J. (2016) On omitting commits and committing omissions: Preventing Git metadata tampering that (re)introduces software vulnerabilities. In *25th USENIX Security Symposium, USENIX Security 16*, Austin, TX, USA, August 10–12, 2016 (pp. 379–395).
59. Tysowski, P. K. (2013). *Highly scalable and secure mobile applications in cloud computing systems*. Ph.D. thesis, University of Waterloo.
60. Whitten, A., & Tygar, J. D. (1999). Why Johnny can't encrypt: A usability evaluation of PGP 5.0. In *Usenix Security* (Vol. 1999).

61. Wilcox-O'Hearn, Z. (2008). Drew Perttula and attacks on convergent encryption [Online]. Available from: https://tahoe-lafs.org/hacktahoelafs/drew_perttula.html. Accessed December 9, 2016.
62. Wu, T. D., et al. (1998). The secure remote password protocol. In *NDSS* (Vol. 98, pp. 97–111).
63. Xue, K., & Hong, P. (2014). A dynamic secure group sharing framework in public cloud computing. *IEEE Transactions on Cloud Computing, 2*(4), 459–470.
64. Yang, G., Yu, J., Shen, W., Su, Q., Fu, Z., & Hao, R. (2016). Enabling public auditing for shared data in cloud storage supporting identity privacy and traceability. *Journal of Systems and Software, 113*, 130–139.
65. Yeo, H. S., Phang, X. S., Lee, H. J., & Lim, H. (2014). Leveraging client-side storage techniques for enhanced use of multiple consumer cloud storage services on resource-constrained mobile devices. *Journal of Network and Computer Applications, 43*, 142–156.

Chapter 16
Secure Elliptic Curves in Cryptography

Victor Gayoso Martínez, Lorena González-Manzano,
and Agustín Martín Muñoz

16.1 Introduction

In 1985, Neal Koblitz [24] and Victor Miller [28] independently suggested using
elliptic curves defined over finite fields for implementing different cryptosystems.
This branch of public-key cryptography is typically known as Elliptic Curve
Cryptography (ECC), and its security is based on the difficulty of solving the Elliptic
Curve Discrete Logarithm Problem (ECDLP), which is considered to be more
difficult to solve than the Integer Factorization Problem (IFP) used by RSA or the
Discrete Logarithm Problem (DLP) which is the basis of the ElGamal encryption
scheme [11, 16, 24]. Since the inception of ECC, elliptic curves have been typically
represented in what is called the short Weierstrass form (which is described in
Sect. 16.2).

One of the most important aspects when working with secure elliptic curves
is how they are generated. Even though some standards include several sample
curves or even the description of the procedures for generating them (e.g., X9.63
[2], IEEE 1363 [21], or NIST FIPS 186 [30]), in most cases the information
contained in those standards has important limitations, such as the lack of clarity
in the selection procedure regarding the seeds and prime numbers involved or the
insufficient explanation for some of the requirements specified in the procedure.

V. Gayoso Martínez (✉) • A. Martín Muñoz
Institute of Physical and Information Technologies (ITEFI), Spanish National
Research Council (CSIC), Madrid, Spain
e-mail: victor.gayoso@iec.csic.es; agustin@iec.csic.es

L. González-Manzano
Computer Security Lab (COSEC), Universidad Carlos III de Madrid, Leganés, Madrid, Spain
e-mail: lgmanzan@inf.uc3m.es

© Springer International Publishing AG 2018

283

K. Daimi (ed.), *Computer and Network Security Essentials*,
DOI 10.1007/978-3-319-58424-9_16

In this scenario, in the early 2000s a working group called ECC Brainpool focused on this topic and completed a first set of recommendations in 2005 [10] for elliptic curves in the short Weierstrass form. Five years later, the Brainpool specification was revised and published as a Request for Comments (RFC) [25]. The Brainpool initiative was considered to be the first international effort with the goal of producing a truly transparent curve generation scheme, and the curves suggested in its specification were initially considered to be secure without any hint of doubt.

Several years later, researchers Daniel Bernstein and Tanja Lange published an analysis in which they reviewed the existing elliptic curve generation mechanisms, including the one devised by Brainpool. In their website SafeCurves [6], they compared not only the strength of the curve parameters and the soundness of what they called "ECC security" (basically the strength against rho attacks [34] and transfers of the ECDLP to other fields where the DLP is easier to solve [18–20, 27], the class number associated with the trace of the curve [12], and the rigidity of the definition of the curve parameters) but also what they termed "ECDLP security," a concept in which they included the resistance to attacks based on the Montgomery ladder [29], the strength of the associated twisted curves [8], the completeness of the addition formulas [5], and the indistinguishability of elliptic curve points from random binary strings [9]. The main result of that analysis was that all the schemes included in the standards overlooked some aspects of the ECDLP security and, for that reason, required to increase the complexity of the implementations in such a way that it opened the door to some types of side channel attacks [6].

As a solution, Bernstein and Lange decided to propose new curves different to those provided by previous specifications. Going one step further, they evaluated 20 curves obtained from different sources, showing that only a small subset of curves fulfilled all their security requirements. That subset is composed by elliptic curves in the Edwards and Montgomery formats (both of them introduced in Sect. 16.2).

However, from the point of view of availability, Montgomery and Edwards curves have not been popular choices so far, and in that respect traditional curves in the short Weierstrass form are the dominant option in both hardware and software implementations. In addition to that, the extra security offered by Edwards and Montgomery curves could affect the performance of the point operations which are the core of the scalar-multiplication operation (the product of a point of the elliptic curve by an integer, an operation needed in any protocol involving elliptic curves).

Regarding the resistance of elliptic curve cryptosystems to quantum computing, at the time of writing this contribution the National Security Agency (NSA) has confirmed the status of transition algorithms to some ECC cryptosystems while new cryptographic systems that are secure against both quantum and classical computers are defined [31, 33]. As it is not clear when those new systems will be available, it is safe to state that ECC will continue to be used at least in the near future.

This chapter presents to the reader the different types of elliptic curves used in Cryptography as well as the Brainpool procedure. The contribution is completed with the examination of the proposal regarding secure elliptic curves represented by the SafeCurves initiative.

16.2 Elliptic Curves

In this section, the reader is presented with the mathematical description of elliptic curves, as well as the specific details of elliptic curves described in the short Weierstrass form and the Edwards and Montgomery formats.

16.2.1 Definition

An elliptic curve defined over a field $\mathbb{F}$ is a cubic, non-singular curve whose points $(x, y) \in \mathbb{F} \times \mathbb{F}$ verify the following equation, known as the Weierstrass equation:

$$E : y^2 + a_1 xy + a_3 y = x^3 + a_2 x^2 + a_4 x + a_6,$$

where $a_1, a_2, a_3, a_4, a_6 \in \mathbb{F}$ and $\triangle \neq 0$, where $\triangle$ is the discriminant of E that can be computed as follows [26]:

$$\triangle = -d_2^2 d_8 - 8d_4^3 - 27d_6^2 + 9d_2 d_4 d_6,$$
$$d_2 = a_1^2 + 4a_2,$$
$$d_4 = 2a_4 + a_1 a_3,$$
$$d_6 = a_3^2 + 4a_6,$$
$$d_8 = a_1^2 a_6 + 4a_2 a_6 - a_1 a_3 a_4 + a_2 a_3^2 - a_4^2.$$

An elliptic curve point is singular if and only if the partial derivatives of the curve equation are null at that point. The curve is said to be singular if it possesses at least a singular point, while it is non-singular if it does not have any such points.

The nonhomogeneous Weierstrass equation can also be expressed in the following homogeneous form [12]:

$$Y^2 Z + a_1 XYZ + a_3 YZ^2 = X^3 + a_2 X^2 Z + a_4 XZ^2 + a_6 Z^3.$$

This equation defines a curve which includes a special point called the point at infinity, which is typically represented as $\mathcal{O} = [0 : 1 : 0]$ and that has no correspondence with any point of the nonhomogeneous form. However, this point is very important as it works as the identity element of the addition operation when working with Weierstrass and Montgomery elliptic curves.

16.2.2 Elliptic Curves Over Finite Fields

Most cryptosystems defined over elliptic curves use only the following finite fields $\mathbb{F}_q$ with $q = p^m$ elements: prime fields $\mathbb{F}_p$ (where p is an odd prime number and

$m = 1$) and binary fields $\mathbb{F}_{2^m}$ (where m can be any positive integer). However, due to a combination of license issues and security concerns [4], prime fields have been favored in the latest specifications at the expense of binary fields (see, for example, Brainpool [25], NSA Suite B [32], or BSI TR-03111 [11]). Following that criterion, this chapter focuses on elliptic curves defined over prime fields. In this type of curves, the term key length must be interpreted as the number of bits needed to represent the prime number p.

When using prime fields, the order (i.e., the number of points) of any elliptic curves is finite. In this context, the order of a point P of the elliptic curve is the minimum nonzero value n such that $n \cdot P = \mathcal{O}$, where $n \cdot P$ is the scalar multiplication of the point P by the number n (i.e., $P + P + \cdots + P$, where P appears n times).

A point G is said to be a generator if it is used to generate either all the points of the curve or a subset of those points. For security reasons, only generators whose order is a prime number are used in Cryptography.

Given a curve and a generator, the term cofactor refers to the result of dividing the number of points of the curve by the order of the generator. Most standards only allow curves whose cofactor is either 1 or a small number like 2, 3, or 4.

16.2.2.1 Weierstrass Curves

The peculiarities of prime fields allow to simplify the general Weierstrass equation, obtaining in the process what is called the short Weierstrass form represented as $y^2 = x^3 + ax + b$, where $4a^3 + 27b^2 \not\equiv 0 \pmod{p}$.

As in the case of the general Weierstrass equation, the identity element of the short Weierstrass form is the point at infinity $\mathcal{O}$, while the opposite element of a point $P = (x, y)$ is the point $-P = (x, -y)$. Adding two points $P_1 = (x_1, y_1)$ and $P_2 = (x_2, y_2)$ such that $P_1 \neq \pm P_2$ produces a point $P_3 = (x_3, y_3)$ whose coordinates can be computed as follows [7]:

$$x_3 = \frac{(y_2 - y_1)^2}{(x_2 - x_1)^2} - x_1 - x_2, \qquad y_3 = \frac{(2x_1 + x_2)(y_2 - y_1)}{x_2 - x_1} - \frac{(y_2 - y_1)^3}{(x_2 - x_1)^3} - y_1.$$

In comparison, when $P_1 = P_2$ it is necessary to use an alternative addition formula, so in this case the point $P_3 = 2P_1$ obtained through the doubling operation has the following coordinates [7]:

$$x_3 = \frac{(3x_1^2 + a)^2}{(2y_1)^2} - 2x_1, \qquad y_3 = \frac{(3x_1)(3x_1^2 + a)}{2y_1} - \frac{(3x_1^2 + a)^3}{(2y_1)^3} - y_1.$$

16.2.2.2 Edwards Curves

Edwards curves were introduced by Harold M. Edwards in [15], though during the last decade slightly different equations have been given that name. The first expression related to Edwards curves was its *normal form* $x^2 + y^2 = c^2(1 + x^2y^2)$, where the neutral element of the addition operation is the point $(0, c)$ [15]. Edwards showed that any elliptic curve could be transformed into its normal form if the finite field used by the curve is algebraically closed. If that was not the case, then only a small fraction of elliptic curves could be transformed into the normal form using the original finite field [5, 8].

A few months later, Bernstein and Lange presented a variant using the *generalized form* $x^2 + y^2 = c^2(1 + dx^2y^2)$, where $cd(1 - dc^4) \not\equiv 0 \pmod{p}$ [5]. The goal of Bernstein and Lange was to increase the number of curves that could be converted into the Edwards form using the original finite field. In addition to the previous definition, an Edwards curve is said to be *complete* if d is not a square in $\mathbb{F}_p$ (i.e., if d is not a quadratic residue in $\mathbb{F}_p$), which allows to use only one addition operation for any pair of points, avoiding the case of exceptional points. If $c \neq 0$, $d \neq 0$, d is not square, and $dc^4 \neq 1$, then the coordinates of P_3 can be computed as follows for any pair of points $P_1 = (x_1, y_1)$ and $P_2 = (x_2, y_2)$:

$$x_3 = \frac{x_1y_2 + y_1x_2}{c(1 + dx_1x_2y_1y_2)}, \qquad y_3 = \frac{y_1y_2 - x_1x_2}{c(1 - dx_1x_2y_1y_2)}.$$

Bernstein and Lange also proved in [5] that all curves in the generalized form are isomorphic to curves defined by means of the equation $x^2 + y^2 = 1 + dx^2y^2$, which is the equation typically associated with regular *Edwards curves* in the literature. If $d \in \mathbb{F} - \{0, 1\}$, then in this type of curves the point $(0, 1)$ is the identity element of the addition operation, the point $(0, -1)$ has order 2, and the points $(1, 0)$ and $(-1, 0)$ have order 4.

One year later, Bernstein et al. proposed another generalization, producing as a result *twisted Edwards curves*, defined by the equation $ax^2 + y^2 = 1 + dx^2y^2$ [8]. The addition formula for twisted Edwards curves is the same as in the case of Edwards curves in the generalized form, where $c = 1$ in the case of twisted Edwards curves.

Theoretically, it would be possible to work with a more general model given by the equation $ax^2 + y^2 = c^2(1 + dx^2y^2)$. However, as the curves defined according to that model are always isomorphic to twisted Edwards curves, in practice it is not used [8].

The most interesting characteristic of complete twisted Edwards curves is that the equations for both adding two points P_1 and P_2 such that $P_1 \neq \pm P_2$ and doubling a point are exactly the same. Moreover, it is not necessary to implement any logic for detecting if the points to be added are such that $P_2 = -P_1$, as the Edwards addition equations also take into account that circumstance. This feature is important when defining the countermeasures to some type of side channel attacks.

16.2.2.3 Montgomery Curves

Montgomery curves conform to the equation $By^2 = x^3 + Ax^2 + x$, where $B(A^2 - 4) \not\equiv 0 \pmod{p}$. As in the case of Weierstrass curves, the identity element in Montgomery curves is the point at infinity $\mathcal{O}$, while the opposite element of $P = (x, y)$ is the point $-P = (x, -y)$. The addition of two points $P_1 = (x_1, y_1)$ and $P_2 = (x_2, y_2)$ such that $P_1 \neq \pm P_2$ is the point $P_3 = (x_3, y_3)$ with the following coordinates [7]:

$$x_3 = \frac{B(y_2 - y_1)^2}{(x_2 - x_1)^2} - A - x_1 - x_2, \quad y_3 = \frac{(2x_1 + x_2 + A)(y_2 - y_1)}{x_2 - x_1} - \frac{B(y_2 - y_1)^3}{(x_2 - x_1)^3} - y_1.$$

Unlike Edwards curves, it is necessary to use different equations for the doubling operation, so in this case the coordinates of the point $P_3 = 2P_1$ can be computed as follows [7]:

$$x_3 = \frac{B(3x_1^2 + 2ax_1 + 1)^2}{(2By_1)^2} - A - 2x_1,$$

$$y_3 = \frac{(3x_1 + A)(3x_1^2 + 2Ax_1 + 1)}{2By_1} - \frac{B(3x_1^2 + 2Ax_1 + 1)^3}{(2By_1)^3} - y_1.$$

16.2.3 Transforming Formulas

Complete twisted Edwards curves $ax^2 + y^2 = 1 + dx^2y^2$ are birationally equivalent to Montgomery curves $By^2 = x^3 + Ax^2 + x$, where two curves are birationally equivalent if their fields of rational functions are isomorphic [15] (when referring to projective non-singular curves, the term *birationally equivalent* simply means that projective non-singular curves are isomorphic). This means that every Montgomery curve can be expressed as a twisted Edwards curve, and vice versa [8]. In order to transform a complete twisted Edwards elliptic curve into a Montgomery elliptic curve as displayed in the previous section, it is necessary to use the following equivalence formulas [6]:

$$A = 2\frac{a + d}{a - d}, \qquad B = \frac{4}{a - d}.$$

In this way, a curve point (x_E, y_E) which belongs to an Edwards curve can be converted to a point (x_M, y_M) of the associated Montgomery curve, where the equations for obtaining (x_M, y_M) are as follows:

$$x_M = \frac{1 + y_E}{1 - y_E}, \qquad y_M = \frac{x_M}{x_E}.$$

Besides, in order to transform a Montgomery elliptic curve into the short Weierstrass form, it is necessary to use the following equivalences [6]:

$$a = \frac{3 - A^2}{3B^2}, \qquad b = \frac{2A^3 - 9A}{27B^3}$$

In this specific case, a curve point (x_M, y_M) belonging to a Montgomery curve can be converted to a point (x_W, y_W) of the associated short Weierstrass curve, where the transforming equations are the following ones:

$$x_W = \frac{x_M + \dfrac{A}{3}}{B}, \qquad y_W = \frac{y_M}{B}.$$

It is important to notice that, given the number and nature of the field operations needed in each case, elliptic curves in the short Weierstrass form have a better performance than the elliptic curves expressed in the Edwards and Montgomery forms when computing scalar multiplications, which is the basic operation when dealing with elliptic curves [14].

16.3 Brainpool

Even though elliptic curve cryptographic protocols are well defined in standards from ANSI [2, 3], IEEE [21, 22], ISO/IEC [23], NIST [30], and other organizations, it is usually the case that the elliptic curve parameters that are necessary to operate those protocols are offered to the reader without a complete and verifiable pseudo-random generation process. Some of the most important limitations detected across the main cryptographic standards regarding the processes for generating elliptic curves suitable for Cryptography are the following [10]:

- The seeds used to generate the curve parameters are typically chosen ad hoc.
- The primes that define the underlying prime fields have a special form aimed at facilitating efficient implementations.
- The parameters specified do not cover in all the cases key lengths adapted to the security levels required nowadays.

In this scenario, a European consortium of companies and government agencies led by the Bundesamt für Sicherheit in der Informationstechnik (BSI) was formed in order to study the aforementioned limitations and produce their recommendations for a well-defined elliptic curve generation procedure. The group was named ECC Brainpool (henceforth simply Brainpool), and, apart from the BSI, some of the most relevant companies and public institutions that took part in the effort were G&D, Infineon Technologies, Philips Electronics, Gemplus (now part of Gemalto), Siemens, the Technical University of Darmstadt, T-Systems, Sagem Orga, and the Graz University of Technology.

In 2005, Brainpool delivered the first version of a document entitled "ECC Brainpool standard curves and curve generation" [10], which was revised and published as an RFC memorandum in 2010, the "Elliptic Curve Cryptography (ECC) Brainpool standard curves and curve generation" [25]. The following sections present the main characteristics of the Brainpool procedure.

16.3.1   Key Length

As mentioned before, the Brainpool procedure only manages elliptic curves defined over prime fields expressed in the short Weierstrass form. The key lengths allowed by Brainpool are 160, 192, 224, 256, 320, 384, and 512 bits [25, p. 6].

16.3.2   Seed Generation

The seeds used in Brainpool are generated in a systematic and comprehensive way. These seeds have been obtained as the first 7 substrings of 160 bits each of the number $\Pi \cdot 2^{1120} = Seedp160||\dots||Seedp512||Remainder$, where $||$ denotes the concatenation operator [25, p. 24].

16.3.3   Seed to Candidate Conversion

Brainpool uses SHA-1 [25, p. 22] during the process of finding candidates for the parameters p, a, and b, as it can be observed in Fig. 16.1, where L represents the bit length of p. Even though it is not recommended to use SHA-1 as a hashing function in security environments (e.g., digital signatures), it is important to note that in this context it is only used for generating candidate values.

As the output of the hashing function SHA-1 is 160 bits, and for different curves the length of the resulting parameters must be necessarily different, Brainpool performs a loop concatenating several SHA-1 outputs until the concatenated number has the proper bit length [25, pp. 22 and 24].

In addition to that, Brainpool uses two functions to generate the candidates, one for p and another for a and b. Those functions are very similar; in fact, the only difference is that the most significant bit of a and b is forced to be 0 [25, p. 24]. Given that another requirement states that the most significant bit of p must be 1 [25, p. 23], this implies that the values a and b generated are such that $a, b < p$.

16.3.4   Validation of Parameters a and b

In Brainpool, once the algorithm has determined the value of p, it starts searching the proper values for the elliptic curve parameters a and b, as it can be seen in Fig. 16.2. When a candidate pair is found, the resulting curve is tested against the security requirements. In case the curve is rejected, both a and b are discarded, starting a new search for a proper pair [25, p. 25].

Fig. 16.1 Generation of
candidates for parameters p,
a, and b in Brainpool

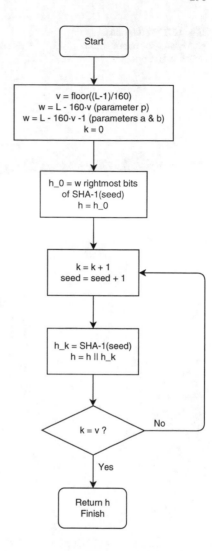

16.3.5   Cofactors

In order to generate cryptographically strong elliptic curves, it is necessary to compute the number of points of the elliptic curve and to determine if that value is a prime number or if it has a small cofactor. In this regard, the Brainpool specifications only accept curves whose number of points is a prime number [25, p. 6]. This means that the Brainpool curves cannot be transformed into the twisted Edwards or Montgomery forms, as in those types of curves the number of points is always divisible by 4.

Fig. 16.2 Validation of candidates for parameters *a* and *b* in Brainpool

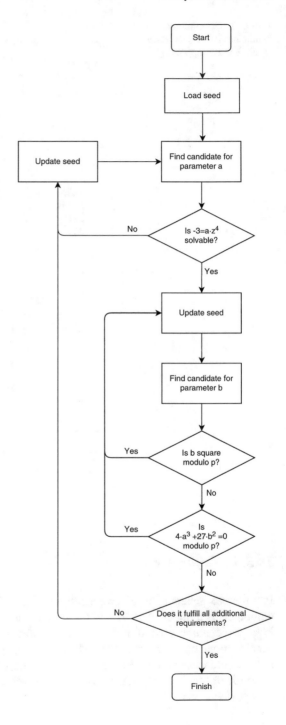

16.3.6   Factorizations

Two of the security requirements defined by Brainpool imply the factorization of integers. In one case, it is necessary to factorize the value $|E| - 1$, where q is the order of the elliptic curve, in order to avoid attacks using the Weil or Tate pairings. Those attacks allow the embedding of the cyclic subgroup of the elliptic curve into the group of units of a degree-l extension field of $\mathbb{F}_p$, where subexponential attacks on the DLP exist [10, p. 5].

In the other case, the specification requests to factorize the value d, which is the square-free factor of $4p - u^2$, where $u = |E| - p - 1$, so it can be checked that the class number of the maximal order of the endomorphism ring of the elliptic curve is larger than 10^7 [10, p. 5].

16.4   SafeCurves

Researchers Daniel Bernstein and Tanja Lange explain in their website SafeCurves [6] how all the standards that include recommended elliptic curves have addressed ECDLP security but not ECC security. The standards and official documents analyzed by SafeCurves are ANSI X9.62 [3], IEEE P1363 [21], SEC 2 [35], NIST FIPS 186 [30], ANSI X9.63 [2], Brainpool [10, 25], NSA Suite B [32], and ANSSI FRP256V1 [1].

Bernstein and Lange state that elliptic curves designed to be ECDLP secure may be attacked if they are not implemented properly, which would allow, for example, to produce incorrect results for some rare curve points, leak secret data when the input is not a curve point, or provide secret data through branch or cache timing attacks. The authors believe that secure implementation of the previously mentioned standard curves is theoretically possible but very hard. In order to avoid those implementation problems, the authors propose to use new curves which allow simple and secure implementations.

SafeCurves includes an evaluation of 20 curves taken from several sources (one SEC 2 curve, one ANSSI curve, two curves provided as examples of bad design, two Brainpool curves, three NIST curves, 5 Montgomery curves, and 6 Edwards curves) showing that some of them do not pass all their security requirements, which are divided into 3 main groups: curve parameters, ECDLP security, and ECC security.

16.4.1   Curve Parameters

In 2006, Bernstein stated that prime fields "have the virtue of minimizing the number of security concerns for elliptic-curve cryptography" [4], citing as two examples [13] and [17]. Later, he affirmed that "there is general agreement that prime fields are the safe, conservative choice for ECC" [6]. Sharing that point of view, other standards and recommendations like Brainpool [25], NSA Suite B [32], or BSI TR-03111 [11] consider only prime fields for all type of applications.

While in all the Brainpool curves the order of the generator G equals the number of points of the curve (i.e., the cofactor is 1), the Montgomery and Edwards examples from SafeCurves have cofactors whose value is either 4 or 8. Small subgroups attacks, which could take advantage of curves with a cofactor greater than 1, are easily deactivated either by implementing the appropriate software checks or by the inherent characteristics of the Montgomery and Edwards curves described in [6].

Regarding the key length, the curves analyzed by SafeCurves have one of the following lengths: 221, 222, 251, 255, 382, 383, 414, 448, 511, and 521 bits.

16.4.2   ECDLP Security

Bernstein and Lange analyzed how resistant are the curves against some well-known attacks to the ECDLP. More specifically, they studied the following characteristics:

- *Rho method*: Bernstein and Lange stated that the rho method [34] breaks ECDLP using, on average, approximately $\sqrt{\Pi|G|/4}$ additions, where $|G|$ represents the order of the generator of the set of elliptic curve points used in the computations. SafeCurves requires curves such that that value is over 2^{100}.
- *Transfers*: In the authors' language, a "transfer" converts the ECDLP into the DLP using linear algebraic groups. Multiplicative transfers were introduced in 1993 and are often referred to as "the MOV attack" [19, 27]. Additive transfers were introduced in 1998 and are sometimes called "the Smart attack" [18, 20]. SafeCurves checks if the elliptic curve is safe against additive and multiplicative transfers by computing a value associated to the parameters of the curve and checking if it is higher than a certain threshold.
- *Complex-multiplication field discriminant*: The number of rational points on an elliptic curve over $\mathbb{F}_p$ is $p + 1 - t$, where t is the trace of the curve, while the order of the generator G is a prime divisor of that value [12]. If s^2 is the largest square dividing $t^2 - 4p$, then it can be affirmed that $(t^2 - 4p)/s^2$ is a square-free negative integer. If $(t^2 - 4p)/s^2 \equiv 1 \pmod 4$, then D is defined as $(t^2 - 4p)/s^2$; otherwise, D is $4(t^2 - 4p)/s^2$. SafeCurves requires the absolute value of this complex-multiplication field discriminant D to be larger than 2^{100}.
- *Rigidity*: SafeCurves checks the degree to which the elliptic curve generation process is explained (e.g., the choice for the seed values, the operations performed on them to derive the parameters, etc.).

16.4.3 ECC Security

In this aspect of their study, Bernstein and Lange analyzed the following list of features:

- *Ladders*: The authors consider that the most important computation in ECC is the single-scalar multiplication. Montgomery curves support a very simple scalar-multiplication method, the Montgomery ladder [29], which is simpler than the standard short Weierstrass scalar-multiplication methods. The Montgomery ladder uses a single standard addition–subtraction–doubling chain, always following a simple, highly efficient double-add pattern.

    SafeCurves requires curves to support simple and constant-time multiplications, avoiding conflicts between efficiency and security. Both the Montgomery and Edwards curves included in the study satisfy this requirement.
- *Twist security*: If the original curve has $p + 1 - t$ points, then any nontrivial quadratic twist has in turn $p + 1 + t$ points. In a generic implementation, programmers have to be careful enough to include different checks (e.g., that the point sent by the other party effectively belongs to the elliptic curve, that the order of the point multiplied by the cofactor is not equal to $\mathcal{O}$, etc.) [8]. This requirement checks if the elliptic curve under analysis is protected against attacks derived from those situations without forcing the programmer to include specific code to handle those situations. In the scope of their contribution, the authors define such a curve as a "twist-secure" curve.
- *Completeness*: This requirement checks if the curve equations allow to use complete single-scalar and multi-scalar multiplications, in the sense that the same single point addition formula must return valid values in all the cases (i.e., when one of the input points is P and the other is $-P$, P, $\mathcal{O}$, or any other point) [5].
- *Indistinguishability from uniform random strings*: Standard representations of elliptic-curve points are easily distinguishable from uniform random strings. This poses a problem for many cryptographic protocols using elliptic curves (e.g., censorship-circumvention protocols, password-authenticated key-exchange protocols, etc.). One of the workarounds for this problem is for the protocol to bounce randomly between a curve and its twist, but according to the authors this is a complicated and error-prone approach [9].

    In 2013, a team of researchers led by Bernstein and Lange proposed a solution to this problem using bijective maps. Hence, this requirement checks the availability of those bijective maps for the curves under analysis.

16.4.4 Results

Using the previously described criteria, the authors of SafeCurves evaluated the aforementioned 20 curves. As it was expected given the content of the requirements, all the curves originally described using the short Weierstrass form do not satisfy at least three of the four ECC security requirements. Apart from that, the SEC 2 curve does not pass the complex-multiplication field discriminant test, while the NIST and the ANSSI curves are considered to be manipulable. Regarding the two Brainpool curves considered in the comparison, they pass all the parameter and ECDLP security requirements.

16.5   Conclusions

Selecting a secure elliptic curve is one of the most important steps when using an ECC algorithm. The Brainpool specification has been analyzed by the scientific community since its first appearance in 2005 and includes clear instructions that allow any interested researcher to replicate the curve generation procedure. However, when it was designed there were requirements that were not taken into account and that are important when deploying an ECC solution.

Compared to the Edwards and Montgomery curves analyzed in SafeCurves, the Brainpool curves do not fulfill some security requirements, though it could be argued that a careful implementation of the curves should avoid the attacks related to those requirements.

In addition to that, it is also important to note that, using standard coordinates, short Weierstrass curves outperform Edwards and Montgomery curves when computing scalar-multiplication operations. As it is usually the case, having a higher level of security by default comes at a cost.

Most software libraries and hardware devices (e.g., smart cards) released during the last years support curves using the short Weierstrass form, so the installed base of those curves is quite large. The implementation pace of Edwards and Montgomery curves is still slow, but it will definitely increase in the new years as new ECC protocols and applications are made public.

Acknowledgements This work has been partly supported by Ministerio de Economía y Competitividad (Spain) under the project TIN2014-55325-C2-1-R (ProCriCiS), and by Comunidad de Madrid (Spain) under the project S2013/ICE-3095-CM (CIBERDINE), cofinanced with the European Union FEDER funds.

References

1. Agence Nationale de la Sécurité des Systèmes d'Information. (2011). Avis relatif aux paramètres de courbes elliptiques définis par l'Etat français. http://www.legifrance.gouv.fr/affichTexte.do?cidTexte=JORFTEXT000024668816.
2. American National Standards Institute. (2001). Public Key Cryptography for the Financial Services Industry: Key Agreement and Key Transport Using Elliptic Curve Cryptography. ANSI X9.63.
3. American National Standards Institute. (2005). Public Key Cryptography for the Financial Services Industry: The Elliptic Curve Digital Signature Algorithm (ECDSA). ANSI X9.62.

4. Bernstein, D. J., & Lange, T. (2007). Curve25519: New Diffie-Hellman speed records. In *Proceedings of the 9th International Conference on Theory and Practice in Public-Key Cryptography (PKC 2006)* (pp. 207–228).
5. Bernstein, D. J., & Lange, T. (2007). *Faster addition and doubling on elliptic curves* (pp. 29–50). Berlin/Heidelberg: Springer.
6. Bernstein, D. J., & Lange, T. (2014). SafeCurves. http://safecurves.cr.yp.to/.
7. Bernstein, D. J., & Lange, T. (2016). Explicit-Formulas Database. https://hyperelliptic.org/EFD/.
8. Bernstein, D. J., Birkner, P., Joye, M., Lange, T., & Peters, C. (2008). Twisted Edwards curves. Cryptology ePrint Archive, Report 2008/013. http://eprint.iacr.org/2008/013.
9. Bernstein, D. J., Hamburg, M., Krasnova, A., & Lange, T. (2013). Elligator: Elliptic-curve points indistinguishable from uniform random strings. In *Proceedings of the 2013 Conference on Computer & Communications Security* (pp. 967–980).
10. Brainpool. (2005). ECC Brainpool Standard Curves and Curve Generation. Version 1.0. http://www.ecc-brainpool.org/download/Domain-parameters.pdf.
11. Bundesamt für Sicherheit in der Informationstechnik. (2012). Elliptic Curve Cryptography. BSI TR-03111 version 2.0. https://www.bsi.bund.de/SharedDocs/Downloads/EN/BSI/Publications/TechGuidelines/TR03111/BSI-TR-03111_pdf.pdf?__blob=publicationFile.
12. Cohen, H., & Frey, G. (2006). *Handbook of elliptic and hyperelliptic curve cryptography*. Boca Raton, FL: Chapman & Hall/CRC.
13. Diem, C. (2003). The GHS attack in odd characteristic. *Journal of the Ramanujan Mathematical Society, 18*, 1–32.
14. Durán Díaz, R., Gayoso Martínez, V., Hernández Encinas, L., & Martín Muñoz, A. (2016). A study on the performance of secure elliptic curves for cryptographic purposes. In *Proceedings of the International Joint Conference SOCO'16-CISIS'16-ICEUTE'16* (pp. 658–667).
15. Edwards, H. M. (2007). A normal form for elliptic curves. *Bulletin of the American Mathematical Society, 44*, 393–422.
16. ElGamal, T. (1985). A public-key cryptosystem and a signature scheme based on discrete logarithm. *IEEE Transactions on Information Theory, 31*, 469–472.
17. Frey, G. (1998). How to Disguise an Elliptic Curve (Weil Descent). http://www.cacr.math.uwaterloo.ca/conferences/1998/ecc98/slides.html.
18. Frey, G. (2001). Applications of arithmetical geometry to cryptographic constructions. In *Proceedings of the 5th International Conference on Finite Fields and Applications* (pp. 128–161). Heidelberg: Springer.
19. Frey, G., & Ruck, H. (1994). A remark concerning *m*-divisibility and the discrete logarithm in the divisor class group of curves. *Mathematics of Computation, 62*, 865–874.
20. Gaudry, P., Hess, F., & Smart, N. P. (2002). Constructive and destructive facets of Weil descent on elliptic curves. *Journal of Cryptology, 15*, 19–46.
21. Institute of Electrical and Electronics Engineers: Standard Specifications for Public Key Cryptography. IEEE 1363 (2000).
22. Institute of Electrical and Electronics Engineers: Standard Specifications for Public Key Cryptography - Amendment 1: Additional Techniques. IEEE 1363a (2004).
23. International Organization for Standardization/International Electrotechnical Commission: Information Technology-Security Techniques-Encryption Algorithms—Part 2: Asymmetric Ciphers. ISO/IEC 18033-2 (2006).
24. Koblitz, N. (1987). Elliptic curve cryptosytems. *Mathematics of Computation, 48*(177), 203–209.
25. Lochter, M., & Merkle, J. (2010). Elliptic curve cryptography (ECC) Brainpool standard curves and curve generation. Request for Comments (RFC 5639), Internet Engineering Task Force.
26. Menezes, A. J. (1993). *Elliptic curve public key cryptosystems*. Boston, MA: Kluwer Academic Publishers.
27. Menezes, A., Okamoto, W., & Vanstone, S. (1993). Reducing elliptic curve logarithms to logarithms in a finite field. *IEEE Transactions on Information Theory, 39*, 1639–1646.

28. Miller, V. S. (1986). Use of elliptic curves in cryptography. In *Lecture Notes in Computer Science* (Vol. 218, pp. 417–426). Berlin: Springer.
29. Montgomery, P. L. (1987). Speeding the Pollard and elliptic curve methods of factorization. *Mathematics of Computation, 48*, 243–264.
30. National Institute of Standards and Technology: Digital Signature Standard (DSS). NIST FIPS 186-4 (2009).
31. National Institute of Standards and Technology: Report on Post-quantum Cryptography (2016). http://nvlpubs.nist.gov/nistpubs/ir/2016/NIST.IR.8105.pdf.
32. National Security Agency: NSA Suite B Cryptography (2009). http://www.nsa.gov/ia/programs/suiteb_cryptography/index.shtml.
33. National Security Agency: Commercial National Security Algorithm Suite (2015). https://www.iad.gov/iad/programs/iad-initiatives/cnsa-suite.cfm.
34. Pollard, J. (1978). Monte Carlo methods for index computation mod *p. Mathematics of Computation, 32*, 918–924.
35. Standards for Efficient Cryptography Group: Recommended Elliptic Curve Domain Parameters. SECG SEC 2 version 2.0 (2010).

Chapter 17
Mathematical Models for Malware Propagation in Wireless Sensor Networks: An Analysis

A. Martín del Rey and A. Peinado

17.1 Introduction

Wireless sensor networks are networks constituted by a large number of sensor nodes that cooperatively collect and transmit data from the environment or other scenarios, allowing interaction between individuals or computers and the surrounding environment [27]. WSNs play a basic role in many new paradigms and scenarios: Internet of Everythings, Industry 4.0, Critical Infrastructures, etc. Due to the nature and the special characteristics of this technology, it is exposed to several security threats ranging from DoS attacks to Sybil, traffic analysis, node replication, privacy, or physical attacks [9]. Furthermore, the increase of computational and storage capabilities of the sensors makes possible the propagation of malicious code in WSNs and, by extension, in the other networks to which they are connected. Consequently, special security and performance issues have to be carefully considered for this type of wireless networks.

This work deals with malware propagation in WSNs; specifically, an analysis of the mathematical models proposed in the scientific literature is performed by focusing the attention on heterogeneous network models.

The rest of the paper is organized as follows: in Sect. 17.2, a description of wireless sensor networks and the major threats to which they are exposed is introduced; a study of the mathematical models to simulate malware propagation in WSNs is shown in Sect. 17.3; in Sect. 17.4, the mathematical models in

A. Martín del Rey (✉)
Department of Applied Mathematics, University of Salamanca, Salamanca, Spain
e-mail: delrey@usal.es

A. Peinado
Dept. Ingeniería de Comunicaciones, Universidad de Málaga, Andalucía Tech,
E.T.S.Ingeniería de Telecomunicación, Málaga, Spain
e-mail: apeinado@ic.uma.es

© Springer International Publishing AG 2018
K. Daimi (ed.), *Computer and Network Security Essentials*,
DOI 10.1007/978-3-319-58424-9_17

heterogeneous networks are detailed focusing the attention on the SIS (Susceptible-Infected-Susceptible) and SIR (Susceptible-Infected-Recovered) models; finally, the conclusions and further work are introduced in Sect. 17.5.

17.2   Wireless Sensor Networks and Their Security

A WSN is usually defined as a network of nodes, usually a large number of them, designed to sense, monitor, control, or interact with the environment, in such a way that the nodes work cooperatively to reach a common objective using wireless communication channels. Hence, a WSN can be studied making the analogy with a group of individuals, where each node partially performs two out of the three vital functions of the human beings: nutrition and interaction; that is, the nodes need energy to work and their behavior depends on the interactions to the other nodes and to the environment, taking into account that the decision adopted by a singular node may affect the whole group. This fact allows one to apply biological models to study the global behavior of the network. However, although they all share common functionalities, it is important not to forget the different features and capabilities of each node, which can be classified into three categories as a function of the microcontroller [21], thus determining the type and capacity of the power unit. As a consequence, a global model does not cover the details provided by the diversity of individuals, while an individual model may require a very complex management.

From a technical point of view, data is transmitted hopping from node to node until the destination is reached, using specifications oriented to the energy saving [16], such as Bluetooth 4.0 Low Energy [2] (in medical applications), IEEE 802.15.4e [14] (for industrial developments), or WLAN IEEE 802.11™[15] (for IoT).

The major vulnerabilities that one may identify in a WSN are derived from the constraints of the nodes because they do not facilitate the implementation of traditional cryptographic algorithms and protocols. Furthermore, it is easy to physically access sensor nodes in order to destroy them or infect them with malware. The wireless feature and distributed nature of the WSNs turn out itself also into a vulnerability since the messages (data and control information) can easily be intercepted, and the status of each node is difficult to monitor, thus keeping the failures and attacks unnoticed. Malware and its propagation in WSNs represent an important threat because it can become a very efficient way to launch a wide range of attacks [22] (*denial of service attack, impersonation attacks, traffic analysis, node replication, and tampering*). For this reason, the study of the propagation of malware is of vital importance in order to predict its effects on the global system, that is, on the own WSN and on the system or application that uses it, e.g., a critical infrastructure or an intelligent transport system.

17.3   Mathematical Models for Malware Propagation
          in WSNs

Mathematical epidemiology [3, 10, 18] is devoted to the study and design of mathematical models for biological agents (virus, bacteria, etc.) propagation. This branch of Mathematics goes back to 1776 when D. Bernoulli was the first to use mathematical tools to understand epidemic processes; specifically, his main contribution was to express the proportion of susceptible (healthy) individuals of an endemic infection in terms of the force of infection and life expectancy [6]. There were no significant advances in this field until nineteenth century, when Hamer and Roos proposed the use of the law of mass action to explain (in a mathematical way) the infection process [12, 31]. However, the most important milestone was due to Kermack and McKendrick who proposed a mathematical model to study the propagation of the plague in London during 1665–1666 [19]. This is a SIR compartmental model where the population is divided into three classes: susceptible, infectious, and recovered individuals, such that susceptible become infectious when the biological agents reach them, and when these pathogens disappear the infectious host becomes recovered. The number of individuals in each of these compartments stands for the time variables of the system of three ordinary differential equations that governs the dynamics of the model. The coefficients of this system are the transmission rate a and the recovery rate $b = \frac{1}{T}$, where T is the length of the infectious period. A fundamental result derived from the work of Kermack and McKendrick is that it substantiated the notion of epidemiological threshold parameter by introducing the *basic reproductive number* $R_0 = \frac{a}{b}N$, where N is the total number of individuals. Roughly speaking, this threshold parameter represents the average number of secondary infections caused by only one infectious individual during its infectious period, in the case that all the members of the population are susceptible. Kermack and McKendrick explicitly computed this coefficient and demonstrated that if $R_0 < 1$, then the infectives will decrease so that the disease will go to extinction, whereas if $R_0 > 1$, then the infectives will increase so that the disease cannot be eliminated and remains endemic.

The Kermack–McKendrick model is a *global* model since the population is homogeneously distributed and randomly mixed, that is, each individual has a small and equal chance of contacting every other individual (in fact, each individual is in contact with the rest of individuals of the population). Consequently, the contact topology is homogeneous. Moreover, all the individuals are considered to be endowed with the same characteristics (individual diversity is not taken into account). The importance of this model lies in the fact that it established a paradigm on which the great majority of subsequently models are based (see [13] and references therein). Global models are usually based on systems of differential equations, and there is a well-established mathematical theory to study the behavior of such systems; this provides us a complete knowledge about the dynamics of the evolution of all compartments (susceptible, infectious, recovered, etc.)

On the opposite side of global models, there are the so-called *individual-based* models. This alternative paradigm considers the particular characteristics of all individuals of the population: not only the local interactions (the number of neighbors varies from one individual to another which yields a particular contact topology) but also the properties in relation to malware: the individuals could have different transmission rates, recovery rates, etc. These models are usually based on discrete mathematical tools such as cellular automata, agent-based models, etc. In this case, both the global evolution of the population and the individual behavior of the system are derived. Nevertheless, the study of the qualitative properties of the dynamics of the model is more difficult, and, in some cases, only the data obtained from empirical simulations can be used to obtain behavioral patterns.

It is important to remark that global models can capture some topological features and consider different classes of individuals according to their contact structures. In this case, *network* (global) models appear. These are also compartmental models where the population is classified into some compartments taking into account both the relation with the malware and the topological structure. For example, infectious compartment is also divided into infectious with 1 neighbor (1-degree infectious individuals), infectious with 2 neighbors (2-degree infectious individuals), etc. The topological structure of the compartment of infectious with k neighbors is defined by means of a degree distribution $P(k)$ whose explicit expression determines the type of the network model: *random* network models, *lattice* network models, *small-world* network models, *spatial* network models, *scale-free* network models, etc.

These paradigms have also been used to design mathematical models for malicious code spreading in different environments (see, for example, [17, 28]). In this sense, several models have appeared in the last years dealing with the simulation of malware propagation in wireless sensor networks [5, 30]. The great majority are global models and usually four compartments are considered: susceptible, exposed (the malware is in latent state), infectious, and recovered. Thus, we can highlight the SIRS model by Feng et al. [7] when reinfection is considered, the SEIRS-V model by Mishra and Keshri [26] where a maintenance mechanism in the sleep mode of nodes is considered in order to improve the antimalware capability, the delayed SEIRS-V model proposed by Zhang and Si [33], or the SCIRS model [23] where a novel compartment constituted by carrier sensor nodes is detailed and studied.

In the particular case of network models, some proposals have appeared: the model by Khayam and Radha [20] which is a topological-aware worm propagation model where a rectangular grid is considered; Ping and Rong [29] proposed a model based on a cluster structure of geographic-adaptive fidelity showing that this type of topological structure could inhibit the malware spreading without considering security countermeasures; Vasilakos introduced the study of malware spreading in WSNs defined by small-world topologies [32]; and a hierarchical tree-based small-world topology for WSNs is considered in [21] in order to design a malware spreading model.

Finally, also individual-based models have been designed to simulate malware propagation in wireless sensor networks. In [25] the individual-based version of the global model proposed by Zhu and Zhao in [34] was introduced, and in [24], a model based on cellular automata is proposed.

Although the use of individual-based models may seem the best option to design a simulation tool, it is also true that network models are also adequate to achieve this goal. However, in no case does it seem appropriate to use global models. Note that although the great majority of sensor nodes in a WSN have similar functionalities, the local topology varies from one to another.

17.4 Network Propagation Models

17.4.1 Mathematical Background on Networks

A *network* $\mathcal{N}$ can be defined as a collection of entities and links that interconnects some of these entities, such that each link represents a certain type of interaction or association between the involved entities. The notion of network corresponds to the mathematical notion of graph $\mathcal{G} = (V, E)$ [11], such that the entities stand for the nodes or vertices, V, and the links represent the edges of the graph, E. In fact, networks can be understood as formal abstractions of physical, logical, or social systems [1].

A *complex network* can be defined as a network that exhibits emergent behaviors that cannot be predicted a priori from known properties of the individual entities constituting the network [17]. Usually complex networks are constituted by nodes with different intelligence and processing capabilities, whereas the links between them depend on this nature (packet routing in communication networks, malware spreading in computer networks, biological agent spreading in human networks, information pathways in cell networks, influence in social networks, etc.) Considering the origin of network formation and operation, complex networks can be classified into natural (transcriptional networks, virus spreading, neuron networks, etc.), human-initiated (social networks, malware diffusion, linguistic networks, etc.), or artificial networks (computer networks, air-traffic networks, power grids, etc.)

Nevertheless, the most important classification of complex networks deals with the topological structure of these networks. It depends on the node degree (i.e., the number of neighbor nodes adjacent to the specific node by means of a link) and the degree distribution, $P(k)$, which stands for the probability that a randomly chosen node has degree k. In this sense, complex networks can be classified into *homogeneous* networks and *heterogeneous* networks.

Homogeneous networks are defined by a uniform degree distribution, that is, all nodes have the same degree, and, consequently, the underlying topology is mathematically defined by a k-regular graph. The paradigmatic example of this class of networks is given by complete networks where all nodes are linked with all nodes: the topology is defined by a $(N - 1)$-regular graph (complete graph).

On the other hand, heterogeneous networks are characterized by a topological structure which follows a non-regular distribution. The most important types of heterogeneous networks are random networks, small-world networks, and scale-free networks. *Random* networks are usually defined by a normal degree distribution; the nodes in a *small-world* network have a small number of neighbors, but the average distance between them also remains small. Finally, the degree distribution of a *scale-free* network follows a power law: $P(k) \sim k^{-\gamma}$, where $2 \leq \gamma \leq 3$. In this type of heterogeneous networks, the highest degree nodes are usually called *hubs*, and they serve for specific purposes in their networks.

Wireless sensor networks follow different topologies depending on the corresponding application and the environment where they are deployed. These topologies can be defined by arbitrary degree distributions (mesh network or multihop network) or a uniform degree distribution (grid topology).

17.4.2 Mathematical Models in Heterogeneous Networks

Taking into account the notation used in the last sections, global models are mathematical models based on homogeneous networks (those whose associated graphs are complete or regular graphs). On the other hand, network models stand for mathematical models based on heterogeneous networks. In this section, the basics of this last type of models are stated detailing the classic SIS and SIR models.

17.4.2.1 General Considerations

Let us suppose that the connection topology of the WSN is defined by a heterogeneous complex network $\mathcal{N} = (V, E)$ constituted by N sensor nodes and M edges. If $P(k)$ stands for the probability that a randomly chosen sensor node has degree k (k-node) and N_k is the number of k-nodes in the WSN, then $P(k) = \frac{N_k}{N}$. If $\langle k \rangle = \frac{2M}{N}$ is the average degree of G, then it is easy to check that

$$\langle k \rangle = \sum_{k=1}^{k_{\max}} kP(k), \qquad (17.1)$$

where $k_{\max}$ is the maximum sensor node degree of the network. As a consequence, the following result holds [4]:

Proposition 1 *In a WSN defined by a non-correlated heterogeneous complex network, the probability that an edge connects to a k-node is given by* $\frac{kP(k)}{\langle k \rangle}$.

Set $I_k(t)$ the number of infectious k-nodes at t; then $I(t) = \sum_{k=0}^{N} I_k(t)$ represents the number of infectious sensors at time t. If $\rho_{I,k}(t) = \frac{I_k(t)}{N_k}$ is the relative density of infectious k-nodes at t, the absolute density of infectious k-nodes, $\rho_I(t)$, is defined as $\rho_I(t) = \sum_{k=0}^{N} P(k) \rho_{I,k}(t)$.

Proposition 2 *In a WSN defined by a non-correlated heterogeneous complex network, the probability that an edge connects to an infectious sensor node at time t, $\Theta(t)$, is given by the following expression:*

$$\Theta(t) = \frac{1}{\langle k \rangle} \sum_{k=1}^{N} kP(k)\rho_{I,k}(t). \tag{17.2}$$

Proof The probability that an edge connects to an infectious sensor node is $\Theta(t) = \sum_{k=1}^{N} P_I(k)$ where $P_I(k)$ is the probability that an edge connects to an infectious k-sensor node. As

$$P_I(k) = \left\{ \begin{array}{l} \text{Probability that an edge} \\ \text{connects to a } k\text{-node} \end{array} \right\} \times \left\{ \begin{array}{l} \text{Probability that a } k\text{-node} \\ \text{will be infectious} \end{array} \right\}, \tag{17.3}$$

and taking into account Proposition 1 and Laplace rule, it yields:

$$P_I(k) = \frac{kP(k)}{\langle k \rangle} \frac{I_k(t)}{N_k}. \tag{17.4}$$

Consequently

$$\Theta(t) = \sum_{k=1}^{N} \frac{kP(k)}{\langle k \rangle} \frac{I_k(t)}{N_k} = \sum_{k=1}^{N} \frac{kP(k)}{\langle k \rangle} \rho_{I,k}(t) = \frac{1}{\langle k \rangle} \sum_{k=1}^{N} kP(k)\rho_{I,k}(t), \tag{17.5}$$

thus finishing.

We refer the reader to the work by Chen et al. [4] for a more detailed description of the theory of complex networks.

17.4.2.2 The SIS Model

In the SIS classic model, the following assumptions are made [8]:

- A susceptible sensor becomes infectious with a certain probability α when there exists an adequate contact with an infectious sensor (i.e., when there exists an edge between the susceptible and the infectious sensors).
- An infectious sensor becomes susceptible again when the malware is successfully detected and removed from it. The recovery rate ν rules this process. Note that in this case the security countermeasures only confer temporary immunity (see Fig. 17.1).
- Population dynamics is not considered, that is, sensor nodes are not removed and new sensors do not appear in the WSN. As a consequence, the total population remains constant over the time: $\rho_{S,k}(t) + \rho_{I,k}(t) = 1$ for every $1 \leq k \leq N-1$, where $\rho_{S,k}(t)$ stands for the relative density of susceptible k-nodes at time t.

Fig. 17.1 Flow diagram
representing the dynamics of
a SIS compartmental model

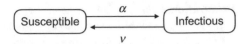

Table 17.1 Notation table for SIS model

Parameter	Description
α	Probability of infection
ν	Recovery rate
k	Number of neighbor devices
$\rho_{S,k}(t)$	Relative density of susceptible k-nodes
$\rho_{I,k}(t)$	Relative density of infectious k-nodes
$\Theta(t)$	Probability of connection with an infectious device

Taking into account these suppositions, the system of ordinary differential equations that governs the dynamics of the model is the following:

$$\begin{cases} \dfrac{d\rho_{S,k}(t)}{dt} = -\alpha k \rho_{S,k}(t)\,\Theta(t) + \nu\rho_{I,k}(t), & 1 \le k \le N-1, \\[2mm] \dfrac{d\rho_{I,k}(t)}{dt} = \alpha k \rho_{S,k}(t)\,\Theta(t) - \nu\rho_{I,k}(t), & 1 \le k \le N-1, \\[2mm] \rho_{S,k}(0) = 1 - \rho_k^0,\ \rho_{I,k}(0) = \rho_k^0, & 1 \le k \le N-1. \end{cases} \quad (17.6)$$

As the total number of sensor nodes remains constant, this system of ordinary differential equations can be reduced to the following:

$$\begin{cases} \dfrac{d\rho_{I,k}(t)}{dt} = \alpha k[1 - \rho_{I,k}(t)]\Theta(t) - \nu\rho_{I,k}(t), & 1 \le k \le N-1, \\[2mm] \rho_{I,k}(0) = \rho_k^0, & 1 \le k \le N-1. \end{cases} \quad (17.7)$$

Note that the addend $\alpha k[1 - \rho_{I,k}(t)]\Theta(t)$ stands for the new infectious k-nodes that appear. Specifically, as $\Theta(t)$ is the probability that an edge connects to an infectious sensor node and every susceptible sensor has k connections, $k\Theta(t)$ depicts the number of connections to infectious sensors of every susceptible k-node. Furthermore, $\alpha k\Theta(t)$ stands for the number of connections to infectious nodes that will lead to a successful propagation. Finally, considering the relative density of susceptible k-nodes, the fraction of new infectious k-nodes will be obtained: $\alpha k[1 - \rho_{I,k}(t)]\Theta(t)$. The quotient $\lambda = \frac{\alpha}{\nu}$ is called the *effective propagation rate*. In Table 17.1, the parameters and variables involved in the SIS model are summarized.

The steady states (or equilibrium points) of this system are obtained when the number of infectious sensor nodes does not change over time, that is, when $\frac{d\rho_{I,k}(t)}{dt} = 0$. A simple calculus shows that there exist two steady states: the *disease-free steady state* $E_0^* = (1, 0)$, where $\rho_{S,k} = 1$ and $\rho_{I,k} = 0$ for every k, and the *endemic steady state*

$$E_2^* = \left(1 - \sum_{k=1}^{N} P(k) \frac{\alpha k \Theta}{\alpha k \Theta + v}, \sum_{k=1}^{N} P(k) \frac{\alpha k \Theta}{\alpha k \Theta + v}\right),$$ (17.8)

where $\Theta = \lim_{t \to \infty} \Theta(t)$. Note that in this case

$$\rho_{I,k} = \frac{\alpha k \Theta}{\alpha k \Theta + v}, \quad 1 \le k \le N-1,$$ (17.9)

$$\rho_I = \sum_{k=1}^{N} P(k) \frac{\alpha k \Theta}{\alpha k \Theta + v}.$$ (17.10)

Moreover

$$\Theta = \frac{1}{\langle k \rangle} \sum_{k=1}^{N} k^2 P(k) \frac{\alpha \Theta}{\alpha k \Theta + v}.$$ (17.11)

The following result holds:

Theorem 1 *The endemic steady state exists if* $\lambda \ge \mu = \frac{\langle k \rangle}{\langle k^2 \rangle}$, *where* $\langle k^2 \rangle = \sum_{k=1}^{N} k^2 P(k)$.

Proof Set

$$\Theta = \frac{1}{\langle k \rangle} \sum_{k=1}^{N} k^2 P(k) \frac{\alpha \Theta}{\alpha k \Theta + v} = \mathscr{F}(\Theta).$$ (17.12)

As $\Theta \in [0,1]$, $\mathscr{F}(\Theta) \in [0,1)$ and $\mathscr{F}(0) = 0$, then the equation $\Theta = \mathscr{F}(\Theta)$ has a non trivial solution if $\left(\frac{d\mathscr{F}}{d\Theta}\right)_{\Theta=0} \ge 1$. As a consequence:

$$\left(\frac{d\mathscr{F}}{d\Theta}\right)_{\Theta=0} = \left(\frac{1}{\langle k \rangle} \sum_{k=1}^{N} \frac{\alpha v k^2 P(k)}{(\alpha k \Theta + v)^2}\right)_{\Theta=0} = \frac{\alpha}{\langle k \rangle v} \sum_{k=1}^{N} k^2 P(k) = \lambda \frac{\langle k^2 \rangle}{\langle k \rangle} \ge 1,$$ (17.13)

thus finishing.

Example 1 Suppose that the contact topology of a WSN is given by an scale-free network. Its average degree is $\langle k \rangle = 2M$, whereas its degree distribution is $P(k) = \frac{2m^2}{k^3}$, where m is the minimum degree of the sensor nodes. Then, from Eq. (17.11) a simple calculus shows that

$$\Theta = m\lambda\Theta \log\left(1 + \frac{1}{m\lambda\Theta}\right),$$ (17.14)

and consequently, the endemic steady state is obtained from the nontrivial solution of this equation:

$$\Theta = \frac{1}{m\lambda} \frac{1}{e^{\frac{1}{m\lambda}} - 1}.$$ (17.15)

Moreover, the total density of infectious nodes is

$$\rho_I = \sum_{k=1}^{\infty} P(k) \frac{\alpha k \Theta}{\alpha k \Theta + \nu} = 2m^2 \alpha \Theta \sum_{k=1}^{\infty} \frac{1}{k^2} \frac{1}{\alpha k \Theta + \nu}$$ (17.16)

$$= 2m^2 \alpha \Theta \int_m^{\infty} \frac{1}{k^2} \frac{dk}{\alpha k \Theta + \nu} = 2m^2 \lambda \Theta \left[\frac{1}{m} + \lambda \Theta \log \left(1 + \frac{1}{m \lambda \Theta} \right) \right],$$

in such a way that in the endemic equilibrium the proportion of infectious sensor nodes will be

$$\rho_I = \frac{2}{e^{\frac{1}{m\lambda}}} \left[1 + \frac{1}{m\lambda} \frac{1}{\left(e^{\frac{1}{m\lambda}} - 1 \right)^2} \right].$$ (17.17)

Note that $\lim_{\lambda \to 0} \rho_I = 0$, and, thus, $\lambda_c = 0$ if the topology of the WSN is defined by means of a scale-free network.

17.4.2.3 The SIR Model

The classic SIR model on non-correlated heterogeneous complex networks is characterized by the following assumptions [8]:

- A susceptible sensor node becomes infectious with probability $0 \le \alpha \le 1$ when there exists an adequate link with an infectious sensor.
- An infectious sensor node becomes recovered when the malware is successfully detected and removed. The recovery rate, $0 \le \beta \le 1$, governs this transition. Note that in this case the security countermeasures on infectious sensors confer permanent immunity.
- As in the previous case, it is supposed that population dynamic is not considered.
- Finally, "vaccination" processes (i.e., the implementation of security counter-measures on susceptible sensor nodes) is not allowed (Fig. 17.2).

Fig. 17.2 Flow diagram representing the dynamics of a SIR compartmental model

Table 17.2 Notation table for SIR model

Parameter	Description
α	Probability of infection
β	Recovery rate
k	Number of neighbor devices
$\rho_{S,k}(t)$	Relative density of susceptible k-nodes
$\rho_{I,k}(t)$	Relative density of infectious k-nodes
$\Theta(t)$	Probability of connection with an infectious device

As a consequence, the system of ordinary differential equations that defines the model is the following:

$$
\begin{cases}
\dfrac{d\rho_{S,k}(t)}{dt} = -\alpha k \rho_{S,k}(t)\,\Theta(t), & 1 \le k \le N-1, \\[2ex]
\dfrac{d\rho_{I,k}(t)}{dt} = \alpha k \rho_{S,k}(t)\,\Theta(t) - \beta \rho_{I,k}(t), & 1 \le k \le N-1, \\[2ex]
\dfrac{d\rho_{R,k}(t)}{dt} = \beta \rho_{I,k}(t), & 1 \le k \le N-1, \\[2ex]
\rho_{S,k}(0) = 1 - \rho_k^0, \ \rho_{I,k}(0) = \rho_k^0, \ \rho_{R,k}(0) = 0.
\end{cases}
\tag{17.18}
$$

In Table 17.2, the parameters and variables involved in the SIR model are introduced.

The steady states are obtained by equating to zero the equations of system (17.18), and a simple calculus shows that there is only one steady state in this case:

$$
(\rho_{S,k}, 0, 1 - \rho_{S,k}), \qquad 0 \le \rho_{S,k} \le 1,
\tag{17.19}
$$

such that the system always evolves to a disease-free steady state:

$$
E^* = \left(\sum_{k=1}^{N} P(k)\,\rho_{S,k}, \ 0, \ 1 - \sum_{k=1}^{N} P(k)\,\rho_{S,k} \right).
\tag{17.20}
$$

Furthermore, the following result holds:

Theorem 2 *Set* $\mu = \dfrac{\langle k \rangle}{\langle k^2 \rangle}$; *then*

(1) If $\lambda \le \mu$, *the system evolves to the disease-free steady state* $E^* \approx (1, 0, 0)$.
(2) If $\lambda > \mu$, *the system evolves to the disease-free steady state with the following explicit expression:*

$$E^* = \left(\sum_{k=1}^{N} P(k) e^{-\alpha k \phi_\infty}, 0, \sum_{k=1}^{N} P(k) \left(1 - e^{-\alpha k \phi_\infty}\right) \right), \tag{17.21}$$

where ϕ_∞ is the nontrivial solution of the equation

$$\phi_\infty = \frac{1}{\beta} - \frac{1}{\beta \langle k \rangle} \sum_{k=1}^{N} k P(k) e^{-\alpha k \phi_\infty}. \tag{17.22}$$

Proof Suppose initially (at $t = 0$) that all infectious nodes are homogeneously distributed: $\rho_k^0 = \rho^0$ for every $1 \leq k \leq N$. From the last equation of the system (17.18), it yields:

$$\rho_{R,k}(t) = \beta \int_0^t \rho_{I,k}(\tau) \, d\tau. \tag{17.23}$$

On the other hand, by integration of the first equation of (17.18) considering the initial conditions of this system, we obtain $\rho_{S,k}(t) = e^{-\alpha k \phi(t)}$, where $\phi(t) = \int_0^t \Theta(\tau) \, d\tau$. Using the explicit expression of $\Theta(t)$ and (17.23), it is:

$$\phi(t) = \int_0^t \Theta(\tau) \, d\tau = \frac{1}{\langle k \rangle} \sum_{k=1}^{N} k P(k) \int_0^t \rho_{I,k}(\tau) \, d\tau = \frac{1}{\beta \langle k \rangle} \sum_{k=1}^{N} k P(k) \rho_{R,k}(t). \tag{17.24}$$

By deriving both members of this equation, we obtain:

$$\Theta(t) = 1 - \frac{1}{\langle k \rangle} \sum_{k=1}^{N} k P(k) e^{-\alpha k \phi(t)} - \beta \phi(t). \tag{17.25}$$

Since $\rho_{I,k} = 0$ in the steady state, then taking limits in the last equation, the following equation holds:

$$\phi_\infty = \frac{1}{\beta} - \frac{1}{\beta \langle k \rangle} \sum_{k=1}^{N} k P(k) e^{-\alpha k \phi_\infty}. \tag{17.26}$$

It has a trivial solution $\phi_\infty = 0$, and taking into account (17.24), we obtain:

$$0 = \frac{1}{\beta \langle k \rangle} \sum_{k=1}^{N} k P(k) \rho_{R,k}, \tag{17.27}$$

that is, $\rho_{R,k} = 0$ and consequently $\rho_{S,k} = 1$, and the disease-free steady state is derived: $E^* = (1, 0, 0)$.

On the other hand, Eq. (17.26) has a nontrivial solution $0 < \phi_\infty \leq 1$ which leads to the disease-free steady state (17.21) when

$$\frac{d}{d\phi_\infty} \left(\frac{1}{\beta} - \frac{1}{\beta \langle k \rangle} \sum_{k=1}^{N} kP(k) e^{-\alpha k \phi_\infty} \right)_{\phi_\infty = 0} > 1. \qquad (17.28)$$

From this inequality, we obtain

$$1 < \frac{\alpha}{\beta \langle k \rangle} \sum_{k=1}^{N} k^2 P(k) = \lambda \frac{\langle k^2 \rangle}{\langle k \rangle}. \qquad (17.29)$$

As a consequence, if $\lambda > \mu = \frac{\langle k \rangle}{\langle k^2 \rangle}$, then the system evolves to the disease-free steady state defined by $\phi_\infty \in (0, 1]$, whereas if $\lambda \leq \mu$, then the system evolves to the disease-free steady state given by $\phi_\infty = 0$.

17.5 Conclusions

In this work, a study of mathematical models to simulate malware propagation in wireless sensor networks has been introduced. It is shown that those models where the contact topologies are based on complete or regular graphs (homogeneous networks) are not suitable for malicious code spreading, whereas the best option is to consider models based on heterogeneous networks defined by different degree distributions. Also individual-based models can be considered as an adequate framework, but, due to the homogeneity of the capabilities of the sensor nodes and the computational resources required for large networks, the use of (heterogeneous) network models seems to be more efficient.

The mathematical analysis of network models allows us to obtain the evolution patterns of the different compartments without the computation of several simulations with different initial conditions. Moreover, these behaviors depend on some threshold coefficients that can be explicitly computed in the case of heterogeneous networks.

These models exhibit two principal drawbacks. The first one is related to the nature of the phenomenon to be simulated (malware propagation over WSNs), whereas the second one deals with the design of mathematical models for malware propagation. Specifically, the first drawback consists in not considering the individual characteristics of the sensor nodes. As is mentioned above, this can be overcome since the great majority of sensor nodes have the same capabilities and it seems not necessary to implement an individual-based model to solve it; moreover, the nodes with different capabilities and functionalities can be included in new compartments (the system would have more equations). The second drawback is related to the definition of the parameters involved in the model (transmission rate, recovery rate, etc.). These parameters are inherited from mathematical epidemiology, and, consequently, the majority are not suitable to be used in malware propagation models. For example, the recovery rate is defined as the inverse of the length of the infectious period T; this is adequate for infectious diseases where the life cycle of biological agents is constrained to rigid statements, but it is unrealistic when the spreading of malicious code is tackled.

Consequently, future work aimed at defining in a proper way the coefficients of network models taking into account the specific characteristics of the malware specimens studied. Moreover, it is also necessary to obtain explicit expressions of the degree distributions of the different topological structures that WSNs can adopt. Finally, improved models would consider some centrality measures of complex networks such as betweenness, eigenvalue centrality, etc.

Acknowledgements We would like to thank the anonymous referees for their valuable suggestions and comments. This work has been supported by Ministerio de Economía y Competitividad (Spain) and the European Union through FEDER funds under grants TIN2014-55325-C2-1-R, TIN2014-55325-C2-2-R, and MTM2015-69138-REDT.

References

1. Barabási, A. L. (2002). *Linked*. Cambridge, MA: Plume.
2. Bluetooth SIG. (2010). *Bluetooth specification version 4*. Kirkland, WA, USA: The Bluetooth Special Interest Group.
3. Brauer, F. (2009). Mathematical epidemiology is not an oxymoron. *BMC Public Health, 9*, S2.
4. Chen, G., Wang, X., & Li, X. (2014). *Fundamentals of complex networks. Models, structures and dynamics*. Chichester, UK: Wiley.
5. De, P., & Das, S. K. (2009). Epidemic models, algorithms, and protocols in wireless sensor and Ad Hoc networks. In A. Boukerche (Ed.), *Algorithms and protocols for wireless sensor networks* (pp. 51–75). Hoboken, NJ: Wiley.
6. Dietz, K., & Heesterbeek, A. P. (2000). Bernoulli was ahead of modern epidemiology. *Nature, 408*, 513–514.
7. Feng, L., Song, L., Zhao, Q., & Wang, H. (2015). Modeling and stability analysis of worm propagation in wireless sensor networks. *Mathematical Problems in Engineering, 2015*, Article ID 129598.
8. Fu, X., Small, M., & Chen, G. (2015). *Propagation dynamics on complex networks. Models, methods and stability analysis*. Singapore: Wiley.
9. de Fuentes, J. M., González-Manzano, L., & Mirzaei, O. (2016). Privacy models in wireless sensor networks: A survey. *Journal of Sensors, 2016*, Article ID 4082084.
10. Grassly, N. C., & Fraser, C. (2008). Mathematical models of infectious disease transmission. *Nature Reviews-Microbiology, 6*, 477–487 .
11. Gross J. L., & Yellen, J. (Eds.). (2004). *Handbook of graph theory*. Boca Raton, FL: CRC Press.
12. Hammer, W. H. (1906). Epidemic disease in England. *Lancet, I*, 733–754.
13. Hethcote, W. H. (2000). The mathematics of infectious diseases. *SIAM Review, 42*, 599–653.
14. IEEE Computer Society. (2012). IEEE 802.15.4e-2012, IEEE Standard for local and metropolitan area networks – Part 15.4: Low-Rate Wireless Personal Area Networks (LR-WPANs) Amendment 1: MAC sublayer.

15. IEEE Computer Society. (2012). IEEE Std 802.11™-2012, Wireless LAN Medium Access Control (MAC) and Physical Layer (PHY) Specifications.
16. International Electrotechnical Commission: White Paper. Internet of Things: Wireless Sensor Network (2014).
17. Karyotis, V., & Khouzani, M. H. R. (2016). *Malware diffusion models for modern complex networks. Theory and applications.* Cambridge, CA: Morgan Kaufmann.
18. Keeling, M. J., & Danon, L. (2009). Mathematical modelling of infectious diseases. *British Medical Bulletin, 92,* 33–42.
19. Kermack, W. O., & McKendrick, A. G. (1927). A contribution to the mathematical theory of epidemics. *Proceedings of the Royal Society of London, Series A, 115,* 700–721.
20. Khayam, S. S., & Rahha, H. (2006). Using signal processing techniques to model worm propagation over wireless sensor networks. *IEEE Signal Processing Magazine, 23,* 164–169.
21. Li, Q., Zhang, B., Cui, L., Fan, Z., & Athanasios, V. V. (2014). Epidemics on small worlds of tree-based wireless sensor networks. *Journal of Systems Science and Complexity, 27,* 1095–1120.
22. López, J., & Zhou, J. (2008). *Wireless sensor network security.* Amsterdam: IOS Press.
23. Martín del Rey, A., Hernández Guillén, J. D., & Rodríguez Sánchez, G. (2016). A SCIRS model for malware propagation in wireless networks. In E. Corchado, et al. (Eds.), *Advances intelligence systems and computation* (Vol. 527, pp. 538–547). Berlin: Springer.
24. Martín del Rey, A., Hernández Guillén, J. D., & Rodríguez Sánchez, G. (2016). Modeling malware propagation in wireless sensor networks with individual-based models. In E. Corchado, et al. (Eds.), *Advances in artificial intelligence.* Lecture Notes in Artificial Intelligence (Vol. 9868, pp. 194–203). Berlin: Springer.
25. Martín del Rey, A., Hernández Encinas, A., Hernández Guillén, J. D., Martín Vaquero, J., Queiruga Dios, A., & Rodríguez Sánchez, G. (2016). An individual-based model for malware propagation in wireless sensor networks. In S. Omatu (Ed.), *Advances in intelligent systems and computation* (Vol. 474, pp. 223–230). Berlin: Springer.
26. Mishra, B. K., & Keshri, N. (2013). Mathematical model on the transmission of worms in wireless sensor network. *Applied Mathematical Modelling, 37,* 4103–4111.
27. Obaidat, M. S., & Misra, S. (2014). *Principles of wireless sensor networks.* Cambridge: Cambridge University Press.
28. Peng, S., Yu, S., & Yang, A. (2014). Smartphone malware and its propagation modeling: A survey. *IEEE Communications Surveys & Tutorials, 16,* 925–941.
29. Ping, S. X., & Rong, S. J. Y. (2011). A malware propagation model in wireless sensor networks with cluster structure of GAF. *Telecommunication Systems Journal, 27,* 33–38.
30. Queiruga-Dios, A., Hernández Encinas, A., Martín-Vaquero, J., & Hernández Encinas, L. (2016). Malware propagation in wireless sensor networks: A review. In E. Corchado, et al. (Eds.), *Advances in intelligence systems and computing* (Vol. 527, pp. 648–657). Berlin: Springer.
31. Ross, R. (1911). *The prevention of malaria* (2nd ed.). London: Murray.
32. Vasilakos, V. J. (2012). Dynamics in small world of tree topologies of wireless sensor networks. *Journal of Systems Engineering and Electronics, 23,* 325–334.
33. Zhang, Z., & Si, F. (2014). Dynamics of a delayed SEIRS-V model on the transmission of worms in a wireless sensor network. *Advances in Differential Equations, 2014,* 1–18.
34. Zhu, L., & Zhao, H. (2015). Dynamical analysis and optimal control for a malware propagation model in an information network. *Neurocomputing, 149,* 1370–1386.

Part IV
Biometrics and Forensics

Chapter 18
Biometric Systems for User Authentication

Natarajan Meghanathan

18.1 Introduction

People are normally verified or identified using one or more of the following three means: (1) With something they have (e.g., ID card, ATM card); (2) With something they know (e.g., passwords) and (3) With something they are (e.g., biometrics). Authentication schemes that are based on ID cards or passwords do not really differentiate between authorized users and persons who are in unauthorized possession. Biometrics includes methods to uniquely recognize humans based on one or more physiological or behavioral identifiers (referred to as biometric traits) using which the users can be authenticated to access data and system resources. Biometric identifiers can be divided into two main classes: (1) Physiological identifiers are those that are related to the body—often unique and can be used for identification as well as verification. Examples are: fingerprint, DNA, palm print, iris recognition, retinal scans, etc.; (2) Behavioral identifiers are those that are related to the behavior of a person—may not be unique for each person and can be used mainly for verification. Examples include: typing rhythm, body mechanics (gait), voice, etc.

In cryptographic systems, the possession of a decryption key is considered sufficient enough to authenticate a user. The cryptographic keys are significantly long and randomly chosen and hence it may not be easy for a user to remember. Hence, the decryption keys are often stored somewhere and are released based on an alternative authentication mechanism (e.g., password). Hence, data protected by a cryptographic system is only as secure as the password (is the weakest link) used to release the correct decryption keys to establish user authenticity. Many people use

N. Meghanathan (✉)
Department of Computer Science, Jackson State University, Mailbox 18839, Jackson,
MS, 39217, USA
e-mail: natarajan.meghanathan@jsums.edu

© Springer International Publishing AG 2018   317
K. Daimi (ed.), *Computer and Network Security Essentials*,
DOI 10.1007/978-3-319-58424-9_18

the same password for multiple applications. Hence, if an imposter can get access to the password, then he can login to several applications as a legitimate user. In multi-user account scenarios, passwords cannot provide non-repudiation. Instead of passwords, biometric systems could be used to protect the strong cryptographic keys [1]. Biometric identifiers are difficult to be lost or forgotten, difficult to be copied or shared, and require the person in question to be physically present while going through authentication. Biometric systems (like fingerprint, iris pattern, retinal images, etc.) employed for user identification (see Sect. 18.4) cannot be forged and a user cannot claim his biometric identifier was stolen and misused!! (i.e., non-repudiation). Moreover, for a given biometric identifier, the level of security is relatively the same for all users—one user's biometrics will not be relatively easy to break or forge than others. Also, there cannot be many users with something like an "easy to guess" biometrics that can be misused to launch intrusion or spoofing attacks.

The rest of the chapter is organized as follows: Sect. 18.2 illustrates the basic building blocks of a biometric system and describes each of them. Section 18.3 outlines the performance metrics used to evaluate biometric systems and identifies the tradeoffs. Section 18.4 presents different biometric systems widely employed for user identification and Sect. 18.5 presents the different biometric systems available for user verification. Section 18.6 compares the biometric systems presented in Sects. 18.4 and 18.5 based on several parameters considered critical for data collection and usage. Section 18.7 presents the different spoofing attacks that could be launched on biometric systems and explains the suitability of multi-biometric systems to prevent these attacks. Section 18.8 describes multi-biometric systems and the different levels of fusion in more detail. Section 18.9 concludes the chapter. For the rest of the chapter, the terms "trait" and "identifier" are used interchangeably; they mean the same.

18.2 Basic Block Diagram of a Biometric System

When an individual uses a biometric system for the first time, it is called enrollment. During the enrollment phase, biometric information from the individual is collected and securely stored in a database. During the subsequent attempts, biometric information is collected from the individual and compared with the information stored at the time of enrollment. The comparison is considered to be successful if the biometric sample collected falls within the threshold values, representing the identifier, in the database. The retrieval of the information from the database must be done in a secured fashion.

Typically, a biometric system for user authentication operates in two modes: (1) *Identification mode*: The biometrics captured of an unknown individual goes through a "one-to-many comparison" with those enrolled in the database and the identity is established if there is a match. (2) *Verification mode*: The biometrics

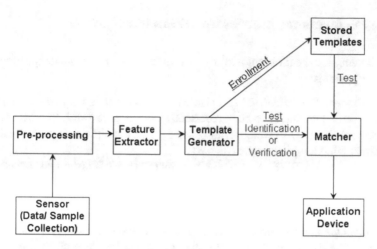

Fig. 18.1 Basic block diagram of a biometric system

captured of an individual, who is already identified through another means of authentication, goes through a "one-to-one comparison" with a stored template in the database to verify whether the individual is the person who claims to be.

Figure 18.1 illustrates a basic block diagram of a biometric system. We now describe the different blocks of the biometric system:

- *First block (sensor)*: The sensor is an interface between the biometric system and the real world. It is used to acquire all the necessary data, depending on the characteristic in consideration. An image acquisition system is the commonly used interface.
- *Second block (pre-processing)*: This block is needed to enhance the input (i.e., remove all the background noise and unnecessary artifacts during data collection) and also to use some kind of normalization, if needed.
- *Third block (feature extractor)*: This block is responsible to extract the necessary features from the pre-processed input in the correct and in the optimal way.
- *Fourth block (template generator)*: The template, typically a vector of numbers or an image, contains the relevant features extracted from the source (characteristic of the enrollee). Features that are not required by the comparison algorithm to establish or verify the identity of the enrollee are discarded from the template to reduce the file size and also for identity protection.

If an enrollment is performed, the template is typically stored in a central database (or sometimes within the biometric reader device itself or on a smartcard owned by the user). To perform a match, the test template is compared with the relevant templates in the databases to estimate the distance between the templates using specific algorithms and the results are returned to the application device, which will then decide how to handle the user being evaluated.

18.3   Performance Metrics for Biometric Systems

The following are the different performance metrics used to evaluate the efficacy of biometric systems:

- *False Accept Rate* (FAR, a.k.a. False Match Rate): It is defined as the percent of invalid inputs that have been incorrectly accepted as valid. In other words, it is the probability with which a biometric system matches an input template to a non-matching template in the database.
- *False Reject Rate* (FRR, a.k.a. False Non-match Rate): It is defined as the percent of valid inputs that have been incorrectly rejected as invalid. In other words, it is the probability with which a biometric system fails to detect a match between an input template and the relevant templates in the database.
- *Relative Operating Characteristic*: It is a curve drawn between the False Accept Rate vs. the False Reject Rate. The shape of the curve depends on the threshold value set for acceptance. If the threshold value (for the difference or the distance between the templates—tolerance to input variations and noise) is too small, the FAR would be low, but the FRR would also be high. If the threshold value is too high, the FAR would be high, but the FRR would be low.
- *Crossover Error Rate (CER)*: The rate at which both the False Accept Rate and the False Reject Rate errors are equal. A lower value for the CER is desired for a biometric system in order to be considered more accurate as well as convenient for its users. Fig. 18.2 illustrates the tradeoff and relationship between FAR, FRR, and CER.
- *Failure to Enroll Rate (FER)*: The rate at which attempts to enroll a user's template into the database turn out to be unsuccessful. This is often attributed either to the low quality inputs that have insufficiently distinct biometric samples characteristic of the trait and the user being enrolled or to poor system design

Fig. 18.2 Relationship and tradeoff between FAR, FRR, and CER (adapted from [2, 3])

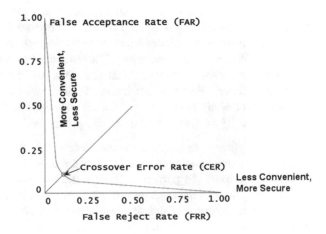

that fails to provide consistent biometric data. Also, it may not be possible to capture distinctive fingerprints from people who do lot of manual labor; Retinal scans require high precision—so people with diabetes, blood pressure cannot be considered for such biometric systems.

- *Failure to Capture Rate (FCR)*: The rate at which a biometric system fails to detect an input when presented correctly. This is mostly applicable for automated systems.
- *Template Capacity*: The number of users who can be uniquely identified based on their biometric templates. From an Information Theory perspective, an n-bit template should support holding the unique feature vectors for 2^n users. However, the capacity of n-bit biometric templates is often far less than 2^n because: Not all bit combinations may be valid as a feature vector representing a particular user. A single user would require more than one combination of bits, since it is not practically feasible to extract 100% identical biometric templates of a user at different instances.

18.4 Biometric Systems for User Identification

In this section, we describe some of the commonly used biometric systems for the purpose of "identification." These biometric systems have low FAR and FRR; but, the tradeoff is in the difficulty associated with data collection and usage. The biometric systems we describe in this section are: (1) fingerprint recognition; (2) iris recognition, and (3) retinal scans. The biometric systems employed for user identification also guarantee non-repudiation as no other user could have the identity (like fingerprint, iris pattern, retinal image, etc.).

18.4.1 Fingerprint Recognition

Fingerprint recognition is an automated method of authenticating an individual by verifying the person's fingerprints with those in the database of enrolled users [4]. Fingerprint patterns of family members have been observed to share the same general patterns, and hence these are often thought to be inherited [5]. Fingerprint recognition involves an analysis of several features of the print pattern (comprising aggregate characteristic of ridges) and minutia points (representing unique features observed within the patterns). The three basic fingerprint ridge patterns include the arch, loop, and whorl; while, the three major minutia features observed within fingerprint ridges include ridge ending, bifurcation, and short ridge (dot). Print patterns and minutia points are critical to the analysis of fingerprints since no two fingerprints (even for the same person) have been proven to be identical. If

Fig. 18.3 Three common fingerprint ridge patterns (adapted from [2, 6])

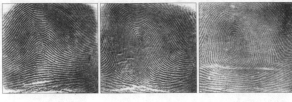

Fig. 18.4 Three common fingerprint minutia patterns (adapted from [2, 6])

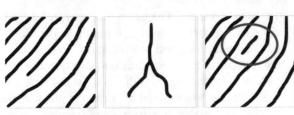

Fig. 18.5 Human eye and the iris (adapted from [1])

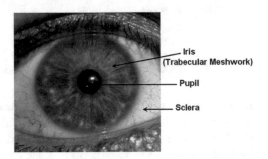

fingerprint recognition can be conducted on an individual, then even a DNA testing need not be conducted on the individual for verification. Figures 18.3 and 18.4, respectively, illustrate the different fingerprint ridges and the minutia patterns.

The "Arch" is a fingerprint pattern in which the ridges enter from one side of the finger, rise in the center, and exit from the other side of the finger; thus, resembling an arch. A "Loop" is the fingerprint pattern in which the ridges enter and leave from the same side of the finger forming a curve. The "Whorl" is a fingerprint pattern in which the ridges form concentric circles around the center of the fingertip. The "Ridge Ending" is a minutia pattern wherein a ridge terminates. "Bifurcation" is a point where a single ridge splits into two ridges. "Short Ridges" (also called dots) are ridges whose length is significantly shorter than the average length of the ridges in the fingerprint.

18.4.2 *Iris Recognition*

Iris (plural: irides) is a thin, circular structure in the eye (refer Fig. 18.5, adapted from [1, 2]), responsible for controlling the diameter and size of the pupil (the black

Fig. 18.6 Visible wavelength
and NIR iris images (adapted
from [1, 2])

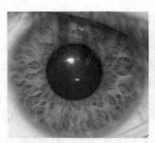 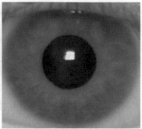

Visible Wavelength (VW)   Near Infra Red (NIR)
Iris Image   Iris Image

hole) as well as the quantity of light entering the pupil. The color of human eye is
classified according to the color of the iris and it can be green, blue, grey, or brown.
The pupil expands or contracts through the muscles attached to the iris; the larger
the pupil, the more light can enter.

The biometric method of iris recognition involves capturing the detail-rich,
intricate structures of the iris, using near infra-red illumination (NIR, 750 nm
wavelength) that would reduce the reflection (would be very noisy if visible light
was used) from the cornea. The images captured through the infra-red camera are
converted to digital templates that provide a unique mathematical representation of
the iris, leading to successful identification of an individual [7].

Although, NIR-based iris images are less noisy and are of good quality, NIR is
insensitive to the melanin pigment (the color pigment) in the iris. On the other hand,
visible wavelength (VW) imaging could also capture melanin and this could provide
more information about the shape patterns in iris. With effective noise-filtering
techniques and feature extraction methods, images captured from both NIR and VW
spectrum could be fused to obtain templates that could provide high accuracies of
recognition in large databanks [8]. Figure 18.6 illustrates the difference between a
VW-based Iris image and a NIR-based iris image.

The following are the advantages of iris and iris recognition systems in the
context of biometrics:

- The iris in an internal organ and is well-protected by the cornea, a highly
  transparent and sensitive membrane. Thus, the biometric templates of the iris
  are considered time-invariant; unlike fingerprints which could change with time
  if the concerned person indulges in lot of manual labor.
- The shape of iris is more predictable (compared to other structures such as face),
  as the iris is mostly flat and its geometric configuration is maintained by two
  complementary muscles that control the pupil's diameter.
- The iris texture (like fingerprints) is formed randomly during embryonic gesta-
  tion (8th month); even genetically identical twins (similar DNA print) will have
  completely independent iris textures.
- Unlike fingerprinting or retinal scanning (wherein the eye has to be brought very
  close to a microscopic lens), the person whose iris image is being captured need

not touch any equipment and can be about 10 cm or even few meters away from the biometric device.

- Iris recognition systems have very low false acceptance rate as well as low false rejection rate [9]. Hence, these systems are considered to be one of the most effective biometric technologies for one-to-many identification.
- The iris texture can be remarkably stable for more than 30 years, barring very few medical and surgical procedures or trauma [10].

Iris recognition systems are vulnerable to the problem of live-tissue verification [5, 11]. To avoid spoofing using manufactured or forged templates, biometric identification systems are often required to ensure that signals acquired and compared have been actually recorded from a live body part of the individual to be authenticated. Iris recognition systems are vulnerable to be deceived when presented with a high-quality photograph of the eye instead of the real eye. As a result, iris recognition systems are often considered not suitable for unsupervised applications (for example, door access-control systems). However, the live-tissue verification problem is of least concern in supervised applications (for example, immigration control), wherein a human operator oversees the entire process including that of taking the picture.

18.4.3 Retinal Scans

Retina is a light-sensitive tissue lining the inner surface of the eye. Retina for an eye is critical like the "film for a camera." The photoreceptor cells ("rods" for black and white vision and "cones" for daytime support and color perception) are responsible for generating the neural signals (transmitted to the brain through the optic nerve) upon incident of light. The network of blood vessels (capillaries) that supply blood to the retina is so complex and is unique for every person, including genetically identical twins. Like iris texture, the retina remains largely unchanged from birth until death and hence a retinal scan image is considered to be very precise and reliable metric with an error rate of one in a million.

A retinal scan is conducted by focusing an unperceivable low-energy infra-red light into a person's eye (as they look through the scanner). As the retina is small and quite internal; the user has to be perfectly still. Thus, retinal scans are considered to be most difficult biometric templates to capture. The retinal blood vessels absorb relatively more infra-red light compared to the rest of the eye and the reflection patterns of the light is captured as a computer code and stored in a database. Retinal patterns are likely to be altered in cases of diabetes, glaucoma, cataracts, or severe astigmatism. Retinal scanners are becoming more popular [12]: used in government agencies like FBI, CIA, and NASA as well as in medical diagnosis and commercial applications like ATM identity verification.

18.5   Biometric Systems for User Verification

In this section, we describe some of the commonly used biometric systems for the purpose of "verification." These biometric systems typically have high FAR and FER rates, but, are favored for ease associated with data collection and usage. The biometric systems we describe in this section are: (1) face recognition systems; (2) speaker recognition systems; (3) hand geometry-based systems, and (4) signature recognition systems. The biometric systems for user verification could also be employed for user authorization (i.e., validating whether a user identified as a valid user is permitted to the access rights claimed by the user). Note that the biological and behavioral characteristics that are extracted from the biometric systems for user verification could also be used to diagnose the well-being of the user (medical biometrics; [13]).

18.5.1   Face Recognition Systems

A facial recognition system is used to authenticate (normally verification) an individual through a digital image or video frame, obtained from a multi-media source. This is often done by comparing selected facial features that are not easily altered with those in the biometric database. Such facial features include the upper outlines of the eye sockets, the areas surrounding the cheekbones, and the sides of the mouth. A key advantage of facial recognition systems is that the biometric template (facial image) can be obtained without the cooperation or consent of the test subject. Hence, facial recognition systems are often considered for use in mass surveillance (for example, in airports, multiplexes, and other public places that need to be monitored), though their correctness and effectiveness is often questionable. Other biometric systems like fingerprints, iris scans, retinal scans, speech recognition, etc. cannot be used for mass surveillance. Some of the weaknesses associated with facial recognition systems are: (1) sensitiveness to facial expressions (a big smile can make the system less effective) and the frontal orientation at which the photo is taken; (2) privacy concerns as it could lead to a "total surveillance society."

18.5.2   Speaker Recognition Systems

Speaker recognition systems distinguish between speakers based on a combination of the physiological differences in the vocal tracts (e.g., the shape of these tracts) and the speaking habits. Speaker recognition systems are mostly passphrase-dependent so that it can provide an added security feature (rather than being text-independent—such systems are also available). During the enrollment phase, a user is required to speak a particular passphrase (like a name, birth date, birth city, favorite color, a sequence of numbers, etc) for a certain number of times. The analog version of this passphrase is transformed to a digital format and a speaker model is established by extracting distinctive vocal characteristics like the pitch, cadence, and tone. This leads to the generation of a biometric template that is stored in the database for future comparisons. Speaker recognition systems are often used when the only available biometric identifier is the voice (e.g., telephone and call centers).

18.5.3 Hand Geometry-Based Biometric Systems

Unlike fingerprints, human hand is not unique. However, hand geometry-based biometrics is not as intrusive as a fingerprint recognition system and hence may be sufficient enough to be used for verification (after the identity of the individual has been established through another mechanism), but, not for identification. The hand geometry features are extracted by computing the length and width of the fingers at various positions in the captured image of the hand palm of the enrollee. The hand geometry metrics constitute the feature vector of the enrollee.

The advantage with this biometric system is that hand geometry is considered to mostly remain the same during the growth period of a human (i.e., from a child to adult). Also, the accuracy of hand geometry systems will not be much affected due to environmental factors like dry weather or individual anomalies such as dry skin. However, there are some limitations: It will not be easy to extract the correct information about the hand geometry in the presence of jewelry (like wedding ring) that people would not like to remove from hand or in the case of people with limited movement of fingers (e.g., arthritis). Hand geometry systems are also physically larger in size and cannot be easily embedded for use with certain devices like laptops.

18.5.4 Signature Recognition Systems

Signature recognition refers to authenticating the identity of a user by measuring handwritten signatures. The user signs his or her name on a digitized graphics tablet or a PDA that can capture handwritten notes. The series of movements made by the user (while signing) and the associated personal rhythm, stroke order, stroke count, acceleration, and pressure flow constitute the unique biometric data characteristic of the user. Such information on the dynamics of the user's signature is encrypted and compressed into a template. Signature recognition systems (for hand signatures) measure how a signature is signed and are different from electronic signatures, which treat a signature as a graphic image.

18.6 Comparison of the Biometric Systems Based on Operating Parameters

The following parameters are considered critical for data collection and usage of biometric systems:

- *Universality*—every individual who needs to be enrolled should have the characteristic
- *Uniqueness*—the biometrics captured for one user should be different from that of the other users
- *Permanence*—the biometrics captured for a user should resist aging and be time-invariant
- *Collectability*—the ease with which the biometrics data could be collected for a user
- *Performance*—the accuracy, robustness, and speed at which the biometric system can be used
- *Acceptability*—the level of approval the technology has with the users of the biometric system
- *Circumvention*—the degree to which the biometric template can be forged with a substitute.

Table 18.1 compares the different biometric systems (discussed in Sects. 18.4 and 18.5) with respect to the above 7 parameters. The terms "Best" (green color cells), "Average" (white color cells), and "Poor" (red color cells) in Table 18.1 represent the suitability and/or usability levels of the biometric systems with respect to the individual parameters.

Table 18.1 Comparison of the biometric systems with respect to data collection and usability parameters (adapted from [1])

Biometric identifier	Data collection and usability parameters to choose a biometric system						
	Universality	*Distinctiveness*	*Permanence*	*Collectable*	*Performance*	*Acceptability*	*No circumvention*
Face	Best	Best	Average	Best	Poor	Best	Poor
Finger print	Average	Best	Best	Average	Best	Average	Average
Hand geometry	Average	Average	Average	Best	Average	Average	Average
Iris	Best	Best	Best	Average	Best	Poor	Best
Signature	Poor	Poor	Poor	Best	Poor	Best	Poor
Voice	Average	Poor	Poor	Average	Poor	Best	Poor

18.7 Spoofing Attacks on Biometric Systems

One common type of attacks that biometric systems are more vulnerable to are the spoofing attacks [6]. In a spoofing attack on biometric systems, an unauthorized person of interest (imposter) typically tries to go through the identification/verification process by posing the biometric traits of an authorized user as his own. The biometric sample of the legitimate user is forged and submitted as the biometric sample of the imposter [14]. In this pursuit, spoofing attacks include capturing and creating a copy of the captured sample. We discuss below the different kinds of spoofing attacks on biometric systems.

18.7.1 Spoofing Fingerprints

In order for an imposter to be able to successfully spoof a fingertip, he must obtain a valid biometric sample (fingerprint) either willingly from an authorized user or obtain the fingerprint without the knowledge of the owner of the biometric sample. The traditional way of capturing a fingerprint is with the use of a powder. Fingerprints can be captured from the residual print on hard surfaces such as metal or glass. Fingerprints can also be captured with the waste toner from a regular copy machine [15]. In addition to the above technical and formal approaches, fingerprints could be very easily obtained by the imposter through some everyday life routines. For example, in public restaurants, people are unaware that their fingerprints are left on things they use such as a drinking glass. An imposter could easily take the glass and obtain the fingerprint by rolling the glass in powder. Once the fingerprint is captured, the imposter uses a camera and transports the image to the computer. The quality of the image depends on the nature and the circumstances under which the object was touched by the person whose fingerprint is stolen.

18.7.2 Spoofing the Face

Compared to other biometric systems, there is an advantage with facial recognition systems because it is not difficult to obtain a facial sample without collusion or cooperation of the users being spoofed. That is, it is possible to get the facial sample without the person knowing about it or without their consent. For example, one can get images by photographing faces of people in public places like bank, mall, grocery stores, etc. Attackers will be able to try several possible means to obtain a facial image without the knowledge of the user being spoofed. Once an image is captured, it could be used to deceive facial recognition systems. Some of the facial recognition algorithms proposed in the literature require users to blink their eye to differentiate live faces from a photograph or painting [15].

18.7.3   Spoofing the Voice

There are two approaches for voice-based authentication. These approaches are categorized as text-dependent and text-independent. In text-dependent authentication, a user speaks fixed phrases, passwords, or other words. In the text-independent category, a user can choose any phrases or words for authentication. A victim's voiceprint is typically obtained through social engineering attacks. For example, the voiceprint could be captured by the imposter calling and asking the victim to repeat different words or phrases as a test [3]. With this, the voiceprint can be successfully obtained without the victim knowing about it.

18.7.4   Transmitter Attacks

When a message is transmitted between a sender and receiver over an insecure channel, a man-in-the-middle attack is likely to occur. With transmission attacks such as the man-in-the-middle attack, data sent to a receiver is intercepted, modified, or corrupted by a third party, thereby giving the attacker the biometric image of the legitimate user. When such an attack occurs on a biometric system, the attacker sends a fake template and disguises as a legitimately enrolled user. Transmission attacks can also result in generating a fake matching score when the spoofed biometric test sample submitted by the imposter matches with the biometric information of the enrolled user in the application system database. The threat of transmission attacks can be reduced by sending data over a secure channel using techniques like encryption and watermarking.

18.7.5   Replay Attacks

Replay attacks are those in which the attacker successfully sniffs the packets transmitted over an insecure channel and retransmits them after a while—thus causing the receiver to accept and process these packets as legitimate packets originating from the authentic sender. For example, if an attacker can capture packets related to login authentication, it gives them the opportunity to play back the username and password at a later time and pose as a legitimate user. An attacker can also obtain genuine data by collecting prints left from a successful authentication process; for example, the fingerprints left on the sensor itself. Replay attacks can be prevented by using encryption and including a way for the sender and receiver to be able to verify the identity of each other. The latter could be accomplished by passing back and forth a "nonce," which is a unique one-time generated number.

18.7.6   Template Attacks

Biometric templates are stored mostly in a database and sometimes at the machine that does the matching or the sensor used to collect the samples. When the templates are attacked, the stored template is modified, existing templates are removed, or new templates are added. Of all the different threats, stolen templates are the most risky. Once a person's biometric data is stolen and compromised, it is compromised forever. Stolen templates can be used to reverse engineer how a biometric system operates. If the template of a fingerprint is reverse engineered, it can be used to instead appear to be a thumbprint. Encrypting the template data can help to reduce the risk of template attacks.

18.7.7   Solution to Mitigate Spoofing Attacks

Multi-biometric systems (discussed in Sect. 18.8) could be a potential solution to prevent or at least effectively reduce the occurrence of spoofing attacks. Most biometric systems use one biometric trait such as a fingerprint or an iris scan as the source of data samples for authentication. Multi-biometric systems use two or more biometric traits to compare (at different levels of fusion) to identify or verify a person. By using the combination of the biometric samples based on two or more traits, the authentication system is expected to be more reliable. A multi-biometric system is basically more secure than single modal systems since it is more difficult to spoof two or more biometric traits than a single biometric trait of a legitimate user. The advantage of using a multi-biometric system for anti-spoofing is that the level of difficulty increases for an attacker because it would require breaking several biometric systems simultaneously, Also, in some situations, the user might find one form of biometric trait or sample to be not enough for authentication with a low false error rate, thus necessitating the need for more than one biometric trait. This can be the case with fingerprints, where at least 10% of the population have worn, cuts or unrecognizable prints [15]. The choice and the number of biometric identifiers are decided by the nature of the application, the computational demand, the cost of recording the identifiers, and the relationship between the identifiers considered.

18.8   Multi-Biometric Systems

Multi-biometric systems use more than one biometric system to identify or verify a user. These multiple biometric systems can be run either one after the other (serial mode) or simultaneously (parallel mode). With the *serial mode*, the output of the matching operation on one biometric identifier could be used to narrow down the records to be searched for to validate the sample corresponding to the other biometric identifier(s). Also, there is no need to simultaneously collect the biometric samples corresponding to all the biometric identifiers. When operated in the *parallel mode*, the sample for each biometric identifier is simultaneously processed to decide on the final recognition. Hence, we need to simultaneously collect the biometric samples corresponding to all the multiple identifiers beforehand.

Though operating in serial mode offers the advantage of collecting only one biometric data at a time and filtering invalid users right away after they fail to go through a particular biometric system in the series, the sequence of biometric systems through which a user goes through the validation tests has to be meticulously selected. For example, if we include a biometric system that has high False Acceptance Rate and/or False Rejection Rate up front in the sequence, then the purpose will not be achieved as either too many invalid users penetrate in and have to be any way validated through the more rigorous biometric systems or too many valid users get invalidated and could not access the system. A valid user has to anyway go through the entire sequence of biometric systems; hence, the access delay (the time between a user starts getting validated through the multi-biometric system and the time the user is actually able to access the application system in question after getting validated) will be quite longer. As a result, serial mode multi-biometric systems are not preferable for real-time delay-sensitive applications.

Operating a multi-biometric system requires lots of simultaneous data collection and processing. However, the access delay for the parallel mode will much lower than that of the serial mode. And also, there will not be too much dependence on each of the constituent biometric systems. If a particular biometric system appears to have errors in its validation (i.e., a high False Acceptance Rate or False Rejection Rate), the data and/or decisions obtained from that particular biometric system can be given a lower weight or preference compared to others. Thus, multi-biometric systems operating in parallel mode are more flexible and are a preferred choice for their accuracy and time-sensitive features, if simultaneous data collection is feasible. The rest of this section explains the different levels of fusion of data/decision that is generally possible when multi-biometric systems are operated in parallel mode.

18.8.1 Levels of Fusion of Multi-biometric Systems Operating in Parallel Mode

There could be four levels of fusion of the data and/or decision, depending on the particular stage of the biometric system at which the fusion takes place. Figures 18.7, 18.8, and 18.9 illustrate fusion at the feature, score, and decision levels.

- *Feature-level fusion*: The feature sets extracted from the raw data for each biometric identifier can be combined to create a new comprehensive feature set characteristic of the individual (see Fig. 18.7). For example, the geometric features of the hand and the Eigen-coefficients of the face may be fused to obtain a new high-dimension feature vector.

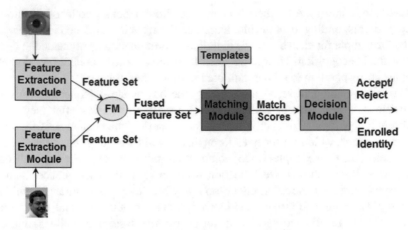

Fig. 18.7 Feature-level fusion: multi-biometric systems operating in parallel mode (adapted from [2])

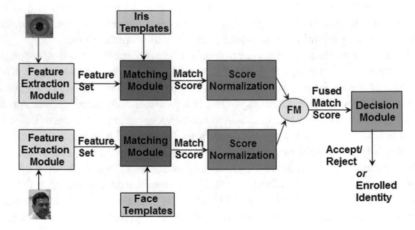

Fig. 18.8 Score-level fusion: multi-biometric systems operating in parallel mode (adapted from [2])

- *Sensor-level fusion*: The raw data gathered from the multiple sensors (see Fig. 18.8), one for each biometric identifiers, can be processed and integrated to generate a new comprehensive dataset from which features characteristic of the individual are extracted.
- *Match score-level fusion*: The match scores obtained from each biometric classifier are normalized and the normalized scores are summed to obtain a new match score—used to make the final decision.
- *Decision-level fusion*: The decisions (accept/reject) made at each biometric system based on the individual scores are then combined (usually a majority voting approach) to arrive at a final decision (see Fig. 18.9).

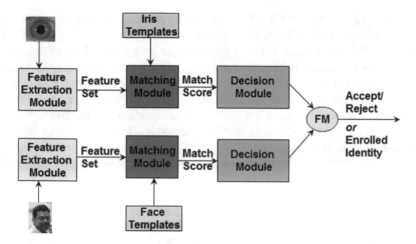

Fig. 18.9 Decision-level fusion: multi-biometric systems operating in parallel mode (adapted from [2])

The sensor-level and feature-level fusion techniques are considered to be tightly coupled as the raw data and the feature vectors convey the richest information about the different identifiers. However, it is more difficult to perform a combination at the sensor-level and feature level, because the raw data and/or feature sets corresponding to the individual biometric identifiers could be of different incompatible formats, unknown representations, units, etc. and may have to be excessively processed to bring them to a common compatible format, scale, representation, etc.; but this could ultimately result in loss of useful information. Also, if the individual biometric systems are developed by different commercial vendors, different feature values may not be accessible and comparable. In such scenarios, loosely coupled fusion systems (at the score or decision level) have to be chosen. The match score-level and decision-level fusion techniques are considered to be loosely coupled as there is little or no interaction among the biometric inputs and the fusion occurs at the output of relatively autonomous biometric authentication systems that independently assess the input from their own perspective.

18.9   Conclusions

In conclusion, biometrics is a valuable tool for information security and can be employed for both user identification and verification. The biometric systems employed for user identification should have very low false error rates (such as the false acceptance rate and false rejection rate) as they are the primary means of user authentication. Hence, the biometric samples in such systems have to be more accurately collected, which may also sometimes cause inconvenience to the users involved. This is a tradeoff and cannot be avoided. Typical biometric traits for user identification include fingerprints, iris images, and retinal scans. With these biometric traits, there are very high chances of a unique biometric template available for each user. Hence, it would be appropriate to match a test biometric template with those stored in a database and authenticate a user (i.e., identify a user) if there is a match. On the contrary, biometric systems employed for user verification are only the secondary means of user authentication (the user being authenticated primarily through some other means) and hence have more flexibility in terms of data collection and accuracy. Typical biometric systems employed for user verification include hand geometry, face recognition, speech recognition, and signature recognition systems. With these biometric systems, more than one user will have the same biometric template; however, when user identification needs to be just verified—we will compare the biometric template of a known user (validated through some primary means of authentication) to the biometric template of that particular user stored in the database. If there is a match, the user is authenticated.

The second half of the chapter discusses the potential issues associated with biometric systems and discusses a specific type of security attacks called spoofing attacks on biometric systems. Spoofing attacks involve an imposter capturing the identity of a legitimate user (either through collusion or without the knowledge of the victim) and spoofing it as his own identity or faking a biometric template as if it is the template of a legitimate user. Multi-biometric systems can be a potentially effective solution to mitigate security attacks, especially those involving spoofing. With the involvement of multiple traits (two or more), it becomes difficult for an imposter to simultaneously forge the different biometric traits and penetrate successfully. Multi-biometric systems can be operated in serial mode or in parallel mode and we discussed the potential pros and cons of each mode. The parallel mode is more preferred for real-time delay-sensitive applications and also for the flexibility available to discard one or more constituent biometric systems with high error rates; but, still be usable for authentication. The data and/or decision of multi-biometric systems operating in parallel mode can be fused at four-different levels: sensor, feature, score, and decision-levels. Accordingly, multi-biometric systems with fusion at the sensor and feature levels are referred to being tightly coupled and those operating with fusion at the score and decision levels are referred to being loosely coupled systems.

References

1. Chandra Murty, P. S. R., Sreenivasa Reddy, E., & Ramesh Babu, I. (2009). Iris recognition system using fractal dimensions of Haar patterns. *International Journal of Signal Processing, Image Processing and Pattern Recognition, 2*(3), 75–84.
2. Jain, A. K., Ross, A., & Pankanti, S. (2006). Biometrics: A tool for information security. *IEEE Transactions on Information Forensics and Security, 1*(2), 125–143.

3. Sabena, F., Dehghantanha, A., & Seddon, A. P. (2010). A review of vulnerabilities in identity management using biometrics. In *Proceedings of the 2nd international conference on future networks* (pp. 42–49). Sanya, Hainan, China.

4. Feng, J., & Jain, A. K. (2009). FM model based fingerprint reconstruction from minutiae template. In *Proceedings of the 3rd international conference on advances in biometrics* (pp. 544–553). Alghero, Italy.

5. Li, S. Z., & Jain, A. K. (2015). *Encyclopedia of biometrics* (2nd ed.). New York: Springer.

6. Cao, K., & Jain, A. K. (2015). Learning fingerprint reconstruction: From minutiae to image. *IEEE Transactions on Information Forensics and Security, 10*(1), 104–117.

7. Solanke, S. B., & Deshmukh, R. R. (2016). Biometrics: Iris recognition system: A study of promising approaches for secured authentication. In *Proceedings of the 3rd international conference on computing for sustainable global development* (pp. 811–814). New Delhi, India.

8. Al-Khazzar, A., & Savage, N. (2011). Biometric identification using user interactions with virtual worlds. In *Proceedings of the 11th international conference on trust, security and privacy in computing and communications* (pp. 517–524). Changsha, China.

9. Zuo, J., & Schmid, N. A. (2013). Adaptive quality-based performance prediction and boosting for iris authentication: Methodology and its illustration. *IEEE Transactions on Information Forensics and Security, 8*(6), 1051–1060.

10. Chen, W.-K., Lee, J.-C., Han, W.-Y., Shih, C.-K., & Chang, K.-C. (2013). Iris recognition based on bi-dimensional empirical model decomposition and fractal dimension. *Information Sciences, 221*, 439–451.

11. Singh, Y. N., & Singh, S. K. (2013). A taxonomy of biometric system vulnerabilities and defenses. *International Journal of Biometrics, 5*(2), 137–159.

12. Sui, Y., Zou, X., Du, E. Y., & Li, F. (2014). Design and analysis of a highly user-friendly, secure, privacy-preserving, and revocable authentication method. *IEEE Transactions on Computers, 63*(4), 902–916.

13. Kostyuk, N., Cole, P., Meghanathan, N., Isokpehi, R., & Cohly, H. (2011). Gas discharge visualization: An imaging and modeling tool for medical biometrics. *International Journal of Biomedical Imaging, 2011*, 196460. 7 pages.

14. Yampolskiy, R. V. (2008). Mimicry attack on strategy-based behavioral biometric. In *Proceedings of the 5th international conference on information technology: New generations* (pp. 916–921). Las Vegas, NV, USA.

15. Akrivopoulou, C. M., & Garipidis, N. (2013). *Digital democracy and the impact of technology on governance and politics: New globalized practices*. Hershey, PA: IGI-Global.

Chapter 19
Biometric Authentication and Data Security in Cloud Computing

Giovanni L. Masala, Pietro Ruiu, and Enrico Grosso

19.1 Introduction

The migration, from local to web applications, is probably one of the most significant advances of the recent years in the arena of the application software: sharing critical data and resources and giving support to multi-user/multi-tenancy scenarios. The development of service-oriented architectures (SOA) and WEB services are key issues in all frameworks. SOAs support designing and developing in terms of services with distributed capabilities, which can be under the control of different ownership domains. These architectures are essentially a collection of services or, in different terms, entities performing single or a limited number of repeatable activities and communicating with each other by simple data passing. Service consumers view a service provider as a communication endpoint supporting a particular request format or contract; this request format (or interface) is always separated from the service implementation.

As a matter of course, security breaches on web applications are a major concern because they can involve both enterprise and private customer data: protecting these assets is then an important part of any web application development. This process usually includes authentication and authorization steps, asset handling, activity logging, auditing. A variety of protection mechanisms has been developed,

G.L. Masala (✉)
School of Computing, Electronics and Mathematics, Plymouth University, Plymouth, UK
e-mail: giovanni.masala@plymouth.ac.uk

P. Ruiu
Istituto Superiore Mario Boella (ISMB), Torino, Italy
e-mail: ruiu@ismb.it

E. Grosso
Department POLCOMING, University of Sassari, Sassari, Italy
e-mail: grosso@uniss.it

© Springer International Publishing AG 2018   337
K. Daimi (ed.), *Computer and Network Security Essentials*,
DOI 10.1007/978-3-319-58424-9_19

for this purpose, like: password management, encryption, intrusion prevention, and vulnerability analysis. The extension of the web application paradigm to the cloud computing model is denoted as software as a service (SaaS). The adoption of cloud computing, in particular leveraging on the public and hybrid models [1], involves many advantages in terms of flexibility, scalability, and reliability, but also implies new challenges on security, data privacy, and protection of personal data.

Literature is vast on this topic, and different risks and vulnerabilities have been extensively studied and highlighted [2, 3]. Attacks to cloud systems are becoming more targeted and sophisticated [4], since attackers know that cloud storage is becoming one of the most adopted ways to archive and share personal information. Incidents of data leakage from the cloud are increasingly frequent and affect also big players like Apple, PlayStation, and others [5–7]. These vulnerabilities are accompanied by collateral legal and reputational risks that should be regulated by national governments. The USA and European Union have enacted regulatory requirements applicable to data stored by cloud providers [8]. The security specific risks of the cloud are primarily derived from the complexity of the architecture, which includes different models of services and distribution. Furthermore there are risks related to the characteristics of multi-tenancy and resource sharing, allowing to allocate the same resources in different times to different users [9].

A first element of risk is related to the failure of the isolation systems for storage and computational resources. When data reside on the same physical infrastructure, a failure of the isolation systems can compromise machines hosted through guest-hopping, SQL injection, and side channel attacks [10]. Individuals and organizations may have different interests and requirements, or even conflicting/competing objectives. To this concern, it is necessary to protect data and systems using methods that guarantee the physical and logical separation of resources and data flows [11]. Moreover, being the cloud a distributed architecture, this implies an increased use of networks and data communication flows, compared to traditional architectures. For example, data must be transferred to the synchronization of images, of the same virtual machine, among various and distributed hardware infrastructures. Or else, simple storage operations can involve communication between central systems and cloud remote clients. Risks are, therefore, those of incurring on sniffing, spoofing, man-in-the-middle, and side channel attacks. An additional element of risk is related to the cloud model adopted. In fact, some cloud models require the user to transfer part of the control over his own data to the service provider. In this case, not only the data are allocated on the provider's servers, but also the user cannot apply specific protection mechanisms like encryption or access control, as the service provider is the sole subject having total control of the cloud resources. Finally, some key roles for managing the cloud infrastructure, such as system administrators and managers of security systems, must be considered. These actors usually have the power to perform all types of activities, within the system, and this would potentially break safety requirements imposed by corporate policies. Yet, the assessment of this kind of fraudulent actions is very complex and there is a lack of certification agencies internationally recognized for the independent evaluation of cloud security.

The "remote user authentication" or "logical access control" is one of the fundamental steps in protecting data and IT infrastructures. Authentication protocols allow to verify that each of the participants in the electronic communication is really who he claims to be. This task is commonly demanded to a specialized architecture denoted as the authentication server (AS). The AS preserves and manages the access keys to the various subsystems. In order to access private services or data, each authorized person must first establish a connection with the AS, declare and prove his own identity, and obtain a session key useful to require further services. Currently, the most common authentication mechanisms of the ASs make use of passwords and private tokens. Passwords are subject to various security threats; for example, they can be easily stolen or intercepted and used fraudulently. Tokens are more difficult to be reproduced and for this reason they are often used in banking services. However, being more expensive and difficult to manage, they are far to be an optimal solution. Moreover, they are usually based on the possession of a physical card or device that can be easily shared with different people.

As reported in the scientific literature [12, 13], the efficient use of multiple biometric features for identity verification is still an open and attracting scientific problem; biometric physical access systems are perceived as reliable [12], then minimizing the typical risks of traditional authentication systems in applications that require a high level of security like border control. On the other hand, the use of biometric data for the logical access to IT services is a more challenging and still unsolved problem. Certainly, the use of biometric techniques can be considered as one way to ensure a significant increase of security in the authentication protocols managed by modern authentication servers.

One of the criticisms of some biometric approach is related to privacy risks. In particular, this has to do with the storage of images or other biometric features in the database of the authentication server, in order to be compared during the recognition phase. These images are considered as sensitive data and should be protected with high secure systems [14]. Hence, according to privacy regulations, it is not possible to outsource these data to cloud services. Authors use often techniques to overcome this problem, as fuzzy biometric templates, based on the fuzzy vault of Jules and Sudan [15], for instance, the Biometric Encryption scheme by Soutar et al. [16], Cancelable Biometrics by Ratha et al. [17], robust bit extraction schemes based on quantization, e.g., of Linnartz and Tuyls [18], of Chang et al. [19], and of Chen et al. [20], and applications of the fuzzy commitment scheme of Juels and Wattenberg [21] to biometric templates, e.g., the constructions of Martini and Beinlich [22] for fingerprints. Authors in [23] propose a solution, using a compact representation of the biometric feature, converted using Scale Invariant Feature Transform (SIFT) representation: only this model is used to recognize the user and stored in the cloud; thus, it is not required to protect sensible data.

In this chapter, we present an example of cloud system [23, 24] that uses biometric authentication based on fingerprints [25]. This advanced access control is combined with a very peculiar fragmentation technique guaranteeing the security of the data residing on the cloud architecture. In Sect. 19.2 some preliminary considerations concerning the cloud platform are introduced while in Sect. 19.3 an example of cloud system is described in detail and the main results on the cloud security are discussed. Section 19.4 draws some conclusions, pointing out issues and problems that will be faced in the near future.

19.2 Preliminaries

19.2.1 Cloud Platform

OpenStack [26] is an open source project that many identify as the first true cloud operating system. OpenStack has to be considered as a basic technology rather than a solution; by analogy is often associated with the Linux kernel.

The example of project [23, 24] described in this chapter has the primary goal of supporting basic web applications shared by small and medium companies; candidate platforms for cloud computing should be, therefore, oriented to scalability, to be implemented according to the public or private cloud models. In this respect, OpenStack has many interesting features; it allows a prompt and elastic control of computing resources such as CPUs, storage, and networks, and includes features for general system management, process automation, and security.

OpenStack consists of several individual sub-components. This modular design improves flexibility because each component may be used alone or in combination with others. Some of these modules marked as cores (such as compute, storage, and networking) represent the essential parts of the platform. Other modules are initially placed in an incubator from which they come only if needed.

The main modules of OpenStack, fully distributable and replicable, are the following: computing (Nova), networking (Neutron), image templates (Glance), block (Cinder) and object storage (Swift), graphical interface platform accessible via the web (Horizon), authentication, the native orchestration module (Heat), and accounting (Keystone). The architecture is based on the concept of "sharing nothing" that makes components independent and self-sufficient, avoiding the sharing of memory or storage. Communications between the different modules are asynchronous and are managed by queue managers (message brokers) that implement the Advanced Message Queuing Protocol (AMQP). The various services communicate with each other through specific Application Programming Interfaces (APIs) that implement the REST model. All these features make OpenStack an ideal tool to be deployed on commodity hardware, with consequent economic benefits and flexibility.

Virtualization is an important element of cloud computing because it guarantees the required elasticity in resource allocation. Virtualization is a technique that allows to run multiple virtual machines on a single physical server and to optimize the available resources. It is possible to provide different levels of abstraction that make the operating system do not see the physical hardware but the virtual hardware.

This abstraction is achieved by a software layer, called *hypervisor,* which is usually integrated into the operating system kernel and it is loaded at system startup. The *hypervisor* does not offer any management capabilities to virtual machines. Like many of the cloud computing platforms also OpenStack is not released with a specific *hypervisor*; the system administrator can choose among a set of supported hypervisors like VMware, Hyper-V, Xen, and KVM. In this project the Kernel-based Virtual Machine (KVM) is used; it is one of the most supported and popular among scientific developers. KVM is a Linux kernel module that allows a user program to use hardware virtualization capabilities of various processors. It supports in particular processors from AMD® and Intel® (x86 and x86_64) having these features (Intel VT or AMD-V). From the point of view of the operating system each virtual machine is seen as a regular Linux process that can use the hardware resources according to what established by the scheduler. A normal Linux process has two execution modes: kernel and user. KVM adds a third mode, a guest mode that has its own kernel and user modes. The main benefit of KVM is that being integrated into the kernel improves performance and reduces the impact on existing Linux systems.

19.2.2   Data Security

A possible solution, to guarantee the security of data residing on distributed cloud infrastructure, is the use of systems for the fragmentation and distribution of data, which allow to split the data into fragments and disperse them on all machines available to the cloud. In this way the recovery and the use of the data is very complex for an unauthorized user. By using fragmentation techniques, it is possible to distribute data on platforms of different providers, and to problems arising from the lack of trust in the service provider. However, in order to achieve a proper fragmentation and distribution of the data in the network, it is necessary to develop support tools to ensure the prompt availability and integrity of these data, without increasing the complexity of the system. In fact, an excessive consumption of resources or performance degradation related to procedures of information retrieval would compromise this approach.

The use of fragmentation techniques to protect outsourced data is not a novel approach in literature. Different solutions have been proposed; however, the most prominent ones use cryptography to obfuscate data [27, 28] and traditional relational databases [29, 30], exploiting sharding functionalities. The approach proposed in this paper is completely different and original, since disclaims these two elements. The solution proposed avoids cryptography, seen as an excessive overhead to data retrieval processes, since encryption makes it not always possible to efficiently execute queries and evaluate conditions over the data. Moreover, another innovative aspect regarding the use of modern database platforms, which embracing the NoSQL paradigm, is characterized by highly scalable distributed architectures.

These platforms include also native management features (redundancy, fault tolerance, high availability) which permit to design simple fragmentation systems without the burden of having to implement these complex control systems.

19.3 An Example of Cloud Platform

19.3.1 General Implementation of the Cloud System

The meaning of the term "node" usually relates to individual machines running the functions of the cloud. In some cases a node corresponds to a physical machine, in other cases it corresponds to an instance of a virtual machine (VM). OpenStack has a distributed nature; therefore, during installation it is necessary to take account of the number of nodes required for the installation of the platform. From the official documentation of OpenStack, the minimum number of nodes to be used in a stable installation is five, at least one for each of the following functions: Horizon, Keystone, Neutron, Nova, and Swift. In particular:

- Neutron is the system that allows to manage the network connectivity and to address the VMs in the cloud. It includes some features of type "networking as a service" that support the use of advanced networking.
- Swift is a distributed storage system that can accommodate data of users of the platform or VMs. It allows to manage the policies of replication and consistency ensuring the integrity, safety, and protection of distributed data in the cloud.
- Keystone manages all security policies, privileges, and authorization for user access to the platform. It provides API client authentication, service discovery, and distributed multi-tenant authorization.
- Horizon is a graphical interface platform accessible via the web, for easy and intuitive management of the cloud.
- Nova is designed to provide power massively scalable, on demand, self-service access to compute resources. It is developed to manage and automate computer resources and can work with several virtualization technologies.
- Glance is the Virtual Machine Image Repository or a catalog of images of the operating system that users can use to instantiate VMs.
- Cinder allows to provide storage that can be used by Nova to serve the VMs. Storage is provided in the form of block storage device and may be required as a service without reference to the real physical allocation.
- Heat is the native orchestration module of processes of the cloud.

In the considered system [23, 24] one module of OpenStack is not installed: Ceilometer, which allows monitoring and billing use of cloud resources. Figure 19.1 (top) highlights the distribution of modules in the nodes; the network configuration of the platform is illustrated in Fig. 19.1 (bottom).

The architecture is divided into two different Italian data centers located in Alghero and Turin. Each server stands on a virtual private LAN: we have a server

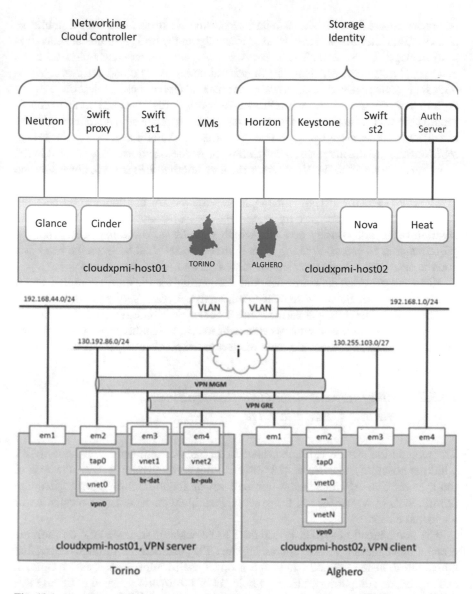

Fig. 19.1 (*Top*) The subdivision of OpenStack functions between our two Italian data centers of Alghero and Turin: services Nova and Heat have a physical machine on the server of Turin and all other services are arranged on virtual nodes. (*Bottom*) The general network configuration of the cloud platform

in Turin, which uses the *em1* interface, while another server, in Alghero, uses the interface *em4*. The other network adapters are used to configure the three networks necessary for the operation of OpenStack. The public network is used to allow the connection of the virtual machines to the outside (Internet). For this network it is necessary to configure a virtual interface for the Neutron node with a public IP address. This interface will then be used to configure the bridge virtual audience (*br-pub*) managed by Neutron. The *management network* interconnects physical hosts and virtual machines, which are the functional nodes of the cloud platforms. These nodes are equipped with the software modules of OpenStack, as described in the bottom part of Fig. 19.1. A Virtual Private Network (VPN) has been set up to ensure secure communication between these nodes (which manage all data transiting in the cloud). The Turin node has been configured as the VPN server, using the bridge *tap0*, attached to the interface *em2*. The host of Alghero and the nodes hosted in the same server connect to the VPN server through another bridge *tap0*, always on the respective interface *em2* (see Fig. 19.1).

The *data network* instead is the channel reserved for communication between virtual machines. OpenStack manages this kind of communication through the creation of ad hoc overlay network, which uses *Generic Routing Encapsulation (GRE) tunnels to encapsulate traffic*. A tunnel is established between the two hosts and the other two tunnels between the same host and the Neutron node.

Keystone provides authentication and accounting for the entire platform, and it is installed on a dedicated virtual machine, on the physical server of Alghero. This is necessary to facilitate its interface with a dedicated biometric authentication, via private network connection; the service is hosted in the authentication server (AS) of the data center, but externally with respect to the platform OpenStack.

19.3.2   Integration of Biometric Recognition with the Cloud Platform

The recognition system is implemented in an isolated authentication server (AS), which exposes the necessary API for ensuring interoperability with the rest of the system. The API includes a minimal set of functions providing registration (enrollment) of a new user in the system, identification of a user, cancellation of a registered user.

The authentication system is designed to be scalable in horizontal on multiple computing nodes and vertical optimizing the CPU performance, through the parallel computation inside the node, in which it operates. To improve processing time, at start-up of the computing node, the whole set of information related to the users is copied directly into RAM to cancel the disk access times. With the current service configuration (1 node with 4 vCPU) the total time of identification is calculable, on average, in 3/10 of a second per registered user.

A VPN is placed between the system and the desktop user application. When the VPN encrypted tunnels are enabled, the user starts the session simply touching the fingerprint scanner. This VPN selectively enables the services that can be accessed by the user: at the start of the process the user only sees the API server while, if authenticated, the system creates a route to the GUI. In this way, communications between the client and the API are always protected and the session ID is never transmitted in clear.

19.3.2.1   Biometric Recognition

The desktop application includes software modules both for the enrollment and the authentication of users. During enrollment, the new user's fingerprint is converted into a compact representation, called model; only this model will be used to recognize the user, thus it is not required to store the fingerprints in the AS database; only the models are recorded.

The features characterizing the model are obtained by using the Scale Invariant Feature Transform (SIFT) representation [31, 32]. Recently SIFT has emerged as a cutting-edge methodology in general object recognition as well as for other machine vision applications [31–35]. One of the interesting features of the SIFT approach is the capability to capture the main local patterns working on a scale-space decomposition of the image. In this respect, the SIFT approach is similar to the Local Binary Patterns method [36, 37], with the difference of producing a more robust view-invariant representation of the extracted 2D patterns.

The matching for the authentication application is performed considering the SIFT features located along a regular grid and matching overlapping patches; in particular, the approach subdivides the images in different sub-images, using a regular grid with a light overlap. The matching between two images is then performed by computing distances between all pairs of corresponding sub-images, and therefore averaging them [34]. A fusion module takes the final decision.

The fingerprint scanner used for the purpose of the project has a 1×1 inch sensor and is certified by FBI according to Personal Identity Verification (PIV) Image Quality Specifications. These technologies ensure a good quality and performance level, currently unreachable with most commercial devices.

19.3.2.2   Performance of the Authentication System

Our authentication system, based on SIFT [23, 24, 34], is tested on a subset of the Biosecure database [38]. More in detail, we used a subset including two different acquisitions (A and B) of the same fingerprint for 50 persons, randomly extracted from the original database of 400 subjects. The dataset contains features extracted in a realistic acquisition scenario, balanced gender, and population distributions. We made the comparison between each fingerprint A, against the fingerprints B of all 50 persons, for a total number of 2500 comparisons. We used normalized scores to express the similarity between two biometric patterns. The higher the score is, the higher the similarity between the biometric patterns.

The access to the system is granted only, if the score for a trained person (identification), or the person that the pattern is verified against (verification), is higher than a certain given threshold. Depending on the choice of the classification threshold, between all and none of the impostor, patterns will be erroneously accepted by the system. The threshold, dependent fraction of the falsely accepted patterns, divided by the number of all impostor patterns, is called False Acceptance Rate (FAR). Again depending on the value of the threshold, between none and all, also a varying number of genuine patterns will be falsely rejected. The fraction of

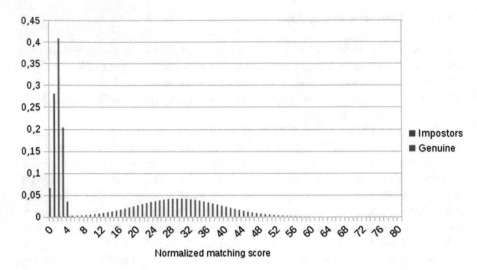

Fig. 19.2 Estimation of the normal distributions for genuine and impostors, in the dataset

Table 19.1 Normal distributions for genuine and impostors

	Mean	St. Dev.
Genuine	29.3	±9.7
Impostors	1.9	±1.0

Table 19.2 Estimations of FAR and FRR, varying the matching threshold

Matching threshold	FAR	FRR
6	4.060E-05	0.00012
7	2.760E-07	0.00016
8	6.430E-10	0.00023
9	5.140E-13	0.00032
10	1.410E-16	0.00044
11	1.320E-20	0.00059

the number of rejected genuine patterns, divided by the total number of genuine patterns, is called False Recognition Rate (FRR). The distributions of the genuine and the impostors scores sometimes overlap, and it is not easy for the choice of the threshold value.

To this purpose, the distributions for genuine users and impostors are estimated in Fig. 19.2. The threshold is tuned in such a way to give suitable FAR and FRR rates. In Table 19.1 are shown the estimated mean and standard deviation for our distributions, while in Table 19.2 are expressed the estimations of FAR and FRR, on such distributions are given, varying the matching threshold.

It is possible to note, in Table 19.2, that with a high threshold (e.g., 10) the FAR is virtually zero (no impostors enter into the system), without causing actual drop in FRR performance. In fact a FRR = 0.00044 corresponds to the above threshold value, which means that only in 44 cases over 100,000 the system rejects genuine fingerprints.

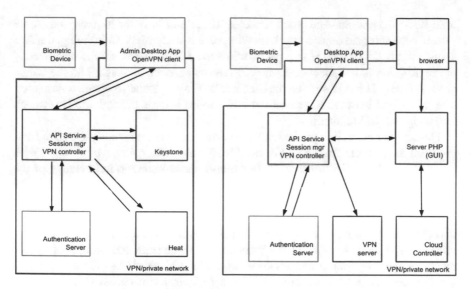

Fig. 19.3 The components and workflow of the registration procedure are shown in the *left* diagram, while figure on the *right* is related to the authentication procedure

19.3.2.3   Automation of the Biometric Access

Registration Process

Before registration process starts, a secure channel is created through a VPN. Next, the client sends new user's data to the system. The API service receives two files in JSON format, containing the meta-data that are generated during fingerprint acquisition. The JSON object also contains the user's company name. After receiving user data properly, the system initiates an automated procedure to set up the virtualized environment that will host the user's services. A general overview of the process is represented in Fig. 19.3.

During registration (Fig. 19.3, left), the service API does the following:

1. Add user to API service list;
2. Add user to AS;
3. Add user to OpenStack Keystone;
4. Add user to OpenVPN Server;
5. Create a new Stack with OpenStack HEAT.

At this stage, automated checks on each component are carried out. Given the complexity and heterogeneity of the resources involved, the system will conduct checks to prevent misalignment between authentication needed configuration service, Keystone, OpenVPN server, and API service. Therefore, the registration process ends when at least one of the operations listed above fails. Initially, the system calculates the new user's ID, password, and a network CIDR. The password is generated by random algorithm, and it is used by the API service to communicate with Keystone and manage cloud services. Therefore, in order to make the whole system even safer, no other component will possess credentials. Continuing, the API service sends the username and password to the AS that registers the new user. If the registration is not successful, the AS returns an error message, and the whole process stops. The API service requires a token in Keystone to create a new user. The interaction between Keystone and API service is done through the OpenStack API endpoint, called Identity.

During the registration process, a VPN certificate is automatically generated and provided to the user. This should be used by the user every time a connection with the cloud services is established. In this phase, the automation is in charge of the OpenVPN server which accepts as input (communication done using a Rest API Interface) the user name, and the network's CIDR, and returns the OVPN certificate, ready to be used on the VPN client. The OpenVPN server setup correct routing rules that allow user access to the network and thus to its services. The rules will take effect only if user is authenticated successfully. OpenStack has images of virtual machines, pre-configured and ready to use. The API service sends a request to create a new stack. A stack consists of a set of virtual machines connected to a new network. The network is then connected to a virtual router. All these operations are carried out automatically by the machine-accessible orchestrator heat. This is the hardest operation of the entire process because it involved almost all OpenStack services: Nova for the creation of virtual machines, Neutron security groups, ports, subnets, and vRouter interfaces. Finally, API service has successfully completed all operations and returns to the desktop client, the Open VPN certificate.

Authentication Process

The procedure for authenticating the user is shown in Fig. 19.3 (right). Before performing any operation, the user connects to the system with its OpenVPN certificate. When the client desktop finishes to acquire and convert the fingerprint, sends the file to the API service through a VPN tunnel. The data is transmitted to the AS and if user is recognized, AS returns a pair of values (that are username and password). The credentials will also be used in this case (as happens in the registration process) by API service, GUI, and OpenStack. According to result of authentication stage, the API service creates or not a new session. When the user is correctly recognized the API service generates new PHP session by creating a session file in PHP session path containing usernames, passwords, token (Open Stack), and stack ID. The username and password parameters are supplied from the AS, while stack ID is obtained by consulting a list on the API service and a token is generated by an automatic procedure. The API service connects to Open Stack Keystone requesting the token that will be used to manage the virtual machines (start, stop, resume, etc.). Finally, the API service has completed its task and returns the generated session ID to desktop client.

At this stage, the user can access to services simply connecting to the URL via browser. When the user makes a request for service management, the GUI server interacts with Nova and other Open Stack services, through the REST API. All the automation layer is run with PHP with a light framework which is able to manage processes quickly. When the user leaves the GUI, the session is destroyed and all environmental variables used for service management are removed.

It is worth to highlight some aspects of the implemented security procedure:

- User and password to access the cloud are never transmitted out of the cloud itself.
- Web GUI, AS, and private cloud controller are not accessible outside the cloud.
- Sensitive data residing on the cloud (fingerprint model file) are compared inside the cloud.
- The data transfer is not related to the user (nobody outside the cloud can associate the model file with some user information).

19.3.3   Data security

Cloud computing services and applications must face various challenges, including latency, unreliability, malicious behavior; most of these challenges are related to the public shared environment in which cloud services are hosted. In particular, security of outsourced data is still one of the main obstacles to cloud computing adoption in public bodies and enterprises. The main reason is impossibility to trust the cloud provider due to the lack of control that the user has over the infrastructure, an issue intrinsic of the public cloud model. Algorithms have to be developed to cope with these challenges and innovative architectures.

In this work is proposed a solution to ensure the data security and high availability of the resources, using an innovative distributed cloud storage architecture. The solution is based on data chunking technique: The basic idea is to share data in small chunks and spread them on different VMs hosted on cloud computing. The complete control of the distributed storage system is delegated to the user who hosts the master node of the system, as shown in Fig. 19.4. The master node maintains the namespace tree and the mapping of blocks to the slaves nodes. Thus, only the user knows the location of the chunks needed to recompose the data. Even if a malicious user can access to one of the nodes which possess the chunks he cannot use it as the information is incomplete.

This solution is a viable countermeasure also for malicious behavior of the cloud provider. Some of the features of the proposed solution are:

- Distributed storage system implemented in cloud, with client–server architecture and partially trusted environment;
- Security granted by chunking data and spreading it on different nodes (VM) possibly hosted by different cloud providers;

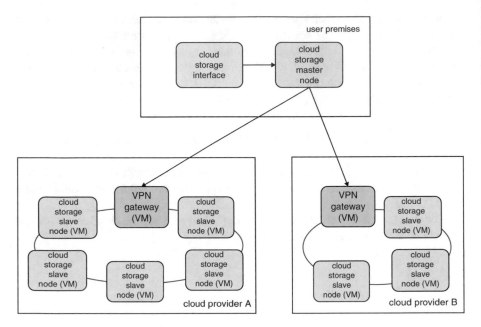

Fig. 19.4 Architecture of the distributed storage system

- Availability and resiliency ensured by the redundancy of nodes and replica of chunks;
- The possibility to use different cloud providers prevents also the so-called vendor "lock-in."

19.3.3.1 Distributed Storage Systems

There are two main categories of distributed storage systems architectures: peer-to-peer and client–server [39]. The latter architecture has been chosen for the implementation because best fit the objectives of the proposed solution. A client–server-based architecture revolves around the server providing a service to requesting clients. The server is the central point, responsible for authentication, sharing, consistency, replication, backup, and servicing requesting clients. In our implementation the master node embraces the server's role and slave nodes the client's role. As slaves nodes are hosted on the cloud, the system operates in a partially trusted environment; users are exposed to a combination of trusted and untrusted nodes [39].

In Distributed Storage Systems data can be replicated across multiple geographical sites to improve redundancy, scalability, and data availability, as shown in Fig. 19.4.

Although these solutions provide the scalability and redundancy that many cloud applications require, they sometimes do not meet the concurrency and performance needs because of the latency due to the network [40]. Some examples of the most known distributed storage systems are HDFS, Ceph, MooseFS, mongoDB.

19.3.3.2   Architecture of the System

The architecture of the solution is comprised of interconnected nodes where files and directories reside. There are two types of nodes: the master node that manages the filesystem namespace and regulates client access to files, and the slave node that stores data as blocks within files. All nodes communicate with each other using TCP-based protocols. The mechanism of data protection does not rely on RAID approaches, but the file content is replicated on multiple slaves for reliability. Master and slave nodes can run in a decoupled manner across heterogeneous operating systems, and on different cloud providers. The complete control of the system is delegated to the master node, which maintains the namespace tree and the mapping of blocks to slave nodes. Slave nodes have little intelligence and not know the location of other slaves or chunks of data.

User applications access the system using a specific client, a library that exports the filesystem interface. When a user wants to perform a reading action on filesystem, the client first asks the master node for the list of *namenodes* that host the chunks of the file. After that, the client contacts a slave node directly and requests the transfer of the desired block. Instead, when a user wants to write on the filesystem, it first asks the master to choose slaves to host chunks of the file. All decisions concerning replication of the chunks are taken by the master node. This ensures the reliability of the data and the fault tolerance of the system.

19.4   Conclusion

A complete system for web applications, and data management over the Cloud, is presented and it is coupled with strong biometric authentication. The system guarantees the identity of the users and makes easy, and secure, the access to data and services. Moreover, the adoption of a data chunking solution is proposed, which is based on a distributed cloud storage architecture. This provides protection of data residing also from provider's administrators and hardware supervisors. A further improvement of the system will extend biometric access to multimodal techniques, thus including face and face + fingerprint authentication. The development of a web server application for the user side, aimed to avoid the installation of local software, will be also pursued.

References

1. Srinavasin, M. K., et al. (2012). State of the art cloud computing security taxonomies: A classification of security challenges in the present cloud computing environment. In *ICACCI 2012 proceedings of the international conference on advances in computing, communications and informatics* (pp. 470–476). ACM.
2. Zissis, D., & Lekkas, D. (2012). Addressing cloud computing security issues. *Future Generation Computer Systems, 28*(3), 583–592.
3. Subashini, S., & Kavitha, V. (2011). A survey on security issues in service delivery models of cloud computing. *Journal of Network and Computer Applications, 34*(1), 1–11.
4. Nelson, C., & Teller, T. (2016). Cloud attacks illustrated: Insights from the cloud provider. In *RSA conference, February 29, 2016–March 4, 2016*. Moscone Center San Francisco.
5. Skokowski, P. (2014). Lessons from Apple iCloud Data Leak. *CSA–Cloud Security Alliance Industry Blog* [Online]. https://blog.cloudsecurityalliance.org/2014/11/19/lessons-from-apple-icloud-data-leak/
6. Gonsalves, A. (2013). Data leakage risk rises with cloud storage services. *Computer world Hong Kong* [Online]. http://cw.com.hk/news/data-leakage-risk-rises-cloud-storage-services
7. Konstantas, J. (2011). What does the Sony PlayStation network breach teach us about cloud security? *Security week* [Online]. http://www.securityweek.com/what-does-sony-playstation-network-breach-teach-us-about-cloud-security
8. Sotto, L. J., Treacy, B. C., & McLellan, M. L. (2010). Privacy and data security risks in cloud computing. *World Communications Regulation Report, 5*(2), 38.
9. European Commission (2012). Exploiting the potential of cloud computing in Europe, September 27, 2012 [Online]. Available: http://europa.eu/rapid/press-release_MEMO-12-713_it.htm
10. Yinqian Zhang, M. K. (2012). Cross-VM side channels and their use to extract private keys. In *CCS'12*. Raleigh, North Carolina, USA.
11. NIST (2013). NIST Cloud Computing Standards Roadmap. NIST
12. Ross, A. A., Nandakumar, K., & Jain, A. K. (2006). *Handbook of multibiometrics* (Vol. 6). Berlin: Springer.
13. Vielhauer, C. (2005). *Biometric user authentication for IT security: From fundamentals to handwriting (advances in information security)* (Vol. 18). New York: Springer.
14. Ratha, N. K., Connell, J. H., & Bolle, R. M. (2001). Enhancing security and privacy in biometrics-based authentication systems. *IBM Systems Journal, 40*(3), 614–634. Chicago.
15. Juels, A., & Sudan M. (2002). A fuzzy vault scheme. In *Proceedings of the 2002 IEEE international symposium on information theory* (p. 408). IEEE.
16. Soutar, C., Roberge, D., Stoianov, A., Gilroy, R., & Kumar, B. V. (1998). Biometric encryption using image processing. In van Renesse, R. L. (Ed.), *Proceedings of the SPIE, optical security and counterfeit deterrence techniques II* (Vol. 3314, p. 178U188).
17. Ratha, N. K., Connell, J. H., & Bolle, R. M. (2001). Enhancing security and privacy of biometric-based authentication systems. *IBM Systems Journal, 40*, 614–634.
18. Linnartz, J.-P., & Tuyls, P. (2003). New shielding functions to enhance privacy and prevent misuse of biometric templates. In *Proceedings of the 4th international conference on Audio- and video-based biometric person authentication (AVBPA'03)* (pp. 393–402). Springer.
19. Chang, Y., Zhang, W., & Chen, T. (2004). Biometrics-based cryptographic key generation. In *Proceedings of the IEEE international conference on multimedia and expo (ICME '04)* (pp. 2203–2206). IEEE Computer Society.
20. Chen, C., Veldhuis, R., Kevenaar, T., & Akkermans, A. (2007). Multibits biometric string generation based on the likelyhood ratio. In *Proceedings of the IEEE conference on biometrics: Theory, applications and systems (BTAS '07)* (pp. 1–6). IEEE Computer Society.
21. Juels, A., & Wattenberg, M. (1999). A fuzzy commitment scheme. In *Proceedings of the 6th ACM conference on computer and communication security* (pp. 28–36). ACM.

22. Martini, U., & Beinlich, S. (2003). Virtual PIN: Biometric encryption using coding theory. In Brömme, A., & Busch, C. (Eds.), *BIOSIG 2003: Biometrics and electronic signatures, ser. Lecture notes in informatics* (Vol. 31, pp. 91–99). Gesellschaft fur Informatik.
23. Masala, G. L, Ruiu P, Brunetti A, Terzo O, & Grosso E (2015). Biometric authentication and data security in cloud computing. In *Proceeding of the international conference on security and management (SAM). The Steering Committee of The World Congress in Computer Science* (p. 9). Computer Engineering and Applied Computing (WorldComp).
24. Ruiu, P., Caragnano, G., Masala, G. L., & Grosso, E. (2016). *Accessing cloud services through biometrics authentication on proceedings of the international conference on complex, intelligent, and software intensive systems (CISIS-2016), July 6–8, 2016.* Japan: Fukuoka Institute of Technology (FIT).
25. Maltoni, D., Maio, D., Jain, A., & Prabhakar, S. (2009). *Handbook of fingerprint recognition* (2nd ed.). Berlin: Springer.
26. OpenStack. OpenStack cloud administrator guide [Online]. Available http://docs.openstack.org/admin-guide-cloud/content/
27. Aggarwal, G., Bawa, M., Ganesan, P., Garcia-Molina, H., Kenthapadi, K., Motwani, R., Srivastava, U., Thomas, D., & Xu, Y.. Two can keep a secret: A distributed architecture for secure database services. In: *Proceeding of the 2nd conference on innovative data systems research (CIDR).* Asilomar, California, USA.
28. Ciriani, V., Di Vimercati, S. D. C., Foresti, S., Jajodia, S., Paraboschi, S., & Samarati, P. (2007). Fragmentation and encryption to enforce privacy in data storage. In *European symposium on research in computer security* (pp. 171–186). Berlin, Heidelberg: Springer.
29. Damiani, E., De Capitani, S., di Vimercati, S., Jajodia, S., Paraboschi, S., & Samarati, P. (2003). Balancing confidentiality and efficiency in untrusted relational DBMSs. In: *CCS03 proceeding of the 10th ACM conference on computer and communications security, Washington, DC, USA, October 2003.* New York: ACM Press.
30. Hacigümüs, H., Iyer, B., & Mehrotra, S. (2002). Providing database as a service. In *ICDE'02 proceedings of the 18th international conference on data engineering, San Jose, California, USA.* Los Alamitos, California: IEEE Computer Society.
31. Lowe, D. (1999). Object recognition from local scale-invariant features. In *International conference on computer vision and pattern recognition* (pp. 1150–1157).
32. Lowe, D. (2004). Distinctive image features from scale-invariant keypoints. *International Journal of Computer Vision, 60*(2), 91–110.
33. Lowe, D. (2001). Local feature view clustering for 3d object recognition. In *IEEE conference on computer vision and pattern recognition* (pp. 682–688).
34. Bicego, M., Lagorio, A., Grosso, E., & Tistarelli, M. (2006). On the use of SIFT features for face authentication. *In CVPRW'06 Conference on computer vision and pattern recognition workshop* (pp. 35–35). IEEE.
35. Ke, Y., & Sukthankar, R. (2004). PCA-SIFT: A more distinctive representation for local image descriptors. In *IEEE conference on computer vision and pattern recognition.*
36. Heusch, G., Rodriguez, Y., & Marcel, S. (2005). Local binary patterns as an image preprocessing for face authentication. *IDIAP-RR 76, IDIAP.*
37. Zhang, G., Huang, X., Li, S., Wang, Y., & Wu, X. (2004). Boosting local binary pattern (lbp)-based face recognition. In *L. 3338, SINOBIOMETRICS* (pp. 179–186). Springer.
38. Fierrez, J., Galbally, J., Ortega-Garcia, J., et al. (2010). BiosecurID: A multimodal biometric database. *Pattern Analysis and Applications, 13,* 235.
39. Placek, M., & Buyya, R. (2006). The University of Melbourne, a taxonomy of distributed storage systems. Reporte Técnico, Universidad de Melbourne, Laboratorio de Sistemas Distribuidos y Cómputo Grid.
40. Assunção, M. D., Calheiros, R. N., Bianchi, S., Netto, M. A., & Buyya, R. (2015). Big Data computing and clouds: Trends and future directions. *Journal of Parallel and Distributed Computing, 79,* 3–15.

Chapter 20
Approximate Search in Digital Forensics

Slobodan Petrović

20.1 Introduction

In digital investigation, many sources of information (such as hard disks from personal computers, tablet computers, and mobile devices) must be analyzed in a relatively short time period in order to find evidence in criminal cases. Very often, the amount of captured information is so large that manual analysis is practically impossible. In addition, data may not be visible at the operating system/application level (e.g., fragments of files deleted at the operating system level may still be possible to recover by means of special tools). Because of that, various systems have been developed that are capable of performing forensic search (Elasticsearch [5], Forensic Toolkit [7], etc.)

In particular, in network forensics, the requirement for analysis in a short time period is very strict—the malware/attack traces are to be detected in real time. To this end, on hosts, various malware detection systems are used (like Host-based Intrusion Detection Systems (HIDS), usually present in antivirus solutions). On networks, Network-based Intrusion Detection Systems (NIDS) (e.g., Snort [19], and Suricata [21]) and network monitoring tools (e.g., Bro [4]) are widely used for this purpose.

The class of problems and algorithms related to finding particular objects and relationships among them in large and heterogeneous data sets is often referred to as *Big Data*. The big data algorithms can be used to solve problems of digital forensics search. However, due to the specific nature of digital forensics, many of these algorithms must be adapted to the application requirements.

S. Petrović (✉)
Norwegian University of Science and Technology (NTNU), Trondheim, Norway
e-mail: slobodan.petrovic@ntnu.no

© Springer International Publishing AG 2018
K. Daimi (ed.), *Computer and Network Security Essentials*,
DOI 10.1007/978-3-319-58424-9_20

We are particularly interested in the following scenario of digital forensics investigation: suppose we have to find pieces of evidence data on a captured media volume under the assumption that the data have been deliberately changed in order to cover traces. The perpetrator might have used a tool to change the incriminating pieces of information automatically. Such a tool has parameters determining the nature of these changes (number of changes, their distribution, and so on).

A similar scenario can be considered in network forensics: we suppose that an attacker has used a tool to modify the attack traffic in order to pass unnoticed by the defense systems (Intrusion Detection Systems, network monitoring tools, etc.)

This scenario introduces variations in the data set that has already been very large, which makes the search task even more complicated. Under such a scenario, an exact search algorithm is not capable of finding evidence or detecting attack traffic. Approximate search must be used for this purpose. Alternatively, all the possible (or acceptable) variations of the original search pattern must be included in the pattern dictionary, which consumes additional time and space. The general-purpose big data search systems mentioned above are capable of finding distorted evidence in large data volumes in a reasonable amount of time. For example, Elasticsearch [5] offers the user "fuzzy queries" to search for strings in text with a given tolerance (i.e., maximum edit distance from the original search pattern). This system is based on Apache Lucene library [11] implementing various search algorithms (including the approximate ones) in Java. The approximate search algorithm implemented in the Lucene library is the one based on so-called Levenshtein automaton [12, 16]. This algorithm, along with other algorithms from the same family, is explained in Sect. 20.2.

The search algorithms that are usually used in network forensics employ exact search and as such they are not capable of detecting small changes in attack traffic. Most of these systems use by default the Aho–Corasick algorithm [1] as a multi-pattern search algorithm that is reasonably efficient and resistant to so-called algorithmic attacks against the very IDS. Attempts to use approximate search in network forensics/intrusion detection have been made (see, e.g., [8]), but in a different scenario where whole strings were inserted in attack traffic.

Even though approximate search can be used to detect distorted evidence or attack patterns, in the present form the available search algorithms do not take into account the distribution of the changes applied on the original patterns. The search tolerance incorporated in these algorithms only takes the number of applied changes into account. The consequence of this is increased probability of false alarms. In a criminal investigation, false alarms may lead to false accusations. In attack detection, false alarms are annoying and the investigator may ignore all the reports from the IDS, even the true positive ones, because of that.

To reduce the number of false positives in approximate search, we propose introducing specific constraints in the approximate search algorithms. The constraints reflect the a priori knowledge about the choice of parameters that the perpetrator/attacker has used in the incriminating/attack pattern modification. This knowledge is a consequence of a limited choice that is at the modifier's disposal: too many changes on strings saved on the media volume reduce their intelligibility too much and make those patterns useless due to impossibility of their recognition even by the intended consumers of information. Too many changes on attack traffic patterns can make the obtained traffic pattern harmless for the victim.

In addition to resolving for the false positives, a big data search algorithm for digital forensics applications must also be as efficient as possible, especially if real-time operation is necessary. So-called bit-parallel search algorithms have proved themselves to be the most efficient of all. They exploit inherent parallelism of computer words (typically 32 or 64 bits at the moment) and process the input characters of the search string by performing bit operations only (typically shift, OR, and AND). This makes them extremely fast if the search pattern length is shorter than the length of the computer word. We consider modifications of the existing bit-parallel algorithms that enable introduction of the constraints explained above. We analyze efficiency of the obtained algorithms together with influence of the introduced constraints on the false-positive rate.

Approximate search algorithms (both constrained and unconstrained), in addition to being capable of detecting modified search patterns, reduce the data set indirectly. Namely, if we want to detect modified search patterns by means of *exact* search algorithms, we have to put all the possible intelligible modifications of the search pattern in the dictionary. In network forensics, we have to include every small modification of the original attack pattern in the attack signature database. These strategies significantly increase the size of the data set besides contributing to the increase of the searching time complexity. By using approximate search, we do not need these large data sets and we can just keep a single representative of each family of attack signatures.

20.2 Bit-Parallel Search

During the last 25 years, bit-parallel techniques in search for patterns have been thoroughly studied (see, e.g., [2, 6, 14]) The goal of their development was to speed up the execution as much as possible in order to make them capable of processing extremely large amounts of data present in modern information systems. A lot of effort has been made to design search algorithms that are fast *on average* [3, 13]. While this approach is useful in general digital forensics applications, in network forensics it opens the way to so-called algorithmic attacks against IDS, where the attacker deliberately produces and launches traffic that forces these algorithms to perform poorly, i.e., to process the input traffic at their worst-case time performance. This means that in real-time operation, where the most efficient search algorithms are needed, we cannot rely on the algorithms that are fast on average. Because of that, specific search algorithms intended for application in network forensics have also been developed. Pure bit-parallel algorithms for this purpose have been investigated, like the one described in [20], but also other techniques using so-called bit-splitting architectures [22] and dynamic linked lists [9].

All the algorithms mentioned above perform *exact* search, and as such they are incapable of producing desired results under the scenario described in the previous section. Approximate search algorithms employing bit parallelism have also been studied (see, e.g., [14, 23]) but only in the unconstrained case. To introduce constraints, modifications of the existing approximate search bit-parallel algorithms are necessary.

20.2.1 Exact Bit-Parallel Search

The bit-parallel search algorithms simulate a Non-deterministic Finite Automaton (NFA) that performs parallel transitions from the current state(s) to some other state(s) (many at a time in general) at each input character from the search string. Such an NFA is assigned to the search pattern. The essence of these NFA simulations is the proven fact [2] that only the status of each string comparator running in parallel (hence NFA) matters for the execution of the search algorithm. The status can be *active* or *inactive*, which is encoded with a single bit. The number of comparators running in parallel in the NFA simulation is equal to the length of the search pattern. If the length of the search pattern is less than or equal to the length of the computer word, then it is possible to place all the status bits in a computer word (so-called *status word*) and update the bits of this word at each input character from the search string. This update is performed by means of the *update formula*. In most cases, the update formula only manipulates the status word by means of the bit operations (shift, OR, and AND). These bit operations consume a very small number of processor cycles on modern computers and, because of that, bit-parallel search algorithms are very fast.

Different algorithms from this family differ in the form and complexity of the update formula. The algorithms, whose average and worst-case time complexities are (approximately) the same, have very simple update formulas. The algorithms capable of skipping the portions of the search string, in which the search pattern cannot be detected, have more complex update formulas. These algorithms have a lower average-case than the worst-case time complexity.

We illustrate the principles of bit-parallel search with an example. Suppose we want to perform exact search for the pattern "gauge" in the search string "omega-gauge." We consider the string comparator assigned to the search pattern. It is a finite state machine capable of making transitions from one state to another after receiving an input character from the search string. Two forms of such a machine are possible to analyze: with or without ε transitions (the transitions without input). These two forms of the machine are equivalent. Without ε transitions, each time a new input character arrives, the comparator tries to match it, i.e., to make a transition from the current state to the state corresponding to the character at the input. The NFA in this case makes a new copy of such a machine for each new input character and tries to match this character on each copy. At success, the corresponding machine remains active; otherwise, it becomes inactive and it is not used for processing the

Fig. 20.1 Two forms of a
string comparator—with and
without ε transitions

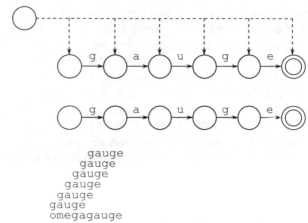

```
                  gauge
                  gauge
                  gauge
                  gauge
                  gauge
                  gauge
              omegagauge
```

Fig. 20.2 Operation of an NFA searching for the pattern "gauge" in the string "omegagauge"

following characters. On the other hand, with ε transitions, we can consider that
several transitions occur at a time and that such a machine matches substrings of the
pattern at once. The two forms of the string comparator are presented in Fig. 20.1.
The final state is double-circled, and, if the machine manages to reach it, it has
recognized the whole pattern.

An infinite parallel string comparator starts from the first character of the search
string and performs the comparison of the whole search pattern of length m with the
first m characters of the search string. The search stops if the complete match is
detected. Otherwise, the algorithm advances one character in the search string,
creates a new string comparator, and continues this procedure until a match is found
(an *occurrence*) or the end of the search string is reached; see Fig. 20.2.

Since it is not possible to have infinite parallelism (i.e., NFA) in practice, we
replace it with a simulated NFA where the number of string comparators operating
at a time is limited to the length of the search pattern m. Each time a new comparator
is created (when a new character from the search string is processed), we remove
the oldest comparator and try to match the new character from the input string
with a newly created comparator. Every comparator incapable of matching the new
character from the search string becomes *inactive* and its status can be encoded
with a "0." Otherwise (i.e., if active), it is encoded with a "1." We place all the status
bits in a computer word of length m, called *the status word*. When processing any
character from the search string, some bits of the status word will have the value
0, indicating the inactive string comparators, and some bits will have the value 1,
indicating active comparators. The fact that the oldest comparator is discarded when
a new comparator is created is encoded in the status word update formula by shifting
the status word to the left by one position. The newly created comparator is always
active before trying to match the new character from the search string, and this is
encoded in the search status word update formula by OR-ing the status word with 1.
What remains is to check which bits of the status word will remain active after

processing the new character from the search string. It is easy to see that each bit in the status word will remain active (i.e., will have the value 1) if it was active previously and if the corresponding string comparator matches the new character from the search string. In addition, each comparator corresponding to a bit of the status word will always try to match the same character. This means that we can pre-compute *bit masks* corresponding to each character contained in the search pattern and AND the bit mask corresponding to the current character from the search string with the status word. That completes the updating process of the status word when processing the current character from the search string.

In our example, the length of the search pattern is $m = 5$, so the search status word D contains 5 bits. Initially, they are all set to 1. The bit mask for each character of the search pattern is obtained by setting a "1" to the position that corresponds to the position of that character in the *inverted* pattern. This is because the most significant bit of the status word corresponds to the oldest string comparator. If the search pattern is "gauge," the bit masks are the following: $B[a] = 00010$; $B[e] = 10000$; $B[g] = 01001$; $B[u] = 00100$. For all other characters, $B[*] = 00000$. Each time a new character from the search string is processed, a new string comparator is created that is active and the oldest string comparator is discarded. In the search status word update formula, this corresponds to shifting the status word to the left by 1 position (discarding the oldest comparator) and OR-ing the obtained status word with 1 (creating of a new active comparator). Suppose that the algorithm has processed the first five characters of the search string "omegagauge" and it is about to process the sixth character ("g"); see Fig. 20.3. The current search status word is 00010, which means that only the string comparator created before processing the fourth character of the search string is still active since it managed to match that character and the next character as well ("ga").

The oldest string comparator must be discarded before the new one is created, so we shift the status word to the left by one position. The value of the status word becomes 00100. After creating the new string comparator, the value of the status word will become 00100 OR 00001 = 00101. Finally, since the next character from the search string is "g," we take the bit mask $B[g]$ and AND it with the status word, which gives the updated value of the status word 00101 AND 01001 = 00001. Note that the bit mask contains ones at the positions where the currently processed character from the search string is located in the inverted search *pattern*. Thus, if the status bit at that location in the status word is active (which means that the corresponding string comparator is waiting for a "g"), it will remain active after

Fig. 20.3 The search status word before processing the next character of the search string

Fig. 20.4 Updating of the status word—shifting left, OR-ing with 1, AND-ing with the bit mask

	j=5	j=6	j=6
		Shift+OR	Shift+OR+AND
ǀ			
ǀg		1	1
gǀa	0	0	0
gaǀu	1	1	0
gauǀg	0	0	0
gaugǀe	0	0	0
gaugeǀ	0		
omegagauge			

AND-ing with the bit mask. The process of shifting to the left, OR-ing with 1, and AND-ing with the bit mask $B[g]$ is illustrated in Fig. 20.4.

The search status word update expression formalizes the procedure described above. Let W be the search pattern of length m. Let S be the search string of length n and let the j-th character of S be denoted S_j. Let the corresponding bit mask be $B[S_j]$ and let D_j be the value of the search status word after processing the first j characters from S. Then the search status word update formula is

$$D_j = ((D_{j-1} << 1) \text{ OR } 1) \text{ AND } B[S_j], \quad j = 1, \ldots n \qquad (20.1)$$

where $D_0 = 1^m$ (all ones). After updating the status word D, the search algorithm has to check whether the most significant bit of D has the value 1. If so, the pattern W has been recognized in the search string S (we have an occurrence). The bit-parallel exact search algorithm described here is called the Shift-AND algorithm and was the first algorithm from the family [2]. A small modification of this algorithm (Shift-OR) eliminates the need to OR with 1 in the update formula, which makes the algorithm even faster (see, e.g., [14]). The modification consists in complementing the bit masks and the search status word (0 is considered active and 1 inactive). These algorithms have the same worst-case and average-case complexities, which makes them suitable for network forensics applications (e.g., IDS) since they are resistant to algorithmic attacks.

Many algorithms from the exact search family of bit-parallel algorithms involve skipping some regions in the search string, where the search pattern cannot be located. This makes those algorithms faster on average than the Shift-AND/Shift-OR algorithm. A typical representative of these so-called *skip algorithms* is the BNDM (Backward Non-deterministic DAWG[1] Matching) algorithm [13]. These algorithms can be used in exact search in digital forensics investigations, but in network forensics applications they are vulnerable to algorithmic attacks.

---

[1]DAWG—Directed Acyclic Word Graph.

20.2.2 Approximate Bit-Parallel Search

To be capable of solving the search problem under the scenario described in Sect. 20.1, we need an approximate search algorithm. Such an algorithm also has to be the fastest possible and, because of that, it is of interest to try to use approximate bit-parallel search algorithms. We assume that the search tolerance k is given in advance. This number is the maximum acceptable number of allowed modifications to the original search pattern, where the modifications can be performed by inserting, deleting, and substituting characters. Bearing this in mind, we can say that k is the maximum allowed *edit distance* from the original search pattern [10]. The problem of unconstrained approximate bit-parallel search was first studied in [23]. The idea was to simulate an extended NFA containing $k+1$ string comparators in a matrix form, where each comparator occupies one row of the NFA matrix. This technique is called Row-based Bit Parallelism (RBP) [14]. Transitions at each input character from the search string can be horizontal (a match), vertical (an insertion), solid diagonal (a substitution), and dashed diagonal (a deletion). A dashed diagonal transition is an ε transition.

As in the exact search case, we illustrate the concept of unconstrained approximate bit-parallel search with an example. Suppose that the search pattern W is "gauge" and the search tolerance is $k = 2$. The NFA capable of performing this kind of search is presented in Fig. 20.5.

A transition from any active state of the simulated NFA can be horizontal, vertical, solid diagonal, or dashed diagonal. At each input character of the search string S, all the active states make all these transitions simultaneously. If the simulated NFA reaches the double-circled (i.e., final) state in some row, then we have an occurrence within the tolerance k. Instead of a status word, we have a *status array* R with the rows $R_0, \ldots, R_k$. If we denote the new state of the status array with R', then the update formula for the automaton of the same type as the one from Fig. 20.5 is

Fig. 20.5 The NFA performing unconstrained bit-parallel approximate search for the pattern "gauge" with the search tolerance $k = 2$

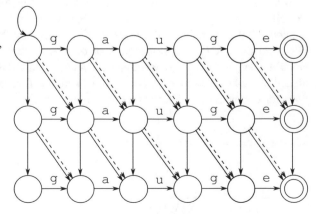

$$R'_0 = ((R_0 << 1) \text{ OR } 1) \text{ AND } B[S_j]$$

$$R'_i = ((R_i << 1) \text{ AND } B[S_j]) \text{ OR } R_{i-1} \text{ OR } (R_{i-1} << 1) \text{ OR } (R'_{i-1} << 1) \quad (20.2)$$

where $i = 1, \ldots, k$ and $B[S_j]$ is the bit mask corresponding to the input character S_j. Note that in Eq. (20.2) the influences of a match, an insertion, a substitution, and a deletion on the status array row bits are taken into account, respectively. Note also that for encoding the influence of deletion on the status array row bits we use the newly computed previous row, unlike the other edit operations (match, substitution, insertion). The reason for this is that a deletion implies an ε transition (dashed diagonal in the NFA array). Finally, we can also notice that the update formula for the 0-th row of the status array R is equivalent to the Shift-AND update formula since the 0-th row of R corresponds to exact search.

Schulz and Mihov [12, 16] use an approach to approximate string matching similar to the one described above. They call the NFA equivalent to the one from Fig. 20.5 the *Non-deterministic Levenshtein Automaton*. They use a deterministic equivalent (Deterministic Finite Automaton—DFA) of this automaton to speed up the search process. Namely, they show that the transitions of the NFA can be pre-computed, which makes the search algorithm faster at the cost of increased memory consumption. This approach can be feasible in a garbled word dictionary search in the cases where the search tolerance k is relatively small. For other applications, the space complexity of the DFA approach is too high. The concrete application determines what is more important—low memory consumption (where NFA simulation is better) or short execution time (where DFA implementation is better). On the other hand, if implemented as a bit-parallel algorithm, the approach from [12, 16] in its NFA variant is shown to be very efficient in any application (including the digital forensic ones), and if k is small, then extreme search speeds can be expected.

20.3   Introducing Constraints in Approximate Search

Approximate search without constraints is capable of detecting modified patterns and this gives a partial solution to the digital investigation problem/scenario described in Sect. 20.1. But the key assumption of the scenario, the fact that the perpetrator/attacker uses a tool with a set of parameters to modify the original pattern, is not taken into account by the unconstrained approach. Namely, the parameters of the tool do not only limit the number of changes (which is correctly addressed in the approximate search algorithms by introducing the search tolerance k) but also the *distribution* of these changes. For example, it is not the same to modify a pattern by using deletions only as to modify the same pattern by using certain (limited) number of deletions, substitutions, and/or insertions. The distribution of pattern modification operations influences the intelligibility of the modified information and the threat level of the modified attack traffic. Because of that, this distribution cannot be allowed to be arbitrary in general.

Using unconstrained approximate search algorithms to detect patterns modified by means of a tool with the properties described above gives rise to situations where the edit distance between the detected string and the original pattern is less than or equal to k, but the distribution of the changes in the detected string does not correspond to the one defined by the perpetrator/attacker at the time of modification, as determined by the set of tool parameters. In these cases, we have *false positives*. The number of false positives may be quite high, which reduces the trust in the search algorithm.

We can solve the problem of taking into account the applied distribution of changes by introducing specific constraints in the approximate search algorithm. These constraints can define tolerances on the numbers of specific edit operations or the lengths of runs of some edit operations. The values of the constraints to be defined in a constrained approximate search algorithm are determined by the a priori knowledge of parameters of the tool used by the perpetrator/attacker. Since the values of these change parameters must be small in order to keep the intelligibility and attack threat level in an acceptable range, the probability of correctly guessing these parameters by the attacked party is high. In the case of a perfect guess of these parameters, the false-positive rate caused by the change distribution parameters can be reduced to very law values, often to zero [17, 18].

The question now arises if we can construct a fast constrained approximate search algorithm and still keep the false-positive rate at an acceptable level. The bit-parallel technique has proved itself to guarantee high efficiency of unconstrained approximate search algorithms. Then we can modify these algorithms in order to introduce constraints. The modification depends on the type of constraints that is introduced. The list of these types includes the constraints on the total numbers of elementary edit operations (insertions, deletions, substitutions), the total number of *indels* (insertions + deletions), and the maximum lengths of runs of insertions and deletions.

In most cases, to take into account the constraints in approximate search, we have to assign *counters* to the bits of the simulated NFA. Each time a transition in the NFA is performed, the counter values are updated for all the active states. Then they are checked against the defined constraints, and if the maximum is reached, the corresponding NFA bit becomes inactive. The time and space complexities of the obtained constrained approximate search algorithm are influenced by the addition of counters and new bit masks defining which bits of the NFA rows become inactive due to the constraints. Bit masks can be implemented efficiently by means of the existing bit operations on most modern CPUs. The counters, on the other hand, influence the space complexity to a great extent, and if any values for the constraints would be allowed, performance of the constrained approximate search algorithm would be significantly reduced. However, bearing in mind the discussion above, our scenario assumes that the changes in the original search pattern are small and their distribution is such that a small number of elementary edit operations are performed at a time. Because of that, the limits for the counters used in the NFA are small (often equal to 1), which reduces the increase in space complexity for a constrained approximate search algorithm compared to the unconstrained one. Thus, in most

---

Algorithm 1 Generic constrained approximate search algorithm

1: **procedure** CONSTRAINEDSEARCH(W, S, con, k)
2:    *Input parameters*:
3:    $W = w_1 w_2 \ldots w_m$ - search pattern,
4:    $S = s_1 s_2 \ldots s_n$ - search string
5:    con - constraints, usually in the form of an array
6:    k - maximum number of errors
7:    k depends on the values of constraints
8:    *Output*:
9:    Positions in S where W was found, allowing up to k errors
10:    *Initialization*
11:    We first define bit masks corresponding to the characters from W
12:    Then we initialize status array R with all 0s, except for $R[0, 0] = 1$
13:    The constraint-related bit masks are set to all 1s for every row of R
14:    *Processing of the search string S*
15:    **for** $j \in 1 \ldots n$ **do**
16:        Update the row 0 of R (the same as unconstrained search)
17:        **for** row $\in 1 \ldots k$ **do**
18:            Generate bit mask for the row based on counter values
19:            Update the row status word and apply the bit mask
20:            (Contributions for match, insertion, substitution and deletion are OR-ed)
21:            Compute the new counters for the row
22:        **end for**
23:        Check the MSB of each row of R; if it is equal to 1 $\rightarrow$ occurrence at j
24:    **end for**

---

cases, we obtain an efficient constrained approximate search algorithm that produces few false positives compared to the original unconstrained case. In [15], a generic constrained approximate search algorithm is proposed for a SPAM filtering scenario, which is very similar to the digital forensics scenario analyzed in this chapter. The algorithm is presented below (Algorithm 1). Its concrete form depends on the set of constraints that models the perpetrator's (or attacker's) behavior.

20.4   Conclusion

In this chapter, we have analyzed the possibilities of using constrained bit-parallel approximate search algorithms to solve problems that appear in digital investigation practice related to the perpetrator's/attacker's ability to modify the incriminating/attack patterns in order to hide traces of criminal activity. We have shown that not only the number of changes in the original search pattern but also the distribution of these changes influences the quality of the results obtained from a search algorithm (the most prominent quality parameter is the false-positive rate). In addition, we have shown that, by using bit-parallel search techniques, we can obtain approximate search algorithms that are both very efficient and produce few false positives. The concrete form of the constrained approximate search algorithm depends on the set of constraints that model the attacker's behavior at changing the search patterns. The a priori probability of guessing the right value for the constraints is high since, due to the intelligibility/attack threat level requirements, the probability space size is relatively small.

References

1. Aho, A., & Corasick, M. (1975). Efficient string matching: An aid to bibliographic search. *Communications of the ACM, 18*, 333–340.
2. Baeza-Yates, R., & Gonnet, G. (1992). A new approach to text searching. *Communications of the ACM, 35*, 74–82.
3. Barton, C., Iliopoulos, C., & Pissis, S. (2015). Average-case optimal approximate circular string matching. In A. Dediu, E. Formenti, C. Marín-Vide, & B. Truthe (Eds.), *Language and automata theory and applications* (pp. 85–96).
4. Bro. https://www.bro.org/. Cited April 25, 2017
5. Elasticsearch. https://www.elastic.co/products/elasticsearch. Cited May 9, 2017
6. Faro, S., & Lecroq, T. (2012). Twenty years of bit-parallelism in string matching. In J. Holub, B. Watson, J. Žďárek (Eds.), *Festschrift for Bořivoj Melichar* (pp. 72–101).
7. Forensic Toolkit (FTK). http://accessdata.com/solutions/digital-forensics/forensic-toolkit-ftk. Cited May 9, 2017
8. Kuri, J., & Navarro, G. (2000). Fast multipattern search algorithms for intrusion detection. In *String processing and information retrieval (SPIRE 2000)* (pp. 169–180).
9. Le-Dang, N., Le, D., & Le, V. (2016). A new multiple-pattern matching algorithm for the network intrusion detection system. *IACSIT International Journal of Engineering and Technology, 8*, 94–100.
10. Levenshtein, V. (1966). Binary codes capable of correcting deletions, insertions and reversals. *Soviet Physics-Doklady, 10*, 707–710.
11. Lucene, A. http://lucene.apache.org/. Cited April 25, 2017
12. Mihov, S., & Schulz, K. (2004). Fast approximate search in large dictionaries. *Journal of Computational Linguistics, 30*, 451–477.
13. Navarro, G., & Raffinot, M. (2000). Fast and flexible string matching by combining bit-parallelism and suffix automata. *ACM Journal of Experimental Algorithms, 5*(4), 1–36.
14. Navarro, G., & Raffinot, M. (2002). *Flexible pattern matching in strings: Practical on-line search algorithms for texts and biological sequences.* New York: Cambridge University Press.
15. Petrović, S. (2016). A SPAM filtering scenario using bit-parallel approximate search. In P. Gomila, & M. Hinarejos (Eds.), *Proceedings of the XIV Spanish Conference on Cryptology and Information Security (RECSI2016)* (pp. 186–190).
16. Shulz, K., & Mihov, S. (2002). Fast string correction with Levenshtein automata. *International Journal on Document Analysis and Recognition (IJDAR), 5*, 67–85.
17. Shrestha, A., & Petrović, S. (2015). Approximate search with constraints on indels with application in SPAM filtering. In V. Oleshchuk (Ed.) *Proceedings of Norwegian Information Security Conference (NISK-2015)* (pp. 22–33).
18. Shrestha, A., & Petrović, S. (2016). Constrained row-based bit-parallel search in intrusion detection. In A. Kolosha (Ed.) *Proceedings of Norwegian Information Security Conference (NISK-2016)* (pp. 68–79).
19. Snort. https://www.snort.org/. Cited April 25, 2017
20. Sung-il, O., Min, S., & Inbok, L. (2013). An efficient bit-parallel algorithm for IDS. In: A. Aghdam, & M. Guo (Eds.) *Proceedings of RACS 2013* (pp. 43–44).

21. Suricata. https://suricata-ids.org/. Cited April 25, 2017
22. Tan, L., & Sherwood, T. (2006). Architectures for bit-split string scanning in intrusion detection. *IEEE Micro, 26*, 110–117.
23. Wu, S., & Manber, U. (1992). Fast text searching allowing errors. *Communications of the ACM, 35*, 83–91.

Chapter 21
Privacy Preserving Internet Browsers: Forensic Analysis of Browzar

Christopher Warren, Eman El-Sheikh, and Nhien-An Le-Khac

21.1 Introduction

Internet security has been a major and increasing concern for many years. Internet security can be compromised not only through the threat of Malware, fraud, system intrusion, or damage, but also via the tracking of Internet activity. Indeed, Internet web browsers are daily used by most individuals and can be found on computers, mobile devices, gaming consoles, smart televisions, in wristwatches, cars, and home appliances. Web browsers store user information such as the sites visited, as well as the date and time of Internet searches [16]. In addition to clearing their browsing history, users can also prevent storage of such information by using 'private browsing' features and tools [3]. Users may choose to use private browsing features and tools for many reasons, including online gift shopping, testing and debugging websites, and accessing public computers. Besides, criminals are using numerous methods to access data in the highly lucrative cybercrime business. Organized crime, as well as individual users, are benefiting from the protection of Virtual Private Networks (VPN) and private browsers, such as Tor, Ice Dragon, and Epic Privacy to carry out illegal activity such as money laundering, drug dealing, and the trade of child pornography. Weak security has been identified

C. Warren
RCMP, Fredericton, NB, Canada
e-mail: cewarren15@gmail.com

E. El-Sheikh
Centre for Cybersecurity, University of West Florida, Pensacola, FL, 32514, USA
e-mail: eelsheikh@uwf.edu

N.-A. Le-Khac (✉)
School of Computer Science, University College Dublin, Belfield, Dublin 4, Ireland
e-mail: an.lekhac@ucd.ie

© Springer International Publishing AG 2018
K. Daimi (ed.), *Computer and Network Security Essentials*,
DOI 10.1007/978-3-319-58424-9_21

and exploited in a number of high profile breaches in recent years. Additionally, the release of National Security Agency (NSA) and Government Communications Headquarters UK (GCHQ) documentation by Edward Joseph Snowden in 2013 further highlighted the need for improved online security. Following the Snowden breach, there was public outrage at the lack of privacy leading to a rise in the number of browsers offering private browsing. News articles advising on Internet privacy assisted in educating the public and a new era of private browsing arose. Although these measures were designed to protect legitimate browsing privacy, they also provided a means to conceal illegal activity.

Mozilla Firefox and Google Chrome have 'private browsing' modes in which the user can browse the Internet without the browser saving Internet histories [3]. On the other hand, there are Internet browsers that are marketed based on its 'total browse privacy'. These browsers claim that, for instance, they do not save cookies, temporary history files, passwords, or cache. In addition to the privacy claims, these browsers do not require a local installation, and can even be launched from removable media, such as an external hard drive. Their features are very appealing for those seeking complete online anonymity, as well as for those users who intend to commit criminal acts online, such as child exploitation. However, there are very few researches on evaluating of private browsing in terms of privacy preserving as well as forensic acquisition and analysis of privacy preserving feature of 'total browse privacy' Internet browsers. Therefore, in this chapter, we firstly review the private mode of popular Internet browsers. Next, we describe the forensic acquisition and analysis of Browzar, a privacy preserving Internet browsers, and compare it with popular Internet browsers. The rest of this chapter is organized as follows: Sect. 21.2 shows the background of this research including related work in this domain. We present the forensic acquisition and analysis methods of Browzar privacy browser in Sect. 21.3. We describe our experimental results and discussion of analysing Browzar privacy browser in Sect. 21.4. Finally, we conclude and discuss on future work in Sect. 21.5.

21.2 Background

Today, with its rise in popularity, the Internet has created a new form of criminal activity, often referred to as 'cybercrime'. Cybercrimes have been defined in various ways, but are generally categorized into one of the following two types of Internet-related crime: 'advanced cybercrime' and 'cyber-enabled crime' [12]. Advanced cybercrimes are attacks against computer hardware and software, whereas cyber-enabled crimes are 'traditional' crimes that have been enabled by and have taken new form with the advent of the Internet. Cyber-enabled crimes include crimes against children, financial crimes, and terrorism. International agencies have identified cybercrimes as an issue. Europol, for example, notes that the Internet is a major source of criminal activity [6]. The obscure nature of the criminal activity makes it very challenging for law enforcement not only to catch criminals using the

Internet, but also to successfully pursue prosecution. In an effort to avoid detection and prosecution by law enforcement agencies, criminals resort to using various tools and tactics to conceal their illegal online activity. These tools and tactics often leave little to no evidence behind.

21.2.1 Forensic Analysis and Web Browser

According to Junghoon et al. [16], a crucial component of a digital forensic investigation is searching for evidence left by web browsing. Each movement taken by a suspect using a web browser can leave a recoverable trace on the computer. Analysing this information can produce a variety of artefacts, such as websites visited, time and frequency of access, and keyword searches performed by the suspect. Junghoon et al. [16] further state that in a Web browser forensic investigation, the simple parsing of information is not enough. Additional forensic techniques may be required to extract more significant information such as 'the keyword searches and the overall user activity'. For example, depending on the possible user activities during a single browser session (i.e. email, online banking, blogging), investigators must analyse the information generated from each browser using the same timeline. Web browser history and analysis has become an increasingly important area of computer forensics [14]. As techniques advance and new tools become available to examine web browsers, investigators are able to reconstruct timelines, identify suspect browsing habits, and recover evidence of crimes. Furthermore, the results uncovered during web browser forensics can determine the objective, methods, and criminal activities of a suspect [13].

21.2.2 Privacy Browsing

Although private browsing has legitimate uses, such as activity on multiple user devices, many individuals are using the shield of anonymity to carry out illegal activity on the Internet. Private browsing is designed in some web browsers to disable browsing history and the web cache. This allows a user to browse the Web without storing data on their system that could be retrieved by investigators. Privacy mode also disables the storage of data in cookies and browsing history databases. This protection is only available to the local device as it is still possible to identify websites visited by associating the IP (Internet Protocol) address at the website. Apple introduced the first Internet privacy features in a browser in 2005 in the web browser Safari (http://www.apple.com/safari). Since this time, 'privacy features' or 'privacy modes' have become standard in most Internet web browsers. Google Chrome, Mozilla Firefox, and Internet Browser each also have private browsing features.

Aggarwal et al. [2] examined private browsing features introduced by four popular browsers: Internet Explorer, Mozilla Firefox, Google Chrome, and Apple Safari. The authors noted that private browsing modes have two goals: (1) to ensure sites visited while browsing in private leave no trace on the user's computer; (2) to hide a user's identity from websites they visit by, for example, making it difficult for websites to link the user's activities in private mode to the user's activities in public mode. The research also identified the inconsistencies when using private mode with the popular browsers and revealed that, although all major browsers support private browsing, inconsistency in the type of privacy provided by each differs greatly. Firefox and Chrome attempt to protect against both web and local attacks while Safari only prevents local issues [20]. Although legitimate reasons do exist for using private browsing, some users take advantage of these features to conceal criminal activity.

Akbal et al. [3] explain that a typical private browsing session is initiated from within the interface of the main browser [3]. Once initiated, it is carried out in its own private browsing session window until the session is terminated. While in a private browsing session, no updates are made to the browser history, and upon terminating the session, the browser will delete any cookies stored during the session. This will also clear the download list. Akbal et al. [3] suggest that while these features do add privacy, they will not completely cover a user's tracks. Hedberg's research led to similar findings and suggests that there are many misconceptions about private browsing [11]. To the end user, private modes and private browsers typically perform as advertised. However, using a series of web browser forensic techniques and tools, Hedberg proved that claims of total browse privacy are false. During his research involving the private modes for the Google Chrome, Mozilla Firefox, and Microsoft Internet Explorer browsers, Hedberg successfully recovered artefacts for each browser from within the hard drive and memory of the system.

Well-known browsers such as Google Chrome, Internet Explorer, Safari, and Mozilla Firefox rely on similar methods to ensure speed and popularity of their product. Web Cache is a popular way of storing data that can be easily and quickly accessed, thereby negating the necessity to find data that has already been used. History databases, thumbnails (small stored images), temporary files, and cookies (user and site specific data) all help to speed up the user experience and, in their path, leave a plethora of artefact evidence for examiners to feast on [21]. Many studies have been carried out in this area and free tools, such as ChromeHistoryView, ChromeCacheView, IECacheView [5], as well as forensic software such as Internet Evidence Finder, are available to automate the examination process. All the above browsers have the option to operate in private mode. Research by Khanikekar [17] indicates that the use of Internet Explorer in 'Protected Mode' runs a 'Low Privilege' process, preventing the application writing to areas of the system that require higher privilege. Besides, there is also research on forensic investigation of privacy and portable modes of web browser in literature [9, 10] but they are not in the scope of this paper. Information commonly stored on a device when using Internet browsers include cache, temporary Internet files, cookie information, search history, passwords, and registry changes. This chapter aims to establish what, if any, data relating to the use of Browzar Privacy Browser is produced during the installation and user interaction with the browser.

21.2.3 Browzar

Browzar Internet browser (http://www.browzar.com/) was designed to offer total browse privacy for users. It was designed and marketed for individuals to browse the Internet in a private way—either on a personal or shared computer. Browzar claims that it does not save any information to the local system—and even includes an automatic purging feature to remove any information left behind upon terminating a browsing session. Browzar is an Internet Explorer (IE) shell browser, so Internet Explorer is required on the host system for it to function. Further, it does not require a standard installation; instead, it can be launched from a removable media device, or even from the vendor website. Working in computer forensics, it is common for investigators to rely on the recovery of Internet evidence to solidify investigations and prosecution. Presently, there is little documented research on Browzar and its privacy claims. This chapter will challenge the key selling points advertised by Browzar. Various experiments will be conducted as part of a full forensic analysis to determine if any artefacts are left behind on a system after using Browzar. Additional comparisons will be made to the Google Chrome and Mozilla Firefox browsers to identify similarities in artefact preservation. More details can be found in the following sections.

21.3 Forensic Acquisition and Analysis of Browzar Privacy Browser

In this chapter, we aim to answer the following research questions:

1. Is Browzar capable of providing complete privacy to users, avoiding all forensic techniques? With claims of total browse privacy, it is important to confirm whether Browzar performs as advertised simply from an end user's stand point, or if it can actually avoid all forensic techniques, including memory analysis, tools specific to web browser forensics, and a full forensic analysis using software such as X-Ways.
2. How does Browzar privacy compare to Google Chrome and Mozilla Firefox privacy modes? This will help determine if the Browzar product works better or worse than most private modes that are now standard in many Internet browsers. This comparison will also help to identify whether Browzar could be deemed an anti-forensic web browser.

3. What evidence was recoverable during post-mortem analysis of Browzar? The most common type of analysis occurs after a device is seized. The post-mortem analysis of Browzar will benefit investigators and offer recommendations on analysing data using specific tools such as web browser forensic tools. Further, post-mortem analysis enables access to database files and information found in unallocated space—information that would otherwise not be available during a live or logical preview.
4. What evidence was recoverable during live state analysis of Browzar? Investigators today are increasingly facing situations that no longer allow them to unplug a system and perform a traditional hard drive acquisition. The information captured in the memory of a system is considered volatile information—which is information that is lost once the host system is powered off. This data is extremely important and useful as it can often help determine the most recent activities on a computer—including browser specific findings such as keyword searches and website history.

21.3.1 Adopted Approach

The following approach was adopted to address the questions posed above. Based on the nature of this research subject, various techniques and tools were used to recover and analyse information.

Change monitoring is a useful technique used to determine the impact a specific program or action has on a system. With respect to web browsers, change monitoring can help determine any modifications that the browser is making to the system, and, in turn, identify the system files and registry keys being changed as a result of a particular action. Further to this, a virtual environment is ideal for testing different versions of a browser and for validating findings.

Live data forensics is an increasingly important aspect of computer forensic investigations. The objective of live data forensics is to minimize impacts to the integrity of data while collecting volatile evidence from the suspect system. The primary source of volatile information is contained within the RAM of a system. Information residing in RAM is stored there on a temporary basis. Once the host system is powered off, the data is cleared from the RAM and lost forever. Forensic tools have been developed that allow investigators to capture and process the data stored in RAM. As part of the experiments performed in this research, the host system will be treated as a real world exhibit. Using forensic techniques and tools, the RAM will be captured in a forensic manner and preserved for further analysis. Various tools will be used to process and analyse the contents of the RAM for web browser specific evidence.

Post-mortem analysis often provides the best opportunity to gather the most evidence in a digital forensic investigation. The post-mortem forensic process begins with a data acquisition. This involves creating bit by bit forensic duplicates of any digital evidence. These forensic duplicates not only protect the integrity of the original data, but also enable additional information to be recovered from deleted and slack space. Software tools, such as X-Ways, are then used to process and categorize the information as part of the forensic analysis. With respect to web browsers, full forensic software like X-Ways include features designed to parse out web browser databases and any other relevant metadata.

21.3.2 Experimental Environments

The results outlined in this chapter were obtained using two separate working environments. The first environment was created using an Apple Mac Mini computer. Apples Boot Camp program was used to configure the system to allow a dual boot configuration—thus enabling an installation of the Windows 7 operating system (service pack 1). All available Windows 7 updates—both security and software—were performed using the Windows Update feature to ensure a complete up to date working environment. The operating system was installed on a 75 GB VMware virtual machine (VMware Workstation 12 Pro). The virtual machine served two purposes: to create an isolated working environment and to allow the system to be reverted back to an earlier date (i.e. snapshot) if necessary.

Change monitoring tools including Process Monitor (Procmon) and Index.dat Analyzer were installed in the virtual environment. Microsoft explains Process Monitor as an 'advanced monitoring tool for Windows that shows real-time file system, Registry and process/thread activity'. Internet Explorer, which is native to Windows 7, also remained as part of the environment. It should be noted that Browzar was tested using Internet Explorer versions 9 through 11. Although compatible with each version, there are instances where results were gathered using only Internet Explorer version 9. Depending on the stability and compatibility, certain results may have been gathered using Internet Explorer versions 10 or 11.

To further analyse the behaviour of each browser, a second testing environment was created using a Dell Latitude E6430 laptop containing a 120 GB OCZ solid state hard drive and 4 GB of RAM. The system was left connected to the Internet for several days, all while actively using various search engines to perform specific web activity. These websites required a successful login with a username and password. After approximately 48 h, the system memory (RAM) was acquired using the dumpit.exe tool, and the system was successfully shut down using a proper shutdown sequence.

21.4 Experiments and Analysis

21.4.1 Change Monitoring

21.4.1.1 Browzar

Contrary to the successful research findings discovered for Google Chrome and Mozilla Firefox, limited information pertaining to Browzar was found. Therefore, the best source of information would be from testing the product itself. Upon completing the initial setup of the virtual machine, Browzar version 2.0 (Windows Style Theme) was downloaded directly from the vendor website. The installation file, as advertised, was around 200 kb in size and took only seconds to download in its entirety. Using Procmon, the file system and registry changes were examined to determine the footprint of the Browzar installation. A Browzar session was initiated in order to visit various websites and to perform the keyword searches outlined above. This activity was designed to populate the system with random Internet artefacts, such as pictures, keyword searches, and cookies. Procmon values were created for browzarwinstyle2000.exe and iexplore.exe to help filter results. Although no results were registered for the iexplore.exe value during the launch, plenty of activity was logged for browzarwinstyle2000.exe (Fig. 21.1).

The Browzar interface, originally modelled after Mozilla Firefox, is similar in appearance to a typical Internet browser. Upon further inspection, differences emerged. Various privacy features such as a separate field for private searches as well as Secure Delete and Force Cleanup options were present. Similar to other Microsoft Internet Explorer shell browsers, much of Browzar's activity involved the registry, specifically with the following keys:

- HKCU\Software\Microsoft\InternetExplorer
- HKCU\Software\Microsoft\Windows\CurrentVersion\InternetSettings
- HKCU\Software\Wow6432Node\Microsoft\InternetExplorer.

A review of the file system activity in Procmon revealed that various files were being created on a temporary basis upon starting Browzar. To illustrate this further,

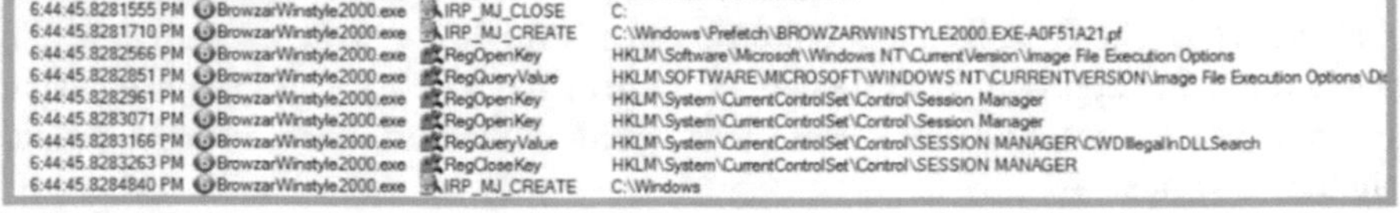

Fig. 21.1 Activity logs of Browzar

8:40:39...	BrowzarWinstyle2000.exe	5700 CloseFile	C:\Users\21\AppData\Local\Microsoft\Windows\History
8:40:39...	BrowzarWinstyle2000.exe	5700 CreateFile	C:\Users\21\AppData\Roaming\Browzar
8:40:39...	BrowzarWinstyle2000.exe	5700 ReadFile	C:
8:40:39...	BrowzarWinstyle2000.exe	5700 CloseFile	C:\Users\21\AppData\Roaming\Browzar
8:40:39...	BrowzarWinstyle2000.exe	5700 CreateFile	C:\
8:40:39...	BrowzarWinstyle2000.exe	5700 QueryNameInfo...	C:\
8:40:39...	BrowzarWinstyle2000.exe	5700 QueryAttributeIn...	C:\
8:40:39...	BrowzarWinstyle2000.exe	5700 CloseFile	C:\
8:40:39...	BrowzarWinstyle2000.exe	5700 CreateFile	C:\Users\21\AppData\Roaming\Browzar\recovery.lck
8:40:39...	BrowzarWinstyle2000.exe	5700 CloseFile	C:\Users\21\AppData\Roaming\Browzar\recovery.lck
8:40:39...	BrowzarWinstyle2000.exe	5700 CreateFile	C:\Windows\SysWOW64\ntdll.dll
8:40:39...	BrowzarWinstyle2000.exe	5700 QueryBasicInfor...	C:\Windows\SysWOW64\ntdll.dll
8:40:39...	BrowzarWinstyle2000.exe	5700 CloseFile	C:\Windows\SysWOW64\ntdll.dll
8:40:39...	BrowzarWinstyle2000.exe	5700 CreateFile	C:\Windows\SysWOW64\ntdll.dll
8:40:39...	BrowzarWinstyle2000.exe	5700 CreateFileMap...	C:\Windows\SysWOW64\ntdll.dll
8:40:39...	BrowzarWinstyle2000.exe	5700 CreateFileMap...	C:\Windows\SysWOW64\ntdll.dll

Fig. 21.2 Procmon results for BrowzarWinstyle2000.exe

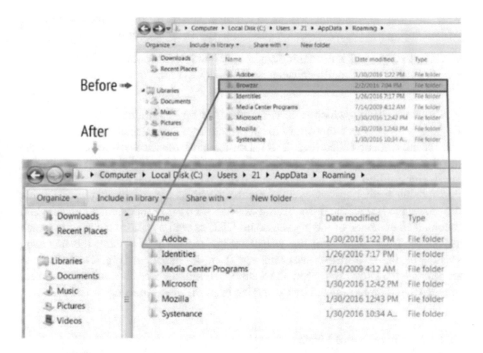

Fig. 21.3 Before and after illustration of deleted temporary Browzar folder

a Browzar folder was created in C:\Users\21\AppData\Roaming which contained a single file called recovery.lck (see Fig. 21.2).

Browzar claims that the recovery.lck contains a date and time stamp of the precise moment the Browzar session began [23]. It states further that this file contains no information about websites visited during a Browzar session. As advertised, any remnants of these *findings* were purged upon closing the browsing session (Fig. 21.3). File operations were also monitored using Process Monitor to determine

2:32:33.1425120 PM	Browzar Winstyle 2000.exe	WriteFile	C:\Users\21\AppData\Local\Microsoft\Windows\Temporary Internet Files\Content.IE5\index.dat
2:32:33.1425464 PM	Browzar Winstyle 2000.exe	WriteFile	C:\Users\21\AppData\Local\Microsoft\Windows\Temporary Internet Files\Content.IE5\index.dat
2:32:33.1425792 PM	Browzar Winstyle 2000.exe	WriteFile	C:\Users\21\AppData\Local\Microsoft\Windows\Temporary Internet Files\Content.IE5\index.dat
2:32:33.1426121 PM	Browzar Winstyle 2000.exe	WriteFile	C:\Users\21\AppData\Local\Microsoft\Windows\Temporary Internet Files\Content.IE5\index.dat
2:32:33.1426458 PM	Browzar Winstyle 2000.exe	WriteFile	C:\Users\21\AppData\Local\Microsoft\Windows\Temporary Internet Files\Content.IE5\index.dat
2:32:33.1426789 PM	Browzar Winstyle 2000.exe	WriteFile	C:\Users\21\AppData\Local\Microsoft\Windows\Temporary Internet Files\Content.IE5\index.dat
2:32:33.1427117 PM	Browzar Winstyle 2000.exe	WriteFile	C:\Users\21\AppData\Local\Microsoft\Windows\Temporary Internet Files\Content.IE5\index.dat
2:32:33.1427854 PM	Browzar Winstyle 2000.exe	WriteFile	C:\Users\21\AppData\Local\Microsoft\Windows\Temporary Internet Files\Content.IE5\index.dat
2:32:33.1428206 PM	Browzar Winstyle 2000.exe	WriteFile	C:\Users\21\AppData\Local\Microsoft\Windows\Temporary Internet Files\Content.IE5\index.dat
2:32:33.1428536 PM	Browzar Winstyle 2000.exe	WriteFile	C:\Users\21\AppData\Local\Microsoft\Windows\Temporary Internet Files\Content.IE5\index.dat
2:32:33.1428866 PM	Browzar Winstyle 2000.exe	WriteFile	C:\Users\21\AppData\Local\Microsoft\Windows\Temporary Internet Files\Content.IE5\index.dat
2:32:33.1429196 PM	Browzar Winstyle 2000.exe	WriteFile	C:\Users\21\AppData\Local\Microsoft\Windows\Temporary Internet Files\Content.IE5\index.dat
2:32:33.1429525 PM	Browzar Winstyle 2000.exe	WriteFile	C:\Users\21\AppData\Local\Microsoft\Windows\Temporary Internet Files\Content.IE5\index.dat
2:32:33.1429855 PM	Browzar Winstyle 2000.exe	WriteFile	C:\Users\21\AppData\Local\Microsoft\Windows\Temporary Internet Files\Content.IE5\index.dat
2:32:33.1430460 PM	Browzar Winstyle 2000.exe	WriteFile	C:\Users\21\AppData\Local\Microsoft\Windows\Temporary Internet Files\Content.IE5\index.dat
2:32:33.1438071 PM	Browzar Winstyle 2000.exe	WriteFile	C:\Users\21\AppData\Local\Microsoft\Windows\Temporary Internet Files\Content.IE5\index.dat

Fig. 21.4 Procmon activity showing create, read, write, and close operations

what information was being written to the disk. The majority of information being written during the Browzar session was to the following locations:

- C:\Users\21\AppData\Roaming\Microsoft\Windows\Cookies
- C:\Users\21\AppData\Local\Microsoft\Windows\TemporaryInternetFiles\ Content.IE5.

The index.dat file, which is located in the Content.IE5 folder, is used by Internet Explorer version 9 as a repository for information such as websites, search queries, and recently opened files (Fig. 21.4). During the Browzar session, the index.dat file showed continuous create, read, and write operation activity. Upon terminating the session, additional close operation activity to the index.dat file was observed. To illustrate this further, the www.nfl.com website was accessed using Browzar. During the browsing session, the index.dat file was accessed using Index.dat Analyzer v2.5. It contained evidence of the www.nfl.com URL as well as additional artefacts such as .png and .jpg files from the website (Fig. 21.5). Upon closing the browsing session, all the information was removed from the index.dat (Fig. 21.6). Prior to terminating the Browzar session, the Secure Delete and Force Cleanup options were initiated and then terminated by exiting the browsing session using the File menu.

21.4.1.2 Chrome and Firefox

Using the Chrome and Firefox private modes, the remaining searches and web activity assigned to each browser. Chrome and Firefox browsers were creating and using SQLite databases to store browsing information, as evident from the Procmon results. These SQLite files were analysed using X-Ways and Oxygen forensic software programs. The Oxygen Forensic software is a mobile forensic software typically used for extracting and analysing cell phones, smartphones, and other mobile devices. However, it also contains a built-in SQLite viewer that parses out and presents the information stored in an SQLite database file. While observing the installation of Chrome and Firefox, it was also documented that user profiles for each browser were generated as part of the installation. Both browsers allowed

Fig. 21.5 Index.dat Analyzer results showing www.nfl.com artefacts

for multiple user profiles using their own SQLite database files, enabling individual users the ability to collect their own bookmarks, store passwords, and history.

21.4.2   Forensic Acquisition

21.4.2.1   Live Data Forensics

To successfully capture the RAM, a 32 GB Lexar USB thumb drive was forensically wiped and loaded with the dumpit.exe utility. Using the dumpit.exe utility, the RAM capture was initiated, storing the output directly to the previously forensically wiped 32 Lexar USB thumb drive.

21.4.2.2   Post-Mortem Data Acquisition

In order to acquire information from the hard drive of the Dell Laptop, a forensic data acquisition was performed. The hard drive was removed from the laptop and connected via a USB tableau write blocker to a forensic analysis machine. Using FTK Imager version 3.4.2.2 forensic software, the data acquisition process was

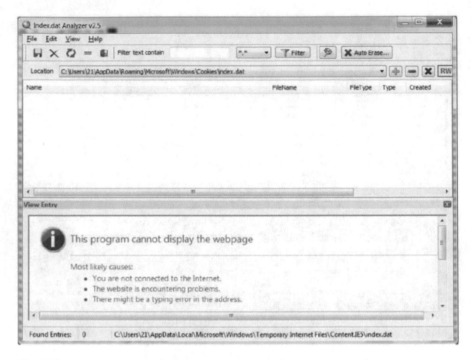

Fig. 21.6 Index.dat Analyzer showing www.nfl.com artefacts were removed

performed. The electronic information contained on computer storage media must be acquired by making a complete physical copy of every bit of data located on the computer media in a manner that does not alter that information. The integrity of the data is authenticated at the end of the acquisition using the MD5 hashing algorithm. This hashing process creates a 128 bit hash value of the data—which is often referred to in computer forensics as the digital fingerprint—to ensure the acquired electronic information is an exact duplicate of the storage media. For this particular acquisition, the forensic duplicate of the hard drive was created using the Encase image file format—known as E01.

21.4.3 Forensic Analysis

21.4.3.1 Browzar: Memory Analysis

During the data processing stage, IEF was used to process the RAM and search for artefacts pertaining to the Browzar specific search strings such as 'Hockey, Baseball, Golf, Football', 'Brazil, South Africa, Canada', 'Beer, wine, Scotch'. IEF recovered several Google Searches performed using Browzar (Fig. 21.7). By further

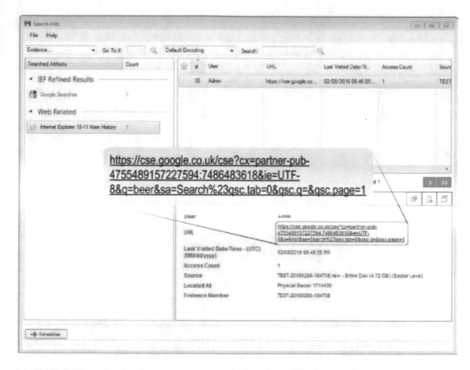

Fig. 21.7 IEF results showing recovered Google Search for 'beer'

analysing the memory using X-Ways, additional keyword searches were identified. For example, through X-Ways, six instances of the term 'beer' were recovered.

Similar results were also identified for the other search terms. Residual artefacts found in RAM such as these are often left behind while the browsing session remains active and stored in RAM. By performing this additional examination, it illustrates the importance for forensic analysts to perform manual data checks and also the importance of validating findings using a second forensic tool.

21.4.3.2   Browzar: Post-Mortem Analysis

Considerable findings were also uncovered during post-mortem analysis for Browzar in the *pagefile.sys*. The *pagefile* is used by the Windows operating system as an extension of random access memory. For example, the *pagefile* is often accessed when a computer system's RAM is completely full. The operating system views this file as actual physical memory, allowing for quicker access to applications stored within it compared to if they remained in their original location. In this instance, *pagefile* findings were recovered using IEF and X-Ways and included keyword searches, web history, and images from webpages. In addition to

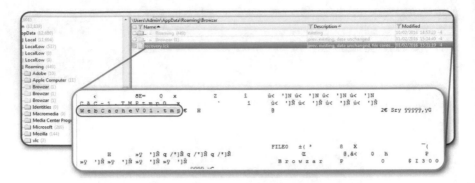

Fig. 21.8 WebCachev01.tmp reference found in recovery.lck file

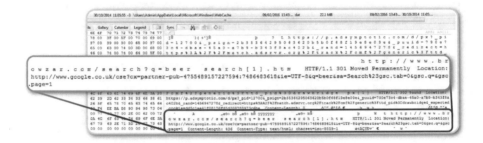

Fig. 21.9 Evidence of Browzar searches found in WebCachev01.dat

the findings in the *pagefile*, several instances of the Browzar folder were discovered. One of the instances containing a recover.lck file was examined. The majority of this file was not legible in X-Ways and it did not contain any stored websites—as claimed by Browzar. However, one section that referenced a WebCachev01.tmp file was discovered (Fig. 21.8). This prompted the researcher to locate and analyse the WebCachev01.dat file. The WebCachev01.dat file is used by Internet Explorer (version 10 and higher) to maintain the web cache, history, and cookies. A review of this file in X-Ways uncovered more references to the keywords used during the Browzar session, confirming that Browzar is leaving traces of activity behind (Fig. 21.9). In addition to these findings, Browzar's private search feature revealed that all searches were performed using a www.google.co.uk address.

To investigate this further, Wireshark tool was used to perform a series of tests. Wireshark is a free and open source network protocol analyser for Unix and Windows. By capturing the browsing session using Wireshark, the data packets sent to and from the test environment during the Browzar session were reviewed and filtered. Based on the results of these tests, it was evident that Browzar was not leaking any session information and that the private search feature found on the Browzar interface was simply a 'Google Custom Search' engine—a feature that was hardcoded into the tool.

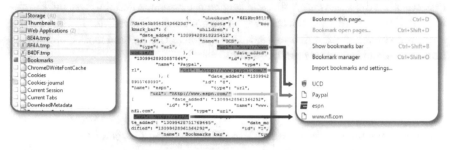

Fig. 21.10   Sample Chrome results found using X-Ways

C:\Users\%username%\AppData\Local\Google\Chrome\User Data\Default\Bookmarks

Fig. 21.11   Illustration showing references to Chrome bookmarks in X-Ways and in Chrome

21.4.3.3   Google Chrome: Memory Analysis

Although minimal, several findings of interest relating to the Chrome private session were recovered during the analysis of the system memory. By filtering the results using the keyword 'football', IEF recovered a user generated Google search involving this word. Similar discoveries were made during a manual examination using X-Ways (Fig. 21.10).

21.4.3.4   Google Chrome: Post-Mortem Analysis

The *pagefile.sys* file contained some of the most valuable information recovered from the Chrome private browsing session. For example, various keyword searches used during the Chrome session were recovered in the *pagefile*. In addition to keyword searches, URLs, bookmarks, and images were all discovered in the *pagefile*. Several bookmarks were created during the private browsing session. It is to be noted that these bookmarks remained available during subsequent normal (not private) browsing sessions. These were easily recoverable using IEF[1] and X-Ways[2] via bookmarks SQLite database file (Fig. 21.11). The remaining Chrome SQLite databases were examined, but did not contain any information related to the private browsing session. Aside from the residual content left behind in memory and the

---

[1] https://www.magnetforensics.com/magnet-ief/.

[2] http://www.x-ways.net/.

spillage found in the *pagefile*, Chrome did a good job of discarding any traces of activity, and certainly performed as advertised from the viewpoint of the end user.

21.4.3.5 Mozilla Firefox: Memory Analysis

IEF was used to process the RAM and search for evidence pertaining to the Firefox. IEF was able to recover several Yahoo searches performed during the Firefox browsing session. In addition to keyword searches, URLs, bookmarks, and images were also identified among the IEF results.

21.4.3.6 Mozilla Firefox: Post-Mortem Analysis

Post-mortem analysis was performed for Firefox using Oxygen.[3] Once again, the *pagefile.sys* contained a wealth of information. From within the *pagefile*, URLs, parsed search queries, and pictures were recovered (Fig. 21.12). To gain a fuller picture of the activity performed during the Firefox session, evidence of a search for 'South Africa' extended beyond the recovered parsed search queries. For example, multiple images—both full and partial—were recovered from the *pagefile* that originated while browsing South African travel websites. With respect to bookmarks, several were created during the private browsing session. It is to be noted that these bookmarks remained available during subsequent normal (not private) browsing sessions (Fig. 21.13). The remaining Firefox SQLite databases were examined, but did not contain any information related to the private browsing session. Aside from the residual content left behind in memory, and the information

☆	#	Search Term	URL
	10	south africa	https://ca.search.yahoo.com/search?p=south+africa&fr=yfp-t-715
	14	south africa	https://ca.search.yahoo.com/search;_ylt=A0LEV2Q24hW2SEA'VbrFAx.?p=south+africa&fr2=sb-top&nojs=1
	21	south africa	https://ca.search.yahoo.com/search;_ylt=A0LEV2Q24hW2SEA'VbrFAx.?p=south+africa&fr2=sb-top&nojs=1
	25	south africa	https://ca.search.yahoo.com/search;_ylt=A0LEV2Q24hW2SEA'VbrFAx.?p=south+africa&fr2=sb-top&nojs=1
	29	south africa	https://ca.search.yahoo.com/search;_ylt=A0LEV2Q24hW2SEA'VbrFAx.?p=south+africa&fr2=sb-top&nojs=1
	75	south africa	https://ca.search.yahoo.com/search;_ylt=A0LEV2Q24hW2SEA'VbrFAx.?p=south+africa&fr2=sb-top&nojs=1
	85	south africa	https://ca.search.yahoo.com/search;_ylt=A0LEV2Q24hW2SEA'VbrFAx.?p=south+africa&fr2=sb-top&nojs=1
	92	south africa	https://ca.search.yahoo.com/search;_ylt=A0LEV2Q24hW2SEA'VbrFAx.?p=south+africa&fr2=sb-top&nojs=1
	98	south africa	https://ca.search.yahoo.com/search;_ylt=A0LEV2Q24hW2SEA'VbrFAx.?p=south+africa&fr2=sb-top&nojs=1

Fig. 21.12 Sample IEF result showing searches for South Africa

---

[3]https://www.oxygen-forensic.com/en/.

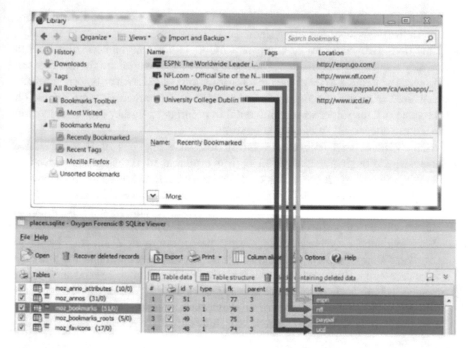

Fig. 21.13 Illustration showing references to Firefox bookmarks in Firefox and Oxygen SQLite Viewer

found in the *pagefile*, Firefox did not store information on the system. From an end user's perspective, the Firefox private browsing feature also performed as advertised.

21.4.4 Discussion

The results of this chapter will serve to fill the gap in knowledge regarding the forensic analysis of the browser, Browzar. Based on the results, it was concluded that private browsing artefacts *are* recoverable. All tests as part of this chapter reduced the amount of information left behind after usage, and were sufficient at minimizing and preventing the amount of data stored on the host system. However, none of the tests were capable of fully preventing or controlling what information remained in the memory of the system and the *pagefile*. In this section, we discuss on the research questions we raised in Sect. 21.3:

1. Is Browzar capable of providing complete privacy to users, avoiding all forensic techniques? To the end user, the product performed as advertised. Upon closing the browsing session, it removed all traces of web browser activity. However, using a combination of forensic tools and techniques, evidence, including

pictures, keyword searches, and URLs, were easily recovered in both the memory and in the *pagefile* of the test system.

2. How does Browzar privacy compare to Google Chrome and Mozilla Firefox privacy modes? Out of the three browsers *forensically* examined for this chapter, Browzar was found to leave the most information behind. Not only were artefacts recovered from various locations, it also left a folder named Browzar behind on the system. This folder was instrumental in pointing to further evidence, leading to a more successful forensic examination.

3. What evidence was recoverable during live state analysis of Browzar? Live analysis, although not always an option available to investigators proved to contain valuable evidence in this case. From within RAM, evidence of keyword searching, websites visited, and pictures were recovered. In certain cases, pictures were not fully recoverable, but they helped demonstrate the interests and activity being performed during the browsing session.

4. What evidence was recoverable during post-mortem analysis in relation to Browzar? Using X-Ways and IEF as primary tools, several additional artefacts were recovered. Files and folders left behind by Browzar were recovered using manual techniques from within X-Ways. The WebCacheV01.dat and *pagefile.sys* files proved to contain a wealth of information—including evidence of Browzar keyword searches, websites visited, and pictures.

21.5   Conclusions and Future Work

In this chapter, we examine Browzar, which claims to be a browser specifically designed for privacy, comparing results to the built-in private browsing modes in Google Chrome and Mozilla Firefox Internet browsers. This chapter also provided elaborate and detailed answers relating to the forensically recoverable information for each web browser tested. The results of this research are useful to, and may be referenced by, forensic experts involved in investigations concerning web activity on both desktop and mobile platforms [7, 8, 22]. Each question posed in the problem statement was addressed and the underlying data presented in subsequent sections should help not only investigators looking for a better general understanding of web browser forensics, but also those seeking advanced techniques and methods for recovering, parsing, and analysing web browser specific data. Some topics for further scientific and practical research are coming up. First of all, investigators can use forensic method proposed in this chapter to examine other privacy Internet browsers such as Epic Privacy Browser (https://www.epicbrowser.com/). Next, this research also assists the studying of website fingerprinting [1, 15]. Besides, we are looking at using clustering methods and tree-based approach [4, 18, 19] to group correlated artefacts from Browzar to deeply analyse the evidence. Moreover, experimental results described in this chapter can assist the researchers who are studying for new methods of preserving the privacy in the next generation of web browser and mobile apps [24].

References

1. Acar, G., Eubank, C., Englehardt, S., Juarez, M., Narayanan, A., & Diaz, C. (2014, November). The web never forgets: Persistent tracking mechanisms in the wild. In *Proceedings of CCS 2014*.
2. Aggarwal, G., Bursztein, E., Jackson, C., & Boneh, D. (2010). An analysis of private browsing modes in modern browsers. In *Proceedings of the 19th USENIX security symposium*, USENIX Association.
3. Akbal, E., Günes, F., & Akbal, A. (2016). Digital forensic analyses of web browser records. *The Journal of Software, 11*(7), 631–637.
4. Aouad, L.-M., An-Lekhac, N., & Kechadi, T. (2009). Grid-based approaches for distributed data mining applications. *Journal of Algorithms & Computational Technology, 3*(4), 517–534.
5. Chivers, H. (2014). Private browsing: A window of forensic opportunity. *Digital Investigation, 11*(1), 20–29.
6. Europol. (2016). *Europol identifies 3600 organised crime groups active in the EU*. Available via https://www.europol.europa.eu/content/europol-identifies-3600-organised-crime-groups-active-eu-europol-report-warns-new-breed-crim. Accessed 10 December 2016.
7. Faheem, M., Kechadi, M. T., & Le-Khac, N. A. (2015). The state of the art forensic techniques in mobile cloud environment: A survey, challenges and current trends. *International Journal of Digital Crime and Forensics (IJDCF), 7*(2), 1–19.
8. Faheem, M., Kechadi, M. T., Le-Khac, N. A.. (2016). Toward a new mobile cloud forensic framework. In *6th IEEE International Conference on Innovative Computing Technology*, Ireland.
9. Flowers, C., Mansour, A., & Al-Khateeb, H. M. (2016). Web browser artefacts in private and portable modes: A forensic investigation. *Journal of Electronic Security and Digital Forensics, 8*(2), 99–117.
10. Ghafarian, A. (2016, May). Forensics analysis of privacy of portable web browsers. In *ADFSL Conference on Digital Forensics, Security and Law*, Daytona Beach, Florida.
11. Hedberg, A. (2013). *The privacy of private browsing* (Technical Report). Available via http://www.cs.tufts.edu/comp/116/archive/fall2013/ahedberg.pdf. Accessed December 2016.
12. Interpol. (2016). *Cybercrime*. Available via http://www.interpol.int/Crime-areas/Cybercrime/Cybercrime. Accessed 30 November 2016.
13. Jones, K., & Rohyt, B. (2005). Web browser forensic. *Security Focus*. Available via http://www.securityfocus.com/infocus/1827. Accessed 10 December 2016.
14. Jones, K. J. (2003). Forensic analysis of internet explorer activity files. *Foundstone*. Available via http://www.foundstone.com/us/pdf/wp_index_dat.pdf. Accessed 15 January 2017.
15. Juarez, M., Imani, M., Perry, M., Diaz, C., & Wright, M. (2016). Toward an efficient website fingerprinting defense. In I. Askoxylakis, S. Ioannidis, S. Katsikas, & C. Meadows (Eds.), *Computer Security – ESORICS 2016. ESORICS 2016, Lecture notes in computer science* (Vol. 9878). Cham: Springer.
16. Junghoon, O., Seungbong, L., & Sangjin, L. (2011, August 1–3). Advanced evidence collection and analysis of web browser activity. In *The digital forensic research conference*, Los Angeles.
17. Khanikekar, S. K. (2010). *Web forensics*. Graduate thesis, A&M University, Texas.
18. Le Khac, NA, Bue, M., Whelan, M., & Kechadi, M. T. (2010, November). A cluster-based data reduction for very large spatio-temporal datasets. In *International conference on advanced data mining and applications*, China.
19. Le-Khac, N. A., Markos, S., O'Neill, M., Brabazon, A., & Kechadi, M. T. (2009, July). An efficient search tool for an anti-money laundering application of an multi-national bank's dataset. In CESRA Press (Eds.), *2009 International conference on Information and Knowledge Engineering (IKE'09)*, Las Vegas, USA.
20. Pereira, M. T. (2009). Forensic analysis of the Firefox 3 Internet history and recovery of deleted SQLite records. *Digital Investigation, 5*(1), 93–103.

21. Satvat, K., Forshaw, M., Hao, F., & Toreini, E. (2014). On the privacy of private browsing – A forensic approach. *Journal of Information Security and Applications, 19*(1), 88–100.
22. Sgaras, C., Kechadi, M. T., & Le-Khac, N. A. (2015). Forensics acquisition and analysis of instant messaging and VoIP applications. In U. Garain & F. Shafait (Eds.), *Computational forensics, Lecture notes in computer science* (Vol. 8915). Cham: Springer.
23. Techdirt. (2016). *According to the government, clearing your browser history is a felony*. Available via https://www.techdirt.com/articles/20150606/16191831259/according-to-government-clearing-your-browser-history-is-felony.shtml. Accessed December 2016.
24. Voorst, R. V., Kechadi, T., & Le-Khac, N. A. (2015). Forensics acquisition of Imvu: A case study. *Journal of Association of Digital Forensics, Security and Law, 10*(4), 69–78.

Part V
Hardware Security

Chapter 22
Experimental Digital Forensics of Subscriber Identification Module (SIM) Card

Mohamed T. Abdelazim, Nashwa Abdelbaki, and Ahmed F. Shosha

22.1 SIM Cards Overview

SIM cards are a subset of smart cards. Smart card's architecture consists of EEPROM, FLASH, and processor that could be used to store information or execute operating system [8, 15]. In addition, the design of smart cards allows user applications to be developed and installed on the cards using SIM API for Java card [5].

SIM cards have been developed for many years. However, there is limited research regarding their security or existing vulnerabilities. They are contact cards used in every mobile phone and GPS device and originally developed for secure communication and GPS devices. Using SIM cards for security means that they can be a potential threat. Mobile phones could be used as a listening device to record the conversations or to steal personal information where SIM cards are one of the keys for remote access control over the device.

SIM card analysis is not a new research; however, it is not a disclosed research and very limited information about it is publicly available. The earliest SIM card forensic analysis research is presented in [12] where the SIM file system structure and data residual in a SIM card were briefly discussed. Other published research proposed set of forensic tools such as [13, 16] and SIMBruch presented in [2]. One of the most recent research explains the different tools in the market and what are the limitations for each tool presented in [9]. However, those tools only focused on extracting the basic information from a SIM card and did not provide clear description about the forensic techniques used to extract evidence from a SIM card.

M.T. Abdelazim (✉) • N. Abdelbaki • A.F. Shosha
Nile University, Cairo, Egypt
e-mail: m.tarek@nu.edu.eg; nabdelbaki@nu.edu.eg; ashosha@nu.edu.eg

© Springer International Publishing AG 2018
K. Daimi (ed.), *Computer and Network Security Essentials*,
DOI 10.1007/978-3-319-58424-9_22

Moreover, the current available public forensic tools for SIM card analysis are not updated or open source, and some of those tools are depending on the presence of specific currently unsupported hardware such as SIM readers compliant to PC/SC specification [14]. Additionally, none of these researches explain the methodology behind data extraction or file system meta-data.

Because of those limitations, we aim to provide an updated implementation that support extraction of digital evidence from modern SIM card models. Also, the proposed implementation is hardware agnostic, and it communicates with a SIM card using the standard communication protocol. This will allow forensic researchers and/or Law Enforcement to use and extend this method without having to rely on specific SIM hardware readers. This research will provide an open-source library to support extracting sound forensic evidence from a SIM card. More importantly, the research will explain what information could be stored in SIM cards, what can be extracted and/or recovered to assist a forensic investigation, how to extract and communicate with SIM cards, and limitations of SIM card forensic analysis.

22.2 Development Environment

SIM card's development environment depends on Java technology known as Java card. Java card's structure is based on four main components [10]:

- Java card virtual machine: used for application installation and defining the features and services.
- Java card runtime: which handles context switching between applications.
- Java card API: which is a set of classes and interfaces to allow the application to use card services.
- Java Card Programming Language: Java cards inherit its security measures from Java programming. This environment allows multi-application to be installed and run, which allow developing application or even malwares [11].

Java card programming language differs slightly from Java programming language. It does not contain some packages and libraries to minimize the size of the developed application due to smart cards' size limitations. For example, java cards do not contain `string` data type. In order to identify string, it is defined as array of `char`.

22.3 SIM Communication Protocols

Smart cards use two types of communication: Application Protocol Data Unit command (APDU) and Attention (AT). APDU executes commands on binary level. AT command is used to communicate with the SIM on the application level using well-defined interfaces.

Fig. 22.1 Smart Card Communication Scheme

Table 22.1 APDU request command description

Code	Name	Length	Description
CLA	Class	1	Class of instruction indicating how to decode the following bytes
INS	Instruction	1	Instruction code
P1	Parameter 1	1	First parameter of the instruction
P2	Parameter 2	1	Second parameter of the instruction
Lc	Length command	0–3	Length of the command data
Data	Data	Lc	String of bytes sent in the data field of the command
Le	Length expected	0–3	Maximum number of bytes expected in the data field of the response to the command

Figure 22.1 shows the communication scheme with smart cards. The figure shows the general representation for smart card communication. The host application can be computer or mobile device. The card acceptance device can be the mobile hardware or smart card reader. The host can initiate a request which will be interpreted to the card through the acceptance device as AT/APDU command. The card processes the command and then responds with the feedback to the host.

To understand the limitations and the differences between these two protocols, we will have to understand each type. We have to understand the command structure, parameters, and the usage.

22.3.1 Using APDU Command

An Application Protocol Data Unit (APDU) contains either a command message or a response message, sent from the interface device to the card or conversely. In a command–response pair, the command message and the response message may contain data.

APDU commands have two formats: Request Command and Response Command. Table 22.1 represents the command structure for the request command. The Request commands are responsible for reading, writing, or updating data on SIM card.

Table 22.2 represents the command structure for the response command. The response commands responsible for the request command acknowledgement either with the requested data or execution statuses either pass or fail.

Table 22.2 APDU response command description

Code	Name	Length	Description
Data	Data	Lc	String of bytes received in the data field of the response
SW1	Status Byte 1	1	Command processing status
SW2	Status Byte 2	1	Command processing qualifier

```
A0A40000023F00

A0A40000027F10

A0A40000026F42

A0DC010428534D532043454E545245FFFFE1FFFFFFFFFFFFFFFFFFFFFFFFFF

07915892020430F4FFFFFFFF0000A9
```

Fig. 22.2 APDU Instructions to update one record file

Figure 22.2 shows an example clarifying this communication method. If we want to update one record at the address "6F42," then we have to follow the following set of instructions.

1. APDU for select 3F00 using the command "A0A40000023F00" where 3F00 is the address for the root directory.
2. APDU for select 7F10 under 3F00 using the command "A0A40000027F10" where 7F10 is the directory holding the requested files.
3. APDU to select 6F42 under 7F10 using the command "A0A40000026F42" where 6F42 is the requested file under the directory.
4. APDU to update record 1 of 6F42 to "534D532043454E545245FFFFE1FFF..."

22.3.2   Using AT Command

GSM communication protocol provides a set of well-defined APIs to extract information from a SIM card. In particular, there are two possible methods for data extraction: using AT command or by accessing a memory address on a binary format. The drawback of the latter method is that not all memory addresses are available to be accessed, as SIM card specification provides a set of restrictions and limitations on the executed command.

By extracting the data using AT commands for example, the command syntax AT+CCID will be executed by the SIM card and return the CCID value as a response to this command, by selecting the address mapping to CCID and then reading the value and returning it. AT commands, like this example, are defined to access a

Table 22.3 CRSM command structure

<command>	176	Read binary
	178	Read record
	192	Get response
	214	Update binary
	220	Update record
	242	Status
<field>	Decimal	Identifiers of an elementary data file on SIM with values in the range
<P1>		0
<P2>		0,1
<data>	Decimal	Number of bytes to read

certain area in the memory to retrieve the information. Generally, AT commands, in the above format, provide user access with no limitations.

Extracting the data using direct memory access depends on identifying the location and how many bytes are required to be accessed. For example, the following syntax AT+CRSM=176,12258,0,0,10 allows accessing memory location 0x2FE2 and retrieves 10 bytes. Memory accessing using this method may allow bypassing some memory restrictions. However, it may also result in memory access exception and blocking command execution.

22.3.2.1 Command Structure

CRSM stands for Command Restricted SIM Access. CRSM command enables accessing "Elementary files" (EF) residual in a SIM card file system. Below is the CRSM syntax:

```
AT+CRSM=<command>[,<filed>[,<P1>,<P2>[,<data>]]]
```

Table 22.3 summarizes CRSM command structure and describes the different parameters and possible values.

22.3.2.2 Response Format

The response of AT command indicates command state, i.e., command executed successfully, command cannot be executed due to permission limitation or incorrect parameters, and data extracted from SIM if any [7]. Following is the command–response syntax:

```
+<command>: <SW1>, <SW2>[, <response>]
```

<command>: Note for executed command.
<SW1, SW2>: Represent the response status which could indicate correct execution or error with the execution. Using both SW1 and SW2, we can identify the

Table 22.4 AT command
response values

Status type	SW1,SW2
Normal execution	9000, 61XX
Warning processing	6 [2|3] XX
Execution error	6 [4|5|6] XX
Checking error	6 [7|D|B|E|F] 00, 6 [8|9|C] XX

error from non-supported commands, incorrect parameters, or denied permission. The size for both SW1 and SW2 is two bytes length. Table 22.4 summarizes the possible different response values for SW1, SW2.

<response>: This field holds the data extracted from the SIM card as a result of command execution.

22.3.3 AT vs APDU Commands

From our experiment, we found some differences between APDU and AT commands.

- APDU command can only be used if the SIM reader supports PC/SC specifications; however, AT commands can be used with standard communication protocols like serial communication.
- APDU and AT commands have the same structure with different implementation; however, APDU commands have more functionality over AT commands.
- AT commands are an abstraction layer for APDU commands where they have limited functionality mapping.
- Using APDU commands will require knowing the SIM behavior, structure, and command sequence in order to extract the needed information which can be achieved with executing multiple commands; however, using AT commands will allow data retrieval with a single command.
- Installing new software "applet" requires the use of APDU commands where there are no AT commands for installing or un-installing applets from SIM card.

As AT commands do not depend on specific types of readers and preexisted libraries for smart card communication and are independent from smart card application, AT commands are used during this research.

22.4 SIM File Identifier

SIM card file system consists of set of files in binary format. Figure 22.3 describes a sample of SIM file system tree where there are three types of files identifiers [6, 7]:

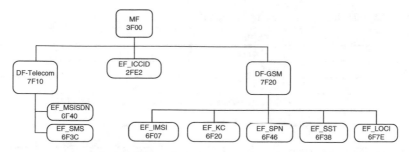

Fig. 22.3 SIM file system structure

- Main Files (MF): reside at memory address $0x3F00$. MF are considered the root directory for SIM file system:
- Dedicated Files (DF): represent the application level for each service provided by SIM such as directories.
- Elementary Files (EF): hold the actual information which needs to be extracted for digital forensics.

All elementary files have file access permissions, which could be summarized as five permissions [13],

- Always: the file is valid without any permissions and always accessible.
- Card Holder Verification 1 (CHV1): the file is protected by user's PIN code.
- CHV2: the file is protected by user's PIN2 code.
- Admin (ADM): the file can only be accessed after having the appropriate administrative authority.
- NEVER: the file cannot be accessed.

The MF can contain DF or EF one or more. However, the DF can contain only EF one or more. Where the EF contains the raw data stored on the SIM card.

Each type of file identifiers consists of a header and a body. All file identifiers have headers which hold the file size, permission access, data storage type, and file ID. However, only EF file has a raw data which is called the file body.

There are no references or official documentation describing the header structure for file system. Through our experiment, we discovered that there are mainly two headers formats: one associated with SIM/USIM cards described in Table 22.5, and the other is for UICC cards described in Table 22.6. The header mapping was a result of a combination between differential analysis technique and experimental trials used by comparing the header with some well-known data size and EF addresses, and through these observations, we concluded these results.

There are three different types of Elementary file identifiers:

- Transparent EF: An EF with a transparent structure consists of a sequence of bytes described in Fig. 22.4a. When reading or updating, the sequence of bytes to be acted upon is referenced by a relative address (offset) and the number of bytes to be read or updated. The total data length of the EF's body is indicated in the EF's header.

Table 22.5 File Header byte mapping SIM/USIM

Byte number	Description
1–2	Null bytes
3–4	File Size
5–6	File ID
14	File data structure which differentiates DF and EF
15	Record size (length)

Table 22.6 File Header byte mapping UICC

Byte number	Description
1..6	Null bytes
7	File type (plain data or records)
8	Record size in case of byte 7 = "00"
9–10	File ID in case of byte 7 != "00"
12–13	File ID in case of byte 7 = "00"
30	File size in case of byte 7 != "00"
15	Record size (length)

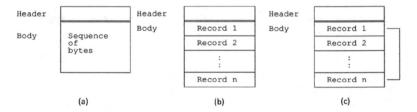

Fig. 22.4 SIM file identifiers types: (**a**) Transparent, (**b**) linear fixed, (**c**) cyclic

- Linear fixed EF: An EF with a linear fixed structure consists of a sequence of records that have the same (fixed) length described in Fig. 22.4b.
- Cyclic EF: Cyclic files are used for storing records in chronological order. When all records are used for storage, then the next storage of data shall overwrite the oldest information described in Fig. 22.4c.

22.5 Data Acquisition

SIM card forensic acquisition can be achieved by two methods: physical or logical. In the physical acquisition, a bit-by-bit forensic image of SIM memory is acquired for analysis. In the logical acquisition, the SIM data structure such as file system is extracted for analysis. In this research, logical forensic acquisition method is used. This will enable extracting forensic data from a SIM card in native format and interpreting these data for forensic analysis.

Using AT commands as described before, we can communicate with SIM card. After being able to communicate with SIM card, we can extract SIM information and stored data. We can extract many information like phonebook, PIN key status, SMS, network provider name, and many others [4].

22.5.1 Short Message Service (SMS) Forensics

SMS is an essential artifact in mobile phone forensics; by developing a solid knowledge about the SMS data structure and how SMS is stored in the SIM card, forensic analyst can use it to reconstruct a human activity in a mobile phone subject for forensic investigation. Each SIM card contains a limited number of memory locations; commonly, it is 40 locations with length size of 176 bytes each for incoming or outgoing SMS. Each location is structured into two parts. First is the record's status that is defined by its first byte, indicating if the record is used or free. The second is Message's data, which holds the rest of the record with maximum size of 175 bytes [3].

Each SMS location initializes the first byte with "00" and the rest of the record with "FF," indicating unused locations. At the point of incoming or outgoing new message, the message is stored in the first available free locations. At the time of message deletion, the assigned record is reset to its initial value removing any trace of the original message.

By knowing that each message is marked by a timestamp and its stored location index, messages' timeline construction can be achieved by providing time frame for received, sent, deleted, or overwritten messages which will be demonstrated later.

22.6 SIM Card Forensic Methodology

We have discussed so far what is SIM card, SIM communication, and data collection. Now we can put all of this information to use and form our methodology and implement our forensic application.

The forensic analysis methodology applied in this research is described in Fig. 22.5. It is a four-stage process. It begins with data acquisition, where the forensic data is extracted from the SIM using serial communication protected by write blocker, to prevent tampering with the forensic data and to ensure a sound forensic analysis where in our case we only use read commands. Second, acquired SIM card data is interpreted in a file system format for analysis. Third, the data analysis stage, where forensic evidence is identified and highlighted. Finally, event reconstruction stage, where digital evidence is presented in timeline representation to assist forensic investigator to develop an understanding of the identified evidence context.

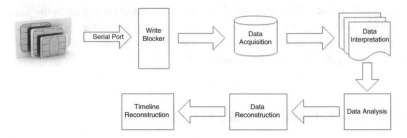

Fig. 22.5 SIM card forensics methodology

22.7 SIM Card Forensics Proof-of-Concept

A PoC to extract artifacts from a SIM card for forensic analysis purposes is developed. It enables extracting the basic information from the SIM subject for analysis, taking into consideration avoiding any usage of write commands. This will ensure a sound forensic analysis and will avoid tampering with a SIM card as digital evidence. The PoC is developed based on the GSM communication protocol standard.

The developed software communicates with SIM card using serial communication. API interfaces are developed to be used to extract certain pieces of information or full data extraction using python programming language. It initiates a communication to the SIM card that allows executing a stream of AT commands [1].

To ensure correctness of the PoC, a number of experiments are conducted using different SIM cards that operate under different service providers, including one SIM card that is un-allocated to any service provider.[1]

We will present some of the case studies where the implemented application is used to extract the information while testing the application availability and stability. The experiments are presented and results are explained, as follows:

22.7.1 Case 1: Old/Fresh SIM Forensic

In this experiment, we used operational SIM card working for several years in an attempt to identify user activities and extract stored SMS. As such, we been able to extract raw data for valuable forensic information and interpret it as follows:

(1) Extracting the EF_LOCI that holds Temporary Mobile Subscriber Identity (TMSI), (2) Location Area Information (LAI), (3) TMSI TIME, and (4) Location update status. The EF_LOCI file size is "11" bytes. Using the AT commands to read the EF_LOCI address AT+CRSM=176,INT(0X6F7E),0,0, 11, it resulted in the following:

---

[1] Service providers have been eradicated.

```
CRSM:144,0,"5CD749D162F2201CDE0000"
```

Another successful attempt to extract SIM service table using AT+CRSM= 176, INT(0x6F38),0,0,14 and SIM card response resulted in the following:

```
CRSM:144,0,"9EEF1F9CFF3E000000FFFFFFFFFF"
```

In addition, we were able to extract the contact list saved on SIM card using AT+CPBR=<param1>, <param2> command, which reads the phone book starting from <param1> to <param2>. SIM cards can hold about 255 records and using AT+CPBR=? we can get phone book properties, such as how many records can be stored on the SIM, the maximum number of digits for each number, and the maximum number of character for each text. Below is a sample of the results:

```
1) Name: "Luck", Number: "01004677080"
2) Name: "Cris", Number: "01006059135"
```

Another extremely valuable forensic information to extract is "pin key status" for PIN1 and PIN2. This is enabled using AT+CPIN? that can be used to return the state of PIN1 key and AT+CPIN2? and get the state of PIN2. SIM operation mode can be extracted using the same method. Below is a sample of the results:

```
Pin 1: READY
Pin 2: Not Found!
Operation Mode: Data
```

Another information that can be extracted is the SIM CCID and IMSI. This is enabled using AT+CRSM=176,int(0x2FE2),0,0,10 for CCID and AT+CIMI for IMSI, as follows:

```
CCID: 89**************385F
IMSI: 602**********385
```

For SMS forensic extraction, saved messages can be extracted using the following AT+CMGL=4 command which will list all the messages including message header information and message content, as follows:

```
Message ID: 4
Message State: received read messages
Length: 30
Content: 0791021197002864640ED0457A7A1E6687E9000851010121...
```

SIM service provider name can also be extracted using the below code: AT+CRS M=176,int(0x6F46),0,0,17.

```
Service provider name: *****
```

Finally, SIM service table is extracted using the below code: AT+CRSM =176, int(0x6F38),0,0,17.

```
SIM Service Table(SST): 9EEF1F9CFF3E000000FFFFFFFFFFFFFFFF
```

Even when fresh SIM is used the SIM CCID, IMSI number, the stored SMS, operation mode, and PIN status were successfully extracted.

22.7.2 Case 2: Blank SIM Forensic

In this experiment, we used a blank card that was not activated with a service provider. We also added some fabricated data, such as some phone book information, SMS information have been inserted and were forensically extracted. Since the SIM was not registered to any service provider, the service provider name was set to the default value "FF," and fabricated data was correctly extracted. Below is a sample of the extracted data:

```
SIM CCID: 898************0529
SIM IMSI: 460**********34

1) Name: "test name ", Number: "545465465"
2) Name: "SOS", Number: "112"

Message ID: 1
Message State: received read messages
Length: 25
Content: 00040681218564000050907231518080A0AF372195407CDDF6F37

Service provider name: None
```

Based on the above experiments, one essential finding can be determined, which is having non-registered SIM card does not refute the potential of having forensic artifacts, and necessity of investigating blank SIM card as it may contain valuable information.

22.7.3 Case 3: PIN Key Enabled/Disabled Forensic

In this experiment, a new SIM card that is locked with PIN key is analyzed. In this case, if the mobile phone restarted or powered off, user will not be able to access the SIM until a PIN key is entered. Below is a sample of the extracted information:

```
PIN Status: SIM PIN
SIM IMSI: Not Found!
SIM Service Table(SST): Not Found!
```

Based on the above output, we can conclude that essential information such as IMSI, SST, phone book, and SMS are protected when PIN key is set. Thus, to allow extracting forensic artifacts from a SIM card, it's essential to disable the set PIN key; otherwise, forensic acquisition using this method will not be successful.

However, when the same SIM card with PIN key disabled is forensically examined, below is sample of the results:

```
PIN Status: READY
SIM IMSI: 262*********205
SIM Service Table (SST): 9E3B140C27FE5DFFFFFFFFFFFF
```

We can conclude that if SIM PIN key was disabled, a forensic analyst will be enabled to extract SIM content for analysis such as IMSI, contact list, and SIM service table.

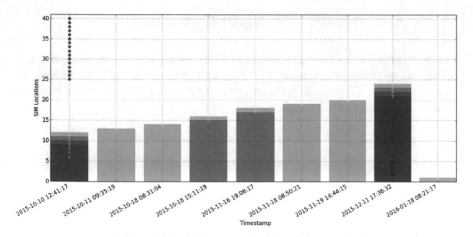

Fig. 22.6 Forensics reconstruction of SMS messages

22.7.4 Case 4: Timeline and Event Reconstruction

In this experiment, we are using the SMS forensic information we have previously examined to construct SIM activities timeline. The experiment shows that each new SMS message is stored at the first empty location, where each SMS has its timestamp which originated with the message. Using these two findings, we can reconstruct some of the SIM activities and provide timeline of the messages received. The experiment was conducted over operational SIM cards and different mobile phones (burner and smart phones) with few SMSs saved in an attempt to build time frame of saved and/or deleted messages.

Figure 22.6 explains PoC of the timeline representation for the received messages where the experiment was conducted using operational SIM card with "40" locations available and few messages saved. The graph shows the number of received messages on a particular day represented by over layered bars, i.e., at timestamp 2015-10-10 12:41:1" which shows that seven messages have been received and stored at locations "6" through "12."

Also the graph represents the possibility of the message deletion at a certain timestamp which is represented by "red circle," i.e., at timestamp 2015-12-11 17:36:32 locations "2" through "5." The rest of the empty locations "26" through "40" at timestamp 2015-10-10 12:41:17 have one of two possible meanings: (1) these locations have never been used and (2) these locations have been freed.

22.7.5   Case 5: SIM/USIM Header Extraction

In this experiment, we are extracting the headers of the files by scanning the SIM address space starting from `0000` to `FFFF` using the `GET RESPONSE` command with instruction ID `192`: `AT+CRSM=192,INT(File_ID),0,0` Below is a sample of extracted raw file header:

```
File ID: 6f38, File Header: 0000000A6F3804001BFFFF01020000
File Size: 10, Raw Data: FF3FFF0F0300F003000C

File ID: 6f4a, File Header: 000000416F4A040011FFFF0102010D
File Size: 65
Record ID: 1, Record Size: 13 ,Raw Data:FFFFFFFFFFFFFFFFFFFFFFFFFF
Record ID: 2, Record Size: 13 ,Raw Data: FFFFFFFFFFFFFFFFFFFFFFFFFF
```

As described in Table 22.5, the byte representation for 15 bytes header. It differentiates between EF, DF, and/or MF files and identifies the data structure for EF files which allows extracting the raw data.

22.8   Findings and Conclusion

In this chapter, the method and protocol to communicate with SIM card for forensic analysis purposes is explained. This includes what information could be extracted and how to extract it. A proof-of-concept implementation was also introduced to automate the investigation and assist forensic analysts. However, the developed software is confronted with some limitations. Most notably, SIM cards that are protected with PIN key. If it is active, extracting information from the SIM card for forensic purposes will not be available using this method. In addition, most of SIM cards have only three attempts to enter the PIN key, which hinder the method of brute forcing the PIN key.

Although SMS messages can be extracted from the SIM card, deleted messages can't be recovered, which is an essential requirement in forensic analysis. Another limitation to the current method is that it is not currently possible to determine or confirm user activities before/after the first/last stored SMS message, where there could be other messages received and deleted. Using SMS timestamps and stored indexes, forensic analyst could construct a timeline for the SIM activities. In particular, it is possible to identify if message(s) have been deleted or overwritten. In summary, the research's contributions are the following:

- Review of SIM card structure and required development environment.
- Review of SIM card file system, identifiers' types, and structure.
- Listing of communication methods with a SIM card.
- Forensic method that allows extracting of digital evidence from a SIM card.
- A proof-of-concept implementation for the proposed forensic methodology .
- Forensic timeline analysis for the recovered messages from SIM card.

References

1. AbdelAzim, M. (2016). *Simanalyzer tool*. https://github.com/Tabaz/SIMAnalyzer. Accessed 30 Jan 2017.
2. Casadei, F., Savoldi, A., & Gubian, P. (2006). Forensics and SIM cards: An overview. *International Journal of Digital Evidence, 5*(1), 1–21.
3. ETSI, T., 100 901 V7.5.0. (2001). Digital cellular telecommunications systems (Phase 2+), technical realization of the short message service (SMS) point-to-point (PP), European Telecommunications Standards Institute, Technical Specification.
4. ETSI, T., 100 916 V7.8.0. (2003). Digital cellular telecommunications system; AT command set for GSM Mobile Equipment (ME), European Telecommunications Standards Institute.
5. ETSI, T., 143 019 V5.6.0. (2003). Digital cellular telecommunications system (Phase 2+); Subscriber Identity Module Application Programming Interface (SIM API) for Java Card; ETSI Standards, European Telecommunications Standards Institute, Technical Specification.
6. ETSI, T., 131 102 V4.15.0. (2005). Universal Mobile Telecommunications System (UMTS); Characteristics of the USIM application, European Telecommunications Standards Institute.
7. ETSI, T., 100 977 v8.14.0. (2007). Digital cellular telecommunications system (Phase 2+); Specification of the Subscriber Identity Module-Mobile Equipment (SIM-ME) Interface. ETSI Standards, European Telecommunications Standards Institute, Technical Specification.
8. Guo, H., Smart cards and their operating systems.
9. Ibrahim, N., Al Naqbi, N., Iqbal, F., & AlFandi, O. (2016). *SIM Card Forensics: Digital Evidence*.
10. I. Oracle. *Java card technology*. http://www.oracle.com/technetwork/java\embedded/javacard/overview/getstarted-1970079.html. Accessed 30 Jan 2017.
11. I. Sun Microsystems. (1998). Java Card Applet Developer's Guide. Palo Alto: Sun Microsystems Inc. Revision 1.10, 17 July 1998.
12. Jansen, W., & Ayers, R. (2006, January). Forensic software tools for cell phone subscriber identity modules. In *Proceedings of the Conference on Digital Forensics, Security and Law* (p. 93). Association of Digital Forensics, Security and Law.
13. Jansen, W.A., & Delaitre, A. (2007, October). Reference material for assessing forensic SIM tools. *Security Technology, 2007 41st Annual IEEE International Carnahan Conference on Security Technology* (pp. 227–234). IEEE.
14. Osmocom. *Osmocom simtrace*, http://bb.osmocom.org/trac/wiki/SIMtrace. Accessed 30 Jan 2017.
15. Rankl, W., & Effing, W. (2004). *Smart card handbook*. New York: Wilcy.
16. Thakur, R., Chourasia, K., & Singh, B. (2012). Cellular phone forensics. *International Journal of Scientific and Research Publications, 2*(8), 233–326.

Chapter 23
A Dynamic Area-Efficient Technique to Enhance ROPUFs Security Against Modeling Attacks

Fathi Amsaad, Nitin Pundir, and Mohammed Niamat

23.1 Introduction

Ring Oscillator (RO) PUFs are one of the most appropriate techniques for the security of silicon chips [12] since it is a proven technique to provide high performance in terms of reliability and uniqueness of its generated response. However, compared to other silicon PUFs (SPUFs), ROPUF can only offer a limited number of challenge–response pairs (CRPs) for generation of secure binary responses. For this reason, ROPUF is categorized as weak SPUF which can be more vulnerable to modeling attacks [2]. Number of CRPs in ROPUF design is related linearly to the number of components that are used to construct a ROPUF design whose behavior depends on the random manufacturing process variations. Thus, to overcome CRPs' limitation, more design components have to be incorporated in ROPUF design. Although, there are other constitutions of ROUF structure using different techniques to integrate more components into ROPUF (i.e., configurable ROPUF), the generation of ROPUF secret keys are primarily based on a single (static) behavior of CRPs which can only extract a non-updated secret keys [4, 14]. Due to these limitations, an adversary may try all challenges and know the corresponding responses that is linear to number of applied challenges, within a certain time [2]. However, a dynamic technique that offers multi-stage ROPUF can be highly unpredictable and less vulnerable to modeling attacks that aims to clone its structure. In this regard, a dynamic technique namely d-ROPUF that offers multiple CRPs behaviors to increase ROPUF unpredictability, which in turn enhances its unclonability against modeling attacks, is proposed. The proposed technique is an area-efficient ROPUF design that utilizes the dedicated FPGA logic (dedicated multiplexers, LUTs, fixed routings) to accommodate four multi-stage structures in

F. Amsaad (✉) • N. Pundir • M. Niamat
University of Toledo, Toledo, OH, USA
e-mail: fathi.amsaad@rockets.utoledo.edu; mohammed.niamat@utoledo.edu

© Springer International Publishing AG 2018   407
K. Daimi (ed.), *Computer and Network Security Essentials*,
DOI 10.1007/978-3-319-58424-9_23

a single CLB using Programmable XOR gates (PXORs) that control the individual RO frequencies. In addition, a reconfiguration mechanism is appropriately designed to automatically alter the behavior of CRPs and reconfigure the design into new structures with different stages (invertors). Hence, the design can generate updated secret keys and consequently becomes highly unpredictable and secure against modeling attacks.

Firstly, a detailed explanation of the significance of the proposed dynamic multi-stage technique is presented and then it is differentiated from prior ROPUF implementations (static CRPs techniques). Secondly, we show how data samples are collected using the proposed techniques from the entire area of 30 Spartan-3E FPGA chips. To quantify the performance of each d-ROPUF structure, the normality of the generated sample RO frequencies and RO loop parameters are studied. In order to do that, RO sample frequencies of each d-ROPUF structure from 30 FPGAs is initially analyzed using two ROPUF normality parameters, namely skewness and kurtosis. In addition, Kolmogorov–Smirnov (K–S) and Shapiro–Wilk (S–W) statistical tests are performed to determine the performance of d-ROPUF structures. ANOVA test is also used to compare the mean values of the average samples frequencies of each d-ROPUF structures. Such a comparison is necessary to ensure that the generated RO frequencies are represented by various data samples. This also reveals the differences in the CRPs behavior of the generated RO frequencies. The result shows that our RO sample frequencies are normally distributed with different CRPs behavior.

Confirming that normality of d-ROPUF structures is the first step toward determining the correlation between RO sample frequencies and their reliability to generate unique secret keys. By selecting d-ROPUF structure with an appropriate number of stages that can generate reliable response bits, the number of RO sample frequencies that can be used to generate unique secret keys at varying operating conditions is maximized [10]. A higher difference between RO sample frequencies will ensure a higher performance in terms of its reliability and uniqueness [8]. In statistics, diverseness and variability are used to indicate the difference between data samples by measuring how far apart they are from each other [3]. For example, having n ROs with frequencies that are very close to each other, their diverseness and variability factors should always be close to 0. However, higher diverseness and variability values specify that data samples vary from each other which is very important for a higher ROPUF performance.

Pearson's correlation factors "r" (-0.93) and (-0.95) indicates a very strong inverse correlation between number of stages in each d-ROPUF structure and the average diverseness, and average variability of their RO sample frequencies, respectively. This shows that d-ROPUF structures with less number of stages exhibits high performance in terms of reliability and uniqueness compared to structure with higher number of stages. Finally, to estimate how effective our d-ROPUF design is, we explore some parameters that are defined by previous researchers including ROPUF loop parameters, diverseness, variability, uniqueness, uniformity, bit-aliasing, and reliability at varying temperature and supply voltage settings.

The book chapter is organized as follows: Sect. 23.2 covers research background. Section 23.3 explains the proposed design. Section 23.4 discusses the experimental results. Lastly, conclusions are drawn in Sect. 23.5.

23.2 Research Background

23.2.1 Basic and Configurable Silicon PUFs

Arbiter PUFs (APUFs) [13] and Ring Oscillators PUFs (ROPUFs) [6] are silicon based PUFs that take into consideration the process variations of Integrated Circuits (ICs) in order to produce random responses. In 2004, Lee et al. [6] used a switch-box structure to create a race between two delay paths with an arbiter at the end. The basic circuit of APUF is presented in Fig. 23.1. Two identical delay paths are formed to produce a response bit based on the fastest path. The arbiter is placed at the end of the circuit (D-latch) to decide the winning signal that reaches first.

Due to their simplicity and high performance, ROPUF is one of the most suitable security solutions for ASICs and FPGAs. Figure 23.2 shows the original ROPUF design for extracting unique signatures. These signatures are produced using challenge–response mechanism and are difficult to clone or predict [13]. As shown in Fig. 23.2, ROPUF circuit compares the two frequencies f_i and f_j, in order to produce "0" or "1" output based on frequency that is higher. The measured process variations should be unique for each RO [13].

In 2007, a 1-out-of-k technique was also proposed by Devadas to improve the reliability factor of RO PUFs [13]. This technique was named the redundancy approach which selects 1-RO pair among k-RO pairs that has the maximum frequency difference. The major disadvantage of this technique lies in its area inefficiency since k times more area is wasted when it is implemented on FPGAs. In 2009, Maiti [3] introduced the notion of configurable ROPUFs (c-ROPUF) for better

Fig. 23.1 An arbiter PUF design

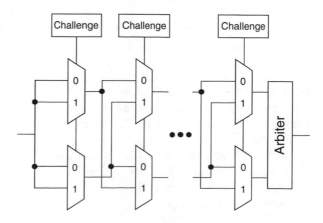

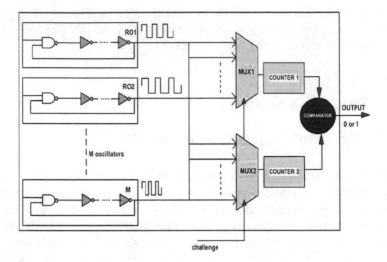

Fig. 23.2 Basic RO based PUF circuit

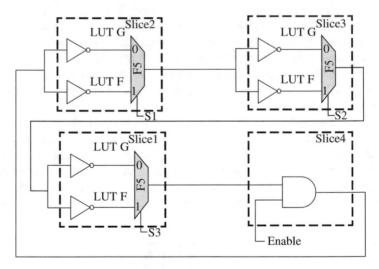

Fig. 23.3 Maiti's design

reliability. As shown in Fig. 23.3, he overcame the 1-out-of-k scheme drawback by offering a configurable design which occupies the same amount of area with more number of ROs. However, Maiti did not propose his method for stronger secret key production.

Besides, the performance of Maiti's configurable design was not fully validated. An improved version of the configurable ROPUF by Maiti [8] that can generate stronger response bits while using the same amount of area is proposed by Xin [14], and Amsaad in Fig. 23.4 [1].

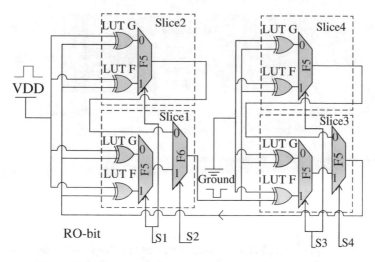

Fig. 23.4 Proposed design

23.2.2 Reconfigurable Silicon PUFs (rPUFs)

In 2005, Lim was the first one to propose the idea of reconfigurable PUF (rPUF) for an Arbiter PUF design [5]. His design was based on floating gate transistors for an Arbiter PUF. However, Lim did not clearly state how his proposed structure can be reconfigured, i.e., how his technique would specifically change the CRPs' behavior.

In 2009, Majzoobi presented a technique for implementing reconfigurable Arbiter PUF [9]. But, his implementation was based on a static PUF that does not satisfy the basic definition and conditions of a secure rPUF.

In 2009, Kursawe first presented the concept of reconfigurable optical PUF [4] as a physical structure with light scattering particles. Kursawe defined reconfigurable PUF (rPUF) as a PUF that is equipped with an appropriate mechanism to automatically convert its structure into a new structure with a new unpredictable challenge–response behavior [4]. In order to achieve this, the reconfiguration mechanism should be separately designed and controlled. Thus, the mechanism can alter the behavior of CRPs without the applied challenge affect. Kursawe proposed to reconfigure his structure with the help of physically state reposition of scattering particles method (polarization). Polarization is defined as the internal structure of the design when it is exposed to a laser beam outside the normal operating conditions [4]. Kursawe laid out a mathematical model and a theoretical example of optical rPUF, but practical aspects and analysis were not specified. As seen in the previous figures for static SPUF designs (APUF, ROPUF, and c-ROPUF), the challenge bits are either totally or partially used to configure their structures which makes the configuration mechanism possibly reversible. Based on the definition of reconfigurable PUFs (rPUF) by Kursawe, neither original SPUFs nor the configurable SPUFs can be considered as reconfigurable challenge–response

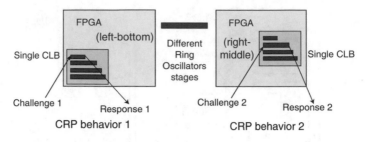

Fig. 23.5 Proposed d-ROPUF general scheme

behavior (rPUFs) [14]. Both basic and c-ROPUFs are FPGA friendly techniques. However, c-ROPUF provides more responses bits for better ROPUF security.

As far as our knowledge goes, prior research is focused on the static ROPUF (basic and configurable ROPUFs) which considers the implementation and performance aspects of static structure. The drawback of static ROPUFs is that, when the design is configured with high number of ROs, due to implementation of the design using a single structure that has a fixed number of stages; the design can only generate RO sample frequencies within a certain range. On the other hand, dynamic ROPUFs can always generate RO sample frequencies within different range of frequencies due to the design's ability to reconfigure itself into a new structure with different number of stages. This also increases CRPs space and updates their behavior which makes it harder for the adversary to try all possible challenges and uncover the corresponding responses (clone or model). Once the behavior of CRPs is known to an attacker, he can easily clone the entire ROPUF structure and store them into fake chips using a memory device in order to hack into a certain system. Having multi-stages ROPUF in the same CLB whose behaviors are automatically altered, not only improves the efficiency of ROPUF in terms of occupied area, but also enhances its unpredictability by enabling the generation of more complex cryptographic keys. As seen in Fig. 23.5, using dynamic ROPUFs, updated secret keys can be randomly extracted from different structures implemented on different FPGA areas with the help of new CRPs behaviors. As a result, the extracted cryptographic keys become more complex and highly unclonable. In the next section, we illustrate the details of implementation and analysis of the proposed technique, namely dynamic ROPUF (d-ROPUF) and show how it can be useful to improve ROPUF security against modeling attacks.

23.3 Implementations of the Proposed Technique on Real Hardware

In this section, we explain how the design is implemented and data is collected from 30 Spartan-3E/100k FPGAs under normal conditions.

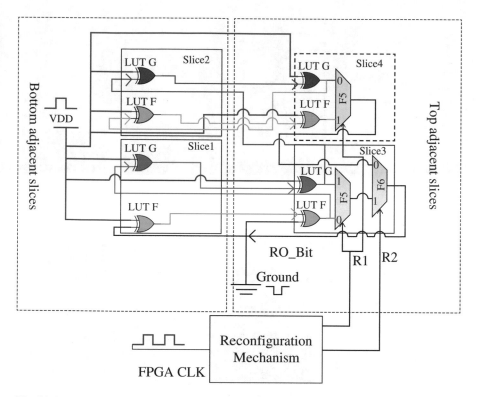

Fig. 23.6 Proposed d-ROPUF in a single CLB of Spartan 3E FPGA

As shown in Fig. 23.6, for area efficiency the proposed design is mapped inside one CLB taking advantage of the identical nature of internal routings that connect LUTs to two levels of dedicated MUXs (F5, F6) with fixed routing delays [1]. Internal routings guarantees identical RO loops in different CLBs and eliminates the local noise caused by dynamic routings [1]. A RO loop in a d-ROPUF structure is a circuit composed of odd number of inverters that are serially connected. As shown in Fig. 23.6, single CLB in Spartan-3E FPGAs contains four slices and eight LUTs (two LUTs per slice). Each of these eight LUTs can be configured as an inverter or a buffer. In order to have odd number of inverters (1, 3, 5, and 7), seven LUTs are configured as inverters and one LUT as a buffer using PXOR gates. These LUTs are connected with the help of internal CLB routings and four different structures are built inside a single CLB. As seen in Fig. 23.6, the red lines connect the enable signal (VDD) to activate the selected CLB and also configure LUTs as controlled inverters (PXORs). There is only one LUT that is configured as a buffer (connected to ground signal) which adds extra delay to keep the generated frequency for one stage inventor RO structure less than 300 MHz, that is the maximum operational value of Spartan 3E FPGA frequency. Likewise, different LUTs are shown in different colors and connected using different colored lines to show how four

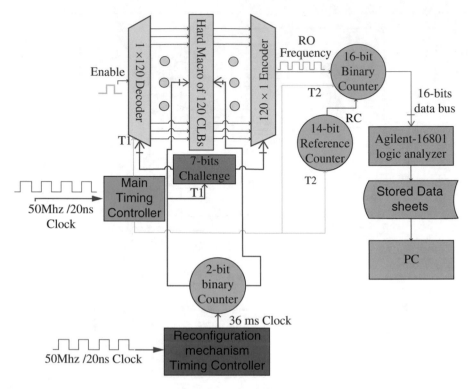

Fig. 23.7 Proposed technique for an FPGA area (120 CLBs)

ROPUF structures are mapped using internal CLB routings. For example, one stage structure is only implemented using the green LUTs, the three stages is implemented using green and blue LUTs. The five stages is implemented using green, blue, and purple LUTs while the seven stages is implemented using green, blue, purple, and brown LUTs. Figure 23.7 shows the proposed technique for an FPGA area. A 0.1 ms delay is provided prior to the activation of each counter, so that the signal is stabilized before the actual RO frequency measurement starts [1, 11].

To implement the design on the entire area of FPGA, we divided each chip into top and bottom area (120 CLBs). To avoid self-heating noise of the neighboring ROs, each RO in our design is activated for a shorter time 0.1 ms (activation period) with a step size of 5000 clock cycles [11]. A deactivation period of 0.1 ms is also allowed before activating the next RO. The timing controller shown in Fig. 23.7 controls the activation and deactivation periods of the individual ROs, the challenge generator, 120 decoder and encoder, the main and reference counters using T1 and T2 signals. Once the T1 signal is received by the challenge generator, input challenge is send to the decoder and encoder, hence a single RO is activated for 0.1 ms. At the end of this period, the main counters receive T2 from the timing controller and starts counting the number of clock cycles that is generated by the activated RO. At the same time, the reference counter receives the T2 signal and counts for 0.1 ms (5000 clock cycles) before it stops the main counter by sending RC signal. At the end of the second 0.1 ms period, the main reference and main counters will be deactivated and a control RC signal is sent by the reference counter to the main counter. Once the main counter receives RC signal, it stops counting and forwards the counted number of clock cycles through a 16-bit data bus to an Agilent logic analyzer. A third 0.1 ms is given before the main counter sends T1 signal to activate another RO.

As a result, 120 sample frequencies for each RO structures are stored for each d-ROPUF structure in the form of data sheets in the logic analyzer before it is collected and analyzed offline. Individual ROs for each d-ROPUF structure is activated from the top and moves down to the bottom of each column of the CLBs. Average frequency of 10 data samples for the entire individual ROs is considered for analysis. This requires a total time for the 10 runs per d-ROPUF structures, which is $10 \times 144 = 1.44$ s.

A synchronous binary counter counts from 0 to $2N - 1$, where N is the number of bits (flip flops) in the counter. The generation of R1 and R2 signals is separately controlled. The reconfiguration mechanism is properly designed as a part of the input challenge using 21-bit up counter that uses 2 ns clock cycle as an input clock and sends a 36 ms signal to the 2-bit up counter to reconfigure d-ROPUF in each CLB. The dedicated Muxs (F5, F6) in each CLB selects different RO structures using the received signals from the reconfiguration lines R1 and R2. Thus, the behavior of challenge–response of each d-ROPUF structure is altered on account of their number of RO stages and provides an updated response bits. To implement the reconfiguration mechanism one flip flop for each 1-bit is needed. Since Spartan 3E FPGA has four dedicated flip flops in each CLB, a proper placement constraint has been applied using VHDL to separately control the implementation of the reconfiguration mechanism within 12 CLBs (23 flip flops). For the sake of simplicity, to collect data from the entire d-ROPUF structure (120 RO frequencies), R1 and R2 are incremented by 1 using the 2-bit up counter (R1R2 = 00, 01, 10, or 11) for every 0.3 ms $\times$ 120 CLBs = 36 ms period.

23.4 Discussion of Experimental Results

23.4.1 The Normality of RO Sample Frequencies

Data samples are collected and analyzed for each d-ROPUF structure (120 ROs in each FPGA region). To collect data samples, the number of clock cycles of an active RO is measured using Agilent-16801 logic analyzer that can correctly recognize the non-uniform pattern of individual RO frequencies. The oth sample RO frequency

$(1 \leq o \leq 10)$ of the nth RO $(1 \leq n \leq 240)$ of the mth FPGA area $(1 \leq m \leq 2)$ in the lth FPGA $(1 \leq l \leq 30)$ is calculated using the following equation [14]:

$$f_{l,m,n,o} = \frac{(CC_{RO} \times 50\,\text{MHz})}{CC_{ref}} \tag{23.1}$$

where CC_{RO} is the number of clock cycles of the active ROs and CC_{ref} is the step size of FPGA clock cycle. After careful assessment, a step size of 5000 cycles was selected for the reference clock cycles. The default FPGA internal clock (50 MHz) is used as a reference clock. To reduce the effect of local noise on RO frequencies, average frequency of 10 data samples for each RO is calculated as follows [7]:

$$F_{avg_{l,m,n,o}} = \frac{1}{o} \sum_{i=1}^{o} f_{l,m,n,o} \tag{23.2}$$

The calculated value of $F_{avg_{l,m,n,o}}$ in Eq. (23.2) is a good measure for the average sample frequency of the individual ROs (d_{avg}) which is a fixed value for all ROs resulting from architectural and technological parameters [7, 10]. Average RO frequencies for an FPGA out of 30 FPGAs are calculated using equation as follows:

$$\text{AVG\_FPGA}_l = \frac{1}{(m \times n)} \sum_{j=1}^{m} \sum_{k=1}^{n} F\_avg_{l,m,n,o} \tag{23.3}$$

The mean and median values are computed with the help of IBM-SPSS software, where n represents RO frequencies of 240 data samples for 30 FPGAs, as follows:

$$\text{Mean} = \frac{1}{l} \sum_{i=1}^{l} \text{AVG\_FPGA}_l \tag{23.4}$$

$$\text{Median} = \frac{1}{2}(N + 1) \tag{23.5}$$

Table 23.1 shows that the value of Mean and Median of all structures is very close to each other which indicates that RO sample frequencies are more likely normally distributed. A distribution is considered normal (symmetric) when the right and left sides of the graph are approximately mirror images of each other. Kurtosis value often provides a measure of sharpness of the central distribution

Table 23.1 Performance parameters of r-ROPUF structures

Statistical measure	ROPUF structures			
	One stage	Three stage	Five stage	Seven stage
RO sample frequencies (N)	240	240	240	240
Mean	264.88	252.34	148.82	133.58
Median	264.80	252.44	148.15	133.55

Table 23.2 Skewness and Kurtosis values of d-ROPUF structures

d-ROPUF parameter	d-ROPUF structures			
	One stage	Three stages	Five stages	Seven stages
Sample RO frequencies	240	240	240	240
Skewness	0.129	−0.146	−0.164	−0.129
Kurtosis	−0.606	0.494	−0.210	−0.350

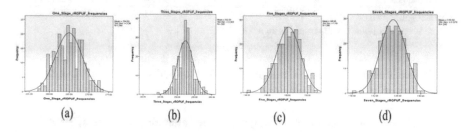

(a)   (b)   (c)   (d)

Fig. 23.8 The distribution of average RO frequencies for r-ROPUF structures

peak. A distribution that has a less sharp peak (more readings near tails) has negative Kurtosis (Kurtosis < 0). On the other hand, a distribution that has a sharper peak than normal (more readings near center) has positive Kurtosis (Kurtosis > 0). For a symmetric distribution, the Skewness and Kurtosis values are equal to zero. As seen in Table 23.2, one stage structure is skewed to the right since it has a positive Skewness value. Furthermore, one and seven stage structures are closest to the symmetric distribution because its Skewness is closest to zero. On the other hand, three and five stage structures appear to have the highest Skewness to the left. As far as Kurtosis is concerned, since three and five stages are closer to zero, they tend to have more readings near the center and are closer to normality when compared to one and seven stages which have more data near the tails (negative Kurtosis) and are far away from the center. Figure 23.8 shows the distribution of average sample frequencies of each structure for all 30 FPGAs. To accurately determine the normality of RO frequencies for each structure, we perform two different statistical tests using IBM-SPSS software, namely: Kolmogorov–Smirnov and (K–S) Shapiro–Wilk (S–W) for each r-ROPUF structure.

For K–S test, the critical p-values = 0.04 at a significant level = 5% are used to determine the normality with the following hypotheses taken into consideration:

H0:   The data follow a normal distribution
H1:   The data does not follow the normal distribution.

The critical region rejects H0 if the statistical value > 0.04. From Table 23.3 it is observed that the five and seven stages d-ROPUF structures are at the lower bound of the significance value which is 0.200. However, the one and three stages reconfigurations are closer to zero (0.003 and 0.29) which indicates that their RO frequencies are more random.

This means that the null hypothesis is not rejected for all d-ROPUF structures, because the p-value for each one is ≤ 0.4, which consequently shows that

Table 23.3 Tests of normality of the d-ROPUF

d-ROPUF structures	Kolmogorov–Smirnov			Shapiro–Wilk		
	Statistic	df	Sig.	Statistic	df	Sig.
One stage	0.074	240	0.003	0.982	240	0.004
Threes stage	0.061	240	0.029	0.991	240	0.142
Five stage	0.043	240	0.200[a]	0.992	240	0.193
Seven stages	0.041	240	0.200[a]	0.994	240	0.413

Critical value = 0.04
[a] This is a lower bound of the true significance

RO frequencies of all r-ROPUF structures are normally distributed. As seen in Table 23.3, Shapiro–Wilk test in the same table shows that at the 98% level we accept the normality of one stage d-ROPUF structures, and at the 99% level we accept the normality of other stages for the remaining structures. IBM-SPSS software is also used to calculate the one-way analysis of variance (ANOVA test) to compare the mean values of frequencies of the d-ROPUF reconfigurations using F distribution. The software uses a significant level of = 0.5. The critical value, as found by the IBM-SPSS software, is p-value = 0.41.

Since $P < \alpha$, this indicates that the group means of d-ROPUF configurations are taken from the data samples with different frequencies. The use of ANOVA test is crucial here since it demonstrates that each r-ROPUF structure can produce different frequencies based on the unique behavior of its CRPs.

23.4.2 Loop Parameters of ROPUF Model

Diverseness of RO is the sum of average diverseness, process variation diverseness, and diverseness due to noise [14]. Average diverseness and variability represent the standard deviation and variance of RO frequencies. Variability is simply the square of diverseness whereas the diverseness of ROs is calculated as follows:

$$\text{Diverseness}_i = \sqrt{\frac{1}{m \times n} \sum_{i=1}^{m} \sum_{j=1}^{n} (F_{\text{avg}_{l,m,n,o}} - f_{l,m,n,o})^2} \qquad (23.6)$$

$$\text{AVG\_Diverseness}_i = \frac{1}{l} \sum_{i=1}^{l} \text{Diverseness}_i \qquad (23.7)$$

As seen in Table 23.4, average diverseness and variability of the individual RO frequencies decrease with increase in the number of stages. This in turn indicates that lesser the number of stages, more reliable is the ROPUF response [7, 10].

Table 23.4 Performance parameters of d-ROPUF structures

Statistical measure	Stages of ROPUFs			
	One	Three	Five	Seven
Avg. diverseness	3.35	2.27	1.18	1.08
Avg. variability	11.22	5.14	1.40	1.16

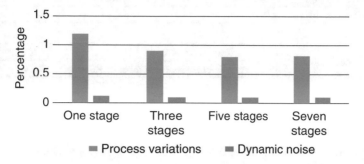

Fig. 23.9 Percentage of static process variation and dynamic noise variation

d_{AVG} is fixed for ROs which represent the average frequency of the identically instantiated RO loops. However, the percentage of intra-die process variation d_{PV} varies from one RO loop to other which is shown in Table 23.4 and computed using the following ratio [14]:

$$d_{PV} = \frac{AVG\_Diverseness_i}{Mean} \times 100\% \tag{23.8}$$

$$d_{NOISE} = \frac{1}{l} \sum_{i=1}^{l} \sqrt{\frac{Diverseness_i}{Mean}} \times 100\% \tag{23.9}$$

As shown in Fig. 23.9, average process variation (dPV) is in the range 0.8–1.18%, with an average of 0.92% and a standard deviation of 0.18%. A higher process variation implies a higher d-ROPUF performance [4, 6, 7]. From the same figure we notice that a d-ROPUF structure with less number of stages has higher static process variation compared to the one with more number of stages. This indicates that a d-ROPUF structure with less number of stages is more reliable to authenticate chips. Furthermore, Table 23.5 depicts the distribution of average noise variation for ROPUF structures which is in the range 0.0089–0.012%, with an average of 0.0095% and a standard deviation of 0.0089%. It is also observed that d-ROPUF structure with less number of stages has lower noise compared to d-ROPUF structures with more number of stages. Table 23.5 also shows a comparison between five stages d-ROPUF loop parameters values and five stages ROPUF [7]. Even though the average frequencies obtained in [7] is higher than our obtained average frequency, the ratio between the average static process variation and average

Table 23.5 Comparison between five stages dynamic and static ROPUF

Comparison	Maiti five stages ROPUF [7]	Proposed five stages d-ROPUF
Avg. RO freq.	205.1 MHz	148.82 MHz
Avg. process Var. (d_{pv})	0.75%	0.81%
Avg. noise Var (d_{Noise})	0.025%	0.0089%
(d_{noise})/(d_{pv}) ratio	0.03	0.01

Table 23.6 Pearson's correlation coefficient "r"

Performance factor	r square	r
Avg. diverseness vs. number of RO stages	0.92	−0.95
Avg. variability vs. number of RO stages	0.87	−0.93

dynamic noise variation is smaller for our five stages d-ROPUF structures. Hence, it is expected that our proposed design exhibits high performance under different conditions [9, 14].

23.4.3 Correlation Between d-ROPUF Structures and Their Performances

In this subsection we study the correlations between the proposed d-ROPUF structures and their performances. Pearson's correlation coefficient is used to study the relationship between each d-ROPUF structure and its performance. Pearson's correlation coefficient "r" describes the type of relationship between a dependent variable V (variability) and an independent variable S (number of stages of each d-ROPUF) as follows:

$$r = \frac{\sum_{i=1}^{n}(S_i - S_i') \times (v_i - v_i')}{\sqrt{\sum_{i=1}^{n}(S_i - S_i')^2} \times \sqrt{\sum_{i=1}^{n}(v_i - v_i')^2}} \tag{23.10}$$

In statistics, variability is always a positive value that is used to estimate how far apart a set of numbers is spread from the mean value. Low variability means that data points are very close to the mean, while a high variance value indicates that data points are quite far off from the mean value. A small variability closer to zero implies that all the values are almost identical. Thus, higher variability indicates that RO frequencies that are generated by a d-ROPUF are far away from each other and can be used to obtain more reliable and unique secret keys. Table 23.6 shows that there is a strong negative linear relationship between d-ROPUF structures and its performance factors in terms of average diverseness and average variability. Average variability (V) as a function of number of stages in a d-ROPUF structure (S) is illustrated in Eq. (23.13) as follows:

$$V = (\beta_0 + \beta_1 S) \tag{23.11}$$

where $\beta_1 = (v2 - v1)/(s2 - s1)$.

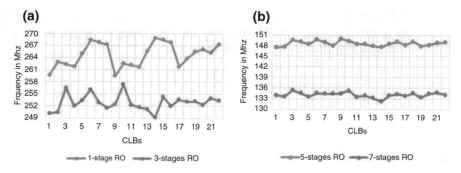

Fig. 23.10 The average frequencies of 22 CLBs

After ensuring that RO frequencies are independently generated and vary according to a normal distribution, we aim to determine the correlation between average variability and number of stages. It is observed that, as the number of stages in our d-ROPUF structure increases, the average variability decreases and vice versa. Figure 23.10a,b shows the average frequencies of 22 CLBs. As seen in Fig. 23.10, 1-stage and 3-stages RO structures have a greater frequency variation compared to 5-stage and 7-stages RO structures. While ROs with five and seven stages vary only by (variability) 0.56 MHz and 0.61 MHz, respectively, one and three stages RO structures have been shown to vary by 4.4 MHz and 9.1 MHz, respectively. A low variability indicates that the RO structures are more vulnerable to bit flip occurrence on account of environmental noise, and therefore generates unreliable responses. However, higher variability in RO structures makes them more appropriate for different applications at varying environmental conditions. For this reason, we subsequently evaluate RO structures at varying conditions to see if they are all suitable for different applications at varying environmental condition.

23.4.4 Quality Metrics of d-ROPUF Structures

In this section, we measure the important RO quality metrics to verify the performance of our d-ROPUF, namely: uniqueness uniformity, bit-aliasing, and reliability. Uniqueness factor shows how unique is the generated output for different chips. Average inter-die Hamming Distances (HDs) is used to evaluate the uniqueness between two chips, u and v with n-bit responses, R_u and R_v, respectively, at normal operating conditions as shown:

$$\text{Uniqueness}_i = \frac{2}{m(m-1)} \sum_{u=1}^{m-1} \sum_{v=u+1}^{m} \frac{\text{HD}(r_u, r'_v)}{d \times c} \times 100 \qquad (23.12)$$

The uniformity metric shows how well distributed are the percentage of 0's and 1's in the generated response bits of a certain chip. Uniformity is calculated using intra-die Hamming Weights (HWs) as follows:

$$\text{Uniformity} = \frac{1}{n} \sum_{i=1}^{n} r_{s,i} \times 100 \tag{23.13}$$

where $r_{s,i}$ is the ith response bit among n response bits. Ideal uniformity of PUF responses is 50% which means that RO responses are well distributed between 0's and 1's in a certain chip.

Bit-aliasing also measures how distributed is the percentage of 0's and 1's in the responses across multiple chips. Ideal bit-aliasing of PUF responses is 50% which means that RO responses are well distributed between 0's and 1's across multiple chips. Bit-aliasing is measured by calculating the inter-chip HW as follows:

$$\text{Bit-aliasing} = \frac{1}{m} \sum_{i=1}^{m} r_{s,i} \times 100 \tag{23.14}$$

Reliability measures how efficient are a PUF structure in reproducing the same response bits using the same RO pair when the operating condition changes. For a certain challenge, reliability can be evaluated by calculating Hamming distance between rs, rt the tth response bit from a chip at room temperature (normal operating condition), and rs, rt the tth response bit from the same chip at varying operating condition such as varying temperature or voltage condition. ROPUF reliability and bit flips percentage are calculated as follows:

$$\text{Reliability} = \frac{1}{n} \sum_{i=1}^{n} \frac{\text{HD}(r_{rt}, r_{s,i})}{n} \times 100\% \tag{23.15}$$

$$\text{Bit-flips} = 1 - \text{Reliability} \times 100\% \tag{23.16}$$

As each FPGA is divided into two areas, top and bottom (120 on each area). The experiment runs two times for each d-ROPUF structure at a certain temperature. For example, in the first run the ROs of the selected d-ROPUF structure (1 or 3 or 5 or 7 stages) are mapped on the top area and the bottom area is the associated logic needed such as reconfiguration mechanism and counters. However, in the second run, the ROs of the selected d-ROPUF structure are mapped on the bottom area and the top area is the associated logic needed such as reconfiguration mechanism and counters.

Response bits from all d-ROPUFs are generated from top (120 CLBs) and bottom area (120 CLBs) of 30 FPGAs to calculate the uniqueness, uniformity, bit-aliasing, reproducibility, consistency, and reliability. Neighbor coding selection technique is used where the neighboring ROs in each CLB are compared [10]. The comparison equation used is shown as follows:

$$R_{\text{bit}} = \begin{cases} 1, & \text{if freq1} \leq \text{freq2} \\ 0, & \text{otherwise} \end{cases} \tag{23.17}$$

Table 23.7 Quality metrics of d-ROPUF percentage under normal conditions

| RO structure | Quality metrics (%) | | | | | |
| | Bottom FPGA area | | | Top FPGA area | | |
	Uniqueness	Uniformity	Bit-aliasing	Uniqueness	Uniformity	Bit-aliasing
One stage	49.7	53.9	44.4	49.4	53.8	49.3
Three stage	48.2	56.6	45.3	47.3	54.6	43.5
Five stage	45.8	55.8	49.6	44.7	55.1	49.8
Seven stage	46.6	54.7	46.0	43.3	52.9	46.77

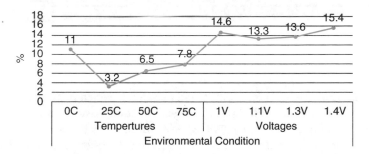

Fig. 23.11 Percentage of bit flips at varying temperature and supply voltage

Table 23.7 shows the obtained result for the top and bottom values of uniqueness, uniformity, and bit-aliasing. As mentioned previously, an ideal value is 50%. For example, for the bottom area of FPGA, the uniqueness of 1-stage ROs have the highest value whereas the 5-stage ROs have the lowest. However, all the values are above 45% which indicates a good uniqueness value. In addition, the average uniformity and bit-aliasing results for all stages taken into account can be considered good since the values are close to 50%. This means the secret bits generated from all RO structures are uniformly distributed between 1's and 0's and imply good randomness of the generated response bits.

To calculate the reliability, response bits are generated at different environmental conditions. In this experiment, responses from 1, 3 and 5-stage 7-stages ROs structures are generated at five different 0 °C, room trumpeter, 25 °C, 50 °C, 75 °C temperature and five voltage conditions 1 V, 1.1 V, 1.2 V (normal), 1.3 V, 1.4 V.

The obtained responses at various temperature and voltage settings are compared with the responses generated at room temperature and normal FPGA supply voltages (1.2 V). Figure 23.11 shows the average total number of bits flipped due to temperature and voltage variations. We notice that the environmental variations have fairly uniform effects on the reliability of d-ROPUF. In addition, voltage variations negatively impact the reliability of d-ROPUF and cause more bit flips in the d-ROPUF response when compared to temperature variations. We also notice from the figure that as the temperature or voltage deviates from the normal value, the number of bit flips increases which in turn affects the ROPUF reliability.

Table 23.8 Reliability of d-ROPUF under varying conditions

| | Reliability (%) | | | | | | | |
| | Temperatures | | | | Voltages | | | |
RO structure stages	0 °C	25 °C	50 °C	75°C	1 V	1.1 V	1.3 V	1.4 V
One	92.2	98.8	95.2	93.45	86.5	89.4	87.2	86.5
Three	89.8	97.3	94.8	92.3	86.1	86.6	87.55	85.2
Five	87.7	96.1	93.2	91.7	84.7	85.8	85.9	83.3
Seven	86.3	95.2	90.8	91.3	84.1	84.9	84.92	83.1
Avg.	89	96.8	93.5	92.2	85.3	86.6	86.5	84.5

As seen in Table 23.8, at 75 °C temperature variation, the average reliability is found to be 93.9% with 14.4 bits flips out of 238 bits which represents the lowest reliability with the highest bit flip percentage. While at 25 °C the average reliability is found to be 97.3% with 6.3 bits flips out of 238 bits which represents the highest reliability with the lowest bit flip percentage. The voltage variation results in lower ROPUF reliability than the temperature variation. In terms of voltage variations, at 1.4 V the average reliability is found to be 92.1% with 19 bits flips out of 238 bits which represents the lowest reliability with the highest bit flip percentage. While at 1.1 V the average reliability is found to be 96.2% with 9.2 bits flipped out of 238 bits which represents the highest reliability with the lowest bit flip percentage.

23.5 Conclusions

In this book chapter, a dynamic ROPUF structure (d-ROPUF) with different challenge response behavior is proposed to enhance ROPUF security against hardware vulnerabilities and attacks. The design is implemented and evaluated on 30 Spartan 3E FPGA chips. Furthermore, the normality and performance aspects of four different d-ROPUF structures are covered in detail. It is shown that the obtained RO frequencies are normally distributed. Using a newly proposed parameter, RO variability, we determine that the performance of d-ROPUF structures can be increased by having less number of stages and vice versa. Our results show that the d-ROPUF exhibits fairly good performance in terms of uniqueness, uniformity, and bit-aliasing at the normal operating conditions. We observe that varying environmental conditions in terms of supply voltage variations have a higher effect on the reliability of d-ROPUF response which causes more bit flips compared to temperature variations. Although the reliability is slightly affected by the environmental variations, a high ratio of static to dynamic variation ensures that the generated d-ROPUF response bits exhibit high reliability in terms of intra-chip Hamming Distance.

References

1. Amsaad, F., Hoque, T., & Niamat, M. (2015). Analyzing the performance of a configurable ROPUF design controlled by programmable XOR gates. In *IEEE 58th International Midwest Symposium on Circuits and Systems (MWSCAS)* (pp. 1–4).
2. Herder, C., Yu, M., Koushanfar, F., & Devadas, S. (2014). Physical unclonable functions and applications: A tutorial. *Proceedings of the IEEE, 102*, 1126–1141.
3. Krishnan, V. (2011). *Statistics for the behavioral sciences* (9th ed.). ISBN-13: 978-1-111-83099-1.
4. Kursawe, K., Sadeghi, A., Schellekens, D., Skoric, B., & Tuyls, P. (2009). Reconfigurable physical unclonable functions: enabling technology for tamper-resistant storage. In *Proceedings IEEE International Conference on Hardware-Oriented Security and Trust* (pp. 22–29).
5. Lim, D. (2004). Extracting secret keys from integrated circuits (Master's thesis, MIT).
6. Lim, D., Lee, J., Gassend, B., et al. (2005). Extracting secret keys from integrated circuits. *IEEE Transactions on Very Large Scale Integration (VLSI) Systems, 13*, 1200–1205.
7. Maiti, A., Casarona, J., McHale, L., & Schaumont, P. (2010). A large scale characterization of RO-PUF. In *IEEE International Symposium on Hardware-Oriented Security and Trust (HOST)* (pp. 94–99).
8. Maiti, A., & Schaumont, P. (2009). Improving the quality of a physical unclonable function using configurable ring oscillators. In *International Conference on Field Programmable Logic and Applications*, Prague (pp. 703–707).
9. Majzoobi, M., Koushanfar, F., & Potkonjak, M. (2009). Techniques for design and implementation of secure reconfigurable PUFs. *ACM Transactions on Reconfigurable Technology and Systems, 2*, 1–33.
10. Mustapa, M., Niamat, M., Alam, M., & Killian, T. (2013). Frequency uniqueness in ring oscillator physical unclonable functions on FPGAs. In *56th International Midwest Symposium on Circuits and Systems (MWSCAS)* (pp. 465–468).
11. Sedcole, P., & Cheung, P. (2006). Within-die delay variability in 90nm FPGAs and beyond. In *IEEE FPT Proceedings* (pp. 97–104).
12. Sklavos, S. (2013). Securing communication devices via physical unclonable functions (PUFs). In *Information security solutions* (Vol. 13, pp. 253–261). Berlin: Springer.
13. Suh, G., & Devadas, S. (2007). *Physical unclonable functions for device authentication and secret key generation* (pp. 9–14). New York: ACM.
14. Xin, X., Kaps, J., & Gaj, K. (2011). A configurable ring-oscillator-based PUF for Xilinx FPGAs. In *14th EUROMICRO Conference on Digital System Design – DSD'11 651–657*.

Chapter 24
Physical Unclonable Functions (PUFs) Design Technologies: Advantages and Trade Offs

Ioannis Papakonstantinou and Nicolas Sklavos

24.1 Introduction

In modern electronic systems, security plays a significant role, Sklavos [13]. Especially in communications the use of cryptography is becoming increasingly essential day by day. Even though there are plenty of software security solutions there is also a vital need to increase the security in the hardware level. An idea for a solution to deal with this matter is to use specific architectures that behave differently each time depending on various parameters. Such architectures are the Physical Unclonable Functions (PUFs) which are units that inherit unique characteristics during their manufacture. These characteristics occur during the manufacturing process due to existing random variations and cannot be controlled. The uncontrolled parameters are a key factor for the security of the PUF since it is the reason that a PUF cannot be cloned. The basic idea is to extract from each device some of their characteristics. Since these are unique and unclonable, we can use them as an ID of the device. In other words PUFs can be considered as the equivalent of fingerprint ID for (electronic) devices.

Some of the earliest references of PUFs were made by Pappu [8] who proposed an optic-based PUF and by Gassend et al. [2] who introduced the notion of PUF and described a fully integrated PUF. Until today, multiple proposals have been made in order to improve various parameters of PUFs that range from cost, area, and run-time, to security and specific implementation designs. The characteristics stated above are some of the reasons why PUFs have emerged in computer systems in the recent years, Sklavos [14]. They can be used as a secret key instead of a key stored in a non-violate memory with multiple advantages over them. They can

I. Papakonstantinou (✉) • N. Sklavos
SKYTALE Research Group, Computer Engineering and Informatics Department,
University of Patras, Patras, Greece
e-mail: papakonsta@ceid.upatras.gr; nsklavos@ceid.upatras.gr

© Springer International Publishing AG 2018
K. Daimi (ed.), *Computer and Network Security Essentials*,
DOI 10.1007/978-3-319-58424-9_24

also be used as cryptographic functions and as a result multiple security protocols can be implemented in hardware. In addition PUFs can be used as a source of randomness and they are an attractive solution for smartcard design. Furthermore, today economic aspects in information security are taken into account, Sklavos and Souras [16].

Despite the above advantages, since it is a relatively new field of research there are some problems that must be dealt with. There have been made multiple studies on PUF security and attacks, Rührmair and van Dijk [11]. Some types of such attacks are modeling attacks that try to mimic the behavior of a PUF or side channel attacks that try to gain information from their operation. These studies indicate that many of the proposed PUFs are not truly unclonable and an adversary could predict their behavior.

This chapter is focusing on silicon type PUFs, presents the basic designs of them and major improvements in various aspect of performance. The chapter is organized in the following way: In Sect. 24.2 general information about PUFs is given and the major criteria of classification are analyzed. Section 24.3 is focusing on silicon PUFs. A further classification of them has been made and some important designs of each type or significant new ideas for further improvement are presented. In this chapter experimental or simulation results are also presented and accompanied by a short analysis of their results. Finally, Sect. 24.4 includes the conclusion and a very short review of the observations that we made during the analysis of various designs.

24.2 Theoretical Background

In this section we will introduce some metrics and explain the terminology found in this topic in order to understand the difference between various designs and architectures. These metrics will help us evaluate the performance of each design and indicate the most appropriate choice for the demands of each case, Sklavos [14]. Then we will discuss about the basic PUFS categories and the criteria that we used to classify them.

24.2.1 Fundamental Terms and Metrics

Challenge is a special input which the verifier gives to the PUFS in order to stimulate the device and to receive an answer for identification. The challenge can have various forms relatively to the PUFS. For example, it can be a laser beam, an electric pulse, or a sequence of binary bits.

Response is the output of a PUF which is generated after a given challenge. The type of response, like the challenge, depends on the technology of the PUFS and can be a laser beam, an electric pulse, a sequence of binary bits or it can have another form. Responses must be different between different PUFs. Also responses must be different between different challenges, even for the same PUFS.

Intra-chip variation is the Hamming Distance between the responses of a PUF for the same given challenge. Each time a challenge is given to a PUF, due to environmental parameters some bits of its final response can be different from those that are expected by the verifier. In order to achieve high level of robustness we want the intra-chip variation to be as less as possible, so the verifier can accept the slightly different response as a valid answer.

Inter-chip variation is the Hamming Distance between the responses of different PUFs for the same given challenge. If the construction process of the PUFs is biased, the sequence of bits of the final response will not be equally distributed. This will compromise the security as an adversary could have more chances to predict the response of the PUF. The best possible inter-chip variation is 50% since this means that the response bits have equal probability to have logic value "0" or "1".

False Acceptance Ratio (FAR) is the percentage of PUF responses that are wrongfully considered valid. This happens when a PUF A has a very similar response to a PUF B and the verifier identifies the PUF A as the PUF B. One reason for this could be high intra-chip variation. If intra-chip variation is high enough, the verifier in order to accept the response in spite of the changed bits must have a tolerance. If this tolerance is too high, the slightly different response that the verifier will accept could match the slightly different response of an other PUF. Having a high FAR in a PUF increases robustness, since the possibility of PUF A to be identified as PUF A is very high. However, a high FAR decreases security since it, at the same time, increases the possibility of PUF B to be identified as PUF A.

False Rejection Ratio (FRR) is the percentage of PUF responses that are wrongfully rejected by the verifier. This can happen as previously referred, if tolerance is very low. The low tolerance could lead to a state where the slightly different response of a PUF has more different bits than the verifier tolerates and because of this, the verifier rejects the response. Having a high FRR increases security since the possibility of PUF B to be identified as PUF A is very low. However, a high FRR decreases robustness since it increases the possibility of PUF A not to be identified as PUF A.

Depending on the application in which a PUF will be used, FRR and FAR must be taken under consideration in order to achieve the best performance for the required demands.

24.2.2 Categories

PUFs can be distinguished with two basic criteria, by their *intrinsic* nature and by their *electronic* nature. Using the intrinsic nature criteria PUF can be distinguished if it is considered *intrinsic* or *non-intrinsic*. PUFs can also be distinguished in two other categories, *Electronic* and *Non-electronic* based on electronic nature criteria.

24.2.2.1 Non-Intrinsic and Intrinsic

Non-intrinsic PUFs are called PUF designs that in order to be identified, the manufacturer is somehow involved. He must give some parameters as an input in the manufacturing process that will make PUF instance unique and able to be recognized. These inputs, for example, can be semitransparent particles on an optical PUF or dielectric particles for a metal coating PUF. In addition, PUFs that cannot be considered as intrinsic are those that there is the need for *an external evaluation*. This means that their responses cannot be used directly as an answer to the verifier and an intermediary transformation must occur before their answer to a given challenge can be considered usable. For example, the response of an optical PUF can be a laser speckle pattern which must be digitized first, so that the verifier can process its response and answer for its id. The most known non-intrinsic PUFs are optical and metal coating PUFs. Since their uniqueness is not based solely on random parameters they can be very secure. Moreover optical PUFs are the only ones that can be considered *true unclonable* [6]. However, they are difficult and expensive to be constructed since they require extra steps in the manufacture process. Also the external evaluation can be source of some significant disadvantages. Having an external system to evaluate them is less accurate and as a result the probability of an error to occur would be higher. Moreover not having the response disclosed to internal systems could result to inner secret leakage that would reduce their security.

On the other hand, *intrinsic* PUFs are PUF designs which inherit their unique characteristics during their manufacturing process due to uncontrolled and random process variations. Furthermore, a PUF to be categorized as intrinsic must internally evaluate itself from an embedded measurement equipment. Having an internal evaluation system means that the PUF has fewer external limitations and its measurements will be more accurate with less influence from environmental parameters. Intrinsic PUFs due to the fact that their variations occur by default during the manufacturing process are less costly and easily produced. Additionally, the fact that the manufacturer cannot control the process variations is another security advantage of them.

24.2.2.2 Electronic and Non-Electronic

Non-electronic PUFs are those whose nature is based on non-electronic technologies. We specifically refer to the basic behavior of the PUF design as non-electronic, since secondary functions of it can occur by the use of electronic subsystems.

In contrast *electronic* PUFs are those which use electronic components for their basic operation. For their identification, resistance, capacitance, or other electronic measurements are used. Naturally they are easier to interconnect with other computer components and easier in processing due to digital responses. A major subclass of electronic PUFs are *silicon* PUFs. Silicon PUFs are the most widely applied among other technologies in computer science. They are designed like integrated circuits and they can be embedded on a silicon chip. By their nature silicon PUFs are directly connected to other digital circuits even in the same chip. This advantage is very important since they can be used in lots of emerging hardware implementations like Field Programmable Gate Arrays (FPGA) or Application-Specific Integrated Circuits (ASIC) and numerous designs of them have been suggested. This is why our analysis will focus on this type of PUFs. Types of silicon PUFs can be distinguished by their operation. Most important types of them are: *delay*-based PUFs, *memory*-based PUFs, and *glitch* PUF. Another significant type is *mixed signal* PUF which is a hybrid type combining both analog and digital components.

24.3   Designs of Silicon PUFs

24.3.1   Delay

Delay-based PUFs exploit delay variations of the design that arose in the manufacture process, Sklavos et al. [15]. Delay variations can occur from uncontrolled differences in wires or gates in silicon. The PUF measures these differences and uses them to decide its response.

Such a PUF is the *Arbiter* PUF which sets up a race condition during which transitions propagate along identical paths. At the end of the paths a module (arbiter) decides the winner and as a result the response of the PUF. In order to have the expected function of a PUF, the paths through the transitions propagate must be symmetrical so the variations are as few as possible, to achieve the unpredictability we are seeking for. An example of an Arbiter PUF is shown in Fig. 24.1, which is constructed from a number of similar switch blocks and each block uses as input one bit from the challenge.

Another interesting delay-based PUF is the *Ring Oscillator* (RO) PUF. For its operation it uses a number of oscillating circuits equally laid-out. Each of these oscillating circuits creates a frequency that is different from the others due to the uncontrollable variations of silicon manufacture. The PUF through the assistance of frequency counters and comparator modules extracts the response for the given challenge.

Another worth mentioning subcategory which we will not cover in this chapter is the *Glitch* PUF. It is composed of combinatorial logic circuits and its operation is based on the transitional effects that occur through changes of their logical values. If a change is made before the circuit output reaches its steady value from a previous change, then the output will be undetermined. In this case the result of the output will be influenced by random processes variations, so this result could be used as a PUF response.

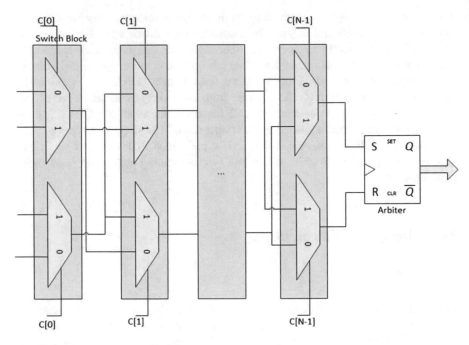

Fig. 24.1 Example of arbiter PUF

Table 24.1 Stochastic
variation relatively to the
FPGA technology [7]

Delay variability in current and future FPGA technologies			
Node (nm)	90	65	45	22
Stochastic var. (%)	3.3	5.5	7.5	11.5

24.3.1.1 Arbiter PUF

Although the simplicity of the arbiter PUFs makes it very appealing for implementation there are some significant drawbacks that must be faced first. Some of these drawbacks can be susceptibility to guessing, reverse engineering, emulation attacks, and sensitivity to environmental variations. Majzoobi et al. [7] proposed and tested some very interesting ideas for solving these problems. They demonstrated how reconfigurability can be exploited to eliminate limitations, the use of FPGA-based PUFs for privacy protection, and new testing techniques. Their idea was to use the delay time measurements for the PUF characterization and store them instead of challenges and responses. After extensive experimentation they found the delay time variations for various designs and technologies. Their results are shown in Table 24.1.

On the other hand, Lao and Parhi [5] proposed several structures for non-FPGA PUFs. Their goal was to overcome drawbacks and limitations of FPGA-based techniques, with a design that is easy to manufacture and to be implemented.

Table 24.2 Inter and intra-chip variation for different PUF structures [5]

Designs	Inter-chip variation		Intra-chip variation	
	Max	Min	Max	Avg
Non-feed-forward	59	22	13	5.8
Feed-forward overlap	66	27	15	8.7
Feed-forward cascade	64	25	20	10.7
Feed-forward separate	65	26	17	9.9
Reconfigurable feed-forward	65	25	19	10.3
Mux and demux	57	23	16	7.1

According to them, FPGA-based solutions require lots of assumptions for the VLSI layout and they are vulnerable to reverse reconfiguration attacks. But most significantly, FPGA platforms are not friendly for PUFs designs that require symmetrical or specific routing, since the designer has not full control of the routing. Because of this they proposed non-FPGA-based designs, targeting high levels of reliability and security, Sklavos [13]. Their basic idea was to add reconfigurability properties to PUFs so that the Challenge–Response Pair (CRP) updates given the different behavior of each reconfiguration, while simultaneously the PUF keeps its original properties. Different approaches were simulated and the extracted results are the following.

Table 24.2 presents inter-chip and intra-chip variations for the different PUF structures that were simulated. Since the minimum inter-chip variation is larger than the maximum intra-chip variation for all of the examples we conclude that randomness in the manufacturing process has a more significant role to the response than environmental conditions. From the above statement we can understand that these PUFs have the requirements to be used as reliable secret keys. Additionally by comparing the results of the different structures we see that feed-forward separate is the most reliable and feed-forward cascade is the least reliable from the feed-forward ones. Moreover, we realize that the reconfigurable feed-forward has similar performance to the other feed-forwards and the MUX and DeMUX is less reliable than the non-feed-forward, due to the increased intra-chip variation.

Table 24.3 shows the reconfigurability of the structures. The numbers represent the Hamming Distance of the response of each PUF for a fixed challenge for different configurations. The output recombination achieves the best results since its response has the largest variation. The MUX and DeMUX is the least reconfigurable. Also the Challenge LFSR has better performance than the Challenge Hash despite their similarities. Finally, the results of the Reconfigurable feed-forward PUF are good enough to be considered as reliable secret key storage, having in mind its advantage of its nonlinear functionality.

At the end the researchers conclude that the reconfigurable feed-forward PUF has the best performance, although the output recombination has the best reconfigurability. This occurs as it has higher security level, making extremely hard to be modeled due to its nonlinear responses.

Table 24.3 Reconfigurability
for different PUF
structures [5]

Designs	Variations (bits)		
	Max	Avg	Min
Challenge LFSR	44	34.6	28
Challenge Hash	42	28.3	19
Output recombination	57	38.9	25
Reconfigurable feed-forward	47	32.4	22
MUX and DeMUX	33	24.7	13

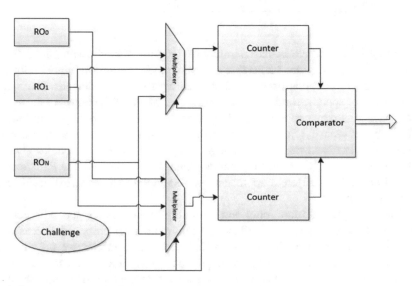

Fig. 24.2 Basic layout of RO PUF

24.3.1.2 Ring Oscillator PUF

Another type of delay-based PUF is the RO PUF as mentioned previously and
the basic layout of this is presented in Fig. 24.2. Such a PUF proposed by
Suh and Devadas [17] for low-cost authentication of IC and generation of secret
keys for cryptographic operations. The design that they describe allows an easier
implementation for ASICs and FPGAs, easier evaluation of entropy and higher
reliability than a simple arbiter PUF. But it has slower response, though, it requires
larger area and consumes more power. The extracted results after experimentation
was 46.15% for average inter-chip variation and in the worst case the intra-chip
variation was 0.48%. Also they computed FAR and FRR for a threshold of 10 bits
out of 128. The computed value for FAR was 2.1×10^{-21} and less than 5×10^{-11}
for FRR.

24.3.1.3 Modeling Attacks and Security

Although major improvements by the years have been made, either due to better technology or new designs, delay-based PUFs are prone to modeling attacks. Responses of delay-based PUFs exhibit some linearity that an adversary can exploit and with the use of machine learning (ML) techniques can predict the response of such a PUF. Rührmair et al. [10] showed how such a PUF can be broken by numerical modeling attacks. They presented an algorithm based on machine learning techniques, according to which knowing a set of CRPs predicts the response of all challenges with very high success rate. Types of PUFs that were tested were arbiter PUFs, Xor arbiter, Lightweight secure PUFs, feed-forward and RO PUFs.

In Table 24.4, the training time shows the amount of time that the algorithm needed to achieve the predict rate. From each PUF's CRP, 5/6 was used as training set (given to the adversary) and the rest as test set (with an exception of Xor

Table 24.4 Results of ML attacks for various PUF designs [10]

Type of PUF	XORs/loops/ oscillators	ML method	Stages	Predict rate (%)	CRPs	Training time
Arbiter	–	LR	64	95	640	0.01 s
Arbiter	–	LR	64	99	2555	0.13 s
Arbiter	–	LR	64	99.9	18,050	0.60 s
Arbiter	–	LR	128	95	1350	0.06 s
Arbiter	–	LR	128	99	5570	0.51 s
Arbiter	–	LR	128	99.9	39,200	2.10 s
Xor Arbiter	4	LR	64	99	12,000	3:42 min
Xor Arbiter	5	LR	64	99	80,000	2:08 h
Xor Arbiter	6	LR	64	99	200,000	31:01 h
Xor Arbiter	4	LR	128	99	24,000	2:52 h
Xor Arbiter	5	LR	128	99	500,000	16:36 h
Lightweight	3	LR	64	99	6000	8.9 s
Lightweight	4	LR	64	99	12,000	1:28 h
Lightweight	5	LR	64	99	300,000	13:06 h
Lightweight	3	LR	128	99	15,000	40 s
Lightweight	4	LR	128	99	500,000	59:42 min
Lightweight	5	LR	128	99	10^6	276 days
Feed-forward	6	ES	64	97.72	50,000	27:20 h
Feed-forward	7	ES	64	97.37	50,000	27:20 h
Feed-forward	8	ES	64	95.46	50,000	46 days
RO	256	QS	–	99	14,060	–
RO	256	QS	–	99.9	28,891	–
RO	512	QS	–	99	36,062	–
RO	512	QS	–	99.9	103,986	–
RO	1024	QS	–	99	83,941	–
RO	1024	QS	–	99.9	345,834	–

and Lightweight). ML method represents the machine learning method that was applied to achieve the result. The methods that were used were Logistic Regression (LR), Evolution Strategies (ES), Lazy Evaluation (LE), Support Vector Machine (SVM), and Quick Short (QS). Stages show the length of the PUF. Predict rate is the percentage of the CRPs that the algorithm can predict.

They conclude that the attack requirement number of CRP grows linear or log-linear in the internal parameter of the PUF, and the training time in most cases is low-degree polynomial (except of Xor Arbiter and Lightweight). But as Xor-based approaches may seem like an excellent solution for tolerance to machine learning algorithms, they also increase exponentially the instability of the PUF. For this reason, it is preferably to increase the number of bits in order to achieve an increased effort for an adversary. At the end the researchers propose ways to increase resilience of delay-based PUFs like increasing the length of the PUF, adding nonlinearity, or combining PUF types that require different ML techniques to be modeled. Additionally, they mention the advantages of using analog parameters, like in mix-signal PUFs, as machine learning of the output of an analog circuit could be very difficult.

24.3.2 Memory

Memory-based PUFs use the memory cells of an integrated circuit (IC) to create a response to a given challenge. The initial state of a memory cell on a starting-up device is unspecified, due to random variations in the manufacturing processes, so the logic value of the memory cell can be used as a PUF response. Since the basic design of such a kind of PUF requires only the existence of an SRAM module (the majority of digital devices) it is very easy for the PUF to be implemented in a wide variety of digital systems without any additional customization.

24.3.2.1 SRAM PUF

Guajardo et al. [3] presented an intrinsic PUF based on the SRAM of an FPGA. Based on previous research the static-noise margin (SNM) that is required for a memory cell to change its logical value is slightly different for each one of them. During the starting-up, the SRAM cells that are composed of cross-coupled inverters do not have a preordered value but the differences of the SNM will determine the value of each memory cell and hence the value of the whole SRAM. The fact that each inverter acts on the output of the other helps to increase the effect described upon. As a result, the SRAM cell will start with the same value almost every time and its behavior will be independent from the other cells. These properties were used to construct an FPGA SRAM PUF. The achieved results were intra-chip variation less than 4% over time and did not exceed 14% over temperature variations. The average intra-chip variation was 3.57% and the inter-chip variation was 49.97%.

Table 24.5 Inter and intra-chip variation for different SRAM PUFs [12]

SRAM	Technology (nm)	Intra-chip var. (%)				Inter-chip var. (%)	
		20 °C avg	±60 °C worst	Normal vdd avg	90–110% vdd worst	20 °C avg	±60 °C worst
Cypress CY7C15632KV18	65	3.8	12.2	3.8	4.5	49.7	48.6
Virage HP ASAP SP ULP 32-bit	90	2.9	19.6	5.5	6.2	49.3	46.8
Virage HP ASAP SP ULP 64-bit	90	3.5	17.6	5.5	6.3	49.2	48.0
Faraday SHGD130-1760X8X1BM1	130	4.5	13.7	4.6	5.4	53.5	65.2
Virage asd-srsnfs1p1750x8cm16sw0	130	4.8	20.5	5.5	6.4	50.1	48.7
Cypress CY7C1041CV33-20ZSX	150	3.5	9.2	3.5	3.9	50.1	48.1
IDT 71V416S15PHI	180	2.8	9.3	1.7	2.2	40.5	40.4

A similar research was made by Schrijen and van der Leest [12]. They investigated properties of SRAM PUFs in a wide variety of SRAM memory technologies and devices ranging from 180 nm to 65 nm. Measurements under different temperatures and supply voltage were conducted to evaluate their reliability.

In Table 24.5, the performance of different SRAMs is shown. We present the average intra-chip variation for normal environmental temperature (20 °C) and normal supply voltage. We also present the worst measurements that were observed in temperature variation from −40 °C to 80 °C and in voltage supply from 90 to 110% of its normal supply, respectively. Additionally, we recorded the average inter-chip variations and the worst inter-chip variations that were measured over temperature variations from −40 °C to 80 °C.

24.3.2.2   Bistable Ring (BR-PUF)

A different approach of memory-based PUF was proposed by Chen et al. [1]. This design does not use memory cells to store its state as other SRAM PUFs. Instead, it's a combination of memory-based PUF and delay-based PUF. Like the RO PUF it uses rings but only from an even number of inverters. In this way the ring can have only two states. As the device is powered-up, each inverter tries to force its own output to the ring but eventually the state of the ring will stabilize in a sequence of alternating "0" and "1". In Fig. 24.3 is shown the basic idea of such a ring and we can see the two possible states of a six stage bistable ring ("010101" or "101010"). The time that the ring takes to stabilize depends on many process variations. The basic idea of bistable rings is used by the Bistable Ring PUF after a series of architectural

Fig. 24.3 Basic idea of bistable rings

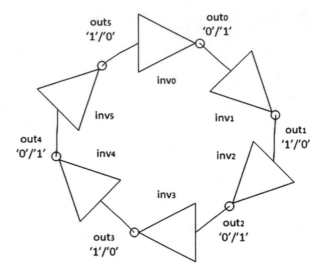

Table 24.6 Max and average settling time various stages of rings [16]

Stages	32	64	128
Max settling time (μs)	8.44	22.25	37.20
Average settling time (μs)	5.26	10.78	23.09

changes so that the PUF will be able to generate an exponential number of CRPs and to be easy to use. Since the stabilization time of the rings is not constant, we can either observe each ring until it stabilizes or set a common upper evaluation time. Choosing a common evaluation time can increase the average measuring time but it also reduces the complexity of the design. Measurements showed that after 22.17 μs, 90.21% of the responses were stabilized. The result of the measurements after the implementation of a BR-PUF on an FPGA is shown on Table 24.6.

In Table 24.6 is presented the settling time for different number of stages for the ring.

From the results of Table 24.6 we can estimate that a 32 stage BR-PUF can be fully measured in 6.28 h making it a risky choice for security applications. To the contrary, a 64 stage BR-PUF needs 6.31×10^4 for the measurement of only 1% of all the CRPs whereas the 128 stage PUF needs even more.

In Table 24.7 are presented some results of identification metrics. In the first measurement the existence of unstabilized responses was ignored. The poor inter-chip variation measurements indicate a layout-biased PUF but the low intra-chip variations make the PUF still operational. For the rest of measurements settling time was taken into account. The experimental results show that there is a trade off between reliability and uniqueness. Choosing a short settling time favors reliability whereas longer settling times favor uniqueness. Since the difference between intra and inter-chip variation increases with longer settling times, the identification and authentication become more efficient. Suggested settling times for this case would be between 35 and 47 μs. Even more, choosing a settling time for the CRPs makes

Table 24.7 Identification performance under different settling time [16]

Measurement	Stages	Settling time (μs)	Avg intra (%)	Max intra (%)	Avg inter (%)	Min inter (%)
1	64	–	0.8	1.2	14.8	6.0
2	64	10	~0	–	~10	–
3	64	40	~2	–	~40	–
4	64	~46	2.19	–	44.1	–

the design of a BR-PUF easier since the symmetry of the layout will not be so necessary. This would be suitable for FPGA-based applications, where the designer doesn't have full control of the layout.

24.3.3   Mix Signal

Mix-signal PUFs are combined from both analog and digital nature components. They behave like analog devices but an analog-to-digital module converts their responses in a digital form wherever needed. Since mix-signal PUFs contain analog devices their manufacture and implementation can be trivial and expensive.

24.3.3.1   MECCA PUF

Many different approaches have been made in this category. Krishna et al. [4] presented a technique based on memory cell PUF. The PUF uses an SRAM array in which inter-chip and intra-chip variations in the device parameters cause mismatches in transistors. These mismatches are exploited to cause failures in the memory cells. Due to random process variations each cell has different limit until its failure. Kinds of failure in memory cell can be *write failure*, *read failure,* or *hold failure.* The MECCA PUF uses the write failure mechanism which occurs when an internal node cannot be discharged through the access transistors during a signal (word line) active duration. To induce such a failure the signals duration was changed. Since write failures can be controlled by a signal duration, we can use them for evaluating the reliability of a cell of the PUF. After simulating the proposed design, the extracted results were: The achieved inter-chip variation was close 50% (49.9%) but for shorter signal duration the average inter-chip variation was 41.32%, while for longer signal duration it was 61.05%. The average intra-chip variation 0.85% with the worst intra-chip variation under temperature variation was 1% and 18% under supply voltage changes. Additionally, compared with a typical RO-PUF we see the difference in overhead of the overall design. The required area for an RO-PUF with a 128-bit key is 3122 μm^2, where the MECCA PUF with 128 cell SRAM array needs 21 μm^2 (0.6%) or 520 μm^2 (16.6%) if we include the memory module.

Finally, we conclude that this design has excellent performance in uniqueness and in reproducibility, and requires very little overhead, especially when compared with delay passed PUFs. Since it is a memory-based PUF it's really easy to be implemented in a wide variety of systems, while the analog nature of the design makes it harder to be cloned by a modeling attack.

24.3.3.2 SHIC PUF

Another approach for a mix-signal PUF was made by Rührmair et al. [9]. Rührmair investigated how irregular I curves can be used in random diodes to construct strong PUFs. After a crystallization-based fabrication method known as ALILE, diodes inherit properties that make them suitable for use in many security applications. Furthermore they are cheap, require very small area, and have a good temperature stability, and thus they making them very desirable for PUF construction. The basic advantage of the Super High Information (SHIC) PUF would be its resistance to machine learning attacks. The two key factors of their proposal are the high density of information and the limited read-out speed. Their hypothesis is that if an adversary gains access even for a limited time to a PUF, he can take measurements that could help him predict CRP of it. Using the above diodes can significantly increase the amount of information that is stored in a standard area (up to 10^{11} bits/cm^2). Furthermore the proposed structure, due to inductive and resistance capacitances, has a slow read-out rate. The above reasons, plus that each diode is independent from the others, means that the adversary needs to extract a very big amount of CRP to compromise the security of the PUF. But considering the amount of information and of the read-out speed, he would need a huge amount of time to achieve a good ratio of already read-out bits vs. the number of overall stored bits.

24.4 Conclusions and Outlook

In conclusion, in this chapter we introduced the concept of PUF and we described ways to measure the performance of different designs. We distinguished the PUFs into major or minor categories based on the operation or their architecture. Then we mention a few designs from various categories that were proposed in recent years which were either improvements of other older designs or completely new ideas. In our days, economic aspects in information security are also considered, Sklavos and Souras [16].

The above analysis indicates that SRAM PUFs require less area and they are easily implemented to FPGAs, Sklavos et al. [15]. Furthermore, already existing SRAM of the device can be used for their construction. There are numerous proposals of delay-based PUFs. Their designs vary from very simple ones to very complex for specific applications. However, their vulnerability to modeling attacks must be taken into account if they are going to be used for high security applications and more research in this area can be done. Finally, mix-signal PUF designs can be very trivial but their achieved performance makes them desirable for the requirements of modern applications.

All in all, we tried to give to the reader an overall picture (overview) about today's PUFs technologies. The state-of-the-art proposals which we presented indicate the research that is carried out in the field. Also the categorization of the various technologies highlights the differences and the advantages of them. For further analysis, the reader can reach to the mentioned papers where more aspects of their design and details are described.

Acknowledgment This work is supported under the framework of EU COST IC 1403: CRYPTA-CUS (Cryptanalysis of Ubiquitous Computing Systems) Project.

References

1. Chen, Q., Csaba, G., Lugli, P., Schlichtmann, U., & Ruhrmair, U. (2011). The bistable ring PUF: A new architecture for strong physical unclonable functions. In *IEEE international symposium on hardware-oriented security and trust—HOST 2011* (pp. 134–141). New York: IEEE.
2. Gassend, B., Clarke, D., van Dijk, M., & Devadas, S. (2002). Silicon physical random functions. In *Proceedings of the computer and communications security conference*.
3. Guajardo, J., Kumar, S. S., Schrijen, G. J., & Tuyls, P. (2007). FPGA intrinsic PUFs and their use for IP protection. In *Lecture notes in computer science (LNCS), Workshop on cryptographic hardware and embedded systems—CHES 2007* (Vol. 4727, pp. 63–80). Berlin: Springer.
4. Krishna, A., Narasimhan, S., Wang, X., & Bhunia, S. (2011). MECCA: a robust low overhead PUF using embedded memory array. In *Lecture notes in computer science (LNCS), Workshop on cryptographic hardware and embedded systems—CHES 2011* (Vol. 6917, pp. 407–420). Berlin: Springer.
5. Lao, Y., & Parhi, K. (2011). Reconfigurable architectures for silicon physical unclonable functions. In *IEEE international conference on electro/information technology—EIT 2011* (pp. 1–7). New York: IEEE.
6. Maes, R. (2013). *Physically unclonable functions*. New York: Springer.
7. Majzoobi, M., Koushanfar, F., & Potkonjak, M. (2009). Techniques for design and implementation of secure reconfigurable PUFs. *ACM Transactions on Reconfigurable Technology and Systems, 2*(1), 1–33.
8. Pappu, R. S. (2001). *Physical one-way functions*. Ph.D. thesis, Massachusetts Institute of Technology (MIT), Cambridge, MA.
9. Rührmair, U., Jaeger, C., Hilgers, C., Algasinger, M., Csaba, G., & Stutzmann, M. (2010b). Security applications of diodes with unique current-voltage characteristics. In *Lecture notes in computer science (LNCS), International conference on financial cryptography and data security—FC 2010* (Vol. 6052, pp. 328–335). Berlin: Springer.
10. Rührmair, U., Sehnke, F., Sölter, J., Dror, G., Devadas, S., & Schmidhuber, J. (2010a). Modeling attacks on physical unclonable functions. In *ACM conference on computer and communications security—CCS 2010* (pp. 237–249). New York: ACM.
11. Rührmair, U., & van Dijk, M. (2013, May 19–22). PUFs in security protocols: attack models and security evaluations. In *2013 IEEE symposium on security and privacy*, San Francisco, CA.

12. Schrijen, G.-J., & van der Leest, V. (2012). Comparative analysis of SRAM memories used as PUF primitives. In *Design, automation and test in Europe—DATE 2012* (pp. 1319–1324). New York: IEEE.
13. Sklavos, N. (2011, April 6–8). Cryptographic algorithms on a chip: Architectures, designs and implementation platforms. In *Proceedings of the 6th design and technology of integrated systems in nano era (DTIS'11)*, Greece.
14. Sklavos, N. (2013, October 22–23). Securing communication devices via physical unclonable functions (PUFs). In *Information security solutions Europe (isse'13)* (pp. 253–261). Brussels: Springer.
15. Sklavos, N., Chaves, R., Di Natale, G., & Regazzoni, F. (2017). *Hardware security and trust*. Cham: Springer.
16. Sklavos, N., & Souras, P. (2006). Economic models and approaches in information security for computer networks. *International Journal of Network Security (IJNS), Science Publications, 2*(1), 14–20.
17. Suh, G. E., & Devadas, S. (2007). Physical unclonable functions for device authentication and secret key generation. In *Design automation conference—DAC 2007* (pp. 9–14). New York: ACM.

Part VI
Security Applications

Chapter 25
Generic Semantics Specification and Processing for Inter-System Information Flow Tracking

Pascal Birnstill, Christoph Bier, Paul Wagner, and Jürgen Beyerer

25.1 Introduction

Distributed usage control (DUC) is a generalization of access control that also addresses obligations regarding the future usage of data, particularly in distributed settings [13]. UC policies are typically specified via events. Events are intercepted or observed by so-called *policy enforcement points (PEP)* as illustrated in Fig. 25.1. PEPs forward events to a *policy decision point (PDP)*, which evaluates them against policies. The PDP replies with an *authorization action*, such as *allow, modify, inhibit*, and *delay*, and triggers *obligations*.

Because data usually comes in different representations—an image can be a pixmap, a file, a leaf in the DOM tree of a website, a Java object, etc.—UC mechanisms have been augmented with information flow tracking technology [5]. One can then specify policies not only for specific fixed representations of a data item but also on *all* representations of that data item. Policies then do not need to rely on events, but can forbid specific representations to be created, also in a distributed setting [6]. In other words, information flow tracking answers the question into which representations within the (distributed) system monitored data have been propagated.

To perform information flow tracking across different applications, across different layers of abstraction of a system, or across systems, a multitude of PEPs, each observing an individual set of information flow-relevant events, have to be integrated

P. Birnstill (✉) • C. Bier • J. Beyerer
Fraunhofer IOSB, Karlsruhe, Germany
e-mail: pascal.birnstill@iosb.fraunhofer.de; christoph.bier@iosb.fraunhofer.de;
juergen.beyerer@iosb.fraunhofer.de

P. Wagner
Karlsruhe Institute of Technology, Karlsruhe, Germany
e-mail: paul.wagner@student.kit.edu

445

Fig. 25.1 Generic UC
architecture with information
flow tracking

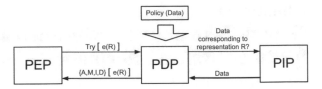

into the information flow tracking system. The so-called *policy information point (PIP)* interprets the information flow semantics of events and accordingly keeps track of new representations of data being created and of information flows between representations. By this means, when evaluating an event concerning a container (such as a file, process, or window), the PDP can ask the PIP whether this container is a representation of a protected data item, for which a policy must be enforced (cf. Fig. 25.1).

This work is also explicitly motivated by the increasing number of mobile apps for accessing video surveillance cameras and systems on the market and by the observation that meanwhile video surveillance is entering highly sensitive areas such as hospitals and nursing facilities. While these apps facilitate the cooperation of control rooms and security personnel on-site, we observe that in comparison to heavily secured control rooms the mobile devices being used fall critically short in terms of security mechanisms for protecting the sensitive data captured by surveillance systems. Obviously, the appearance of leaked surveillance footage showing a patient in an emergency situation on a video sharing portal on the Internet is in the interest of neither the patient nor the hospital. We thus instantiate our approach for protecting video data provided by a video surveillance server against illegitimate duplication and redistribution by mobile clients after receipt.

We address the following problems: We generalize an approach to inter-layer information flow tracking introduced by Lovat [9] to additionally cover inter-system flows so as to enable monitoring of flows of protected data between systems equipped with UC enforcement mechanisms. This approach is suitable for proof-of-concept implementation since it is lightweight. Yet it is prone to over-approximations requiring an extension with monitoring technology of higher precision in future work (cf. Sect. 25.6). Plugging new PEPs into existing UC infrastructures requires information flow semantics of the intercepted events to be deployed at the PIP. We introduce a generic set of primitives for specifying information flow semantics in a uniform syntax to be used by developers of monitors (PEPs). These primitives are derived from analyses of various scenarios in which information flow tracking has been instantiated for UC, such as [5, 14, 15]. Across system boundaries, information flows have to be handled asynchronously, triggered by different events on the particular machines. For this, we specify a protocol for processing inter-layer and inter-system flows based on our semantics description primitives. We thus facilitate UC enforcement on the granularity of representations in distributed settings.

This work is structured as follows. After explaining the formal information flow model of Harvan and Pretschner [5] in Sect. 25.2, we introduce our information flow semantics primitives in Sect. 25.3. In Sect. 25.4, we extend the model so as to allow uniform processing extension of inter-layer and inter-system information flows. We present an instantiation of our approach for protecting video data streamed to a client on behalf of the originating video surveillance server in Sect. 25.5. Eventually we discuss related work in Sect. 25.6 and conclude in Sect. 25.7.

25.2 Information Flow Model

Our approach to information flow modeling is based on works of Harvan and Pretschner [5, 14]. An information flow model is a transition system that captures the flow of data throughout a system. Transitions of the state are triggered by events that are observed by monitors, such as PEPs of a UC infrastructure. A system's information flow tracking component, the PIP, interprets events given information flow semantics provided by monitors when being deployed.

The state of the information flow model comprises three aspects. It reflects which data units are in which container, where a container may be a file, a window in the graphical user interface, an object in a Java virtual machine, a network connection, etc. The state also captures *alias relations* between containers, which express that a container is implicitly updated whenever some other container is updated. This happens, for instance, when processes share memory. Finally, the state comprises different *names* that identify a container, e.g., a file may not only be accessible by its file name but also by a file handle.

25.2.1 Formal Model

As introduced by Pretschner and Harvan in [5, 14], the formal information flow model is a tuple (D, C, F, Σ, E, R). D is the set of data for which UC policies exist. C is the set of containers in the system. F is the set of names. $\Sigma = (C \to 2^D) \times (C \to 2^C) \times (F \to C)$ is the set of possible states, which consists of the *storage function* $s : C \to 2^D$, the *alias function* $l : C \to 2^C$, and the *naming function* $f : F \to C$. Chains of aliases are addressed using the reflexive transitive closure l^* of the alias function. The initial state of the system is denoted as $\sigma_I \in \Sigma$, where the state of the storage function s is given by the initial representation of a data item a UC policy refers to. *Events E* are observed actions that trigger changes of the storage function s, the alias function l, or the naming function f. These changes are described in a (deterministic) transition relation $R \subseteq \Sigma \times E \times \Sigma$. We describe updates to the functions s, l, and f using a notation introduced in [5]: Let $m : S \to T$ be any mapping and $x \in X \subseteq S$ a variable. Then $m[x \leftarrow \text{expr}]_{x \in X} = m'$ with $m' : S \to T$ is defined as

$$m'(y) = \begin{cases} \text{expr} & \text{if } y \in X \\ m(y) & \text{otherwise.} \end{cases}$$

25.3 Generic Primitives for Information Flow Semantics

For any PEP, R is specified in an *information flow semantics*, which the PEP deploys on the PIP when being added to a UC infrastructure. For each event intercepted by a PEP, an information flow semantics specifies the state changes of the functions s, l, and f using generic primitives as introduced in the following. When processing an event given an information flow semantics (e.g., Listing 2), the PIP picks the action description for the event, converts event parameters to match the signatures of the contained semantics primitives (i.e., it implicitly applies f or s on a given parameter: $F \xrightarrow{f} C \xrightarrow{s} D$), and finally modifies its state according to the given primitives.

25.3.1 Primitives for Updating the Storage Function

The storage function keeps track of representations, i.e., mappings between data units and containers. We employ it for modeling the actual information flows.

$$\text{flow}(\text{container } c, \text{ data } \{d_i\}_{1 \leq i \leq n \in \mathbb{N}}) : \; s[c \leftarrow s(c) \cup \{d_i\}] \quad (25.1)$$

The *flow* primitive [cf. Eq. (25.1)] indicates an information flow of a set of data units $\{d_i\}_{1 \leq i \leq n \in \mathbb{N}}$ into the container c. This primitive is used to model that a process creates a new file, a child process, or that a file is copied. Data will then also flow into containers of processes that have a read handle on this file.

$$\text{flow\_to\_rtc}(\text{container } c, \text{ data } \{d_i\}_{1 \leq i \leq n \in \mathbb{N}}) : \; \forall t \in l^*(c) : s[t \leftarrow s(t) \cup \{d_i\}]$$
$$(25.2)$$

The *flow_to_rtc* primitive [cf. Eq. (25.2)] models a flow into containers of the reflexive transitive closure $l^*(c)$ of container c. It is used for processes reading from a file, writing to a file, or getting data from the system clipboard.

$$\text{clear}(\text{container } c) : \; s[c \leftarrow \varnothing] \quad (25.3)$$

We employ the *clear* [cf. Eq. (25.3)] primitive whenever a container is deleted, such as when deleting a file, closing a window, killing a process, etc.

25.3.2 Primitives for Updating the Alias Function

The alias function maintains relations between containers that lead to implicit flows. Whenever data items flow to container c_{from}, they also flow into the aliased container c_{to}.

$$create\_alias(\text{container } c_{from}, \text{ container } c_{to}) : \; l[c_{from} \leftarrow l(c_{from}) \cup c_{to}] \quad (25.4)$$

The primitive *create_alias* shown in Eq. (25.4) adds a unidirectional alias from container c_{from} to container c_{to} to the alias function of c_{from}. We use unidirectional aliases for memory-mapped file I/O, if a process has read-only access to the file (cf. mmap system call on POSIX-compliant UNIX and Linux systems).

$$create\_bidir\_alias(\text{container } c_{from}, \text{ container } c_{to}) :$$
$$l[c_{from} \leftarrow l(c_{from}) \cup c_{to}], \quad (25.5)$$
$$l[c_{to} \leftarrow l(c_{to}) \cup c_{from}]$$

We add bidirectional aliases using the primitive *create_bidir_alias* [cf. Eq. (25.5)]. Examples to be modeled with bidirectional aliases include creating a new window, or a process having read and write access to a file.

$$rm\_alias\_locally(\text{container } c_{from}, \text{ container } c_{to}) : \; l[c_{from} \leftarrow l(c_{from}) \setminus c_{to}] \quad (25.6)$$

The primitive *rm_alias_locally* removes a unidirectional alias from c_{from} to c_{to}, e.g., aliases added using the primitive *create_alias* [cf. Eq. (25.4)].

$$rm\_alias\_globally(\text{container } c_{to}) : \; \forall c \in C : l[c \leftarrow l(c) \setminus c_{to}] \quad (25.7)$$

In some cases, we also need to remove a unidirectional alias from all containers in C, e.g., in case c is a file, which is deleted. For this, we employ the primitive *rm_alias_globally* as shown in Eq. (25.7).

$$rm\_bidir\_alias\_locally(\text{container } c_{from}, \text{ container } c_{to}) :$$
$$l[c_{from} \leftarrow l(c_{from}) \setminus c_{to}], \quad (25.8)$$
$$l[c_{to} \leftarrow l(c_{to}) \setminus c_{from}]$$

Bidirectional aliases as added using the primitive *create_bidir_alias* [cf. Eq. (25.5)] are removed using the primitive *rm_bidir_alias_locally* as shown in Eq. (25.8).

$$clear\_aliases(\text{container } c) : \; l[c \leftarrow \varnothing] \quad (25.9)$$

clear_aliases removes all aliases with the given container as source from the state of the alias function [cf. Eq. (25.9)], e.g., to clean up if a container is deleted.

25.3.3 Primitives for Updating the Naming Function

The naming function maps different names to the same container, e.g., files can be addressed via file names and also via file handles or hard links; in the Windows operating system, we can identify a window via a window handle and also via a window name.

$$add\_naming(\text{naming } n, \text{ container } c) : f[n \leftarrow c] \qquad (25.10)$$

A new name n for a container c is added using the primitive add_naming [cf. Eq. (25.10)] and removed via rm_naming:

$$rm\_naming(\text{naming } n) : f[n \leftarrow \text{nil}] \qquad (25.11)$$

25.4 Inter-Layer and Inter-System Flows

So far, our primitives do not capture *inter-layer* and *inter-system* information flows. When using the term *inter-layer*, we refer to flows between different layers of abstraction, e.g., between an application and the operating system. *Inter-system* flows take place whenever data is exchanged between systems over a network connection. We introduce an information flow model extension for monitoring such flows, which requires that an event indicating an *incoming* flow is matched to a preceding *outgoing* event on another system or layer of abstraction.

25.4.1 Extended Information Flow Model

As an example, consider the transfer of video data from a streaming server to a client. Assume further that both server and client are equipped with PEPs that are capable of intercepting outgoing and incoming events as well as with local UC infrastructures. The server side PEP observes an outgoing event indicating a flow from a local container to another container representing the network connection to the client. When receiving data of the video stream via this connection, the client side PEP observes a related incoming event. Finally, when either the client disconnects from the video stream or the server closes the connection, a third event is observed, which terminates the flow. Initially, these events are independent from the perspective of both PIPs. Detecting an inter-system flow requires that both events are interpreted at both PIPs requiring appropriate *remote* information flow semantics, which are provided by the respective PEPs and exchanged between PIPs.

Within an information flow semantics, a so-called *scope* specification indicates that an event is related to an event on another system (or layer of abstraction). The particular events are matched to a scope by means of a *scope name* parameter, which is a label for a flow mutually known by two systems (or layers of abstraction). We thus extend the information flow model with a set of scopes *SCOPE* and the state with the following two mappings: The *intermediate container function* $\iota : SCOPE \rightarrow C$ maps each scope to an intermediate container $c_\iota \in C$. The *scope state function* $\varsigma : SCOPE \rightarrow \{\text{ACTIVATED}, \text{DEACTIVATED}\}$ indicates currently open scopes. Intermediate containers of different systems are distinct containers, which are mapped on each other by means of scopes and virtually represent the connection. Each event belongs to at most one inter-layer (XLAYER) or inter-system (XSYSTEM) scope. In the initial state σ_I of the system, there is one intermediate container c_ι for each scope ι and $\varsigma(sc)$ is DEACTIVATED for all $sc \in SCOPE$. Three attributes of a scope define how the model state is modified when processing an according event:

$$X_{SCOPE} : \Sigma \times E \rightarrow SCOPE \times BEHAVIOR \times DELIMITER \times INTER$$

$$DELIMITER = \{\text{OPEN}, \text{CLOSE}, \text{NONE}\}$$

$$BEHAVIOR = \{\text{IN}, \text{OUT}, \text{INTRA}\}$$

$$INTER = \{\text{XLAYER}, \text{XSYSTEM}\}$$

The *DELIMITER* of a scope describes whether an event indicates a new XLAYER or XSYSTEM flow. The delimiter OPEN changes the state of the scope within which the event is processed to ACTIVATED. The *BEHAVIOR* describes whether the event indicates an outgoing flow to (OUT), or an incoming flow (IN) from another system or layer of abstraction. The *BEHAVIOR* of a scope affects the processing of semantics primitives when handling XLAYER/XSYSTEM flows as will be described in Sect. 25.4.3 (INTRA is the default behavior, i.e., a flow within a layer of abstraction, which does not affect the interpretation of primitives). *INTER* differentiates between XSYSTEM and XLAYER flows.

25.4.2   Selecting the Appropriate Scope Semantics for an Event

For each event type, a PEP's information flow semantics contains *action descriptions*, which specify its interpretation in terms of information flow using semantics primitives (cf. Sect. 25.3). An action description also includes an ordered list of all scope specifications that possibly apply for this event type. The event notification only contains the scope (as a name–value pair, where the value is the scope itself). When processing an event, the PIP needs to check, in the given order of the action description, which scope specification is applicable. For each scope specification, the PIP evaluates the following three conditions:

1. Does the scope name in the scope specification match the name of a parameter provided in the parameter list of the event notification?
2. If *DELIMITER* = OPEN in the scope specification: scope deactivated?

3. If *DELIMITER* = NONE or *DELIMITER* = CLOSE: scope activated?

If only one of the conditions is not fulfilled, the respective scope is skipped. The ordered list is processed until the matching scope specification X_{SCOPE} is found.

25.4.3 Scope Processing

The transition relation R is modified when processing a scope specification. Algorithm 1 describes how R is modified to obtain R_{mod}, i.e., the transition relation for XLAYER or XSYSTEM flows. $R[\text{left} \overset{\text{subst.}}{\Longleftarrow} \text{right}]$ denotes that the term of R on the left is substituted with the term on the right in R_{mod}. If the delimiter of the scope equals OPEN, the scope is activated (cf. line 6); if the delimiter equals CLOSE, the scope is deactivated after handling the event (cf. line 17). In between (cf. line 8 ff.), depending on the scope's behavior, either the left argument (target) (cf. line 12 ff.)

Algorithm 1 Processing an XSYSTEM scope

1: **procedure** $R_{inter}(\sigma, e)$
2: $\quad (scope, behav, delim, inter) \longleftarrow X_{SCOPE}(\sigma, e)$
3: $\quad$ **if** $scope \neq \emptyset$ **then**
4: $\qquad ic \longleftarrow \iota(scope)$
5: $\qquad R_{mod} \longleftarrow R$
6: $\qquad$ **if** $delim =$ OPEN **then**
7: $\qquad\quad \sigma \longleftarrow \varsigma[scope \leftarrow \text{ACTIVATED}]$
8: $\qquad$ **if** $behav =$ OUT **then**
9: $\qquad\quad R_{mod} \longleftarrow R_{mod}\big[s[c \leftarrow s(c) \cup \{d_i\}] \overset{\text{subst.}}{\Longleftarrow} s[ic \leftarrow s(ic) \cup \{d_i\}]\big]$
10: $\qquad\quad R_{mod} \longleftarrow R_{mod}\big[\forall t \in l(c) : s[t \leftarrow s(t) \cup \{d_i\}] \overset{\text{subst.}}{\Longleftarrow} \forall t \in l(ic) : s[t \leftarrow s(t) \cup \{d_i\}]\big]$
11: $\qquad\quad R_{mod} \longleftarrow R_{mod}\big[\forall t \in l^*(c) : s[t \leftarrow s(t) \cup \{d_i\}] \overset{\text{subst.}}{\Longleftarrow} \forall t \in l^*(ic) : s[t \leftarrow s(t) \cup \{d_i\}]\big]$
12: $\qquad$ **if** $behav =$ IN **then**
13: $\qquad\quad R_{mod} \longleftarrow R_{mod}\big[s[c \leftarrow s(c) \cup \{d_i\}] \overset{\text{subst.}}{\Longleftarrow} s[c \leftarrow s(c) \cup s(ic)]\big]$
14: $\qquad\quad R_{mod} \longleftarrow R_{mod}\big[\forall t \in l(c) : s[t \leftarrow s(t) \cup \{d_i\}] \overset{\text{subst.}}{\Longleftarrow} \forall t \in l(c) : s[t \leftarrow s(t) \cup s(ic)]\big]$
15: $\qquad\quad R_{mod} \longleftarrow R_{mod}\big[\forall t \in l^*(c) : s[t \leftarrow s(t) \cup \{d_i\}] \overset{\text{subst.}}{\Longleftarrow} \forall t \in l^*(c) : s[t \leftarrow s(t) \cup s(ic)]\big]$
16: $\qquad \sigma \longleftarrow R_{mod}(\sigma, e)$
17: $\qquad$ **if** $delim =$ CLOSE **then**
18: $\qquad\quad \sigma \longleftarrow \varsigma[scope \leftarrow \text{DEACTIVATED}]$
19: $\qquad\quad \sigma \longleftarrow s[ic \leftarrow \emptyset]$
20: $\quad$ **else**
21: $\qquad \sigma \longleftarrow R(\sigma, e)$
22: $\quad$ **return** σ

---

or the right argument (source) of the storage function primitives *flow* or *flow_to_rtc* is substituted with the scope's intermediate container. R_{mod} is then applied on the state σ (cf. line 16).

In case of an XSYSTEM flow, the PIP needs to enable its remote counterpart to process the given event. As described in Sect. 25.5.1, this is achieved by forwarding a remote information flow semantics to the remote PIP.

25.5   Instantiation

We implemented XSYSTEM information flow tracking for a scenario concerning video data provided by a streaming server. It enables us to enforce the UC requirement of preventing redistribution of the video data after receipt on client systems, such as mobile apps for video surveillance systems as mentioned earlier. For this, we need to (1) deploy an according policy at the UC infrastructure of the client, (2) track the flow of video data from the server to the client, and (3) monitor the video data at the client so as to inhibit further representations of the data to be created. We achieve (1) and (2) in the following protocol steps:

1. Intercept an event signaling the *outgoing* data at the server side PEP
2. Evaluate the event against an according policy at the server side PDP
3. Deploy a policy for the data at the client side PDP
4. Process the event at the server side PIP
5. Create a new representation of the video data at the client side PIP
6. Process the outgoing event also at the client side PIP
7. Intercept an event signaling the *incoming* data at the client side PEP
8. Evaluate the event at the client side PDP
9. Process the event at the client side PIP
10. Intercept an event signaling *close* of the connection at the server side PEP
11. Process the event at the server side PIP
12. Process the close event also at the client side PIP

We explain the details of the protocol steps by means of Fig. 25.2, where an additional component, a system's *policy management point (PMP)*, takes care of policy shipment and deployment.

25.5.1   Inter-System Information Flow Tracking

Video streaming is triggered by a request from the client, which is intercepted on the server side. The according *outgoing* event triggers the steps 1–6 (cf. Listing 1). It indicates an outgoing flow from the local container *L1*, i.e., the actual server process providing the video stream, to a container *C1* representing the network connection to the client from the perspective of the server.

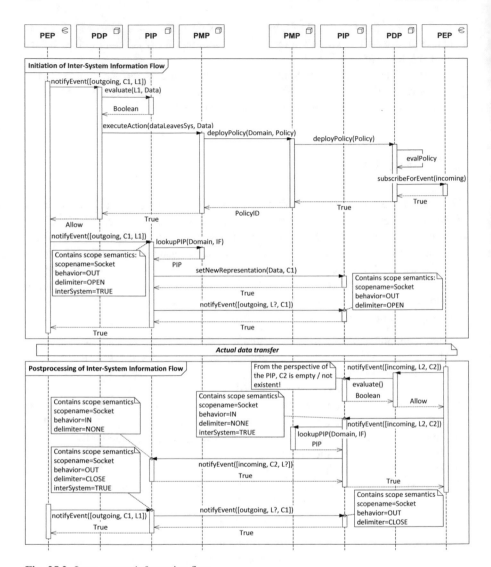

Fig. 25.2 Inter-system information flow

```
<event action="outgoing" timestamp="2015—05—30T09:30:10">
    <parameter name="network" value="192.168.0.2:80;192.168.0.1:49152"> <!-- C1 -->
    <parameter name="process" value="2a26af9d—f565—4775—87b5—8eb1fb987ad5"> <!-- L1 -->
    <parameter name="currentscope" value="192.168.0.2:80;192.168.0.1:49152">
</event>
```

Listing 1 Outgoing event at server side

In step 2, a policy deployed at the server side PDP grants access to the video stream under the condition that a policy for protecting the requested video data is deployed at the client side (step 3): Both policies refer to the video data by means of a unique dataID *data*, which represents the video data within PIPs. Thus, when evaluating the *outgoing* event concerning the local container *L1* against the local policy, the server side PDP queries the local PIP whether *L1* contains *data* (cf. *evaluate*-call to the server side PIP in Fig. 25.2). As the PIP returns *true*, the policy matches the outgoing event and evaluates to *allow* under the condition that the policy deployment at the client is successful. This policy demands that no further representations of the protected video data must be created, which includes that it must not be saved and that the screen must not be captured while the video data is accessed.

In step 4, the outgoing event is handled by the server side PIP. The PIP holds a *local semantics* and a *remote semantics* for this event type. The local semantics for the outgoing event is shown in Listing 2. The scope attribute *interSystem* = TRUE

```
<ifsemantics>
  <params>
    <param name="network" type="CONTAINER"/>
    <param name="process" type="CONTAINER"/>
  </params>
  <actions>
    <action name="outgoing">
      <scope behavior="OUT" delimiter="OPEN" interSystem="TRUE">currentscope</scope>
      <operation name="SF_FLOW">
        <left>
          <operand>network</operand> <!-- C1 -->
        </left>
        <right>
          <operand>process</operand> <!-- L1 -->
        </right>
      </operation>
    </action>
    <action name="close">
      <scope behavior="OUT" delimiter="CLOSE" interSystem="TRUE">currentscope</scope>
      <operation name="SF_CLEAR">
        <left>
          <operand>network</operand> <!-- C1 -->
        </left>
        <right>
        </right>
      </operation>
    </action>
  </actions>
</ifsemantics>
```

Listing 2 Local semantics of outgoing and close event

in the *outgoing* action description is equivalent to *inter* = XSYSTEM in the formal model and activates an XSYSTEM scope. The action description indicates a flow from the local container *L1* into the network container *C1*. Due to *behavior* = OUT of the scope, *C1* is substituted by the scope's *intermediate container* at the server side PIP: As the PIP knows that *L1* contains *data*, it models this flow by mapping *data* to the intermediate container.

The scope specification in the semantics also triggers the server side PIP to signal the upcoming data transfer to the client side PIP. So far, the client side PIP neither knows that this data exists nor that the client requested it. Step 5 takes care of the first part: The server side PIP creates a new representation of the data at the client side PIP, i.e., we add an initial mapping between the dataID *data* of the video data and the remote network container *C1* to the client side information flow model. The server side PIP then forwards the event to the client side PIP. In case the remote semantics for this event has not yet been deployed at the client side PIP, it is attached to this notification (cf. Listing 3).

```
<ifsemantics>
  <params>
    <param name="network" type="CONTAINER"/>
    <param name="process" type="CONTAINER"/>
  </params>
  <actions>
    <action name="outgoing">
      <scope behavior="OUT" delimiter="OPEN">currentscope</scope>
      <operation name="SF_FLOW">
        <left>
          <operand>process</operand> <!--L?-->
        </left>
        <right>
          <operand>network</operand> <!--C1-->
        </right>
      </operation>
    </action>
    <action name="close">
      <scope behavior="OUT" delimiter="CLOSE">currentscope</scope>
      <operation name="SF_CLEAR">
        <left>
          <operand>network</operand> <!-- C1 -->
        </left>
        <right>
        </right>
      </operation>
    </action>
  </actions>
</ifsemantics>
```

Listing 3 Remote semantics of outgoing and close event

In step 6, the client side PIP processes the outgoing event from the server side given the remote semantics (cf. Sect. 25.4.2). Due to *delimiter* = OPEN, the client side PIP also creates a new scope. The semantics indicates a flow from the network container *C1* into the container *L?*, which is a wildcard for an unknown container receiving the flow at the client (the local container at the server included in the event is ignored at the client). According to *behavior* = OUT of the scope, the client PIP replaces *L?* with the scope's *intermediate container* (cf. Sect. 25.4.3). Together with the fact that *C1* contains *data*, we obtain a flow of *data* from *C1* into the *intermediate container* at the client. Now the server starts sending video data to the client.

Steps 7–9 are triggered by an *incoming* event intercepted by the client side PEP when receiving data over the network connection with the server (cf. Listing 4). The *incoming* event refers to the same *scope* as the *outgoing* event. It indicates a flow from a network container *C2* representing the network connection from the perspective of the client side PEP into a local container *L2*, i.e., the process accessing the video stream. The client side PDP evaluates this event against the policy that has been deployed in step 3. This requires the PDP to query the PIP whether this flow involves a representation of the protected video data (step 8, cf. *evaluate*-call to the client side PIP in Fig. 25.2). As *C2* is either empty, i.e., it has been created during a prior connection to the server, or does not yet exist, the PIP returns *false*, and the PDP will *allow* the incoming event.

```
<event action="incoming" timestamp="2015-05-30T09:30:11">
  <parameter name="process" value="16820cec-18c7-49a2-a443-cd94f0fec3e0"> <!--L2-->
  <parameter name="network" value="192.168.0.1:49152;192.168.0.2:80"/> <!-- C2 -->
  <parameter name="currentscope" value="192.168.0.2:80;192.168.0.1:49152">
</event>
```

Listing 4 Incoming event at client side

In step 9, the *incoming* event is processed at the client side PIP, which holds a local semantics for this event type. The semantics is shown in Listing 5. It contains

```
<ifsemantics>
  <params>
    <param name="process" type="CONTAINER"/>
    <param name="network" type="CONTAINER"/>
    <param name="process_id" type="CONTAINER_NAME"/>
  </params>
  <actions>
    <action name="incoming">
      <scope behavior="IN" delimiter="NONE">currentscope</scope>
      <operation name="SF_FLOW">
        <left>
          <operand>process</operand> <!--L2-->
        </left>
        <right>
          <operand>network</operand> <!--C2-->
        </right>
      </operation>
      <operation name="NF_ADD_NAMING">
        <left>
          <operand>process_id</operand>
        </left>
        <right>
          <operand>process</operand> <!--L2-->
        </right>
      </operation>
    </action>
  </actions>
</ifsemantics>
```

Listing 5 Local semantics of incoming event

a scope specification with *behavior* = IN and *delimiter* = NONE. It further signals
a flow from the network container *C2* into the local container *L2*. The *delimiter* =
NONE indicates that the event belongs to an already activated inter-system scope.
Due to *behavior* = IN, *C2* is replaced by the scope's *intermediate container* within
the client side PIP. Together with the state after steps 5 and 6, the client side PIP
observes a flow of *data* from the remote container *C1* via the *intermediate container*
into *L2*, i.e., as of now, the PIP knows that *L2* contains the video data *data* protected
by our policy. Furthermore, a naming is added to the state of the naming function in
order to make *L2* accessible via the PID of the process receiving the video data.

Once the client disconnects from the video stream, the established inter-system
state is no longer needed, i.e., we deactivate the scopes and delete the intermediate
containers at both PIPs. In our example, the termination of the network connection
is observed by the server side PEP (step 10). The according event is processed
at the server side PIP (step 11) and forwarded to the client side PIP. In line with
Algorithm 1, this event is processed with scope delimiter CLOSE at the server and
the client according to the scope specification of the local semantics (cf. Listing 2)
and the remote semantics deployed in step 5 (cf. Listing 3). As *C1* was replaced by
the *intermediate container* at the server in step 4, the event has no effect except for
closing the scope locally. Tracking of this XSYSTEM flow terminates after the close
event is interpreted at the client side (step 12).

25.5.2   Client Side Policy Enforcement

In terms of enforcing our policy to inhibit the redistribution of video data at the client
(3), the PIP is queried each time a user triggers an event indicating an according
information flow, e.g., when trying to take a screenshot. The event of taking a screen
shot is intercepted by a PEP on the client (Android platform, cf. [4] for further

details) and is only allowed if no application in the foreground has access to the video stream protected by our policy. For this, the PIP can be queried using the PIDs of questionable processes (cf. Sect. 25.5.1, step 9). For the PID of the application accessing the video stream, the PIP will answer that this container is a representation of the data for which our policy applies. Thus the event, i.e., the screen shot, is inhibited.

Security in terms of the reliability of distributed UC enforcement is based on the following assumptions: The integrity and the correctness of policies and components of the UC infrastructure are ensured. The infrastructure is up and running and not tampered with, i.e., users do not have administrative privileges on their devices.

25.6 Related Work

This paper addresses specification and processing of information flow semantics depending on intercepted events—including inter-system and inter-layer flows.

Park and Sandhu [12] introduced the first UC model, which has not been combined with information flow tracking. The distributed usage control (DUC) model by Pretschner et al. in [13] has been extended with information flow tracking in [5, 14] to enable the enforcement of policies depending on the state of an information flow model. The aspect of distributed enforcement of UC policies is considered in greater detail in [1, 7], also focusing on efficient PDP-PIP communication.

Our work builds on and extends [5, 6, 9, 14]. We unify information flow semantics specifications of monitoring components and generalize the information flow model to cope with inter-system flows. Since we were up to a lightweight proof-of-concept implementation, we did not yet consider monitoring technology with higher precision such as the following. Lovat et al. [9, 10] proposed approaches to handle implicit flows [11] and to address the issue of over-approximations of such simple taint-based information flow tracking systems, which we do not cover.

Information flows towards operating system resources and in-between processes are addressed by taint-based information flow tracking frameworks such as Panorama [16] and TaintDroid [3]. SeeC [8] also covers inter-system taint propagation. With Neon [17], Zhang et al. provide a virtual machine monitor for tainting and tracking flows on the level of bytes, which does not require the modification of applications and operating systems. Demsky's tool GARM [2] tackles data provenance tracking and policy enforcement across applications and systems via application rewriting.

25.7   Conclusion

We described and implemented a generic, extensible, and application-oriented approach for dynamic information flow modeling and processing of explicit flows, also across the boundaries of systems equipped with usage control (UC) technology. By this means, we can enforce UC requirements on representations of protected data items on remote systems after the initial access to the data has been granted. Our proof-of-concept implementation shows how video footage from a surveillance system can be protected against duplication and redistribution even if it is accessed via mobile applications, which are increasingly used for cooperation between control rooms and security personnel on-site (given that the mobile device is equipped with UC technology, otherwise it would not be granted access in the first place).

Our generic primitives for specifying information flow semantics enable engineers to develop information flow monitors (PEPs), which can easily be plugged into existing UC infrastructures. This approach eliminates the interdependency between event capturing and information flow tracking at development time and thus improves the practical application of state-based UC enforcement.

References

1. Basin, D. A., Harvan, M., Klaedtke, F., & Zalinescu, E. (2013). Monitoring data usage in distributed systems. *IEEE Transactions on Software Engineering, 39*(10), 1403–1426.
2. Demsky, B. (2011). Cross-application data provenance and policy enforcement. *ACM Transactions on Information and System Security, 14*(1), 6.
3. Enck, W., Gilbert, P., Han, S., Tendulkar, V., Chun, B., Cox, L. P., Jung, J., McDaniel, P., & Sheth, A. N. (2014). Taintdroid: An information-flow tracking system for realtime privacy monitoring on smartphones. *ACM Transactions on Computer Systems, 32*(2), 5.
4. Feth, D., & Pretschner, A. (2012). Flexible data-driven security for android. In *2012 IEEE Sixth International Conference on Software Security and Reliability (SERE)* (pp. 41–50). New York: IEEE.
5. Harvan, M., & Pretschner, A. (2009). State-based usage control enforcement with data flow tracking using system call interposition. In *Proceedings of NSS* (pp. 373–380).
6. Kelbert, F., & Pretschner, A. (2013). Data usage control enforcement in distributed systems. In *Proceedings of CODASPY* (pp. 71–82).
7. Kelbert, F., & Pretschner, A. (2014). Decentralized distributed data usage control. In *Proceedings of CANS* (pp. 353–369).
8. Kim, H. C., Keromytis, A. D., Covington, M., & Sahita, R. (2009). Capturing information flow with concatenated dynamic taint analysis. In *Proceedings of ARES* (pp. 355–362).
9. Lovat, E. (2015). Cross-layer Data-centric Usage Control. Dissertation, Technische Universität München, München, Germany. Dissecting scanning activities using ip gray space.
10. Lovat, E., & Kelbert, F. (2014). Structure matters - A new approach for data flow tracking. In *Proceedings of SPW (IEEE)* (pp. 39–43).
11. Lovat, E., Oudinet, J., & Pretschner, A. (2014). On quantitative dynamic data flow tracking. In *Proceedings of CODASPY* (pp. 211–222).
12. Park, J., & Sandhu, R. S. (2004). The ucon$_{abc}$ usage control model. *ACM Transactions on Information and System Security, 7*(1), 128–174.

13. Pretschner, A., Hilty, M., & Basin, D. A. (2006). Distributed usage control. *Communications of ACM, 49*(9), 39–44.
14. Pretschner, A., Lovat, E., & Büchler, M. (2011). Representation-independent data usage control. In *Proceedings of DPM* (pp. 122–140).
15. Wüchner, T., & Pretschner, A. (2012). Data loss prevention based on data-driven usage control. In *Proceedings of ISSRE (IEEE)* (pp. 151–160).
16. Yin, H., Song, D. X., Egele, M., Kruegel, C., & Kirda, E. (2007). Panorama: Capturing system-wide information flow for malware detection and analysis. In *Proceedings of CCS (ACM)* (pp. 116–127).
17. Zhang, Q., McCullough, J., Ma, J., Schear, N., Vrable, M., Vahdat, A., Snoeren, A. C., Voelker, G. M., & Savage, S. (2010). Neon: system support for derived data management. In *Proceedings of VEE* (pp. 63–74).

Chapter 26
On Inferring and Characterizing Large-Scale Probing and DDoS Campaigns

Elias Bou-Harb and Claude Fachkha

26.1 Introduction

Cyberspace is the electronic world created by interconnected networks of Information Technology (IT) and the information on those networks. It can be defined as the interdependent network of IT infrastructure, including the Internet, telecommunication networks, computer systems, and embedded industrial processors and controllers. Cyberspace is a global commons where more than 1.7 billion people are linked together to exchange ideas and services [1]. Moreover, it underpins almost every facet of a modern society and provides critical support for the economy, civil infrastructure, public safety, and national security.

However, recent events have indeed demonstrated that cyberspace could be subjected, at the speed of light and in full anonymity, to severe attacks with drastic consequences. One particular research revealed that 90% of corporations have been the target of a cyber attack, with 80% suffering a significant financial loss [2]. In addition, the cyber security report [1] elaborated that in a recent 1 year period, 86% of large North American organizations had suffered a cyber attack where the loss of intellectual property as a result of these attacks doubled between 2011 and 2015. Moreover, the report alarmed that more than 60% of all the malicious code ever detected, originating from more than 190 countries, was introduced into cyberspace solely in 2016.

E. Bou-Harb (✉)
Cyber Threat Intelligence Lab, Florida Atlantic University, Boca Raton, FL, USA
e-mail: ebouharb@fau.edu

C. Fachkha
University of Dubai, Dubai, United Arab Emirates
e-mail: cfachkha@ud.ac.ae

© Springer International Publishing AG 2018
K. Daimi (ed.), *Computer and Network Security Essentials*,
DOI 10.1007/978-3-319-58424-9_26

461

To this end, generating effective cyber threat intelligence is indeed an effective approach that would aid in preventing, inferring, characterizing, analyzing, and mitigating various Internet-scale malicious activities. Thus, in this chapter, we aim to generate such cyber threat intelligence related to two specific types of malicious actives, namely probing and DDoS events, and their corresponding orchestrated campaigns, by analyzing the darknet IP space.

26.1.1 Background

In this section, we provide brief yet relevant background information related to the concerned topics.

Probing Activities Probing or scanning events [3] could be defined by the task of executing reconnaissance activities towards enterprise networks or Internet-wide services, searching for vulnerabilities or ways to infiltrate IT assets. Such events are commonly the primary stage of an intrusion attempt that enables an attacker to remotely locate, target, and subsequently exploit vulnerable systems [4]. They are basically a core technique and a facilitating factor of various subsequent cyber attacks. Readers that are further interested in inner details related to probing activities are kindly referred to the following surveys [5, 6].

DDoS Activities Denial of Service (DoS) attacks are characterized by an explicit attempt to prevent the legitimate use of a service. Distributed DoS (DDoS) attacks employ multiple attacking entities (i.e., compromised machines/bots) to achieve their intended aim. DDoS attacks could be related to flooding attempts, in which the bots directly attack the victim, or they could be rendered by amplification attempts, where the attacker employs third party servers known as open amplifiers to indirectly launch the attack towards the victim. Readers that are interested in more details related to DDoS activities are kindly referred to [7].

Darknets A network telescope, also commonly referred to as a darknet or an Internet sink [8], is a set of routable and allocated yet unused IP addresses [9]. It represents a partial view of the entire Internet address space. From a design perceptive, network telescopes are transparent and indistinguishable compared with the rest of the Internet space. From a deployment perspective, it is rendered by network sensors that are implemented and dispersed on numerous strategic points throughout the Internet. Such sensors are often distributed and are typically hosted by various global entities, including Internet Service Providers (ISPs), academic and research facilities, and backbone networks. The aim of a darknet is to provide a lens on Internet-wide malicious traffic; since darknet IP addresses are unused, any traffic targeting them represents a continuous view of anomalous unsolicited traffic.

Orchestrated Campaigns A number of malicious activities could operate within the context of large-scale campaigns. These render a new era of such malicious events, since they are distinguished from previous independent incidents as (1) the population of the participating bots is several orders of magnitude larger, (2) the target scope is generally the entire IP address space, and (3) the bots adopt well-orchestrated, often botmaster coordinated, stealth scan strategies that maximize targets' coverage while minimizing redundancy and overlap. Readers that are further interested in inner details related to large-scale orchestrated malicious campaigns are kindly referred to [10].

In this chapter, we aim to infer and characterize probing and DDoS orchestrated campaigns by uniquely analyzing darknet traffic.

26.1.2   Organization

The remaining of this chapter is organized as follows. In the next section, we address the problem of inferring independent and orchestrated probing events while in Sect. 26.3, we focus on inferring and characterizing DDoS events and large-scale campaigns. In Sect. 26.4, we review some literature work to demonstrate the uniqueness of the presented work. We conclude this chapter in Sect. 26.5 by summarizing the offered contributions and pinpointing several topics that are worthy of being investigated in the future.

26.2   Probing Campaigns

In this section, we present methods to infer independent and orchestrated probing events by scrutinizing darknet data. Further, we present some results characterizing such events.

26.2.1   Inferring Probing Events

Motivated by recent cyber attacks that were facilitated through probing [11], limited cyber security intelligence, and the lack of accuracy that is provided by scanning detection systems, this section presents a new approach to fingerprint Internet-scale probing activities. The rationale of the proposed method states that regardless of the source, strategy, and aim of the probing, the reconnaissance activity should have been generated using a certain literature-known scanning technique (i.e., TCP SYN, UDP, ACK, etc. [5]). We observe that a number of those probing techniques demonstrate a similar temporal correlation and similarity when generating their corresponding probing traffic. In other words, the observation states that we can cluster the scanning techniques based on their traffic correlation

Fig. 26.1 Sessions
distribution

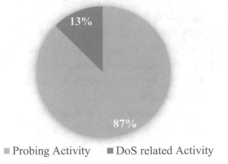

■ Probing Activity   ■ DoS related Activity

statuses. Subsequently, we can differentiate between probing and other darknet malicious traffic based on the possessed traffic correlation status. We can as well attribute the probing traffic to a certain cluster of scanning techniques (i.e., the probing activity, after confirmed as probing, can be identified as being generated by a certain cluster of techniques that possess similar traffic correlation status). To identify exactly which scanning technique has been employed in the probing, we statistically estimate the relative closeness of the probing traffic in comparison with the techniques found in that cluster. To enable the capturing of traffic signals correlation statuses, the proposed method employs the Detrended Fluctuation Analysis (DFA) technique [12]. Elaborative details about the modus operandi of the proposed inference method could be found in [13].

Empirical Results We employ around 10 GB of real darknet data to evaluate the inference approach. We first applied the approach to attempt to differentiate between scanning and darknet backscattered traffic (i.e., DoS related activity). Figure 26.1 represents how the 700 sessions were distributed and fingerprinted. It is shown that probing activity corresponds to 87% (612) of all the sessions. This scanning to backscattered traffic ratio is somehow coherent with other darknet studies [14]. To evaluate the scanning fingerprinting capabilities of our approach, we experimented with Snort's sfPortscan preprocessor using the same 612 sessions that were fingerprinted as probing. Snort's sfPortscan detected 590 scans. After a semi-automated analysis and comparison that was based on the logged scanning traffic flows, we identified that all the 612 scans that our approach fingerprinted as probing activity include sfPortscan's 590 scans. Therefore, relative to this technique and experimenting with this specific data set, we confirm that our approach yielded no false negative, with only 2% as false positives.

26.2.2 Inferring and Characterizing Probing Campaigns

To infer orchestrated probing campaigns, for each of the previously inferred probing event, we generate their feature vectors as summarized in Table 26.1. The machinery that would generate such vectors is summarized in [15].

Table 26.1 Probing feature vectors

Employed probing technique
Probing traffic (random vs patterns)
Employed pattern
Adopted probing strategy
Nature of probing source
Type of probing (targeted vs dispersed)
Signs of malware infection
Exact malware type/variant
Probing rate
Ratio of destination overlaps
Target port

To automatically infer orchestrated probing events, the approach leverages all the previously extracted inferences and insights related to the probing sessions/sources to build and parse a Frequent Pattern (FP) tree. In such a tree, each node after the root represents a feature extracted from the probing sessions, which is shared by the sub-trees beneath. Each path in the tree represents sets of features that co-occur in the sessions, in non-increasing order of frequency of occurrences. Thus, two sessions that have several frequent features in common and are different just on infrequent features will share a common path in the tree. The proposed approach also employs the FP tree-based mining method, FP-growth, for mining the complete set of generated frequent patterns. As an outcome, the generated patterns represent frequent and similar probing behavioral characteristics that correlate the probing sources into orchestrated probing events.

Empirical Results We evaluate the proposed approach using 330 GB of darknet data. We visualize the outcome of the feature vectors as depicted in Fig. 26.2. Such "flower-based" result intuitively and creatively illustrates how the FP-tree is constructed. Recall that the tree depicts frequent probing features that co-occur in the probing sessions, which are generated by the probing behavioral analytics. One can notice several groupings or clusters that depict probing events sharing various common machinery. For the sake of this work, we have devised a parsing algorithm that automatically build patterns from the FP-tree that aim at capturing orchestrated probing events that probe horizontally; probe all IPs by focusing on specific ports.

The proposed approach automatically inferred the pattern that is summarized in Table 26.2. The pattern permitted the detection, identification, and correlation of 846 unique probing bots into a well-defined orchestrated probing event that targeted the VoIP (SIP) service. It is shown that this event adopted UDP scanning, probed around 65% of the monitored dark space (i.e., 300,000 dark IPs) where all its bots did not follow a certain pattern when generating their probing traffic. Further, the results demonstrate that the bots employed a reverse IP-sequential probing strategy when probing their targets. Moreover, the malware responsible for this event was shown to be attributed to the Sality malware.

Fig. 26.2 Visualization of the outcome of the probing behavioral analytics in the FP-tree

Table 26.2 The inferred pattern capturing a large-scale orchestrated probing event	
	Employed probing technique: UDP
	Probing traffic (random vs patterns): Random
	Employed pattern: Null
	Adopted probing strategy: Reverse IP-sequential
	Nature of probing source: Bot
	Type of probing (targeted vs dispersed): Dispersed
	Signs of malware infection: Yes
	Exact malware type/variant: Virus.Win32.Sality.bh
	Probing rate: 12 pps
	Target port: 5060

26.3 DDoS Campaigns

In this section, we present techniques to infer distributed and orchestrated DDoS events by analyzing real darknet data. In addition, we present some results characterizing these large-scale activities.

26.3.1   Inferring DDoS Events

This section leverages darknets to identify independent DDoS attacks. To achieve its aim, our approach adopts three steps: (1) selecting backscattered packets from victims' replies; (2) extracting session flows corresponding to malicious activities; and (3) inferring DDoS attacks by employing a detection algorithm. First, in order to select backscattered packets, we adopt the technique from [16] that relies on flags in packet headers, such as TCP SYN+ACK, RST, RST+ACK, and ACK. However, this technique might cause misconfiguration as well as scanning probes (i.e., SYN/ACK Scan) to co-occur within the backscattered packets. In order to filter out the misconfiguration, we use a simple metric that records the average number of sources per destination darknet address. This metric should be significantly larger for misconfiguration than scanning traffic [17]. Second, in order to filter out scanning activities, we split the connections into separate session flows, each of which consists of a unique source and destination IP/port pair. The rationale for this is that DDoS attempts possess a much greater number of packets sent to one destination (i.e., flood) whereas portsweep scanners have one or few attempts towards one destination (i.e., probe). Third, we aim to confirm that all the extracted sessions in fact reflect real DDoS attempts. To accomplish this, we employ a modified version of the DDoS detection parameters from [18] to label a session as a single DoS attack. Algorithm 1 displays our detection mechanism. We proceed by merging all the previously extracted sessions that have the same source IP (i.e., victim) to extract DDoS attacks.

Empirical Results Similar to the probing analysis, the data is based on the previously darknet data set collected during the same period. We inferred thousands of DDoS attacks, as per Fig. 26.3a, where the majority were shown to abuse TCP services (62%), ICMP (21%), and UDP (17%). Furthermore, as shown in Fig. 26.3b, these attack types are distributed as follows: 82% for TCP flooding, 14% for DNS flooding, and the remaining are ICMP flooding events.

26.3.2   Inferring and Characterizing DDoS Campaigns

In the previous sections, we elaborated on the components of the systematic approach for inferring DDoS activities targeting a unique organization. In this section, we extend the approach by proposing a clustering approach to infer orchestrated DDoS campaigns that target multiple victims. This permits the fingerprinting of the nature of such campaigns. For example, it could be identified that a specific DDoS campaign is specialized in targeting financial institutions while another campaign is focused on targeting critical infrastructure. Further, such clustering approach allows the elaboration on the actual scope of the DDoS campaign to provide cyber security situational awareness; how large is the campaign and what is

---

Algorithm 1 DDoS detection engine

1: In the algorithm:
2: Each flow *f* contains packet count (*pkt_cnt*) and rate (*rate*)
   Tw: Time Window
   p_th: Packet Threshold
   r_th: Rate Threshold
   Tn: Time of packet number *n* in a flow
   pkt: Packet

3: **Input:** A set of darknet flows *F* where each *f* in *F* is composed of a pair of <source IP,
   destination IP> leveraging a series of consecutive packets that share the same source IP
   address.
4: **Output:** DDoS attack flows
5:
6: **for** each *f* in *F* **do**
7:     *attack_flag* = 0
8:     *pkt_cnt* = 0
9:     *T1* = pkt_gettime(1)
10:    *Tf* = *T1* + *Tw*
11:    **while** *pkt* in *f* **do**
12:        *Tn*= pkt_gettime()
13:        **if** *Tn* < *Tf* **then**
14:            pkt_cnt++
15:        **end if**
16:    **end while**
17:    $rate = \frac{pkt\_cnt}{Tw}$
18:    **if** *pkt_cnt* > p_th & *rate* > *r_th* **then**
19:        attack_flag = 1
20:    **end if**
21: **end for**

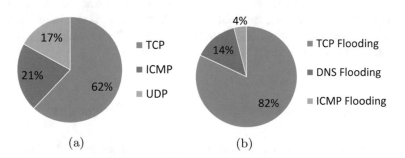

Fig. 26.3 DDoS: major protocols and distribution. (**a**) Abused protocols, (**b**) Attack distribution

its employed rates, when attacking the various victims. Additionally, the proposed
approach could be leveraged to predict the campaign's features in terms of rate and
number of involved machines.

To achieve this task, our approach employs the following statistical-based mechanism. First, backscattered sessions are extracted as previously discussed. Second,

the notion of fuzzy hashing [19] between the different sessions is applied. Fuzzy hashing is advantageous in comparison with typical hashing as it can provide a percentage of similarity between two traffic samples rather than producing a null value if the samples are different. This popular technique is derived from the digital forensics research field and is typically applied on files or images [19]. Our approach explores the capabilities of this technique on backscattered DDoS traffic. We select the sessions that demonstrate at least 20% similarity. We concur that this threshold is a reasonable starting point and aids in reducing false negatives. Third, from those similar sessions, we employ two statistical tests, namely the Euclidean and the Kolmogorov–Smirnov tests [20] to measure the distance between the selected sessions. We extract those sessions that minimize the statistical distance after executing both tests. The rationale of the latter approach stems from the need to cluster the sessions belonging to multiple victims that share a maximized similar traffic behavior while minimizing the false positives by confirming such similarity using various tests. Note that, we hereafter refer to the use of the previous two techniques as the fusion technique. The outcome of the proposed approach are clustered diverse victims that are inferred to be the target of the same orchestrated DDoS campaign.

Empirical Results   In this section, we present the empirical evaluation results. We employ the DDoS campaign clustering model as discussed in the previous section to demonstrate how multiple victims could be modeled as being the target of the same campaign.

26.3.2.1   TCP SYN Flooding on Multiple HTTP Servers

To demonstrate the effectiveness of the approach, we experiment with a 1 day sample retrieved from our darknet data set. We extract more than 600 backscattered DDoS sessions and apply fuzzy hashing between the sessions, by leveraging deep-toad, a fuzzy hashing implementation. The outcome of this operation is depicted in Fig. 26.4a, where the victims are represented by round circles while directed arrows illustrate how the various victims were shown to be statistically close to other targeted victims. It is important to note that we anonymize the real identity of the victims due to sensitivity and legal reasons. Subsequently, the Euclidean and the Kolmogorov–Smirnov tests are executed to exactly pinpoint and cluster the victims that demonstrate significant traffic similarity. Figure 26.4b shows such result while Table 26.3 summarizes the outcome of the proposed DDoS campaign clustering approach. From Fig. 26.4b, one can notice the formation of root nodes, advocating that the approach is successful in clustering various victims that are the target of the same DDoS campaign.

In general, the approach yielded, for 1 day data set, 13 unique campaigns where each campaign clusters a number of victims ranging from 2 to 125 targets.

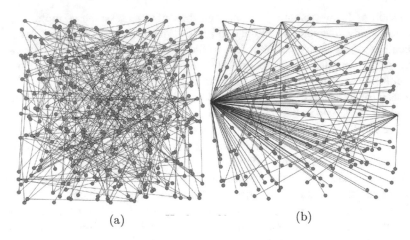

(a)   (b)

Fig. 26.4 DDoS clustering. (**a**) Fuzzy hashing clustering. (**b**) Fusion technique clustering

Table 26.3 Summary of the DDoS campaign clustering approach

Technique	Unique campaign count	Campaign of 2 victim machines	Campaign of 3 victim machines	Campaign of 4 victim machines	Campaign of 5 victim machines	Campaign of 6 victim machines	Campaign of 125 victim machines
Euclidean	16	6	2	3	3	1	1
KS	16	6	2	3	2	2	1
Fusion	13	6	1	2	2	1	1

26.4 Related Work

In this section, we review the related work on various concerned topics.

Extracting Probing Events Li et al. [21] considered large spikes of unique source counts as probing events. The authors extracted those events from darknet traffic using time-series analysis; they first automatically identified and extracted the rough boundaries of events and then manually refined the event starting and ending times. At this point, they used manual analysis and visualization techniques to extract the event. In an alternate work, Jin et al. [22] considered any incoming flow that touches any temporary dark (grey) IP address as potentially suspicious. The authors narrowed down the flows with sustained suspicious activities and investigated whether certain source or destination ports are repeatedly used in those activities. Using these ports, the authors separated the probing activities of an external host from other traffic that is generated from the same host. In contrast, in this work, we propose a method that exploits a unique observation related to the signal correlation status of probing events. By leveraging this, we are able to differentiate between probing and other events and subsequently extract the former from incoming darknet traffic.

Analyzing Probing Events The authors of [22, 23] studied probing activities towards a large campus network using netflow data. Their goal was to infer the probing strategies of scanners and thereby assess the harmfulness of their actions. They introduced the notion of gray IP space, developed techniques to identify potential scanners, and subsequently studied their scanning behaviors. In another work, the authors of [21, 24, 25] presented an analysis that drew upon extensive honeynet data to explore the prevalence of different types of scanning. Additionally, they designed mathematical and observational schemes to extrapolate the global properties of scanning events including total population and target scope. In contrary, we aim at inferring large-scale probing and DDoS campaigns rather than focusing on analyzing probing events.

Probing Measurement Studies In addition to [26, 27], Benoit et al. [28] presented the world's first Web census while Heidemann et al. [29] were among the first to survey edge hosts in the visible Internet. Further, Pryadkin et al. [30] offered an empirical evaluation of IP address space occupancy whereas Cui and Stolfo [31] presented a quantitative analysis of the insecurity of embedded network devices obtained from a wide-area scan. In a slightly different work, Leonary and Loguinov [32] demonstrated IRLscanner, a tool which aimed at maximizing politeness yet provided scanning rates that achieved coverage of the Internet in minutes. In this work, as previously mentioned, we strive to infer large-scale campaigns rather than solely providing measurements of particular events.

Botnet Detection Frameworks A number of botnet detection systems have been proposed in the literature [33–36]. Some investigates specific channels, others might require deep packet inspection or training periods, while the majority depends on malware infections and/or attack life-cycles. To the best of our knowledge, none of the proposals is dedicated to tackle the problem of inferring large-scale probing and DDoS campaigns. Further, in this work, we aim to achieve that task by analyzing the dark IP space and by focusing on the machinery and netflow characteristics of the received darknet traffic, without requiring content analysis or training periods.

26.5   Concluding Remarks

This chapter aims at generating effective cyber threat intelligence to aid in proactive and defensive protection of cyberspace. To this end, several techniques to detect and identify large-scale orchestrated probing and DDoS campaigns by leveraging real darknet data were elaborated. On one hand, we presented approaches that addressed the problem of inferring probing activities, which are typically the precursors of future cyber attacks. In particularity, we discussed an approach rooted in time-series fluctuation analysis to identity probing activities as well as attribute such events to a certain technique. Further, we leveraged this inference approach to detect orchestrating probing events, by proposing a feature generation and clustering approach. The latter is based on a set of behavioral data analytics and the employment of a data mining method. On the other hand, we designed and developed darknet-based techniques to infer and characterize independent DDoS attacks. Additionally, we addressed the problem of DDoS campaigns by exploiting fuzzy and statistical methods. Empirical evaluations based on real darknet data demonstrated that the devised techniques, methods, and approaches are effective in inferring and characterizing such stealthy and devastating events.

26.5.1 Considerations and Research Gaps

Developing and deploying cyber security capabilities to combat contemporary threats in general, and cyber campaigns in particular, require several considerations. First, characterizing security information requires access to real attack data sets, which is relatively difficult to access or obtain. Second, developing techniques might not be as simple as deploying them. For instance, deploying darknet-based models require access to real hardware devices. Furthermore, deploying such techniques must be approved by authorities (network administrators, Internet Service Providers, etc.) and therefore necessitate significant collaborative effort and coordination. Our future plan is to deploy our models in real-time and leverage such capabilities to develop an Internet-scale situation awareness system, working closely with our partners and affiliations.

From the conducted research, we can extract the following points/research gaps:

- Inferring and attributing botnets or malicious campaigns by solely monitoring the dark IP space is very challenging due to the passive nature of such IP space. Therefore, other interactive techniques such as honeypots could be used in conjunction with darknet analysis to enhance botnet investigation.
- Packet analysis is the only technique employed on darknet data to investigate spoofing activities. This method is rendered by inspecting ICMP packets and TTL values. Minimal research has been executed to study spoofing events through darknet analysis. Therefore, spoofing is still a noteworthy malicious activity that needs more attention from the security research community.
- Despite the existence of few collaborative darknet projects, more darknet resources and information sharing efforts should emerge to infer and attribute large-scale cyber activities. Indeed, establishing a worldwide darknet information exchange is a capability that requires collaboration and trust; however, this collaboration necessitates the implementation of numerous global policies and undoubtedly would raise serious privacy concerns.
- There exists a need to explore darknet data to generate cyber threat intelligence for other evolving paradigms, include the Internet-of-Things (IoT) and Cyber-Physical Systems (CPS).

Acknowledgements The authors would like to acknowledge the computer security lab at Concordia University, Canada where most of the presented work was conducted. The authors are also grateful to the anonymous reviewers for their insightful comments and suggestions.

References

1. Government of Canada. (2010). Canada's cyber security strategy report, http://www.capb.ca/uploads/files/documents/Cyber_Security_Strategy.pdf.
2. Hinde, S. (2003). The law, cybercrime, risk assessment and cyber protection. *Computers & Security, 22*, 90–95.
3. Bou-Harb, E., Debbabi, M., & Assi, C. (2013). A statistical approach for fingerprinting probing activities. In *2013 Eighth International Conference on Availability, Reliability and Security (ARES)* (pp. 21–30), Sept 2013.
4. Bou-Harb, E., Lakhdari, N. -E., Binsalleeh, H., & Debbabi, M. (2014). Multidimensional investigation of source port 0 probing. *Digital Investigation, 11*(Supplement 2), S114–S123; Fourteenth Annual {DFRWS} Conference.
5. Bhuyan, M. H., Bhattacharyya, D. K., & Kalita, J. K. (2010). Surveying port scans and their detection methodologies. *The Computer Journal, 54*(10), 1565–1581.
6. Bou-Harb, E., Debbabi, M., & Assi, C. (2014). Cyber scanning: A comprehensive survey. *IEEE Communications Surveys & Tutorials, 16*(3), 1496–1519.
7. Rossow, C. (2014). Amplification hell: Revisiting network protocols for DDoS abuse. In *NDSS*.
8. Fachkha, C., & Debbabi, M. (2016). Darknet as a source of cyber intelligence: Survey, taxonomy, and characterization. *IEEE Communications Surveys & Tutorials, 18*(2), 1197–1227.
9. Moore, D., Shannon, C., Voelker, G. M., & Savage, S. (2004). Network Telescopes: Technical Report. Department of Computer Science and Engineering, University of California, San Diego.
10. Bou-Harb, E., Assi, C., & Debbabi, M. (2016). Csc-detector: A system to infer large-scale probing campaigns. *IEEE Transactions on Dependable and Secure Computing, PP*(99), 1
11. Bou-Harb, E., Debbabi, M., & Assi, C. (2013). A systematic approach for detecting and clustering distributed cyber scanning. *Computer Networks, 57*(18), 3826–3839
12. Peng, C. -K., Buldyrev, S. V., Havlin, S., Simons, M., Stanley, H. E., & Goldberger, A. L. (1994). Mosaic organization of DNA nucleotides. *Phys. Rev. E, 49*, 1685–1689.
13. Bou-Harb, E., Debbabi, M., & Assi, C. (2014). On fingerprinting probing activities. *Computers & Security, 43*, 35–48.
14. Wustrow, E., Karir, M., Bailey, M., Jahanian, F., Huston, G. (2010). Internet background radiation revisited. In *Proceedings of the 10th Annual Conference on Internet Measurement* (pp 62–74). New York, NY: ACM.
15. Bou-Harb, E., Debbabi, M., & Assi, C. (2014) Behavioral analytics for inferring large-scale orchestrated probing events. In *2014 IEEE Conference on Computer Communications Workshops (INFOCOM WKSHPS)* (pp. 506–511). New York, NY: IEEE.
16. Moore, D., Voelker, G. M., & Savage, S. (2001). Inferring internet denial-of-service activity. Technical Report, DTIC Document.
17. Li, Z., Goyal, A., Chen, Y., & Paxson, V. (2011). Towards situational awareness of large-scale botnet probing events. *IEEE Transactions on Information Forensics and Security, 6*(1), 175–188.
18. Moore, D., Shannon, C., Brown, D.J., Voelker, G.M., & Savage, S. (2006). Inferring internet denial-of-service activity. *ACM Transactions on Computer Systems (TOCS), 24*(2), 115–139
19. Kornblum, J. (2006). Identifying almost identical files using context triggered piecewise hashing. *Digital Investigation, 3*(Supplement), 91–97; The Proceedings of the 6th Annual Digital Forensic Research Workshop (DFRWS'06).

20. Lilliefors, H. W. (1967). On the Kolmogorov-Smirnov test for normality with mean and variance unknown. *Journal of the American Statistical Association, 62*(318), 399–402.
21. Li, Z., Goyal, A., Chen, Y., & Paxson, V. (2011). Towards situational awareness of large-scale botnet probing events. *IEEE Transactions on Information Forensics and Security, 6*(1), 175–188
22. Jin, Y., Simon, G., Xu, K., Zhang, Z.-L., & Kumar, V. (2007). Gray's anatomy: Dissecting scanning activities using IP gray space analysis. In *Usenix SysML07*.
23. Jin, Y., Zhang, Z. -L., Xu, K., Cao, F., & Sahu, S. (2007). Identifying and tracking suspicious activities through IP gray space analysis. In *Proceedings of the 3rd Annual ACM Workshop on Mining Network Data, MineNet'07* (pp. 7–12). New York, NY: ACM.
24. Li, Z., Goyal, A., Chen, Y., & Paxson, V. (2009). Automating analysis of large-scale botnet probing events. In *Proceedings of the 4th International Symposium on Information, Computer, and Communications Security, ASIACCS'09* (pp. 11–22). New York, NY: ACM.
25. Yegneswaran, V., Barford, P., & Paxson, V. (2005). Using honeynets for internet situational awareness. In *Proceedings of ACM Hotnets IV*.
26. Dainotti, A., King, A., Claffy, K., Papale, F., & Pescapé, A. (2014). Analysis of a "/0" Stealth Scan from a Botnet. *IEEE/ACM Transactions on Networking, 23*, 341–354.
27. Internet Census 2012-Port scanning /0 using insecure embedded devices, http://tinyurl.com/c8af8lt.
28. Benoit, D., Trudel, A. (2007). World's first web census. *International Journal of Web Information Systems, 3*(4), 378.
29. Heidemann, J., Pradkin, Y., Govindan, R., Papadopoulos, C., Bartlett, G., & Bannister, J. (2008). Census and survey of the visible internet. In *Proceedings of the 8th ACM SIGCOMM conference on Internet measurement, IMC'08* (pp. 169–182). New York, NY: ACM.
30. Pryadkin, Y., Lindell, R., Bannister, J., & Govindan, R. (2004). *An empirical evaluation of ip address space occupancy.* USC/ISI Technical Report ISI-TR, 598.
31. Cui, A., & Stolfo, S. J. (2010). A quantitative analysis of the insecurity of embedded network devices: Results of a wide-area scan. In *Proceedings of the 26th Annual Computer Security Applications Conference, ACSAC'10* (pp. 97–106). New York, NY: ACM.
32. Leonard, D., & Loguinov, D. (2010). Demystifying service discovery: Implementing an internet-wide scanner. In *The 10th ACM SIGCOMM Conference on Internet Measurement.* New York, NY: ACM.
33. Gu, G., Porras, P., Yegneswaran, V., Fong, M., & Lee, W. (2007). Bothunter: Detecting malware infection through ids-driven dialog correlation. In *Proceedings of 16th USENIX Security Symposium on USENIX Security Symposium, SS'07* (pp. 12:1–12:16). Berkeley, CA: USENIX Association.
34. Goebel, J., & Holz, T. (2007). Rishi: Identify bot contaminated hosts by irc nickname evaluation. In *Proceedings of the first conference on First Workshop on Hot Topics in Understanding Botnets (USENIX HotBots)*, Cambridge, MA (pp. 8–8).
35. Wurzinger, P., Bilge, L., Holz, T., Goebel, J., Kruegel, C., & Kirda, E. (2009). Automatically generating models for botnet detection. In M. Backes, & P. Ning, (Eds.), *Computer security – ESORICS 2009. Lecture notes in computer science* (Vol. 5789, pp. 232–249). Berlin: Springer.
36. Tegeler, F., Fu, X., Vigna, G., & Kruegel, C. (2012). Botfinder: Finding bots in network traffic without deep packet inspection. In *Proceedings of the 8th International Conference on Emerging Networking Experiments and Technologies, CoNEXT'12* (pp. 349–360). New York, NY: ACM.

Chapter 27
Design of a Secure Framework for Session Mobility as a Service in Cloud Computing Environment

Natarajan Meghanathan and Michael Terrell

27.1 Introduction

Cloud computing is a rapidly growing technology and offers a paradigm shift in Internet computing as an online utility computing. Cloud computing [1] negates the need for purchasing software licenses, operating system licenses, and hardware platforms for individual users, and instead enables them to access software applications (Software-as-a-Service, SaaS) as well as system and hardware resources as a service (Platform-as-a-Service, PaaS and Infrastructure-as-a-Service, IaaS) offered by several organizations (Amazon, Google, etc.) over the Internet. Service providers deliver applications via the Internet, which are accessed through web browsers, desktop, and mobile apps; the software and data are stored on the servers at a remote location. Cloud computing services are typically run in multiple servers, distributed throughout the Internet, with the servers sometimes collaborating with one another to offer a particular service and sometimes capable of individually offering the service.

Mobile phones, originally conceived purely for communication, are now transitioning to smart mobile devices that can do communication as well as computation. Several billions of such devices are connected to the Internet and access the cloud services. The two major factors that impact the performance of applications delivered as a service over the Internet are the mobility of the end-user clients

---

N. Meghanathan (✉)
Jackson State University, Mailbox 18839, Jackson, MS, USA
e-mail: natarajan.meghanathan@jsums.edu

M. Terrell
Software Developer II, Century Link, Monroe, LA, USA

© Springer International Publishing AG 2018
K. Daimi (ed.), *Computer and Network Security Essentials*,
DOI 10.1007/978-3-319-58424-9_27

and the bandwidth constraint of the intermediate networks connecting the client to the servers in the cloud. With mobility and bandwidth constraints, it becomes imperative to develop a framework that will facilitate seamless transfer of ongoing client sessions from one server to another server in the cloud to provide the desired quality of service.

While commercial offerings of cloud computing may be required to meet service-level-agreements (SLAs), clients accessing cloud computing applications for free are often provided only the best-effort service, as is the case of the Internet—the networking medium through which the clients access the cloud. The route through which these clients access the server(s) in the cloud is often determined on-demand and bandwidth/delay is still a performance bottleneck. When a client moves far away from the server to which it is initially assigned in the cloud or the route through which the packets between the client and server are sent is congested, it would be better to handoff the ongoing session between the client and server to another server in the cloud that is relatively closer to the current location of the mobile client as well as employ a route that is relatively less congested, and provide the requested service with a better performance.

Session mobility has been addressed in the literature from two perspectives: network or client/end-device [2]. With the former approach, migration is handled largely by the network; the user/client merely asks the network to initiate and take care of the session migration, from one end-device to another, by just providing some preferences, and the network is responsible for choosing the best target destination for the session. Network-centric session mobility is very complex and expensive to implement and is usually confined to small, localized session mobility-based systems. On the other hand, device-centric session mobility is more scalable and is often realized with information gathered from the network to initiate session transfer. Whether it is network or client/device-initiated, the session transfer has to happen in a secure fashion without affecting the confidentiality and integrity of the session (including the end-user privacy information) as well as the availability of the service.

27.2 Secure Session Mobility as a Service (SMaaS) Framework

In this research, our objective is to develop a framework for secure transfer of an ongoing client session from one server to another server in the cloud such that the users (on the client side) are unaware of the congestion and the resulting session transfer with different servers in the cloud. We call such a framework as Session Mobility-as-a-Service (SMaaS) and it could be used in conjunction with the Software-as-a-Service (SaaS) cloud computing paradigm, especially for clients who are on the move.

Through this research, we address the problem of securely transferring client–server sessions in a cloud computing environment and providing a mobile user seamless access to the servers in the cloud and run the SaaS applications, while being on the move. The proposed SMaaS framework is perfectly suitable for thin clients because a client has to maintain an active session with only one server in the cloud at any point of time for the requested service. A salient characteristic of the SMaaS framework is that it is security-aware and has appropriate encryption, authentication, and anti-spoofing features incorporated during both session transfer and typical client–server interactions.

We will adopt a client/device-centric approach wherein we will let the server (in the cloud) to sense an impending congestion in the path to the client and/or the moving of the client far away from the server and notify the client through a session handoff message; the client then handles the transfer of the session from the current server to another appropriate server that can continue providing the particular service. The user working at the client is totally unaware of this session transfer process among the servers in the cloud. The entire session transfer will occur in a secure manner with no scope for any denial of service or spoofing attacks that will affect the confidentiality, integrity, and availability. We use the terms "session transfer" and "session migration," and "session mobility" as well as "message" and "packet" interchangeably. They mean the same.

27.2.1 Network Model

There are four entities (see Fig. 27.1 for a topological illustration) involved in the SMaaS framework.

1. SMaaS Server Cluster—A group of service servers, each located in different networks across the cloud. Note that only a subset of the servers in the cloud might provide a specific service (e.g., file download) and this information resides in the SMaaS Gateway Server.
2. SMaaS User—A user running from a client machine to obtain the SMaaS service (e.g., download a file from a server in the SMaaS Server Cluster).
3. SMaaS Client—A client machine through which a user interacts with the SMaaS Server Cluster and obtains the service.
4. SMaaS Gateway Server—The public face of the SMaaS framework. The SMaaS Clients first contact the SMaaS Gateway Server to learn about the servers providing a particular service (e.g., servers hosting a particular file for download). We assume the connection from the SMaaS Client to the SMaaS Gateway Server is uninterruptible. If needed, there can be multiple Gateway Servers for load balancing.

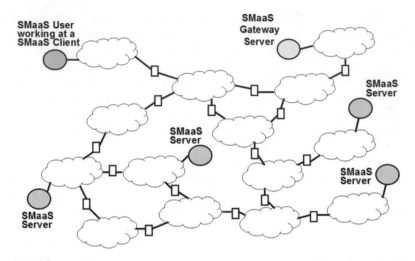

Fig. 27.1 Topological illustration of the entities involved in the SMaaS framework

27.2.2 Sequence of Steps

1. *Exchange of Public-Key Certificates between SMaaS Client and the SMaaS Gateway Server:* A SMaaS Client contacting the SMaaS Gateway Server for the first time initiates the Internet Key Exchange (IKE) protocol through which the client and the Gateway Server exchange each others' public-key certificates. The two sides then decrypt the public-key certificate (received from the other side) to validate each others' identity (IP address). This way, we prevent IP spoofing from either side.

2. *SMaaS Request Message:* The SMaaS Client contacts the SMaaS Gateway Server by sending a *SMaaS Request* message containing the service ID (a unique identification for the service requested), service usage scenario index, and service related parameters (includes an index field identifying a usage scenario of the requested service): all of which could be used for applications that require a certain quality of service, as well as the username and password for the SMaaS User (requesting the particular service). An example for the Service Parameters could be the name of the file, a version number (suitable in case of downloading the .iso file of a particular operating system), and any path information (suitable in case the user wants to download a file stored in a particular directory). As noticed in the above case of file downloading, there could be various parameters associated with different usage scenarios of a particular service. Hence, we have included a *Service Usage Scenario Index* field to identify the set of associated parameters that correspond to a particular usage scenario of the requested service. The contents of the SMaaS Request message (see Fig. 27.2) are encrypted with the public key of the SMaaS Gateway Server so that they can be decrypted and processed only by the

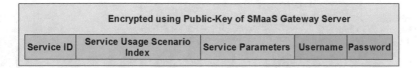

Fig. 27.2   SMaaS request message (sent from the Client to the Gateway Server)

latter (using its private key). Note that for all the messages, illustrated through figures in this chapter, we only show the payload portion of the message; we do not show the standard IP header (containing the source and destination IP addresses) and other headers of the TCP/IP protocol stack, unless needed to highlight the use of some specific field in the protocol headers.

3. *SMaaS Gateway Server Response and the SMaaS Ticket:* The Gateway Server validates the <username, password> tuple in its database and extracts the list of SMaaS servers that could provide the requested service (corresponding to the Service ID and the Service Usage Scenario Index). The SMaaS servers in the cloud are ranked in the order of the number of hops from the network of the requesting client.

   The Gateway Server creates a SMaaS Ticket that contains the username, IP address of the client machine, the time of contact, Service ID, Service Usage Scenario Index, and the Service Parameters as well as the Sequence Number of the latest Acknowledgment packet received from the Client and the Session State Information (both initialized to NULL). The time of contact information (with reference to the Gateway Server) is included to avoid any replay attack. To detect any attempts of a replay attack, we require the clocks of all the SMaaS Servers to be loosely synchronized. SMaaS Tickets lose their validity beyond a certain time after their creation. All of the above information in the SMaaS Ticket is encrypted using a secret key that is shared by all the SMaaS Servers capable of providing the requested service (identified using the Service ID) and the Gateway Server. Along with this information, the Gateway Server also includes the set of IP addresses of the SMaaS Servers in the increasing order of the hop count from the client network. For security purposes, the IP address list of the candidate SMaaS Servers is encrypted through a key that is derived (using a Key Derivation Function agreed upon by the user while creating an account at the Gateway Server) based on the user password. Figure 27.3 illustrates the contents of the SMaaS Ticket along with the SMaaS Server IP address list.

4. *Selection of the SMaaS Server:* The client decrypts the SMaaS Server List and pings the top three servers in the list by sending four short "Echo Request" messages to each of these servers. The client measures the Round Trip Time (RTT) of the "Echo Reply" ping messages. The SMaaS Server that responds with the Reply message at the earliest (i.e., incurred the lowest RTT) is selected. Ties are broken down by the lowest hop count and other predefined criteria.

5. *Session Initiation Request Message:* The SMaaS Client attempts to establish a TCP Session with the chosen SMaaS Server and sends a *Session Initiation*

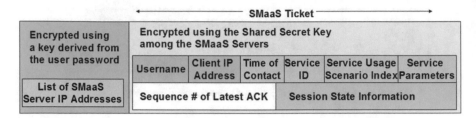

Fig. 27.3 Structure of the SMaaS ticket along with the list of server IP addresses

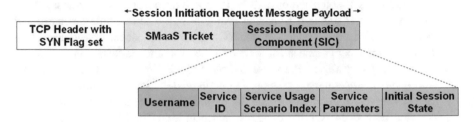

Fig. 27.4 Structure of session initiation request message (SMaaS Client to SMaaS Server)

Request message (structure shown in Fig. 27.4) with the SYN flag set in the TCP header; the payload of the message includes the SMaaS Ticket and a Session Information Component containing the Username, Service ID, Service Usage Scenario Index, values for the Service Parameters, and the Initial Session State (in the case of file download, the Service Parameters could be the file/path name, version #, etc. and the Session State could be the sequence number and byte number of the file downloaded so far in-order, initialized to zero). The SMaaS Server first decrypts the SMaaS Ticket using the secret key shared among the servers in the SMaaS Cloud as well as the Gateway Server. If the extracted contents of the Ticket match with those in the Session Initiation Request message as well as the IP address of the client machine, then the SMaaS Server accepts the TCP connection request (sends a TCP SYN/ACK message) if it can allocate the required resources for the session. Otherwise, the SMaaS Server sends a "Connection Request Reject" message.

6. *Trials in Selecting a SMaaS Server:* If a contacted SMaaS Server denies the TCP connection request, the SMaaS Client includes the SMaaS Server in the *Overloaded/Congested List of SMaaS Servers* and then tries to establish a TCP session with the SMaaS Server that responded with the next lowest RTT. If all the three first-choice SMaaS Servers deny the connection request, the SMaaS Client chooses the next three SMaaS Servers in the list sent by the Gateway Server and pings them. This procedure is repeated until the SMaaS Client manages to successfully find a SMaaS Server; otherwise, the SMaaS Client returns an error message to the user indicating that the requested service cannot be accessed.

7. *IPSec Security Association between the SMaaS Client and the Chosen SMaaS Server:* Once a SMaaS Server has accepted for the TCP session, the SMaaS Client and Server establish a bi-directional IPSec security association (IPSec SA) to prevent IP spoofing from either side. In this pursuit, they go through an IKE session to exchange each others' public-key certificates. All subsequent session-specific messages are encrypted, at the sender, using the public key of the receiver and decrypted at the receiver using its private key. Actually, an IPSec SA could be established between the SMaaS Client and a SMaaS Server before the former sends a TCP SYN Session Initiation Message (prior to Step 5). However, if the SMaaS Server does not accept the client's request, then the client would have to go through the SMaaS Server selection process (Step 6). It would be too much of an overhead and delay incurred to require the client to establish an IPSec SA with a prospective server whose willingness to start or continue a session is not yet confirmed. Hence, we suggest an IPSec SA be formed between a SMaaS Client and Server after the latter has confirmed to start/continue a session. Nevertheless, the identity of the user, the client, and session state information contained in the Session Information Component— all of these are validated by the SMaaS Server through the SMaaS Ticket. As the key to decrypt the SMaaS Ticket is known only to the SMaaS Servers, it is not possible for someone to forge the contents of the SMaaS Ticket. The two sides can further validate each other's identity and the server can commit its resources only if it could validate the client through its public-key certificate and the IPSec SA.

8. *SMaaS Session and Acknowledgment:* After establishing an IPSec SA with the chosen SMaaS Server, the Client accesses/obtains the requested service (e.g., to download the contents of the file requested by the user). We run the standard TCP protocol on the top of IPSec/IP. After receiving a data packet corresponding to the requested service, the SMaaS Client acknowledges for the same (e.g., in the case of a file download session, the SMaaS Client acknowledges for all the packets that have been received in-order and not acknowledged yet). The SMaaS Server measures the RTT for the Acknowledgment packets received from the SMaaS Client. If the RTTs start increasing significantly for every acknowledgment received (the actual rate of increase of the RTT is an implementation issue), then the SMaaS Server decides to handoff the session to another peer SMaaS Server.

9. *SMaaS Session Handoff:* To handoff the session, the SMaaS Server sends a *Session Handoff* message (see Fig. 27.5) to the SMaaS Client and includes the sequence number of the latest acknowledgment that has been received by the server and the current state information of the session corresponding to the service provided (in case of file download, the Session Handoff message would include the sequence number of the last byte whose acknowledgment has been received by the Server and the position of this byte—byte number— in the actual file that is being downloaded). The SMaaS Server also updates the SMaaS Ticket with the sequence number of the latest acknowledgement

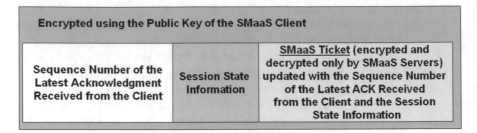

Fig. 27.5 Structure of session handoff message (SMaaS Server to SMaaS Client)

received and the current state information of the session. The SMaaS Server encrypts the updated SMaaS Ticket using the secret key that is shared only among all the servers in the SMaaS cloud. The updated SMaaS Ticket along with the Session Handoff message is sent to the SMaaS Client.

10. *Successful SMaaS Session Transfer:* After receiving the Session Transfer message, the SMaaS Client confirms about the sequence number of the last session related message that was received in-order from the previous SMaaS Server (which is now added to the Overloaded/Congested List of SMaaS Servers that is locally maintained by the SMaaS Client). Unlike the previous procedure adopted (i.e., to look for prospective SMaaS Servers in the increasing order of the number of hops), the SMaaS Client randomly permutes the list and pings all the Servers in the Cloud, except those in the locally maintained Overloaded/Congested List.

The SMaaS Server that responds with an "Echo Reply" at the earliest is chosen as the next server to transfer the session. The SMaaS Client sends a *Session Transfer Request* message that includes a TCP header with the SYN flag set and the payload of which includes the SMaaS Ticket received from the previous server as well as the *Session Information Component* containing the following information (same as those forwarded to the first SMaaS Server): Username, Service ID, Service Usage Scenario Index, values for the Service Parameters, and the Current Session State (in the case of file download, the Service Parameters could be the file/path name, version #, etc. and the Current Session State could be the sequence number and byte number of the file downloaded so far in-order).

Once the newly chosen SMaaS Server receives the SMaaS Ticket and the Session Information Component as part of the Session Transfer Request message (see Fig. 27.6), it decrypts the SMaaS Ticket using the secret key for the SMaaS Server Cluster and compares the contents of the SMaaS Ticket with those in the Session Information Component. If everything matches and it is ready to allocate the required buffer space for this session, the new SMaaS Server agrees to continue with the session and sends a TCP SYN/ACK message; otherwise, it sends a Connection Request Reject message.

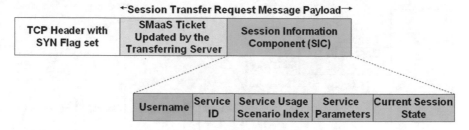

Fig. 27.6 Structure of session transfer request message (SMaaS Server to SMaaS Client)

If the new SMaaS Server accepts the session transfer request, the SMaaS Client establishes an IPSec SA with the new SMaaS Server (see Step 7) before continuing the session. After a while, if the new SMaaS Server decides to handoff the session, Steps 9 and 10 are again followed.

11. *Successful SMaaS Session Transfer:* If the SMaaS Client receives a *Connection Request Reject* message from the last chosen SMaaS Server, the SMaaS Client adds the SMaaS Server to the Overloaded/Congested List. The SMaaS Server that responded with the next lowest RTT is contacted and this procedure is repeated until a new SMaaS Server to transfer the session is found. If unsuccessful over the entire SMaaS Server List, the SMaaS Client quits and reports an error message to the user.

A high-level illustration of the sequence of message exchanges under the proposed SMaaS framework is shown in Fig. 27.7.

27.3   Qualitative Comparison of SMaaS with Related Work

27.3.1   SMaaS vs. Standard FTP

In this section, we compare the advantages of using SMaaS for downloading a large file (whose size is in the order of hundreds of Mega bytes or even Giga bytes, like the .iso files) vis-à-vis the use of the standard File Transfer Protocol (FTP). FTP does not permit transfer of an ongoing session to a new server during the middle of the session. If at all, the client or the server experience frequent timeouts and/or packet loss due to network congestion, the TCP session running as part of FTP has to be discontinued and a new TCP session has to be established. Nevertheless, we cannot be sure whether the new TCP session would be of any remedy to the network congestion problem as packets are more likely to be again routed through the same set of congested routers (and networks) as long as the server and client remain the same. SMaaS (when used for File Download) handles the network congestion problem by initiating the transfer of a session to another server in the cloud. This

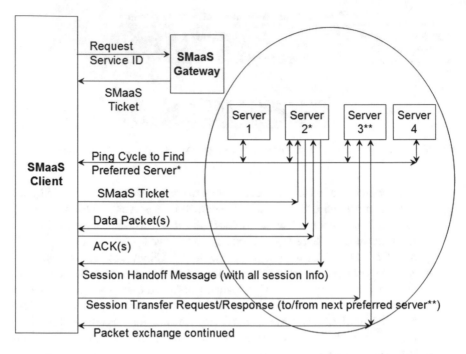

Fig. 27.7 A high-level overview of the sequence of message exchanges under the SMaaS framework

transfer is done in a secure fashion, through the encrypted SMaaS Ticket and the use of an IPSec SA, in order to avoid the scenarios wherein an attacker initiates the transfer without the consent or knowledge of the actual server or the client.

There could be some delay involved in transferring a session from one server to another server as well as to establish an IPSec SA before the actual data packet transfer. However, the transfer delay is expected to be smaller enough to offset the delay incurred if the data packets are continued to be sent on a congested route without any session transfer. The procedure designed for the transfer process sincerely attempts to avoid session thrashing. In other words, we do not want a session to be transferred to a server *I* and the server *I* quickly initiating the transfer of the session to some other server *J*. Note that in Step 10, the SMaaS Server receiving the Session Transfer Request message accepts the message only if it can allocate resources and offer the download service as requested. In other words, only if the parameters are acceptable, the contacted server agrees to become the new alternate server for the session. However, there is still no guarantee that there will not be any congestion on the route between the new alternate server and the client. Since, IP is best effort based and routes are chosen on a per-packet basis depending on the best route available to the routers at the particular moment, it would be difficult to estimate the congestion of a route without any packets being sent on that route.

Hence, from a network congestion point of view, we cannot guarantee that session thrashing will be totally avoidable. It is possible that after the session transfer, one or more networks on the route between the client and the alternate server get congested and the session has to be again transferred to some other server within the set of clusters.

27.3.2 SMaaS vs. Parallel/Mirror Server Schemes

SMaaS is perfectly suitable for thin clients, as at any time—there is a TCP connection/session between the client and only one server in the cloud. On the other hand, as described below, the parallel/mirror server-based schemes proposed for file transfer in the Internet require the client to establish and maintain multiple TCP connections at a time, one with each of the mirror servers, as well as be sometimes constrained to download a pre-assigned portion of the file from a particular mirror server. We provide below a qualitative comparison of SMaaS for file download vs. some of the parallel/mirror server schemes for file download proposed in the literature.

Sohail et al. [3] propose a Parallelized-File Transfer Protocol (P-FTP) that facilitates simultaneous downloads of disjoint file portions from multiple file servers distributed across the Internet. The selection of the set of parallel file servers is done by the P-FTP gateway when contacted by a P-FTP client. The number of bytes to be downloaded from each file server is decided based on the available bandwidth. We observe the following drawbacks with the P-FTP process and the use of multiple mirror file servers for simultaneous download: (1) The P-FTP client would be significantly overloaded in managing multiple TCP sessions, one with each of the parallel file servers. Thus, P-FTP cannot be run on thin clients that are limited in the available memory and resources to run concurrent TCP sessions for downloading a single file. (2) If the path to a particular file server gets congested, the P-FTP client is forced to wait for the congestion to be relieved and continue to download the remaining bytes of the portion of the file allocated for download from the particular file server. The Quality of Service realized during the beginning of the download process may not be available till the end due to the dynamics of the Internet. (3) P-FTP has no security features embedded in it. Hence, it is open for spoofing-based attacks on the availability of the parallel file servers by unauthorized users/clients who simply launch several parallel download sessions that appear to originate from authentic users/IP addresses.

Rodriguez and Biersack [4] propose a Dynamic Parallel Access (DPA) scheme that is also based on downloading a file in parallel from multiple servers, but different from P-FTP in the sense that the portion of the file and the number of bytes to be downloaded from a particular file server are not decided a priori; but done dynamically based on the response from the individual servers. In this scheme, the client chooses the set of parallel servers to request for the file. The download is to be done in blocks of equal size. Initially, the client requests one block of the file from every server. After a client has completely received one block from a server, the client requests the particular server for another block that has not yet been requested from any other server. Upon receiving all the blocks, the client reassembles them and reconstructs the whole file. Unlike the P-FTP scheme, the DPA-scheme appears to be less dependent on any particular mirror server as the latter scheme requests only one block of the file from a server at a time and does not wait for several blocks of the file from any particular server. However, with the DPA-scheme, the client cannot close its TCP connections with any of the mirror file servers until the entire file is downloaded. This is because, if a client fails to receive a block of the file from a particular mirror server and has waited for a long time, then the client has to request another peer mirror server for the missing block. In order to avoid opening and closing multiple TCP connections with a particular mirror server, the client has to maintain the TCP connection with each of the file servers until the entire download is completed. The client has to keep sending some dummy packets to persist with the TCP connections. On the other hand, the P-FTP scheme does not have the necessity for the client to persist with TCP connections for the entire download session; the P-FTP client can close the TCP connection with a P-FTP server once the required portions of the file are downloaded as initially allocated from the particular mirror server. Like P-FTP, DPA also does not have any security features embedded in it.

27.3.3 SMaaS vs. Peer-to-Peer File Sharing Schemes

More recently, peer-to-peer file sharing through the BitTorrent protocol [5] has gained much attention for distributing large amounts of data over the Internet. BitTorrent does not follow the traditional single server or multiple mirror server techniques. Instead, the contents of a file are broken down to pieces and spread across the peer computer nodes (BitTorrent clients); Metadata about the distribution of the multiple pieces is stored in a torrent file that also contains information about a tracker, the computer that coordinates the distribution of the particular file. Any client interested in downloading a particular file should first download the torrent file and connect to the specified tracker, which guides the client to connect to the multiple peers to download the different pieces of the file. As a client download the pieces of a file, they may be shared with one or more clients that are also downloading the file during the same time. As it takes time to form the multiple peer-to-peer connections for complete downloading, BitTorrent is considered to be relatively non-supportive of progressive downloads, real-time, or streaming applications and preferred mainly for applications that support non-contiguous downloads ([6]). Also, if the content that is being distributed does not get popular among the users in the Internet, there will not be sufficient peers hosting the file or downloading a file after a certain time. Hence, BitTorrent is also considered to be not supportive for distributing unpopular contents [7].

27.3.4   SMaaS vs. Kerberos and Anycasting

At the outset, there may appear some similarities between the proposed SMaaS framework and implementation of Kerberos protocol [8] over anycast addressing [9] in the Internet. However, we list below the key differences between the two implementations and the unique features of the SMaaS framework, which could also be considered as significant improvements over the existing technologies:

1. The distinguishing feature of the proposed SMaaS framework is that it supports a seamless secure session transfer in the cloud in the presence of client mobility and Internet congestion, with minimal impact on the service received by the client. After the session transfer, there is no need to start over a session from the beginning with the new chosen server. The session can be continued from where it was left with the previous server.

2. With an implementation of Kerberos over anycast addressing, the client would merely find the server that is closest to its current location in the Internet and get serviced from it. However, such an implementation cannot handle mobility of the client and the congestion in the Internet. Without session transfer, a client has to continue getting serviced from a server even if the route between them is congested and packets suffer significant delay. If a disgusted client wants to get serviced from another server, it has to start all over again going through the Kerberos authentication mechanism, obtain a new ticket to start a session, and cannot continue the session from where it was stopped at the current server.

3. The Kerberos ticket will not include any information about the list of authenticated servers the client can contact to start a session. With Kerberos, if the client likes to get serviced from a server (other than the one that is currently serving it), the client has to again contact the gateway and obtain a server specific ticket. The Kerberos gateway will then sooner become a bottleneck and a single point of failure. With the SMaaS framework, the client contacts the gateway only at the beginning. Moreover, the SMaaS Ticket sent by the gateway to the client will have a list of authenticated service servers the client can contact and this information would be encrypted with a secret derived from the user password (that was initially sent in an encrypted form from the client to the gateway). So, a prospective intruder cannot see the encrypted list of servers included in the SMaaS ticket. Further, the client uses this list of authenticated servers to securely transfer the session each time it wants to.

4. In the proposed SMaaS framework, unlike anycasting, the client does not necessarily choose the nearest server to start or transfer a session. The client waits for the servers (to which it had sent a ping request) to respond, and the server whose reply is received at the earliest is chosen with the presumption that the path to the server, whose response reaches the client at the earliest, is the least congested. So, if the path between a client and server is congested, and even though the client and server machines may be located physically closer to each other, the client may indeed choose a different server—the one that responds at the earliest.

5. Both the client and server play an active role in the session transfer. We require the servers to keep track of the round trip times taken by the acknowledgment packets sent by the client, and initiate a session transfer. However, the choice of the next server to get serviced from is made by the client—because the selection of a prospective server depends on the mobility pattern of the client and the delay (a measure of the number of hops as well as the bandwidth available) on the routes between the client and the candidate servers. The server initiates the session transfer, which is completed by the client by finding a new server.

27.3.5   Other Related Work

Barisch et al. [10] propose a device-centric session mobility framework to transfer ongoing service sessions from one end-device (say a Home TV) to another end-device (Mobile Phone) with minimal involvement from the communication partner on the other side (i.e., a cloud service provider). In a similar context, Johansson [11] has proposed a light-weight session mobility framework that supports migration of multi-media sessions from stationary and semi-stationary devices to smart phones. Shanmugalingam et al. [12] present a web-based communication system architecture that supports session continuity in the presence of user mobility.

Cisco has recently proposed a Location/ID Separation Protocol (LISP; [13]) for virtual machine mobility. With LISP, one can deploy virtual machines as servers (PaaS) anywhere regardless of their IP addresses and can freely move them across the data centers of a cloud to different locations. While LISP focuses on handling mobility of the virtual machine servers in the cloud, our proposed SMaaS framework is focused towards handling client mobility. Once SMaaS is developed, we conjecture that it could be integrated with LISP and we can develop a secure framework that can support session transfer resulting from mobility of clients and/or servers in a cloud computing environment. Dell has developed a virtual desktop solution called Mobile Clinical Computing (MCC; [14])—a cloud-based service aimed at providing full single sign-on and session mobility, and eliminating the need for local storage, thereby reducing the risk of data loss or security breaches.

27.4   Simulation Plan

We are currently in the process of implementing the proposed SMaaS framework as a prototype in a cloud environment setup using the Xen virtualization platform [15]. The client and server applications will be run on virtual machines instantiated in the cloud environment. Our simulation plan is to let the client to download a huge GB file that is replicated across three servers in a cloud of virtual machine servers hosted as an overlay network in the Computer Networks and Systems Security (CNSS) lab at Jackson State University. Somewhere in the middle of the downloading process, a server that is currently serving the client notices variations in the round-trip-time in the acknowledgments sent by the client, and decides a session transfer to another server is needed to better service the client. As a result, the server hands off the session to the client by including the details of the file transfer in the SMaaS Ticket. The client contacts the other two servers for continuing the session, and selects the server that responded at the earliest. The client forwards the SMaaS Ticket to the chosen server and the downloading process would be continued.

We will evaluate the effectiveness of our SMaaS paradigm by measuring the total time taken for the entire downloading process spread over the three servers as well as the average inter-packet arrival time and the standard deviation of the same. For comparison purposes, as a control, we will run a simple file transfer session involving a single client–server pair, wherein the client has to endure downloading the entire file from the same server, without any session transfer, even in the presence of congestion. We will investigate different techniques that are available to simulate congestion in an overlay network and run both the simulation experiments (the SMaaS simulation involving session transfers across servers and the single client–server file download session) under identical congestion scenarios. We will also research on leveraging one or more features (like Live migration, Distributed virtual switching, Live memory snapshot and revert, etc.) of the Xen virtualization platform to enhance the proposed SMaaS framework.

27.5 Conclusions

The proposed Session Mobility as a Service (SMaaS) framework caters to the needs of the mobile computing as well as the cloud computing communities to securely and efficiently transfer an ongoing client–server session to another server, necessitated due to unavoidable factors like client mobility, surge in network traffic on a particular path, etc. A client running the SMaaS framework could uninterruptedly obtain the requested service (e.g., file download) from a different server, without requiring to start from scratch for every session handoff. The user is completely unaware of the session transfers. The constituent mechanisms of session initiation, session handoff, and session transfer are coordinated among the client, SMaaS Gateway Server and the Service Servers through a SMaaS Ticket (that contains all user and client validity information as well as the session state information) that can be encrypted and decrypted only by the Gateway Server and the Service Servers. The use of an encrypted ticket, IPSec SA, and other security features in the SMaaS framework ensures that there cannot be any spoofing-based denial of service attacks to disrupt the services rendered by the cloud servers to genuine clients.

The SMaaS framework requires a client to maintain only one active TCP connection at a time (with a server in the cloud) and hence the framework can be very much suitable for thin client devices (like Android phones) that do not have as much resources as a desktop or laptop computer. The SMaaS framework is a preferred solution for providing services that would require several message exchanges between a client and server (e.g., download of a huge .iso file) either in one or both directions, as well as for providing reliable secure services in environments wherein peer-to-peer services and parallel/mirror services are unavailable or not secure. The SMaaS framework is also portable to the new IPv6 standard as the use of IPSec Security Association is a core required component of its design.

References

1. Faynberg, I., Lu, H.-L., & Skuler, D. (2016). *Cloud computing: Business trends and technologies* (1st ed.). New York City: Wiley.
2. Mate, S., Chandra, U., & Curcio, I. D. D. (2007). Movable-multimedia: Session mobility in ubiquitous computing ecosystem. In *Proceedings of the 5th international conference on mobile and ubiquitous multimedia (# 8)*. Stanford: ACM.
3. Sohail, S., Jha, S. K., & Kanhere, S. S. (2006). QoS driven parallelization of resources to reduce file download delay. *IEEE Transactions on Parallel and Distributed Systems, 17*(10), 1204–1215.
4. Rodriguez, P., & Biersack, E. W. (2002). Dynamic parallel access to replicated content in the internet. *IEEE/ACM Transactions on Networking, 10*(4), 455–465.
5. Huang, W., Wu, C., Li, Z., & Lau, F. (2014). The performance and locality tradeoff in bittorrent-like file sharing systems. *Peer-to-Peer Networking and Applications, 7*(4), 469–484.
6. Yang, Z., Xing, Y., Chen, C., Xue, J., & Dai, Y. (2015). Understanding the performance of offline download in real P2P networks. *Peer-to-Peer Networking and Applications, 8*(6), 992–1007.
7. Menasche, D. S., Rocha, A. A. A., Li, B., Towsley, D., & Venkataramani, A. (2013). Content availability and bundling in swarming systems. *IEEE/ACM Transactions on Networking, 21*(2), 580–593.
8. Garman, J. (2003). *Kerberos: The definitive guide*. Sebastopol: O'Reilly Media.
9. Oki, E., Rojas-Cessa, R., Tatipamula, M., & Vogt, C. (2012). *Advanced internet protocols, services, and applications* (1st ed.). New York City: Wiley.
10. Barisch, M., Kogel, J., & Meier, S. (2009). A flexible framework for complete session mobility and its implementation. In *Proceedings of the 15th open European summer school and IFIP TC6.6 workshop on the internet of the future* (pp. 188–198). Barcelona: ACM.
11. Johansson, D. (2011). Session mobility in multimedia services enabled by the cloud and peer-to-peer paradigms. In *Proceedings of the 5th workshop on user mobility and vehicular networks* (pp. 770–776). Bonn: IEEE.
12. Shanmugalingam, S., Crespi, N., & Labrogere, P. (2010). User mobility in a web-based communication system. In *Proceedings of the 4th international conference on internet multimedia services architecture and application* (pp. 1–6). Bangalore: IEEE.
13. Raad, P., Colombo, G., Chi, D. P., Secci, S., Cianfrani, A., Gallard, P., et al. (2012). Demonstrating LISP-based virtual machine mobility for cloud networks. In *Proceedings of the 1st international conference on cloud networking* (pp. 200–202). Paris: IEEE.
14. Curran, K. (2014). *Recent advances in ambient intelligence and context-aware computing*. Hershey: IGI Global.
15. Binu, A., & Santhosh Kumar, G. (2011). Virtualization techniques: A methodical review of XEN and KVM. In *Proceedings of the 1st international conference on advances in computing and communications* (pp. 399–410). Kochi: Springer.

Part VII
Security Management

Chapter 28
Securing the Internet of Things: Best Practices for Deploying IoT Devices

Bryson R. Payne and Tamirat T. Abegaz

28.1 Introduction

The Internet of Things (IoT) is a catch-all term for the rapidly growing number of non-traditional Internet-enabled devices being added to traditional computer networks. Enterprises that originally installed IP (Internet Protocol) networks to connect employees' computers, printers, and servers to the worldwide web now host innumerable wireless devices, video surveillance, access control systems, automated lighting, smart thermostats, VoIP phones and video teleconferencing systems, paging/speaker systems, smart sensors and alarms, automated manufacturing and industrial control systems, Internet-connected vehicle fleets, RFID inventory scanners, touch-screen building directories and room signage, even programmable window shades. Almost any new (non-computer) physical device with an IP address can be considered part of the Internet of Things. Unfortunately, in many businesses, these devices may share the network with systems that process credit card transactions or other payments, human resources data, medical and/or insurance information, customer databases, research and development, and more.

As an example of another IoT-rich environment, university residence halls feature access-controlled doors that open with a swipe of a student's college ID smartcard. This same card can be used in connected vending machines to buy a soda down the hallway. And often, these share the same network as the on-demand cable TV and Internet-connected laundry machines that alert students when their clothes are dry, via an app delivered over the free wireless network in the building. These devices, along with standard alarm systems, security cameras, and energy-saving smart building controls, all send data streaming through the same network equipment and cabling originally designed simply to provide Internet access to

B.R. Payne (✉) • T.T. Abegaz
University of North Georgia, Dahlonega, GA, 30597, USA
e-mail: bryson.payne@ung.edu; tamirat.abegaz@ung.edu

© Springer International Publishing AG 2018
K. Daimi (ed.), *Computer and Network Security Essentials*,
DOI 10.1007/978-3-319-58424-9_28

493

students' computers in their rooms. And inside those rooms, we might find each student's smartphone and tablet, Bluetooth speakers, fitness-tracker watch, virtual reality gaming headset, handheld e-book reader, smart alarm clock, wireless video projector, or the toy drone or bot they brought from home. The same university network can easily house data centers and student health clinics, along with most of the enterprise IoT device types noted in the previous paragraph, financial aid data, community devices like weather stations and air quality sensors, cell phone towers or range extenders, and classified research systems funded by government grants.

Among home users, not everyone owns a smart refrigerator yet, but many homeowners have installed entryway video surveillance cameras, Amazon Echo/-Google Home or similar voice-enabled speakers, smart thermostats, automated door locks, baby monitors, Wi-Fi-enabled bathroom scales, remote-controllable electronic appliances, digital video recorders (DVRs) and app-enabled TVs, lawn sprinkler systems, and hybrid/electric car charging stations. Similarly, these everyday things may sit on the same Wi-Fi network as a medical device worn by a family member, connected to a single household router configured by the local Internet service provider (ISP) with a default username, password, and wireless network name. Add to this home network a few assorted smartphones, tablets, laptops, printers, video game consoles, and a desktop computer with all the family's financial information and tax returns, and an awareness of the importance of securing the Internet of Things begins to emerge.

This chapter presents the case for securely deploying and maintaining IoT devices by examining two major IoT security incidents, the late-2013 Target Corporation breach involving 40 million stolen bank card numbers, and the late-2016 Mirai botnet's record-breaking Internet service disruption spanning two continents. A survey of current best practices for securing computer networks follows, along with an overview of special challenges and considerations regarding IoT installation. A specific IoT cyber attack scenario is presented in detail, to further highlight both the application of common security best practices and the difficulty inherent in IoT utilization. Finally, we provide a robust framework for securely deploying IoT devices both in industry and within the home.

28.2 Background

In today's connected world, it is harder than ever for individuals and companies to protect their confidential information. In recent news, major data breaches make headlines on a regular basis. The International Standards Organization (ISO) defines a data breach to be a "compromise of security that leads to the accidental or unlawful destruction, loss, alteration, unauthorized disclosure of, or access to protected data transmitted, stored or otherwise processed" [9]. For the first time in modern history, people can't rely solely on law enforcement officials, government, or financial institutions to protect their confidential information from the hands of criminals. We live in an age in which individuals must think twice before exposing their personal data, and where hacking has become a multi-billion-dollar business spanning the globe [23, 26]. Similar to legitimate large-scale enterprises, hackers drive cyber attacks on an industry scale, operating at times with the support of rogue governments and worldwide underground criminal organizations [23, 26].

Hackers have become skilled at utilizing various tactics and exploit kits to evolve and conceal their attacks from security intrusion prevention and detection systems, as well as from law enforcement officials [2]. In today's world, cyber threats are unprecedented, unrelenting, and arguably existential [2]. Cyber threat actors have been successful in targeting and compromising companies and organizations that store critical personal information, trade secrets, intellectual property, and other sensitive information.

The fact that broad classes of cyber attacks have matured into organized criminal enterprises is only part of the motivation for taking network, computer, and IoT device security seriously. In addition to criminal hacks, we now have state-sponsored actors or advanced persistent threats from rogue and enemy government agencies, hacktivists, industrial espionage from competitor companies, random cyber vandalism, and insider threats from disgruntled workers or former employees. Threats to information security are manifold and require the investment of effort both in deploying and in maintaining devices on both corporate and home networks.

28.3 Major IoT Security Incidents

A direct analysis of two major IoT security incidents will provide further motivation to support development of a framework for safe IoT deployment both at work and at home. First, we will examine the 2013 Target Corporation hack, in which an IoT building control system (the heating, ventilation and air conditioning, or HVAC, controller) served as the launch pad for a breach that resulted in the loss of tens of millions of credit card numbers and customer transaction records. Second, we will analyze the record-breaking Mirai botnet's distributed denial-of-service (DDoS) attack on the domain name system (DNS) that interrupted Internet services on two continents in late 2016.

28.3.1 Target Corporation's IoT Breach

Target Corporation's December 2013 data breach exposed more than 40 million credit card and debit card numbers, and over 70 million customer records [8]. The hackers entered the network through a small, Linux-based computer that allowed a contractor to remotely monitor and control the heating, ventilation, and air conditioning (HVAC) system. It is worth noting here that small, Linux-based computers power millions of IoT devices, from webcams to DVRs and door locks, as well. The extent of the breach was due to the fact that Target didn't sufficiently insulate its point-of-sale (POS) and payment system network from other network components, like the HVAC system, via mechanisms such as virtual local area networks (VLANs), or by providing physically separate network infrastructure.

The breach impacted Target's earnings gravely, resulting in a 50% drop in profit for the following year, and an 11% drop in stock price. The breach resulted in the resignations of several top leaders including the CEO. It also cost financial institutions over $200 million for card replacement [8]. Apart from Target, 2013 data breaches cost an estimated $400 billion to the global economy [21]. In addition, by 2019, the global economy is forecasted to lose more than $2 trillion to breaches and cybercrime [15].

28.3.2   The Mirai IoT Botnet

On October 21, 2016, an estimated 100,000 compromised IoT devices launched a record-breaking 1.2 Tbps (1.2 Terabits per second) of malicious traffic attacking a key component of Internet infrastructure, the domain name system (DNS) [11, 12, 27]. The hack was the largest distributed denial-of-service (DDoS) attack in history, disrupting service for users of popular web sites like Twitter, Amazon, PayPal, and Netflix on two continents for most of a day [11, 12].

The Mirai botnet's source code was released just a month before the attack, revealing 62 "horribly insecure default passwords, starting with the infamous admin:admin" [14] used to gain control of hundreds of thousands of endpoints that had been installed without changing the default usernames or passwords. Just 62 hard-coded username and password pairs were needed to compromise as many as 1.5 million IoT devices, primarily digital video recorders (DVRs)— more than 80% of the infected bots were DVRs as of October 2016 according to Level3 Communications—along with IP cameras and similar Linux-based devices [5, 6, 13].

The Mirai Botnet was responsible for two previous record-setting attacks over the preceding two months, and while these attacks used only around 100,000 IoT devices, various hackers have begun offering the botnet as a service known as DDoS for hire with as many as 400,000 infected hosts [13, 25], meaning that a malicious actor with sufficient funding could turn four times the power of this massive IoT botnet against a company, service provider, or continent of their choosing. In fact, a variant of the Mirai code resulted in several days of Internet service loss to more than 900,000 customers of German ISP Deutsche Telekom in late November 2016 [11, 12] by adapting the Mirai code to attack certain vulnerable home and small business Internet routers from two manufacturers used by the German ISP. Just days later, in December 2016, another cybercriminal claimed to have taken control of 3.2 million home Internet routers by exploiting a similar flaw across dozens of brands and models worldwide [5, 6].

The sheer numbers of these infected devices should be sobering, but take into account estimates from Gartner [7] that over 6.4 billion connected "things" were in use in 2016, and the case to take action becomes even more dire. Further, Gartner predicts 20.8 billion devices (excluding smartphones, tablets, and computers) will be connected to the Internet by 2020 [7]. If the largest Internet disruption in history used perhaps as few as 100,000 IoT devices, what would an attack across 20 billion devices look like?

28.4   General Security Best Practices

The proliferation of IoT devices creates a unique security challenge to our society [7]. Although it is arguably impossible to prevent every cyber intrusion, the following are some of the present best-practice mitigation strategies to prevent most attacks from being successful: (1) Network management, (2) Device management, (3) Patch management, (4) Anti-malware applications, (5) Identity management, (6) Access control and data protection, and (7) User education.

With the implementation of IPv6, the scalability of IoT networks is practically limitless (2^{128} addresses possible). Implementing network security though network segmentation and segregation is extremely important to reduce direct access to IoT devices from the Internet. Techniques such as Network Address Translation (NAT), establishing Demilitarized Zones (DMZs), or utilizing virtual private networks (VPNs) are important [11, 12]. However, although network isolation is useful to reduce IoT attacks, many IoT devices can use other methods, including UPnP (Universal plug & play), Bluetooth, Zigbee, and other wireless and wired peer-to-peer (P2P) networks enabled by default, to establish connections and communicate with one another, spreading malware across cities like wildfire [19].

Regarding device management, it is important to treat each IoT device as a network endpoint, similar to smartphones and computers. IoT devices run their own firmware, and each device has an IP address and is capable of connecting to the Internet. Most devices have onboard computers and are capable of performing complicated tasks. Unless properly managed, the devices could easily become infected with malware and used by hackers for coordinated attacks. Therefore, understanding of the number, type, and capability of IoT devices on your network and managing them is crucial to reducing vulnerabilities.

Since IoT devices contain their own OS (operating system) and applications, implementing security patches both at the OS/firmware and application levels is extremely important. For desktop computers, PDF readers, web browsers, Microsoft Office, and Java software systems are among the top applications with serious vulnerabilities. An enterprise security report released by HP indicated that 68% of successful exploits in 2015 were patched more than two years prior [4]. The report indicated that timely patching alone could have blocked hackers from exploiting 68% of the vulnerabilities, saving companies' assets, and reputation. Similarly, the

Australian Signals Directorate (ASD) indicated that OS and application patches are among the top four mitigation strategies (application whitelisting, patch applications, patch operating system vulnerabilities, and restrict administrative privileges) that could prevent more than 85% of cyber attacks [1].

The importance of having an enterprise-base anti-malware application is evidenced by the Internet Security Threat Report [10] that Symantec releases annually. In one report, Symantec discovered more than 430 million new malware samples in 2015 alone, many targeting IoT devices. The report also indicted that the number of zero-day vulnerabilities increased by 125% from the year before. Both network-based, and where possible, endpoint-based anti-malware software, running all the time on all connected devices is a necessary defense to discover, remove, and report newly created malware.

Similarly, due to a large-scale growth in IoT devices joining the Internet, identity and access management becomes a complex task. IoT device manufactures generally create extremely simple generic accounts for devices and enable online management (usually web-based). Unfortunately, many manufacturers do not encourage customers to create their own username and password when configuring the device for the first time, as seen in the Mirai botnet's success. To further complicate matters, some IoT manufactures do not synchronize credentials for both web and command line (like telnet and/or SSH) interfaces. Therefore, it is important for IoT manufactures to encourage or enforce strong passwords, and make sure to synchronize credentials to reduce attacks using manufacturers' default logins.

Data protection using various advanced encryption mechanisms and utilizing role-based access controls are essential to managing and protecting sensitive information from unauthorized use. And, encryption should be standardized and strong (like AES-256, where possible) for data both at rest (on the device) and in transit (wirelessly or wired).

Finally, creating user awareness of common security threats, from phishing emails to bringing unauthorized IoT devices onto the network, is an essential component and should be a front line of defense in any security strategy.

28.5 IoT Challenges

A central issue in securing IoT devices is that the Internet of Things involves connecting new kinds of devices to "old" networks. The introduction to this chapter detailed a number of different IoT applications superimposed on IP networks previously established for communication between desktop computers and local servers and peripherals like printers. These networks were later connected to the Internet, primarily for web access. As building controls, video surveillance and wireless systems began to be added to legacy networks, however, new security concerns multiplied accordingly, above and beyond the vulnerabilities already found in traditional networks, servers, and desktops.

Some of the factors that make IoT devices an especially attractive target are insecure default passwords, lack of encryption, inexperience by users at segmenting networks, the lack of patches available or applied, and a set-it-and-forget-it mentality applied to the deployment of most IoT devices. As mentioned previously in the discussion of the Mirai botnet, most manufacturers use poor default passwords, which are then posted online in PDF user manuals for easy access by hackers intent on compromising your networks. Further, few users bother to change the default password when they install an IoT device—sometimes because the task is unnecessarily difficult due to the manufacturer's web management console [20].

Lack of encryption enables hackers to use network "sniffers" to capture IoT device commands and even user credentials [3, 24]. In one DEF CON 2016 presentation, 12 of 16 IoT locks tested, or 75%, were vulnerable to relatively simple hacks using under $200 of off-the-shelf hardware [3]. Further, a SchmooCon 2016 presentation revealed that only 6 out of 33 Z-Wave IoT devices tested (18%) used encryption by default, and only 9 of 33 (27%) supported encryption at all [24], meaning that many were susceptible to simple plaintext sniffing and replay attacks, as just one example vulnerability.

The best-practices section above listed network segmentation as an important strategy in defending networks from compromised IoT devices. However, most home and small business users have little to no knowledge of effective network segmentation techniques. VLANs are often used to segment mid-sized and larger enterprise networks, overlaying multiple VLANs (for example, one VLAN for VoIP phones, another for desktop computers and printers, and yet another for video surveillance cameras or keycard access doors) onto a single physical network, but small businesses and home users rarely have the equipment or the know-how to separate different types of devices from one another on the same network. This ability may become an important consideration for home network routers supplied by ISPs in the near future.

In addition, IoT security differs from standard computer security in many ways, not least of these is the basic human psychology involved. We're often reminded to change our desktop and email passwords, and prodded to update our operating system, browser, and desktop applications, but we rarely or never think to update the firmware on our Blu-Ray player, thermostat, smart TV, or front door security camera, let alone the broadband router that connects all these systems to the Internet.

This set-it-and-forget-it mentality may be further exacerbated by the fact that many IoT devices are relatively inexpensive—we tend to protect a $1500 MacBook more than a $25 doorbell webcam or a $40 smart coffee maker. However, as Schneier points out, computers and smartphones actually gain security from the fact that we replace them every couple of years—older phones and computers are often not exposed to ongoing threats due to their obsolescence, because we replace them frequently with newer versions [20]. This is not the case for most IoT devices, which may last for years or even decades. As Schneier notes, we may replace a DVR after five or ten years, a refrigerator after 25, and a thermostat "approximately never" [20]. Unfortunately, a neglected IoT webcam or smart building device could remain vulnerable for generations.

28.5.1 Cyber-Physical Attacks

A major point of distinction between traditional computing platforms and IoT devices is the potential for cyber-physical attacks: computer-based and network-based attacks that have a physical effect. Similar to supervisory control and data acquisition (SCADA) and industrial control systems (ICS), IoT devices are often connected to, controlling, or composed of physical systems, like programmable light bulbs, smart thermostats, automated door locks, toy motors, or other physical devices. In SCADA systems, a cyber-physical attack could mean opening a dam or shutting down a section of the power grid; in ICS, a cyber-physical attack could include stopping an assembly line, or releasing chemicals from a storage facility.

For IoT devices in a smart building, a cyber-physical attack could mean that an intruder can enter the building by tricking the smart lock on a door, or damage the air conditioning system and burn out certain types of light bulbs by repeatedly turning them on and off. In one case, hackers were able to destroy industrial florescent bulbs rated for 30,000 h of normal use (approximately 14 years of 9-to-5 business office use) in a single night, some in as little as four hours, by turning them on for one second and off for three seconds repeatedly [24].

In the old days, criminals had to break through windows or pry open doors with a crowbar, meaning they had to be physically present. But, the Internet makes a hacker, whether down the street or halfway around the world, just a ping and a few milliseconds away from our IP address, and often able to enter with little or no trace of evidence that were being attacked.

28.6 IoT Cyber Attack Proof of Concept

While newer cyber attacks aimed specifically at IoT devices are rapidly being developed, like the recently revealed proof-of-concept Thermostat Ransomware [5, 6], many older, traditional cyber attacks from popular hacker toolkits like Metasploit can be carried out on newer devices with little or no customization by the user. Unpatched operating systems, lax or missing encryption, and default passwords make simple replay attacks, brute force attacks, and other hacker staples easy to apply to the newest IoT tools and toys.

Shodan (shodan.io) and Censys (censys.io) are two well-known search engines for IoT devices. Unlike web search engines like Google and Bing that crawl for data on the web pages, IoT databases like Shodan query ports and grab the resulting TCP banners, revealing identifying information about the kind of device at that address [17]. To utilize Shodan's features effectively, a user simply needs to create an account, or use an existing social media account [22].

Both Shodan and Censys allow users to apply filters to narrow down the search results to a manageable level. Common filters include *after/before* (to limit the search result by date), *country, hostname, port, operating system*, and *net* (to limit search by specific IP range or subnet) [22].

Fig. 28.1 Search results for the query "webcam, country:us"

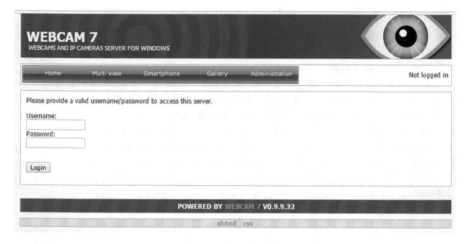

Fig. 28.2 The web management console login page for a particular IoT device

The screenshot in Fig. 28.1 displays a Shodan search result that contains banner information for IoT devices. For this demo, the search query "webcam, country:us" was used to display the results that are registered to US IP addresses that contain the word "webcam" in their banner. The search result displays over 1100 hosts, categorized by city, service, organization, operating system, and product type. More importantly, hyperlinked IP addresses allow the user to access the listed device with a single click. (IP addresses in the figure are blurred intentionally.)

Clicking a hyperlinked result will usually directly open the web management console login page for that device, as shown in Fig. 28.2. Unfortunately, most IoT devices keep the default username and password as discussed earlier [11, 12]. A quick Google search for "user manual pdf webcam 7" or similar yields the default manufacture's username and password with minimal effort. The potential to exploit some or most of the 1100 devices in the query took less than 1 min, and required no risk on our part. *Testing the default username and password, however, should only be done on devices you personally own or have explicit, written permission to access.* As an ethical analogy, consider the login screen in Fig. 28.2 to be a door that appears to be propped open—seeing the open door is not unethical. Attempting a username and password combination, however, is analogous to entering without permission, and is not only unethical, it is unlawful in most countries.

If the above scenario seems artificial or academic in any way, you can find your own IP address by asking Google "what is my IP address" from virtually any computer or smartphone, then search Shodan or Censys for IoT devices listed at similar addresses, or even at your own IP address. If you find devices at addresses that share all but the last few digits of your IP address, they may be located in your business or building, or simply at other customers of the same Internet service provider. If listed devices share your exact IP, they may either be devices currently in your home or business, or they may refer to devices hosted by other customers of your ISP from days ago or longer, because the IP address is dynamically served and changes from time to time. If, however, you find a device listed that you know is actually one of your own, change the default password and follow the other best practices given in the next section.

As bonus proofs-of-concept in attacking IoT devices, Nichols, Yang, and Yuan [16] offer a simple, straight-forward, yet elegant application of a series of Metasploit tools in tricking smart home devices into giving up usernames and passwords, gaining full control of IoT items in an easy, step-by-step fashion. Furthermore, security and hacking conferences alike, including the presentations described by Smith [24] and Franceschi-Bicchierai [5, 6], often include step-by-step instructions to enable white-hat penetration testers, as well as black-hat hackers, to test for vulnerabilities in the devices in your own home or business network.

28.7 Best Practices for Deploying IoT Devices

While IoT devices bring additional security concerns to networks, all hope is not lost. It is possible to deploy IoT in your home or business environment with security in mind to minimize the likelihood of IoT-related attacks. The following 10 best practices are presented as a framework for introducing IoT devices to your network more securely:

1. *Consider whether an IoT device is needed in the first place, and where IoT use is justified, plan for the cost of maintaining it.* A smart thermostat may save hundreds per year in heating and cooling costs, but factor in all of the following costs, as well: the value of the time and materials required to safely install the device and separate it from other important network devices; the annual maintenance time for updating, patching, and testing the device; and the risk posed to other devices and data in the same home or building as the device ages.

2. *Change the default username and/or password to make it difficult to guess, and do not reuse passwords across multiple types of devices.* After changing the password, test to make sure the default combination in the user's manual no longer works. If it does, the credentials may be hard-coded—return the device or select a different product. Also make sure any hardware reset mechanism is protected and not readily accessible on the device, as these may restore default manufacturer's passwords.

3. *Disable unnecessary services and ports on IoT devices.* Some IoT devices allow remote management both through a web interface *and* by command-line tools like Telnet or SSH. Disable these services at the device if possible (see the user's manual), or block ports 22, 23, 2222, and 2323 at the router or firewall. If a device only needs internal access within the building, block it entirely at the firewall.

4. *Segment IoT devices from the rest of your network.* In businesses, this may mean separate network cabling and switches for critical systems, or at least separate VLANs to contain threats to logical networks of similar devices. At home, this may take the form of a set of reserved IP or NAT addresses that you can limit and monitor from your broadband router's management console.

5. *Keep an inventory of IoT devices as you add them to your network.* Include information about the manufacturer, the date, the purpose of the device, and its location. Review your IoT document list annually as a homeowner, or more often as a business user, when updating or patching your devices, and remove devices that are no longer needed.

6. *Enable strong encryption (AES-128 or AES-256 if available).* If a device does not support encryption, consider purchasing a different device, or separating the device from the rest of your network. Once again, the value of the device is not the question, it is the value of all other assets on the same network as a potentially hacked device.

7. *Patch devices regularly.* For home users, annually is a minimum rule of thumb; for business users, check for updates more often based on the value of other assets on your network. Check the manufacturer's site for updates, and when patches are no longer available, consider replacing the device, again depending on the value of other assets on the network.

8. *Turn off or disable devices when not in use (at night or after hours).* Parents of teenagers may already employ this advice by turning off Internet access at the router after a bedtime or curfew hour; business users would do well to follow the same example. Consider which devices could be turned off after business hours, during vacations, and the like, including desktop computers, Wi-Fi routers, DVRs, and more. For home users, plugging your main router into an outlet with an automatic timer, or connecting certain devices like smart TVs, game consoles, and DVRs to a power strip with an on/off switch may suffice.

9. *Train users.* At work, regular security training (every six months, if possible) should include phishing awareness, applying software and OS updates, avoiding unauthorized devices including IoT, and using strong passwords or

passphrases as part of a well-rounded security awareness program. At home, make sure all family members talk about similar dangers, and that they understand the kinds of devices being used, and how they might be misused by an attacker over the Internet.

10. *Prioritize devices, and protect all endpoints appropriately.* Protect every device on your network, even "old" computers, especially when IoT devices are in use. Employ firewalls and antivirus/anti-malware software on all servers, desktops, and laptops, and apply appropriate security configuration on tablets and smartphones (strong passcodes, two-step authentication, disable automatic Wi-Fi connections, be selective when installing apps), in addition to the IoT-specific protections described above.

Perhaps equally important to implementing the above 10 best practices is choosing IoT vendors who follow best practices, as well. The OWASP project's Manufacturer IoT Security Guidance project [18] details security recommendations and considerations to greatly improve the security of IoT products.

28.8 Conclusion

The Internet of Things changes the security landscape of our networks and impacts our daily lives substantially. By applying traditional security best practices, adapted and updated for IoT, we can deploy valuable new devices to our networks more securely and responsibly.

Thoughtfully considering the need for, and full cost of, IoT device deployment and maintenance is the first step. Changing default passwords and disabling unneeded services and ports are important for IoT, just like they are for desktops and other computing equipment. Separating IoT devices from other resources on your network and inventorying IoT devices as they are added are important next steps. Enabling encryption, patching devices regularly, and turning off devices when not in use round out the IoT-specific recommendations. Last but not least, training users and applying appropriate protection across all network endpoints will maximize the security of your entire home or business.

It is worth noting that it is especially important to emphasize security at setup or deployment, because IoT devices aren't likely to be updated or maintained as often as standard desktop or laptop computers. Furthermore, unlike smartphones or traditional desktop computers, the useful life could span decades for various categories of IoT, like Internet-enabled TVs and DVRs, security cameras, smart thermostats, and building automation devices.

Finally, while not all IoT may be equal, and we should prioritize devices based on their type and use, we must remember that the cost of a seemingly disposable IoT device should not be the main consideration in the level of security we apply when deploying the device, but rather the *total value of all systems*, *resources*, and *data* both *on* and *connected to* our networks. If a $25 security webcam sits on the same network as your financial and health records, or if a DVR has Internet access over a Gigabit fiber connection making it capable of spewing network-crippling DDoS attacks against other homes, businesses, governments, or continents, we have a moral and ethical imperative to apply all appropriate security controls to protect ourselves, our families, coworkers, and humanity from rogue or compromised IoT devices on our networks.

Information security requires equal parts of optimism and paranoia. We can make our networks more secure by following relatively simple best practices, but the vulnerabilities are more complex and the threats more destructive than ever. Ethical security professionals everywhere can and must configure and maintain IoT devices responsibly, and the framework presented in this chapter can be the starting point for safe IoT deployment and use.

Acknowledgments This work was supported in part by National Security Agency and National Science Foundation GenCyber grant project #H98230-16-1-0262.

The authors also wish to thank colleagues and security experts Rob Cherveny and Dr. Markus Hitz for thoughtful input and feedback throughout this chapter.

References

1. ASD Australian Signals Directorate. (2014). *Strategies to mitigate targeted cyber intrusions*. http://www.asd.gov.au/infosec/top-mitigations/top-4-strategies-explained.htm. Accessed 5 December 2016.
2. Charney, S. (2010). *Collective defense: Applying public health models to the Internet*. White paper. Redmond, Wash: Microsoft Corporation. http://www.microsoft.com/security/internethealth. Accessed 30 December 2016.
3. Coldewey, D. (2016). 'Smart' locks yield to simple hacker tricks. *TechCrunch*. https://techcrunch.com/2016/08/08/smart-locks-yield-to-simple-hacker-tricks/. Accessed 8 January 2017.
4. Cyber Risk Report. (2016). *HPE security research*. https://www.thehaguesecuritydelta.com/media/com_hsd/report/57/document/4aa6-3786enw.pdf. Accessed 5 January 2017.
5. Franceschi-Bicchierai, L. (2016a). Hacker claims to push malicious firmware update to 3.2 million home routers. Motherboard.com. http://motherboard.vice.com/read/hacker-claims-to-push-malicious-firmware-update-to-32-million-home-routers. Accessed 3 January 2017.
6. Franceschi-Bicchierai, L. (2016b). Hackers make the first-ever ransomware for smart thermostats. Motherboard.com. http://motherboard.vice.com/read/internet-of-things-ransomware-smart-thermostat. Accessed 2 January 2017.
7. Gartner. (2015). *Gartner says 6.4 billion connected "things" will be in use in 2016, Up 30 Percent from 2015*. http://www.gartner.com/newsroom/id/3165317. Accessed 3 January 2017.
8. Greene, C., Stavins, J. (2016). *Did the target data breach change consumer assessments of payment card security?* (Research Data Reports No. 16-1). Federal Reserve Bank of Boston.
9. ISO. (2015). *IT-security techniques-storage security (ISO/IEC Standard No. 27040)*. Retrieved from https://www.iso.org/obp/ui/#iso:std:iso-iec:27040
10. ISTR: Internet Security Threat Report. (2016). https://www.symantec.com/content/dam/symantec/docs/reports/istr-21-2016-en.pdf. Accessed 5 January 2017.
11. Krebs, B. (2016a). *Hacked cameras, DVRs powered today's massive internet outage*. https://krebsonsecurity.com/2016/10/hacked-cameras-dvrs-powered-todays-massive-internet-outage/. Accessed 20 December 2016.

12. Krebs, B. (2016b). *Who makes the IoT things under attack.* http://krebsonsecurity.com/2016/10/who-makes-the-iot-things-under-attack/. Accessed 3 January 2017.
13. Level 3 Research Labs. (2016). *How the grinch stole IoT.* http://blog.level3.com/security/grinch-stole-iot/. Accessed 2 January 2017.
14. MalwareTech. (2016). *Mapping mirai: A botnet case study.* https://www.malwaretech.com/2016/10/mapping-mirai-a-botnet-case-study.html. Accessed 31 December 2016.
15. Morgan, S. (2016). *Cyber crime costs projected to reach $2 trillion by 2019.* http://www.forbes.com/sites/stevemorgan/2016/01/17/cyber-crime-costs-projected-to-reach-2-trillion-by-2019/#216e8d33bb0c. Accessed January 09 2017.
16. Nichols, O., Yang, L., & Yuan, X. (2016, October 4). Teaching security of internet of things in using raspberry Pi. In *KSU conference on cybersecurity education, research and practice*.
17. O'Harrow, Jr. R. (2012, June 3). Cyber search engine Shodan exposes industrial control systems to new risks. *The Washington Post, 6.*
18. OWASP (2016). Manufacturer IoT security guidance. *Open web application security project.* https://www.owasp.org/index.php/IoT_Security_Guidance. Accessed 5 January 2017.
19. Pauli, D. (2016). IoT worm can hack Philips Hue lightbulbs, spread across cities. *The Register.* http://www.theregister.co.uk/2016/11/10/iot_worm_can_hack_philips_hue_lightbulbs_spread_across_cities/. Accessed 5 January 2017.
20. Schneier, B. (2016). We need to save the internet from the internet of things. *Motherboard.* https://motherboard.vice.com/read/we-need-to-save-the-internet-from-the-internet-of-things. Accessed 7 January 2017.
21. Shields, K. (2015). Cybersecurity: Recognizing the risk and protecting against attacks. *North Carolina Banking Institute, 19,* 345.
22. Simon, K. (2016, November 14). Vulnerability analysis using google and shodan. In *International conference on cryptology and network security* (pp. 725–730). Springer International Publishing.
23. Slay, J., & Miller, M. (2007, March 19). Lessons learned from the maroochy water breach. In *Conference on critical infrastructure protection* (pp. 73–82). New York: Springer.
24. Smith, M. (2016). EZ-Wave: A Z-Wave hacking tool capable of breaking bulbs, abusing Z-Wave devices. *Network World.* http://www.networkworld.com/article/3024217/security/ez-wave-z-wave-hacking-tool-capable-of-breaking-bulbs-and-abusing-z-wave-devices.html. Accessed 8 January 2017.
25. Vernon, P. (2016). The Mirai botnet: what it is, what it has done, and how to find out if you're part of it. HackRead.com. https://www.hackread.com/mirai-botnet-ddos-attacks-brief/. Accessed 2 January 2017.
26. Wheatley, S., Maillart, T., & Sornette, D. (2016). The extreme risk of personal data breaches and the erosion of privacy. *The European Physical Journal B, 89,* 1–2.
27. Woolf, N. (2016). DDoS attack that disrupted internet was largest of its kind in history, experts say. *The Guardian.* https://www.theguardian.com/technology/2016/oct/26/ddos-attack-dyn-mirai-botnet. Accessed 20 December 2016.

Chapter 29
Cognitive Computing and Multiscale Analysis for Cyber Security

Sana Siddiqui, Muhammad Salman Khan, and Ken Ferens

29.1 Cyber-Threat Landscape

The pervasive role and contribution of the Internet in our daily lives have made us more susceptible to cyber-attacks. From malware, web based attacks, botnets, phishing to ransomware, and cyber-espionage, the field of cyber-threats is multi-faceted. In 2015 alone, a zero-day vulnerability was found every week on average and nine breaches with over 10 million records per breach were reported. Also, more than one million users became the target of web attacks daily while in general, a large organization observed at least 3.6 successful average attacks; and the deadly ransomware increased by 35% from the previous year [1]. Moreover, malware which has consecutively topped the list of cyber-threats within past few years has observed an increase of one million samples per day [2]. Also, in 2016 alone, average 58 thousand new malicious URLs were found every day [2] which are major source of web based attacks. Further, an extreme volumetric surge in denial of service (DoS) attacks was observed reaching bandwidths of 100 Gbps [2]. Exploit kits (e.g., rootkits) based malicious activities have increased by almost 67% [2]. Overall, there was a reported increase of 64% in security related incidents [3] with varying motives including but not limited to financial gain, physical harm, intellectual property theft, and political damage. It is worth-noting that healthcare was the most frequently targeted industry in addition to government, financial services, educational organization, and manufacturing industries [3, 4]. Also, the biggest risks for any organization remain the insider-threat which was the source of

S. Siddiqui (✉) • M.S. Khan • K. Ferens
Electrical and Computer Engineering, University of Manitoba, Winnipeg, MB, Canada
e-mail: siddiqu5@myumanitoba.ca; muhammadsalman.khan@umanitoba.ca; ken.ferens@umanitoba.ca

© Springer International Publishing AG 2018   507
K. Daimi (ed.), *Computer and Network Security Essentials*,
DOI 10.1007/978-3-319-58424-9_29

60% of all cyber-attacks either in the form of direct adversary or as an inadvertent malicious actor. Therefore, change is the only constant in cyber-threat landscape which is expanding tremendously in terms of its complexity, velocity, magnitude, and targets [2, 4].

Adversaries are constantly innovating, refining, and advancing techniques to develop intelligent obfuscation methods that aim to outsmart state-of-the-art defenses, thus remaining steps ahead of the latest data orchestration and security methods employed to detect malicious activities. Threat actors have improved their attack tactics, techniques, and procedures; they can exploit social media to fulfill their malevolent objectives of scamming, re-direction, and downloading of malicious executables; they can exploit network protocols to launch web exploits without any user interaction; they can generate multiple waves of attacks including but not limited to launching various denial of service attacks to divert the attention of security professionals and gaining access to vulnerable and unattended network resources; they can piggy back on phishing attacks to register malicious domain or infect legitimate servers; they can abuse network protocols and human behavior to steal confidential information, etc. [2]. Thus, cyber-warfare is multi-dimensional, not only in getting unprecedented success, but also in concealing its direct source and involved resources. Additionally, a continuous morphing and divergence in attack tactics have been observed. One such example is the introduction of bi-faced malware, which behaves as normal or benign under surveillance, but transforms into malicious code when scrutiny is removed. Another tactic to evade detection is the application of ghost-ware, which erases its attack-indicators, thus making it extremely difficult for forensic experts to track the extent of damage caused by the attack [5]. Therefore, it can be concluded that threat-agents have outthought and outpaced security leaders, experts, and researchers by deceiving even state-of-the-art protection measures.

Proactive development of policy and technological strategies is required to understand and detect the cyber-attack landscape, which has evolved by the following two extreme approaches. The first methodology used by the threat actors to ensure effectiveness and robustness is the use of simplified tactics. This is attained by maneuvering low-tech procedures to device reliable infection methods. Simultaneously, complexity is introduced in threats by using advanced morphing techniques to generate intelligent attacks [2]. The fusion of complex and simple methods have successfully produced threats which are extremely difficult to detect using contemporary security methods. Therefore, there is a need to perform research and development to introduce, apply, and further improve cognition based computationally intelligent systems, which take into account complexity measures for the detection of advanced anomalies. Such an approach will be effective for modern cyber-threats.

29.2   Cognitive Computing, Complexity, and Complexity Measures

Cognitive computing is an emerging research field capable of outperforming artificial intelligence based state-of-the-art anomaly detection systems. It consists of tools and techniques that mimic or follow human cognition abilities including but not limited to decision making, reasoning, comprehension, perception, inference, and the human thought process. It constitutes probabilistic and statistical knowledge; theories of signal processing and dynamical systems; machine learning; natural language processing, and different aspects of multiscale analysis. The basic research motivation in this area is to improve traditional learning algorithms by mimicking human brain and introducing models, which not only learn through historical data (experience), but also improve as a result of stimulus-response mechanism, which adapts with the changing environment and context (evolution). On the one hand, this is achieved by utilizing the concepts of complexity, which is defined as an important aspect of human's cognitive operations based on many interactions among the system's components which cannot be simplified further [6]. On the other hand, cognitive operations that involve fewer and limited relationships among components are considered simple [6]. So, ease in comprehending refers to simplicity in terms of human cognition, whereas difficulty in doing so is termed as complexity. It is note-worthy to understand that the simplest level of complexity cannot be further decomposed into simpler sub-systems without actually causing a damage to the system's dynamics [7]. Therefore, these notions are utilized to achieve tractability and robustness in the cognition process [8].

Complexity has been defined in the literature from different aspects. The author in [9] has categorized objects, systems, and structures in two classes based on their interactions among themselves; mainly whether (1) they contain any type of order, which refers to the presence of a definite pattern or, (2) they depict complete randomness, which indicate a lack of pattern and order. These patterns can be further classified as simple or complex. If the observed patterns involve interaction with fewer components, then it is considered as simple. Alternatively, complex systems have many components each of which interact with others in a variety of ways and may contain some form of order or hierarchy among themselves which further leads to the emergence of new patterns or behaviors. Similarly, from the perspective of dynamical system theory, the aggregate interactions in complex system give rise to the evolutionary behavior which is responsible for new and unknown patterns, and this system cannot be further decomposed into independent components [10]. In the context of cognitive computing, complexity is regarded as (1) static due to the fixed limited patterns in the system or object, (2) dynamic which refers the temporal behavior, (3) functional which is attributed to the system's functional components, (4) organization which refers the degree of interaction among different components, and (5) design which stems from the structural properties of the underlying system [11]. The author in [11] has discussed at length different complexity measures based on various system models. It can be categorized based on the scope, scale, statistics, and fractality. Some of these include (1) mono-scale or multiscale, (2) algorithmic or probabilistic, (3) local or global, (4) average or asymptotic, (5) arithmetic or logical, (6) absolute or differential.

In this chapter, the authors will explain the utilization of scale based complexity measures in computationally intelligent systems to detect advance threats. To elucidate further, single scale (mono-scale) based learning algorithms use Euclidean measure of complexity at single scale to minimize learning error between the target and the observed samples, whereas, multiscale based intelligence systems improve the cognitive perception by providing detailed insight of the relationships at different scales. Thus, the latter takes into account both the local and the global changes in the system revealing the complexity of a system.

29.3   Cyber-Threats and Need for Complexity Analysis

In the field of cyber security, the massive evolution of different characteristics of cyber-attacks have posed researchers with an unprecedented challenge of extracting unique set of features from available data attributes, which are able to cope with the dynamic and incessantly evolving nature of cyber-threats. Moreover, it is getting extremely difficult to determine a unique feature set or algorithms which can be used to detect a variety of cyber-attacks, simultaneously. The primary rationale for this problem is attributed towards the intelligence introduced by adversaries in the latest threats, which can not only morph in real time, but they are also multifaceted. Traditional cyber security measures like firewalls, antivirus systems, and encryption techniques rely heavily on pre-known signatures for the detection of cyber-threats. However, these techniques are proving to be futile in the face of advance internet threats, which are able to change their signatures in merely 15 seconds [12].

Increasingly, machine learning based approaches for the behavioral analysis of anomalous and legitimate traffic are being utilized. These techniques are based on determining the deviation of an observable object from a defined normal behavior, and further analysis is required to evaluate the deviated object as legitimate or false anomaly. Therefore, one of the limitations of machine learning is to know behavior of interest a-priori that can be further used to find deviation of objects from that known behavior. In addition, human intelligence, expertise, and feedback are also required to analyze the alerts produced by these machine learning engines to identify false positives/negatives and to activate blocking mechanisms, when appropriate.

Further, these threat samples, which are often encrypted and persistent in nature, render obfuscation by masquerading the behavior of a normal or legitimate data sample. It implies that on a feature space, the malicious and benign samples lie on the same co-ordinates making it difficult to detect the anomaly. For instance, phishing emails are used to deliver a malicious payload to the end user. The content of these phishing emails is generated using known normal patterns of user behavior and thus are able to dupe them to open those emails which in turn deliver the malicious payload without letting the user know. Therefore, in this case, the positive and negative samples are inseparable on typical single scale feature space. Also, email application cannot be blocked due to their widespread use. Further, the advanced phishing emails contain web links which can only be detected by the intrusion detection systems if the signatures are available in their database. Unless we know the domain name, source IP, or the regular expression of the email content, we cannot detect phishing attacks. Thus, this has become a signature based detection problem, which requires a human expert to analyze the unique signatures and then update them manually. This example demonstrates that threats are able to successfully masquerade within normal flows of traffic, and they are able to change signatures so quickly that, if there was initially an actionable classification feature space, it is merely a matter of time that the feature space would become overlapped, and the data samples would become inseparable and indistinguishable, thus, making the feature space selection an NP-hard problem [13].

29.4 Inseparability Problem in Cyber Security Through Examples

In order to emphasize the role of complexity and dynamically changing threat landscape, Figs. 29.1–29.3 provide a visual interpretation of how malware detection using traditional machine intelligence and learning mechanisms is becoming difficult. As shown in Fig. 29.1, a two dimensional feature space is depicted where blue cross (**x**) represents normal network traffic and red dot (**o**) represents attack traffic. As can be seen, it is a linear classification feature space and any properly configured machine learning system can easily classify these samples with sufficient reliability and accuracy.

However, as shown in Fig. 29.2, the feature space depicts that the attack and normal packets are not separated across a linear classification boundary but still occupy discrete space which separates attack and normal samples. The boundary

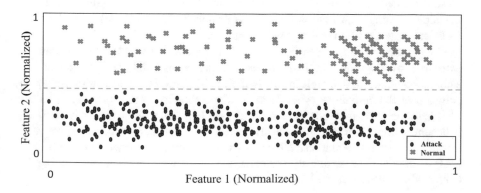

Fig. 29.1 Malware samples having discrete (non-overlapping) and linear classification boundary

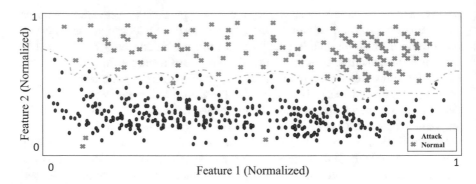

Fig. 29.2 Samples of malware variant showing discrete (non-overlapping) and highly nonlinear classification boundary

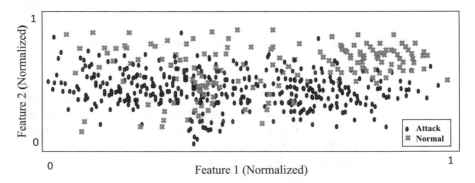

Fig. 29.3 Samples of malware variant showing overlapping classification feature space

is highly nonlinear and classification for such data set is difficult to achieve using existing Euclidean metric based machine learning systems which tend to suffer from under or over generalization issues. Available literature discusses in detail about these 2 classification cases and the efforts are concentrated in finding a suitable machine learning algorithm that can classify non-linearly separable samples with sufficient generalization and regularization [14].

On the other hand, it can be observed in Fig. 29.3 that the feature space is extremely overlapping. Both the attack and the normal data samples lie on the same co-ordinate and therefore are indistinguishable. These scenarios are quite common in the domain of cyber security where threats try to mimic the behavior of a legitimate internet flow and are successful in concealing themselves. Thus, it becomes imperative for the security researchers to consider the cognitive relationship between different data attributes by introducing the concepts of multiscale analysis in feature space and algorithms as well. This will help in understanding the details which otherwise are hidden over single scale.

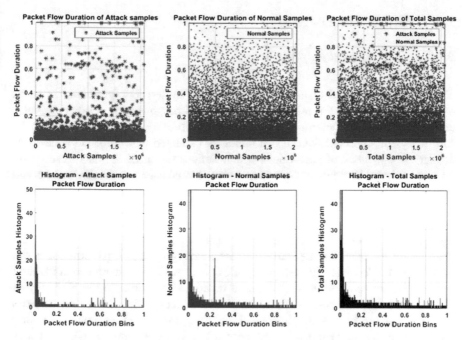

Fig. 29.4 Attack samples, normal samples, and total samples distribution for packet flow duration [34]

One such data set in the field of cyber security which depicts the inseparability characteristic is the UNSW-NB-15 data set captured and processed in the year 2014 at Cyber Range Lab of the Australian Centre for Cyber Security (ACCS) [15]. The authors of this chapter performed HTTP flow based analysis for this data set as it is one of the widely exploited protocols to launch attacks due to its wide range application in internet applications. Out of 49 different attributes and features listed in the data set, the authors of this chapter assessed only four features which include duration, round trip time, total packet inter-arrival time, and total number of bytes exchanged between the source and destination. It is worthwhile to mention that all of the selected features followed heavy tailed distribution to evade detection methods. It was found that the distribution of the attack samples was indistinguishable from that of benign data samples which is due to the masquerading behavior of threats which are following the patterns of normal data. Figure 29.4 [34] can be referred to get a pictorial concept of data distribution for both normal and attack sample for packet flow duration. As evident, it is impossible to classify this data using any linear or non-linear learning algorithm. This is a challenging situation in the context of machine learning although fairly normal for state-of-the-art attacks. Therefore, running a Euclidean measure based neural network which expects at least symmetrical distribution is not feasible [16]. Hence, complexity based cognitive capabilities are needed to be introduced in learning algorithm which are discussed in the next section.

29.5   Multiscale Analysis

29.5.1   Significance of Scale

The quantifiable extent of an object, process, or system is represented in terms of a scale. It signifies the relative fragileness and roughness of various attributes and features. It also determines the resultant selectivity of patterns that can be formed by the observed data. It can therefore be stated that the scale behaves as a filter, or a window of perceptions which defines the basis for data observation and its analysis. The larger the window size, the coarser the details will be and vice versa. Moreover, scale is responsible for setting the bounds for the analysis of an observation thus expressing the extent to which a data set can be modelled and analyzed consequently. For example, in the field of cyber security, a set of network related alerts produced by a learning algorithm can be extended to incorporate host related alerts. It can be scaled to assess emerging attack patterns, or it can be further divided into different categories of attacks to find related mitigation vectors. The relationship among different details with respect to the hierarchy of scales determines the complexity of an object. It depicts the emergence between structures and processes through interactions across a variety of spatial and temporal scales. For instance, the change in the behavior of a system under a cyber-attack is influenced by interactions at finer scales of total flow duration and the total amount of bytes exchanged. Typically, a malware tends to stay low and slow to evade detection which implies that only finer scales can reveal this type of interactions which otherwise cannot be observed on coarser scale levels. The concept of scale is useful for the modelling of an observation by scaling the observation at different levels or it can be considered as the characteristics of a process itself and thus aiding in the system analysis. For example, packet inter-arrival time and total packets exchanged in a flow are two features which can be observed at multiple scales. Moving from the coarser to the finer scales, the data samples reveal the delicate differences in benign and malicious samples of each attribute which help in modelling them. Otherwise, over single scale analysis, the distribution of the attack samples is indistinguishable from the distribution of normal data samples as seen in Fig. 29.4. This is due to the fact that attack samples follow the same behavior as of normal samples. This is natural because threat actors try to hide their existence using the same characteristics as those of normal flows. Moreover, this information can be used to deduce inference about the malware detection process itself which is based on cross-scale interactions.

29.5.2   Single Scale vs Multiscale

Traditional single scale based analysis of a system or process involves measuring the nature of a system at a global or macro scale. In contrast, multiscale analysis takes into account micro level details from various localized temporal, spatial, and regularity scales. Multiscale analysis combines information within and across different scales of a set of data features. The primary importance of this approach is attributed towards the characteristics of finer scales, which provide granular level details, while coarser scale lacks in providing the sufficient details required for the substantial analysis and modelling of a complex system. Multiscale analysis is computationally more expensive than single scale analysis; however, in the domain of cyber security, the former is critical for the successful classification of advanced threats, whose features overlap each other in feature space, and are, therefore, non-classifiable on a single scale using any linear or non-linear learning algorithm. In addition, multiscale analysis is significant for the measurement of the nature of internet data, which is known to be fractal in nature that is characterized by having the scaling property, thus requiring analysis on multiple levels to describe its complexity, faithfully.

29.5.3   Multiscale Analysis Using Fractals

Fractals are characterized as objects which are non-differentiable everywhere. They have unique and fascinating scaling properties such that the portion of an entire object like a geometrical figure, a process, or a time series is a scaled down version of the whole. In other words, fractals are invariant to magnification along multiple scales which can be either symmetrical or asymmetrical and hence, are referred to as self-similar or self-affine, respectively. Formally, Mandelbrot [17] who introduced the subject of Fractal Theory defined it as "a set for which the Hausdorff-Besicovitch dimension strictly exceeds the topological dimension." Another similar description by Robert L. Devaney involves the self-similarity property of an object in n-dimension Euclidean space such that its fractal dimension exceeds its topological dimension. Therefore, the notion of fractal dimension and self-similarity constitutes the basic building blocks in the analysis of fractals.

Typically, Euclidean dimension signifies the required number of co-ordinates to embed the object in Euclidean space while topological dimension indicates the inherent topology or form of an object under any transformation. To elaborate, consider a line which has a topological dimension of 1 but its embedding dimension is based on the space in which it is present. So, a line in 1D has an embedding dimension of 1, in 2D it is 2, and so on. Contrary to the integer dimensions associated with the Platonic objects, fractals have a non-integer dimension which is a measure of the complexity of these structures and hence, is a constant. For example, the fractal dimension of the coastline of Britain is 1.24 compared to its topological dimension of 1. This represents that the coastline of Britain is not a smooth straight line rather, it is irregular and complex. Moreover, higher the complexity of an object, greater will be the fractal dimension [18].

Fractal dimension computation involves finding the exponent of similarity for an object at different magnification scales. For a regular object like a box, a circle, or a line, this exponent corresponds to the topological dimension but exceeds it when the object has a lack of smoothness or is irregular [19]. It can be calculated by taking the logarithmic quotient between a measure of object N_k at a given measurement scale $1/r$, and the measurement scale itself when the scale tends to zero. These measures can be deterministic e.g. morphology or probabilistic e.g. correlation, information and etc. Mathematically, it is expressed as [20],

$$D_s = \lim_{r_k \to 0} \frac{\log (N_k)}{\log \left(\frac{1}{r_k}\right)} \tag{29.1}$$

The significance of multiscale measure is depicted through the use of subscript k which indicates that the measurements are taken at various scales. Fractals provide a way of measuring complexity of objects using multiscale analysis and are mathematically elegant in the sense that fractal dimensions are always embedded within the topological dimensions of the multiscale object under study. As it is already established that finding a feature set where data can be uniquely classified for threats is becoming difficult due to advancement in threat landscape, therefore, fractal analysis is a promising candidate in detecting advanced threats. This is because fractals are able to find a long term correlation based relationship among various scales of an object and thus provide a unique way to analyze patterns of an attack which is both mathematically tractable and convenient algorithms are available for implementation. For example, the authors in [21] illustrated the fractal dimension of the DNS time series in which attack pattern clearly follows a change in fractal dimension and hence complexity of the time series. Also, it is possible for humans to visually analyze this pattern but mono-scale machine learning tools are not able to capture this trend efficiently. Another approach of multiscale analysis is adopted by researchers in [22–24] where self-constructing and re-organizing neural networks having fractal structures have been discussed. This particular property has resulted in better learning capabilities compared to the traditional neural networks as it is based on the connection in biological systems. Further, a unique multiscale based k-Nearest Neighbor algorithm was proposed in [25] which was able to detect latest advanced persistent threats. Hence, fractal based techniques and algorithms are proving to be more robust and reliable to perform tasks that require cognitive capabilities either because of the presence of multiscale distribution of information or highly complex data analysis requirements.

29.5.4 Multiscale Analysis Using Wavelets

The statistical non-stationarity in multiscale objects requires a window estimation technique to achieve temporal and spatial stationarity. Therefore, wavelet analysis is used as a tool because of its inherent ability to consider the space and scale effects simultaneously, and as a result, the stationarity issues are addressed adequately [26–28]. This is achieved by a choice of suitable wavelet. However, it can be noted that this idea is contrary to the Fourier transform, which works on the assumption of stationarity of the signal, and instead wavelet transform finds the frequency content as a function of time. The choice of wavelets as an analysis tool is based on its characteristics of being a mathematical microscope that describes the hierarchal distribution of singularities in fractal measures. Moreover, unlike global analysis tools like Fourier analysis which tend to mask the singularities, wavelets get rid of the possible regular behaviors thus, revealing the irregularities. Therefore, it is equally applicable for non-differentiable measures and functions like fractals [29].

Wavelet analysis has been extensively used in cyber security field related to biometric fingerprinting, digital image steganography, forensic, and cryptography [30]. The concept has been exploited in the context of anomaly detection systems for cyber-threats [31, 32]. For example, if the anomaly detection problem is modelled as a time series as shown by the authors in [21] for the DNS DDoS attack captured from PREDICT data set available at [33], then Wavelets based multiscale analysis is used to find correlation among various samples at different magnification levels which reveals a relationship that otherwise is not possible through a mono-scale analysis. It is important to note that in mono-scale analysis, security threats are either mapped over a 2 or 3 dimension feature space or are either analyzed using time series or through frequency spectrum separately. However, time series has various practical constraints in order to be mathematically valid including but not limited to statistical stationarity and the Nyquist sampling criterion, to name a few [20].

29.6   Conclusion

This chapter is focused on incrementally relating the abstract concept of cognitive complexity with anomaly detection methodologies in cyber security. This is achieved by first briefly reviewing the current threat landscape and then explaining the essence of cognitive computing and the required complexity analysis. Using examples from real-world data set, the need for cognition based detection methods has been discussed to cater the problem of class overlap and indistinguishable classification boundary over a feature space as observed in advanced cyber-attacks. Also, the role of scale has been argued from the perspective of single and multiscale analysis to extract finer details obfuscated by the threat actors. Two key approaches for multiscale analysis are discussed including fractals and wavelets which are proving to be much more robust and reliable. Concluding, this chapter serves as an introductory tutorial to develop deeper understanding of the multiscale analysis from the viewpoint of accurately modelling and analyzing real-world cyber security data using fractals and wavelets.

References

1. Wood, P., et al. (2016). *Internet security threat report*. Symantec Corporation.
2. Marinos, L., Belmonte, A., & Rekleitis, E. (2016). *ENISA threat landscape 2015*. Greece: The European Union Agency for Network and Information Security (ENISA).
3. Bradley, N. (2016). *Reviewing a year of serious data breaches, major attacks and new vulnerabilities*. IBM X-Force® Research.
4. Lee, N. (2016). *Exploits at the endpoint: SANS 2016 threat landscape survey*. SANS Institute.
5. Vijayan, J. (2016 Dec 19). *5 ways the cyber-threat landscape shifted in 2016, Dark Reading* [Online]. Available: http://www.darkreading.com
6. Rauterberg, M. (1992). A method of a quantitative measurement of cognitive complexity. In *proceedings of the 6th European conference on cognitive ergonomics*, ECCE'92.
7. Bennet, C. H. (2003). *How to define complexity in physics, and why* (Vol. 8, pp. 34–47). Oxford: Oxford University Press.
8. Brasil, L. M., Azevedo, F. M. de, Barreto, J. M., & Noirhomme-Fraiture, M. (1998). Complexity and cognitive computing. In *proceeding of 11th international conference on industrial and engineering applications of artificial intelligence and expert systems*.
9. Kinsner, W. (2008). Complexity and its measures in cognitive and other complex systems. In *Proceedings of the IEEE international conference on cognitive informatics and cognitive computing*.
10. Edmonds, B. (1999). *Syntactic measures of complexity*. Dissertation, University of Manchester, Manchester, UK.
11. Kinsner, W. (2010). System complexity and its measures: How complex is complex. *Advances in Cognitive Informatics and Cognitive Computing Studies in Computational Intelligence, 323*, 265–295.
12. Belcher, P. (2016). *Hash factory: New cerber ransomware morphs every 15 seconds* [Online]. Available: https://www.invincea.com
13. Virendra, M., Duan, Q., & Upadhyaya, S. (2012). Detecting cheating aggregators and report dropping attacks in Wireless Sensor Networks. *Journal of Wireless Technologies: Concepts, Methodologies, Tools and Applications, 1*(3), 565–586.
14. Wozniak, M., Grana, M., & Corchado, E. (2014). A survey of multiple classifier systems as hybrid systems. *Information Fusion - Special Issue on Information Fusion in Hybrid Intelligent Fusion Systems, 16*, 3–17.
15. Moustafa, N., & Slay, J. (2014). *ADFA-NB15-Datasets - UNSW-NB15 network packets and flows captures, cyber range lab of the Australian centre for cyber security*. New South Wales: University of New South Wales, Australia.
16. Fan, J., Li, Q., & Wang, Y. (2017). Estimation of high dimensional mean regression in the absence of symmetry and light tail assumptions. *Journal of the Royal Statistical Society: Series B (Statistical Methodology), 19*(1) 247–265.
17. Mandelbrot, B. B. (1977). Fractals, Form, Chance and Dimension (1st ed.). W. H. Freeman.
18. Mandelbrot, B. B. (1967). How long is the coast of Britain? *Science, 156*(3775), 636–638.
19. Gouravaraju, S., & Ganguli, R. (2012). Estimating the Hausdorff–Besicovitch dimension of boundary of basin of attraction in helicopter trim. *Applied Mathematics and Computation, 218*(21), 10435–10442.
20. Khan, M. S., Ferens, K., & Kinsner, W. (2015). A polyscale based autonomous sliding window algorithm for cognitive machine classification of malicious internet traffic. In *Proceeding of the international conference on security and management (SAM'15), WordComp'15*, Nevada, USA.
21. Khan, M. S., Ferens, K., & Kinsner, W. (2015). Multifractal singularity spectrum for cognitive cyber defence in internet time series. *International Journal of Software Science and Computational Intelligence (IJSSCI), 7*(3), 17–45.
22. Kim, E.-S., San, M., & Sawada, Y. (1993). Fractal neural network: Computational performance as an associative memory. *Progress of Theoretical Physics, 89*(5), 965–972.

23. Bieberich, E. (2002). Recurrent fractal neural networks: a strategy for the exchange of local and global information processing in the brain. *Biosystems, 66*(3), 145–164.
24. Zhao, L., Li, W., Geng, L., & Ma, Y. (2011). Artificial neural networks based on fractal growth. In *Advances in automation and robotics*, (Vol. 123, pp. 323–330), Springer, Berlin.
25. Siddiqui, S., Khan, M. S., Ferens, K., & Kinsner, W. (2016). Detecting advanced persistent threats using fractal dimension based machine learning classification. In *Proceedings of the 2016 ACM on International workshop on security and privacy analytics, CODASPY'16*, New Orleans, LA.
26. Khan, M. S., Ferens, K., & Kinsner, W. (2015). A cognitive multifractal approach to characterize complexity of non-stationary and malicious DNS data traffic using adaptive sliding window. In *Proceedings of IEEE 14th international conference on cognitive informatics & cognitive computing (ICCI*CC)*.
27. Houtveen, J. H., & Molenaar, P. C. M. (2001). Comparison between the Fourier and Wavelet methods of spectral analysis applied to stationary and nonstationary heart period data. *Psychophysiology, 38*(5), 729–735.
28. Fryzlewicz, P., Bellegem, S. Van, & Sachs, R. von (2003). Forecasting non-stationary time series by wavelet process modelling. *Annals of the Institute of Statistical Mathematics, 55*(737), 737–764.
29. Jaffard, S., Abry, P., Roux, S., Vedel, B., & Wendt, H. (2010). The contribution of wavelets in multifractal analysis. In Damlamian, A., & Jaffard, S. (Eds), *Wavelet methods in mathematical analysis and engineering*. Singapore :World Scientific.
30. Gupta, B., Agrawal, D. P., & Yamaguchi, S. (2016). *Handbook of research on modern cryptographic solutions for computer and cyber security*. Hershey, PA: IGI Global.
31. Boukhtouta, A., Mokhov, S. A., Lakhdari, N.-E., Debbabi, M., & Paquet, J. (2016). Network malware classification comparison using DPI and flow packet headers. *Journal of Computer Virology and Hacking Techniques, 12*(2), 69–100.
32. Ji, S.-Y., Jeong, B.-K., Choi, S., & Jeong, D. H. (2016). A multi-level intrusion detection method for abnormal network behaviors. *Journal of Network and Computer Applications, 62*, 9–17.
33. PREDICT USC-Lander, DoS_DNS_amplification (2013). *Scrambled internet measurement, PREDICT ID USC-Lander/DoS_DNS_amplification-20130617 (2013-06-17) to (2013-06-17) provided by the USC/Lander Project*.
34. Siddiqui, S., Khan, M. S., Ferens, K., & Kinsner, W. (2017). Fractal based cognitive neural network to detect obfuscated and indistinguishable internet threats. In *Proceedings of the IEEE 16th International Conference on Cognitive Informatics and Cognitive Computing (ICCI×CC)*.

Chapter 30
A Comparative Study of Neural Network Training Algorithms for the Intelligent Security Monitoring of Industrial Control Systems

Jaedeok Kim and Guillermo Francia

30.1 Introduction

A special publication, NIST SP 800-137 [1], by the National Institute of Standards and Technology (NIST) was released in 2011 to provide guidelines for the continuous monitoring of information security on federal information systems and organizations. That publication addresses the assessment and analysis of security control effectiveness and security posture of an organization. In early 2015, the US Department of Energy's Office of Electricity Delivery and Energy Reliability published a document titled "Energy Sector Cyber Security Framework Implementation Guidance" [2] in response to NIST's Framework for Improving Critical Infrastructure Cyber Security [3]. Almost invariably, the North American Electric Reliability Corporation (NERC) continues to update and enforce a suite of Critical Infrastructure Protection (CIP) [4] standards related to the reliability of cyber security.

Clearly these guidelines and standards underscore the importance of protecting our critical infrastructures which are mostly operating through automated controlled systems. In an almost daily basis, this national need for cyber security-related CIP becomes more pronounced in light of intrusions and attempted attacks on internet-facing control systems. As documented in [5], the Industrial Control Systems Cyber Emergency Response Team (ICS-CERT) responded to 245 incidents reported by asset owners and industry partners in 2014.

In response to this national need, we embark of a research study on continuous security monitoring of industrial control systems using artificial intelligent neural systems. This chapter presents some preliminary work towards developing a complex neural system that classifies, in real-time, industrial control system behaviors.

J. Kim (✉) • G. Francia
Jacksonville State University, 700 Pelham Rd N, Jacksonville, AL, 36265, USA
e-mail: jkim@jsu.edu; gfrancia@jsu.edu

© Springer International Publishing AG 2018
K. Daimi (ed.), *Computer and Network Security Essentials*,
DOI 10.1007/978-3-319-58424-9_30

30.2 Facilities and Resources

In May 2010, The National Science Foundation (NSF) funded the Critical Infrastructure Security and Assessment Laboratory (CISAL) project [6] at JSU. The system architecture of CISAL is shown in Fig. 30.1. CISAL is designed to simulate the operations of a typical SCADA system serving a small community. The laboratory is equipped with Remote Terminal Units (RTUs), main control panels, a Human Machine Interface (HMI), and a System Historian. Central to the CISAL architecture is the ControlLogix™ controller backplane. We purposely designed the laboratory with this backplane to accommodate various communication modules operating different control system protocols such as Ethernet/IP, ControlNet, and DeviceNet. The Remote Terminal Units (RTUs) that are integral components of the laboratory are equipped with three wireless communication modes: cellular broadband, Wi-Fi, and 900 MHz radio. Current CISAL project outcomes include the design and implementation of the control systems laboratory that mimics a water distribution system and a nuclear power generation facility, the development of various curriculum modules [7] for a critical infrastructure protection course, and the investigation of various wireless technology applications on control systems.

30.3 Neural Networks

A neural network is an information processing system that is inspired by the human brain's nervous system. It consists of a large number of highly interconnected processing components called neurons. It can be used to solve problems in a wide range of applications such as function approximations, linear and nonlinear regressions, pattern recognitions, and data classifications. The popularity of neural network is primarily due to its ability of learning from big data set and utilizing

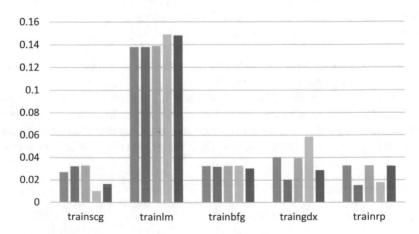

Fig. 30.1 Performance with continuous tank level w/o temperature

the knowledge gained from such learning in many decision-making processes. Techniques from neural networks are broadly applied to solving problems in many industries, and they have become standards in many areas of businesses. For example, a great number of the larger banks have developed sophisticated neural network systems in an effort to build more robust and efficient techniques for credit risk management.

The biggest challenge in implementing a neural network to conduct classification and prediction is on training the network. Network training is the process of finding the values for the weights and biases so that, for a set of training data with known inputs and outputs, the computed outputs closely match the desired training outputs. To train a neural network, we need to measure the error between computed outputs and the target outputs of the training data. The mean squared error, which is the mean of squares of the difference between two sets of outputs, is the most common measure of error. However, some of the recent research results by de Boer [8] and McCaffrey [9] suggest that using a different measure, so called Cross Entropy (CE) error, is better performing in rare event simulations, combinatorial optimization problems, and more. Consequently, the neural network algorithm involving CE error, called Cross Entropy Method, is getting more attention from various areas as it can be developed into a versatile tool for finding optimal solutions to many problems.

In the following discourse, we provide an overview of the process of designing a neural network model for system event classification. The linear models for classification problems are based on linear combination of fixed nonlinear basis functions ϕ_j and take the form

$$y(x, w) = f\left(\sum_{j=1}^{M} w_j \phi_j(x)\right),$$ (30.1)

where $w = [w_j]$ is a vector of parameters (or weights) and f is a nonlinear activation function. Neural network uses this model by updating selections of parameters in each layer so that it maximizes the likelihood of generating the outputs $y(x, w)$ matching the target outputs.

The basic neural network model can be described in terms of a series of functional transformations. First, we construct M linear combinations of the input variables $x_1, \ldots, x_D$ in the form

$$a_j^{(1)} = \sum_{i=0}^{D} w_{ji}^{(1)} x_i,$$ (30.2)

where $j = 1, \ldots, M$, and the superscript (1) indicates that the corresponding parameters are in the first layer of the network. The additional parameter $w_{j0}^{(1)}$ is the bias, and the additional input variable x_0 is the fixed value at $x_0 = 1$ for notational convenience. The quantities $a_j^{(1)}$ are called activations.

Next we transform each of activations using a differentiable and nonlinear activation function h to yield $z_j = h(a_j)$. These quantities are called hidden units. The sigmoid function $\left(\frac{1}{1+e^{-a}}\right)$ and the hyperbolic tangent function $\left(\tanh(a) = \frac{e^a - e^{-a}}{e^a + e^{-a}}\right)$ are generally chosen for the nonlinear function h. Now these hidden units are linearly combined to give output unit activations

$$a_k^{(2)} = \sum_{j=0}^{M} w_{kj}^{(2)} z_j, \tag{30.3}$$

where $k = 1, \ldots, K$, and K is the total number of outputs.

If we combine the two layers of neural network into a single formula, we can write it as follows:

$$y_k(x, w) = f\left(\sum_{j=0}^{M} w_{kj}^{(2)} h\left(\sum_{i=0}^{D} w_{ji}^{(1)} x_i\right)\right) \tag{30.4}$$

The choice of the activation function f in (30.4) is determined by the type of problem. For example, if it is linear standard regression, the activation function is the identity. If we work on a probabilistic regression problem, a popular choice for the activation function is the logistic function. If it is a binary classification problem, the activation function is the nonlinear sign function: $f(x) = 1$ if $x > 0$ and $f(x) = 0$ $(or - 1)$ if $x < 0$.

In order to perform training algorithms, first we are given a training set comprising a set of input vectors $\{x_n\}$, where $n = 1, \ldots, N$, together with a corresponding set of target vectors $\{t_n\}$. Then we can construct an error function (or a performance index) to be minimized to determine the set of parameters that may provide the maximum likelihood of target vectors $\{t_n\}$ matching network output vectors $\{y(x_n, w)\}$.

We now consider two performance indexes that are most commonly used in different algorithms. The first performance index is the mean square error (MSE) which measures the expected value of squares of difference between network output vectors and target vectors.

$$E_1(w) = \frac{1}{N} \sum_{n=1}^{N} \|y(x_n, w) - t_n\|^2 \tag{30.5}$$

In a single layer regression problem MSE is quadratic, the Hessian matrix [10] (or the second derivative) of the MSE function is constant. As a result, the analytic behaviors of the MSE depend heavily on the Hessian matrix. For example, if the Hessian matrix is positive definite, then the MSE has a global minimum.

The second performance index that is constructed by providing probabilistic views to the network outputs is defined as follows:

$$E_2(w) = -\sum_{n=1}^{N} \{t_n \ln y\,(x_n, w) + (1 - t_n) \ln (1 - y\,(x_n, w))\} \qquad (30.6)$$

This error function is called Cross Entropy (CE) [11]. CE measures the negative log likelihood.

30.4 Neural Network Training Functions

The following step in the neural network training process is to choose a network training function (or training algorithm) that minimizes the chosen error function. Among many well-known training algorithms available in the Matlab® Neural Network Toolbox are the following: Levenberg–Marquardt [12], BFGS Quasi–Newton [13], Resilient Backpropagation [14], Scaled Conjugate Gradient [15] [16], Gradient Descent with Momentum [16], and Gradient Descent with Momentum and Adaptive Learning Rate [16].

30.4.1 Conjugate Gradient Method (trainscg)

The conjugate gradient method uses the following steps for determining the optimal value (minimum) of a performance index $E(\mathbf{w})$. Given an initial point ($\mathbf{w}_0$) to start from, it chooses a direction (p_0). A line search is then performed to find the optimal distance to move along the search direction.

$$\mathbf{w}_1 = \mathbf{w}_0 + \alpha_0 p_0 \qquad (30.7)$$

The next search direction is determined so that it is orthogonal to the difference of gradients.

$$\Delta g_1^T p_1 = (g_1 - g_0)^T p_1 = (\nabla E\,(\mathbf{w}_1) - \nabla E\,(\mathbf{w}_0))^T p_1 = 0 \qquad (30.8)$$

Repeating the two steps, we have the algorithm of conjugate gradient methods:

$$\mathbf{w}_{k+1} = \mathbf{w}_k + \alpha_k p_k \qquad (30.9)$$

$$\Delta g_k^T p_k = (\nabla E\,(\mathbf{w}_k) - \nabla E\,(\mathbf{w}_{k-1}))^T p_k = 0 \qquad (30.10)$$

The most common first search direction (p_0) is the negative of the gradient:

$$p_0 = -g_0 = -\nabla E\,(\mathbf{w}_0) \qquad (30.11)$$

A set of vectors $\{p_k\}$ is called *mutually conjugate* with respect to a positive definite Hessian matrix H if

$$p_k^T H p_j = 0 \text{ for } k \neq j \tag{30.12}$$

It can be shown that the set of search direction vectors $\{p_k\}$ obtained from Eq. (30.10) without the use of the Hessian matrix is mutually conjugate. The general procedure for determining the new search direction is to combine the new steepest descent direction with the previous search direction:

$$p_k = -g_k + \beta_k p_{k-1} \tag{30.13}$$

The scalars $\{\beta_k\}$ can be chosen by several different methods. The most common choices are

$$\beta_k = \frac{\Delta g_{k-1}^T g_k}{\Delta g_{k-1}^T p_{k-1}}, \tag{30.14}$$

which is due to Hestenes and Stiefel [17], and

$$\beta_k = \frac{\Delta g_{k-1}^T g_k}{\Delta g_{k-1}^T g_{k-1}}, \tag{30.15}$$

which is due to Fletcher and Reeves [17].

Unlike other conjugate gradient algorithms that use linear optimization methods, scaled conjugate gradient method does not perform a line search when updating the vector.

30.4.2 Newton's Method and Other Variations (trainbf and trainlm)

One of the fastest (second-order training speed) training algorithms is Newton's method.

$$\mathbf{w}_{k+1} = \mathbf{w}_k - H^{-1} g_k \tag{30.16}$$

However, one drawback of the Newton's method is it requires the computation of the Hessian matrix (H), which can become very costly in neural network problem if the number of attributes (or variables) is large. Two variations of Newton's method are available for which we can avoid the computation of the Hessian matrix: Quasi–Newton algorithm (trainbf) and Levenberg–Marquardt algorithm (trainlm).

The idea of quasi–Newton is to replace H^{-1} with a positive definite matrix which is updated at each iteration without matrix inversion.

The second variation Levenberg–Marquardt algorithm works very well with neural network training where the performance index is MSE. The Levenberg–Marquardt algorithm was designed to approach second-order training speed without having to compute the Hessian matrix. When the error function is in the form of a sum of squares such as MSE, the Hessian matrix can be approximated by

$$H = J^T J \tag{30.17}$$

and the gradient can be computed as

$$g_k = \nabla E (\mathbf{w_k}) = J^T (\mathbf{w_k}) \, \mathbf{e} (\mathbf{w_k}) \tag{30.18}$$

where J is the Jacobian matrix that contains first derivatives of the network errors with respect to the weights and biases, and e is a vector of network errors. If we substitute (30.17) and (30.18) into (30.16), we obtain the following algorithm, known as Gauss–Newton method.

$$\mathbf{w}_{k+1} = \mathbf{w}_k - \left[J^T (\mathbf{w_k}) J (\mathbf{w_k}) \right]^{-1} J^T (\mathbf{w_k}) \, \mathbf{e} (\mathbf{w_k}) \tag{30.19}$$

One problem with the Gauss–Newton is that the matrix $H = J^T J$ may not be invertible. This problem can be resolved by using the following modification:

$$G = H + \mu I \tag{30.20}$$

Since the eigenvalues of G are translation of the eigenvalues of H by μ, G can be made positive definite by finding a value of μ so that all eigenvalues of G are positive. This leads to the Levenberg–Marquardt algorithm.

$$\mathbf{w}_{k+1} = \mathbf{w}_k - \left[J^T (\mathbf{w_k}) J (\mathbf{w_k}) + \mu_k I \right]^{-1} J^T (\mathbf{w_k}) \, \mathbf{e} (\mathbf{w_k}) \tag{30.21}$$

Levenberg–Marquardt algorithm is known to work fast and stable for various forms of neural network problems with MSE performance index.

30.4.3   Resilient Backpropagation Algorithm (trainrp)

Multilayer networks typically use sigmoid transfer functions in the hidden layers. Sigmoid functions are characterized by the fact that their slopes must approach zero as the input gets large. This causes a problem when you use the steepest descent to train a multilayer network with sigmoid functions because of possibility of a very small change of magnitude of gradient that could result in slow convergence or no convergence of the algorithm.

The purpose of the resilient backpropagation algorithm (trainrp) is to eliminate those possible issues caused by small magnitude of gradient change. The idea of resilient backpropagation algorithm is to use the size of the weight change that is determined by a separate update value. The update value for each weight and bias is increased by a prescribed constant whenever the derivative of the performance index with respect to that weight has the same sign for two successive iterations. The update value is decreased by the constant whenever the derivative changes sign from the previous iteration. If the derivative is zero, the update value remains the same. If the weight changes are switching signs frequently, then the magnitude of weight change is reduced. If the weight continues to change in the same direction for several iterations, the magnitude of the weight change increases. An elaborate description of the resilient backpropagation algorithm is given in [14].

30.4.4   Gradient Descent with Momentum (traingdm) and Gradient Descent with Momentum and Adaptive Learning Rate (traingdx)

The steepest descent method is an algorithm that uses the gradient of the error function at each updated point when it chooses a new search direction. The training algorithm associated with the steepest descent method is called the steepest descent backpropagation (SDBP). The SDBP uses the chain rule in order to compute the derivatives of the squared error with respect to the weights and biases in the hidden layers. It is called backpropagation because the derivatives are computed first at the last layer of the network, and then propagated backward through the network to compute the derivatives.

When introduced, SDBP was a major breakthrough in the development of neural networks. However, it possesses some drawbacks. The biggest drawback is its slow convergence for most practical applications. In general, when the gradients at various points are calculated, they tend to show sharp turns at each updated point which can be possible causes of such slow convergence. As a remedy to this problem, momentum filter can be applied to SDBP in an attempt to smoothen the sharp oscillatory turns of gradients. The training algorithm that incorporates the momentum filter into the backpropagation is called the gradient descent with momentum (traingdm) [16]. However, as remarked in the following section, the training algorithm traingdm shows still slow convergence for every set of training data we tested.

The gradient descent with momentum and adaptive learning rate backpropagation (traingdx) is a network training function that is invoked the same way as traingdm, except that it uses adjusted learning rate based on the direction and magnitude of the gradient vector.

30.5   Comparative Analysis of Neural Network Training Algorithms

The research study involves the monitoring of a subset of a collection of complex operational data from a power generating system. We monitor the operation of three pumps, the tank water level, and the operational status of water-cooled equipment such as a reactor or a generator. We also test a different set of operational data by including the water temperature as an additional input data element. The operational and pump statuses are binary values (0-OFF, 1-ON). The water temperatures are real numbers ranging from 0 to 400 $^\circ$C. The tank level is either continuous or discrete. The continuous values range from 0.0 to 99.99 reflecting the fill percentage of the tank. The five discrete values are indicative of the following level statuses: 1 (critically low), 2 (low), 3 (normal), 4 (high), and 5 (critically high). The goal of the classification system is to take an input vector x and to assign it to a two dimensional output vector y which equals either $[1, 0]^T$ (normal) or $[0, 1]^T$ (abnormal). The following table summarizes the training parameters (Table 30.1).

In an effort to increase the statistical stability of results of network training, we run each algorithm five times on the same set of input data for both continuous format and discrete format. We make comparisons of five different training algorithms (traincgf, trainlm, trainbfg, traingdx, trainrp) on four neural network system metrics: network error performance, success rate, run time, and number of epochs. We also run comparative analysis investigating the statistical differences using the following different pairings:

1. Continuous (tank level) without water temperature vs discrete without water temperature
2. Continuous with water temperature vs discrete with water temperature
3. Without water temperature vs with water temperature.

Table 30.1 Training parameters

Parameter	Value
Number of input attributes	5
Number of input records	30,000
Number of output attributes	2
Number of output records	30,000
Number of neural network hidden layers	10
Allocation of data for training	70%
Allocation of data for validation	15%
Allocation of data for initial testing	15%
Maximum number of training epochs	5000
Performance index (Cross Entropy (CE) or mean square error (MSE))	MSE or CE

30.5.1 Network Error Performance

30.5.1.1 Comparisons of Training Algorithms

The network error performance, the first point of comparison in this comparative study, is measured by the amount of error generated at each run of the training algorithm. However, a word of caveat is in order: Levenberg–Marquardt (trainlm) uses the mean square error (MSE) as a performance index while the other algorithms use Cross Entropy (CE) as a performance index. Thus, the performance comparison results must be considered judiciously (Figs. 30.1, 30.2, 30.3, and 30.4).

30.5.1.2 Continuous vs Discrete

The performance values are overall better (or smaller) when the input data set is given in discrete tank level (Figs. 30.5 and 30.6).

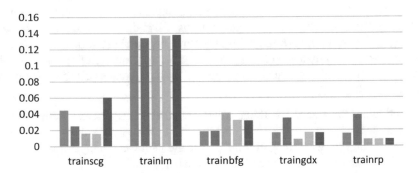

Fig. 30.2 Performance with discrete tank level w/o temperature

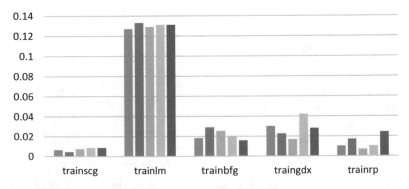

Fig. 30.3 Performance with continuous tank level w/temperature

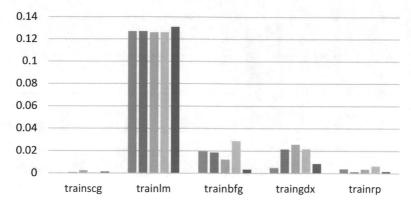

Fig. 30.4 Performance with discrete tank level w/temperature

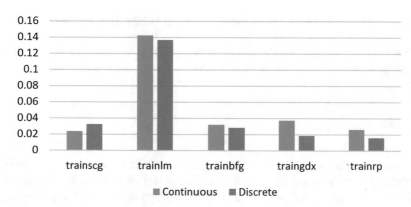

Fig. 30.5 Continuous vs discrete (performance w/o temperature)

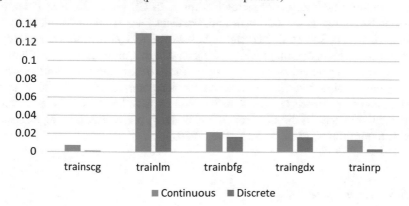

Fig. 30.6 Continuous vs discrete (performance w/temperature)

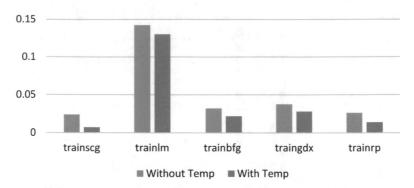

Fig. 30.7 Without temperature vs with temperature in performance

30.5.1.3   Without Water Temperature vs With Water Temperature

It turns out that the addition of the element of water temperature provides significant advantage in the classification of abnormality of the system (Fig. 30.7).

30.5.2   Success Rate

The next comparison point is on the success rate, i.e., how well each training algorithm correctly classifies normal and abnormal system behaviors. Figure 30.3 depicts the average classification success rate for each algorithm. The advantage of using discrete values in both without and with temperature cases in the classification process of the Levenberg–Marquardt (trainlm), BFGS Quasi–Newton (trainbfg), Resilient Backpropagation (trainrp), and Gradient Descent with Momentum and Adaptive Learning Rate (traingdx) can be best explained by the coarse granularity of the given data which, in turn, provides an easier determination of class membership. The Scaled Conjugate Gradient (trainscg) performs better with the continuous data, but it yields a bigger success rate when the temperature value is added to the input data. The evidence is clear that the discrete tank level with the addition of temperature element in the data set provides much better accuracy regardless of training algorithms (Figs. 30.8, 30.9, and 30.10).

30.5.3   Run Time

We observed a very poor performance from the Gradient Descent with Momentum (traingdm) in every metric we measured. One rationale for the problem is its inability of making adjustment when the performance index is changing very little. A similar but more flexible algorithm, Gradient Descent with Momentum

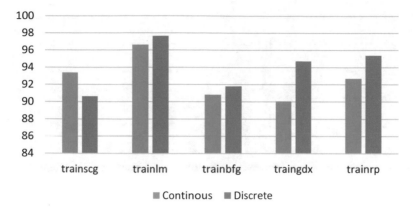

Fig. 30.8 Percentage of success rate w/o temperature

Fig. 30.9 Percentage of success rate w/temperature

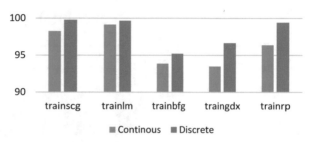

Fig. 30.10 Without temp vs with temp (success rate in percent)

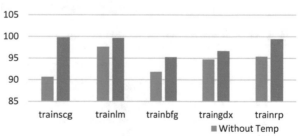

and Adaptive Learning Rate (traingdx), was chosen to replace traingdm, and it outperforms not only traingdm but also the other algorithms in run time. One interesting thing to note is that run time with discrete tank level data values especially with temperature case takes longer than continuous one. There is no immediate clear answer why such things occur with most of the training algorithms (Figs. 30.11, 30.12, 30.13, and 30.14).

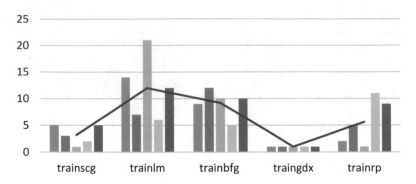

Fig. 30.11 Run time (in seconds) continuous w/o temp

Run Time with Discrete Tank Level (in seconds)

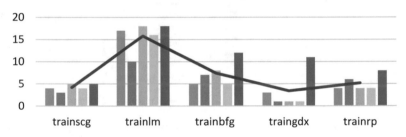

Fig. 30.12 Run time (in seconds) discrete w/temp

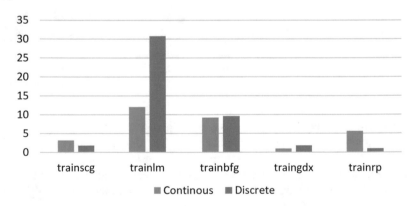

Fig. 30.13 Run time (in seconds) w/o temp

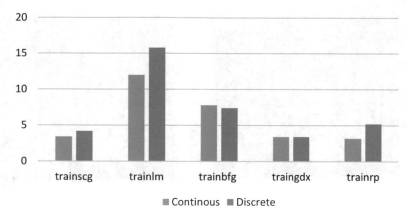

Fig. 30.14 Run time (in seconds) w/temp

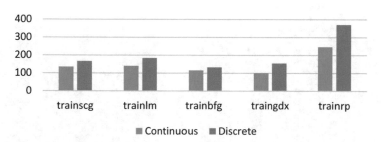

Fig. 30.15 Number of epochs needed

30.5.4 Epochs

The results involving epochs closely replicate those in the run-time comparison.
Two superior algorithms (trainlm and trainrp) in performance and success rate finish
their testing with bigger number of epochs than the other algorithms (Fig. 30.15).

30.6   Comparative Analysis of Neural Network Training Algorithms

The confusion matrix shown in Fig. 30.16 summarizes the result of each development stage of the neural system using the resilient backpropagation training
algorithm. The consistency of classifying normal and abnormal behaviors of the
industrial control system within an acceptable accuracy rate provides enough
evidence that building an intelligent behavior predictor system using this training
algorithm is feasible.

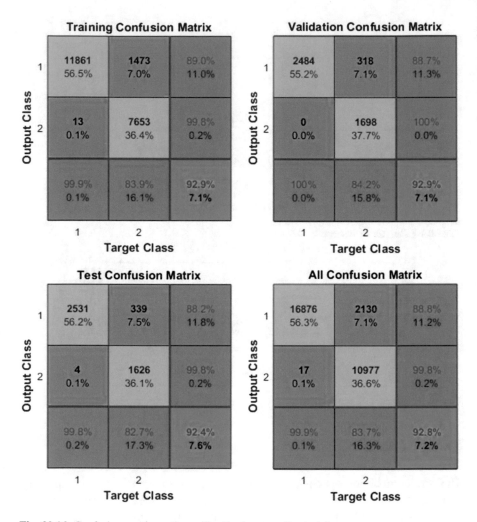

Fig. 30.16 Confusion matrix on the resilient backpropagation training

30.7 Conclusion and Future Work

The work we presented herein is part of a much larger research project on continuous security monitoring of control systems infrastructure. Industrial control systems (ICS) such as those that operate our nation's critical infrastructure are beset by challenges that affect the security of their real-time or near real-time operations. We envision to develop a mechanism to apply machine learning techniques to classify, in real-time, the behavior of control systems by capturing disparate data from multiple sources and feeding those into a neural system for analysis. However, before we can embark on a full scale development process, we need to determine the best performing neural network training algorithm that suits our data and system environment. With this realization, we made a preliminary comparative study of various training algorithms that are available in the Matlab® Neural Network Toolbox. The results of this study are summarized in the following observations:

- Except for the Gradient Descent with Momentum (traingdm) training algorithm, the other algorithms perform at a reasonable success rate of 90% and over. It is obvious that the discrete data set provides improved success rates in most runs than the continuous cases. Also, it is noteworthy to point out that the addition of temperature variable to the input data set increases the success rate considerably.
- Acceptable error rates are highly subjective and we tend to stay away from that issue. Most of the time it is not only subjective but also application dependent. For instance, an acceptable error rate would be very small with regard to an application that may involve the loss of life. However, it may be a bit lenient when predicting weather patterns or climate change.
- The run time and the number of iterations (epochs) displayed by the Resilient Backpropagation (trainrp) and Gradient Descent with Momentum and Adaptive Learning Rate (traingdx) training algorithms are generally better than the other algorithms. More importantly, these results provide us with the confidence in choosing the two methods in the development of an expanded research study on the application of machine learning techniques for enhancing the security of industrial control system security.
- After further experiments on Gradient Descent with Momentum (traingdm) training algorithm, we observed that overall performance can be significantly improved by adjusting learning rate. The default learning rate in Matlab Neural Network Toolbox for traingdm is .01, but much better learning occurs when the learning rate is set between 2 and 3. The average number of epochs is 9362 and the success rate is 97.4%. Varying the momentum constant produces little impact on the performance of the algorithm.
- One rationale that can explain such improvement in the traingdm algorithm performance with bigger learning rate is as follows. The performance surface (graph of error CE function) is relatively flat even when the minimum is still far apart so that in early stage of searching for optimal value it needs to take bigger steps before small magnitudes of gradients start slowing down the search. However, this reasoning can't be applied if the performance surface contains more hills and valleys with larger gradient magnitudes.

As this is an on-going study, we offer the following future research extensions:

- The utilization of artificial intelligence approaches such as Clustering, Principal Component Analysis (PCA), Genetic Algorithms, Deep Learning, and Support Vector Machine (SVM) to enhance the performance of the machine learning classification system;
- The investigation of using the Mean Square Error (MSE) function to further enhance a classification system that was trained with the Cross Entropy criterion as suggested in [18].

Acknowledgement This work is supported in part by a Center for Academic Excellence (CAE) Cyber Security Research Program grant (Grant Award Number H98230-15-1-0270) from the National Security Agency (NSA). Opinions expressed are those of the authors and not necessarily of the granting agency.

References

1. National Institute of Standards and Technology (NIST). (2011). Special Publication 800-137: Information Security Continuous Monitoring (ISCM) for Federal Systems and Organizations, NIST.
2. U.S. Department of Energy. (2015). *Energy sector cybersecurity framework implementation guidance*. DOE.
3. National Institute of Standards and Technology (NIST). (2014). *Framework for improving critical infrastructure cybersecurity*. National Institute of Standards and Technology (NIST).
4. North American Electric Reliability Corporation (NERC). (2015). *Critical infrastructure protection (CIP) standards*. North American Electric Reliability Corporation (NERC).
5. ICS-CERT. (2014). *Incident response/vulnerability coordination in 2014*. ICS-CERT Monitor, Industrial Control Systems Cyber Emergency Response Team (ICS-CERT).
6. Francia, G., Bekhouche, N., & Marbut, T. (2011). Design and implementation of a critical infrastructure security and assessment laboratory. In *Proceedings of the security and management 2011 conference*, Las Vegas, NV.
7. Francia, G. (2011). Critical infrastructure curriculum modules. In *Proceedings of the 2011 information security curriculum development (INFOSECCD) conference*, Kennesaw, GA.
8. De Boer, P., Kroese, D. K., Mannor, S., & Rubinstein, R. Y. (2005). A tutorial on the cross-entropy method. *Annals of Operations Research, 134*, 19–67.
9. McCaffrey, J. (2014, November 4). Neural network cross entropy error. *Visual Studio Magazine*.
10. Magnus, J. R., & Neudecker, H. (1999). *Matrix differential calculus*. Chichester: John Wiley & Sons, Ltd..
11. Bishop, C. M. (2007). *Patern recognition and machine learning*. Heidelberg: Springer.
12. Marquardt, D. (1963). An algorithm for least-squares estimation of nonlinear parameters. *SIAM Journal on Applied Mathematics, 11*(2), 431–441.
13. Dennis, J. E., & Schnabel, R. B. (1983). *Numerical methods for unconstrained optimization and nonlinear equations*. Englewoods Cliffs, NJ: Prentice-Hall.
14. Riedmiller, M., & Braun, H.. (1993). A direct adaptive method for faster backpropagation learning: The RPROP algorithm. In *Proceedings of the IEEE international conference on neural networks*.
15. Moller, M. (1993). A scaled conjugate gradient algorithm for fast supervised learning. *Neural Networks, 6*, 525–533.
16. Hagan, M. T., Demuth, H. B., & Beale, M. H. (1996). *Neural network design*. Boston, MA: PWS Publishing.
17. Scales, L. (1985). *Introduction to non-linear optimization*. New York: Springer-Verlag.
18. Golik, P., Doetsch, P., & Ney, H. (2013). Cross-entropy vs. squared error training: a theoretical and experimental comparison. In *Proceeding of the 14th annual conference of the international speech communication association*, Lyon.

Chapter 31
Cloud Computing: Security Issues and Establishing Virtual Cloud Environment via Vagrant to Secure Cloud Hosts

Polyxeni Spanaki and Nicolas Sklavos

31.1 Introduction to Cloud Computing

According to National Institute of Standards and Technology (NIST), Cloud Computing is a model for enabling pervasive, suitable, on-demand network access to a shared pool of resources such as networks, servers, storage, applications, and services [1]. Cloud Computing delivers services over the internet regarding hardware, software, data resources, and several applications stored in data centers.

Based on the definition provided by NIST, Cloud Computing is composed by some essential characteristics such as on-demand self-service, network access, resource grouping, rapid elasticity, and measured service [1].

Through the past years, Cloud Computing has been growing in popularity since many Information Technology (IT) organizations have started using cloud platforms. This expansive use of the cloud derives from the vast variety of services provided, as well as the many advantages of its use. Cloud, over the past years, has become a popular area of research as well as designing, developing, and deploying several sets of applications [2].

P. Spanaki (✉)
Open University of Cyprus, Nicosia, Cyprus
e-mail: spanakipolyxeni@gmail.com

N. Sklavos
SKYTALE Research Group, Computer Engineering and Informatics Department,
University of Patras, Patra, Greece
e-mail: nsklavos@ceid.upatras.gr

© Springer International Publishing AG 2018
K. Daimi (ed.), *Computer and Network Security Essentials*,
DOI 10.1007/978-3-319-58424-9_31

Some of the important benefits of the use of the "cloud" for companies and users are:

1. Low cost: Organizations save total expenses in equipment and other costs.
2. Increased data resources: Large amount of data can be stored and accessed through the internet.
3. Recovery: Since all of the data are stored in the cloud, recovery is an easy task for users and organizations, compared to storage in physical devices.
4. Accessibility: Any device with Internet Connection can easily access stored data.
5. Improved services: With the use of the cloud, organizations improve their IT services, customer satisfaction as well as productivity [3, 4].

Along with the many advantages of the use of the "cloud" there are some disadvantages:

1. Accessibility: The access of the stored data is only possible with the use of the Internet.
2. Security issues: Security in the cloud, as well as privacy and integrity of the stored data, is a major issue in cloud environments.

In the following chapter, these security issues will be discussed and some of the main attacks will be demonstrated. With the use of a virtual environment created by automation tools, remote hosts are configured with the use of configuration management tools in order to be secured. Section 31.2 analyzes the service and deployment Cloud Computing models. Section 31.3 discusses the main security issues and Sect. 31.4 presents automation tools used for the demonstration. Finally Sects. 31.5 and 31.6 conclude the chapter.

31.2 Cloud Computing

Cloud Computing is composed by a set of service models according to National Institute of Standards and Technology (NIST). These service models are designed in order to be selected and customized by organizations and users for the satisfaction of their organizational requirements [3]. They cover all the different needs of each organization through virtualization. This process includes creating a virtual version of the cloud resources, divided into multiple environments [5]. The major distinctions for Cloud Computing service models according to NIST are Software-as-a-Service (SaaS), Platform-as-a-Service (PaaS), and Infrastructure-as-a-Service (IaaS), as it is represented in the following figure (Fig. 31.1) [1, 3–6].

1. Software-as-a-Service (SaaS): The service providers offer the capability to the users to use applications and services running on a cloud infrastructure. A cloud infrastructure is the hardware and software which implements the five essential characteristics of Cloud Computing and contains a physical and an abstraction layer. The physical layer includes the hardware resources that are used such as

Fig. 31.1 Cloud service models

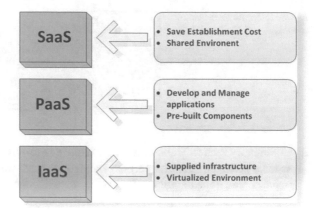

servers, storage, and network components. The abstraction layer consists of the software which is deployed across the physical layer [1]. These applications are accessible from various client devices. Some examples for SaaS's cloud are IBM, Oracle, Salesforce, and Microsoft [5].

2. Platform-as-a-Service (PaaS): The user has the ability to deploy applications which are created using programming languages, libraries, services and tools supported by the provider. The consumer does not control the basic cloud infrastructure including network, servers, operating systems, or storage, but can manage the deployed applications and possible configuration settings [1]. Some examples for PaaS's cloud are Google App Engine, Microsoft Azure, Heroku [3].

3. Infrastructure-as-a-Service (IaaS): The consumer is supplied with storage, networks, and other fundamental computing resources where then is able to deploy and run software, such as operating systems and applications. The consumer does not have the right to manage the cloud infrastructure but has control over operating systems, storage, applications, and networking components in a small amount. Some examples for Infrastructure as a Service cloud are Amazon EC2, Eucalyptus, Rackspace, and Nimbus [3].

The services described can be deployed on one or more of the four models that compose the cloud, according to NIST. Each model can change the way systems are connected and how any amount of work is done in an organization. The usage of any deployment model can contribute in the development of the applications, platforms, infrastructure, and any other resources and services used within the cloud [5].

According to NIST the four deployment models which compose the cloud are Private, Community, Public, and Hybrid cloud as represented in Fig. 31.2 [1].

1. Private cloud: The cloud infrastructure is provisioned for private use by a single organization containing multiple consumers. It may be owned and managed by a single organization, a third party, or both and it may exist on or off organization's bounds.

Fig. 31.2 Deployment cloud
models

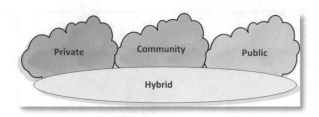

2. Community cloud: The cloud infrastructure is provisioned for private use by a
   specific group of consumers from organizations that share security requirements,
   policies, and considerations. It may be owned, controlled, and managed by the
   organizations, a third party, or both and it may exist on or off organization's
   premises.
3. Public cloud: The cloud infrastructure is provisioned for public and open use. It
   may be owned, controlled, and managed by a business, academic, or government
   organization and exists only on the premises of the cloud provider.
4. Hybrid cloud: The cloud infrastructure is structured by two or more cloud
   infrastructures, as described above. Although they remain unique entities they
   are actually bound together by standardized technology. Hybrid cloud also
   implements data and application portability.

31.3   Cloud Computing Security Issues

Cloud Computing is composed of a large-scale distributed infrastructure, which
delivers endless and dynamically scalable resources such as data storage, hardware
platforms, and web applications [7]. Although the many advantages and benefits
derived by the cloud, since its environment is based on the interaction via the
Internet, the risks and security issues are eminent. According to an IDC Asia/Pacific
Cloud Survey (2009), the major concern within the cloud environment was the issue
of security [7].

Computer security is the protection provided to an automated information system
in order to preserve objectives such as the integrity, availability, and confidentiality
of information system resources. System resources include hardware, software,
firmware, and information data [8]. In computer security, vulnerability is a flaw
or a weakness in the system which allows the potential attacker to exploit it and
violate the basic principles of Information Security (availability, privacy, integrity,
confidentiality).

Regarding a cloud environment, vulnerabilities are cloud-specific if they [9]:

1. Are frequent in a Cloud Computing technology
2. Have their main cause in one or more of National Institute of Standards and
   Technology cloud characteristics

3. Are caused when the cloud novelty hardens the control of security in the cloud environment
4. Are common in cloud offerings.

By exploiting these cloud-specific vulnerabilities, potential attackers may be able to launch attacks with important and dangerous consequences to the cloud environment. Some of the attributes of the cloud could help them evolve into exploitable vulnerabilities, in order to launch an attack against the cloud environment. Some of them are [9] [10]:

1. Pervasive Network Access: Cloud consumers can access all the services and resources by using the Internet with the help of conventional devices.
2. Measured Service: Users are able to monitor their usage of data resources online through a measured service.
3. Multi-tenancy: Multi-tenancy is a part of the public cloud that offers the possibility to provide services to multiple users simultaneously. This introduces reduced overhead and a larger amount of applications [5].
4. Off-premise Infrastructure: In a public cloud, the infrastructure is owned and controlled by a third party and does not give the possibility to be accessed by the user's organization.
5. Rapid elasticity: Resources can easily and quickly be scaled up and down.

The cloud environment should have specific characteristics in order to be considered as a safe environment and in order to preserve the integrity, availability, confidentiality, and privacy of the stored information. Corporate information is not only competitive asset, but it often contains information of customers and employees, so its protection information is crucial [11].

The security personnel should be able to understand attacks that are specifically targeted towards the cloud platform. The responsibility of securing the system must be assigned to someone with great experience and skills in information security. The most important skill required is the ability to design, configure, and penetrate security of systems [6].

In a cloud environment some of the main security issues that we may encounter are [12]:

1. Privileged access control: Sensitive data that are processed outside the organization contain a high level of risk, because outsourced services can find a way to ignore the physical and logical controls.
2. Data recovery: Since data is not stored in the organization's or user's infrastructure, in the event of a disaster the availability of the stored data is compromised.
3. Data security: Security refers to the confidentiality, integrity, and availability of the data, which can be a major security issue for cloud providers.
4. Regulatory compliance: Customers are ultimately responsible for the security and integrity of their own data, even when it is stored in a service provider's data center.

5. Data location: Service providers need to commit to storing and processing data and must provide privacy and security to their customers, since their customers don't know the exact location of the data centers.

Some of the most common attacks occurred in the cloud are [10]:

1. Distributed Denial of Service Attacks: In a DDoS attack, the attacker blocks the access in a machine by exploiting a number of compromised machines called bots in order to force a computer or network to not be able to provide services. DDoS attacks can also target cloud machines, since the attacker can also exploit virtual machines as internal bots in order to "flood" malicious requests to the target's VM.

2. VM Denial of Service attacks: A Virtual Machine Denial of Service (VM DoS) attack occurs when the adversary who is the owner of a VM in the cloud exploits several vulnerabilities that exist in the hypervisor to employ all the resources stored in the machine that the VM is running on.

3. Keystroke Timing Attacks: Keystroke Timing Attacks are attacks that refer to the stealing of private and confidential information, such as login passwords by the keystrokes of the victim. Timing information of keystroke may give out information about the keys' sequence types. While the potential victim is typing a password, the attacker measures the time between keystrokes [12].

4. Side-Channel Attacks: The attribute of multi-tenancy in public clouds gives the possibility to enable multiple VMs to run on the same physical machine. However, a user's VM could be running on the same server as their potential attacker. This may allow the attacker to infiltrate the target's VMs and in the end, compromise the users' confidentiality [13]. If the potential attacker can place the malicious virtual machine on the same physical machine of the target, then the adversary performs a cross-VM attack in order to extract confidential information.

5. Hypervisor Attacks: With the hypervisor attack, a cloud administrator who has privileged access to the hypervisor can penetrate into a guest's virtual machine without any direct privileges on the particular machine.

6. Cloud Malware Injection Attacks: In a cloud malware injection attack, the attacker tries to inject a malicious VM into a cloud environment with the purpose to launch attacks such as eavesdropping and blockings [10, 14].

7. Fraudulent Resource Consumption Attacks: In this cloud-specific attack the adversary's main target is to exploit the pricing model of the cloud. This occurs with an attack very similar to the Distributed Denial of Service attack. The attacker's main interest is to deprive the target of their economic benefits by consuming the user's cloud resources. The main difference of Fraudulent Resource Consumption Attack and DDoS attack is that FRC attack aims to make cloud resources economically unsustainable for the target victim, whereas DDoS attack only consumes a large amount of resources.

31.4   Virtual Cloud Environment

31.4.1   Automation Tools

Security threats regarding managing sensitive data, vulnerabilities in the virtualized environment, and network security are some of the many challenges faced in a cloud and non-cloud environment.

The introduced design methodology, which will be demonstrated, is by the use of a virtual environment. With the use of software automation tools, a specific methodology is followed. The remote hosts are configured, within the cloud in order to be secured.

Some of the major benefits of IT automation are:

1. Less time consumption.
2. Improvement of productivity and collaboration.
3. Elimination of repetitive tasks.
4. Reduction of errors.
5. Degradation of complexity.
6. Accretion of innovation resources.

In order to create a Virtual Cloud Environment, the proper software must be used. There are several open-source software products which are used for creating virtual environments, such as Docker; however, the main interest in this chapter is Vagrant by HashiCorp. Vagrant is a tool used to build complete development environments with a low setup time. The project started in January 2010 by Mitchell Hashimoto.

Vagrant is used to create and configure lightweight, portable development virtual environments which can easily be reproduced. This software product can be used in most operating systems and is written in Ruby; however, it can support the development in almost all major languages. It manages all the important configurations, in order to avoid all the unnecessary maintenance and setup time and is controlled by a single consistent workflow, in order to help maximize the productivity and flexibility. Vagrant is designed to run on almost any VM tool such as VirtualBox and VM Ware. However, default support is only provided for VirtualBox. For any of the other providers the installation of plugins for each one is necessary.

For developers, Vagrant is a practical tool to isolate dependencies and their configuration within a single environment. A very important attribute is that it won't interfere with any of the tools used in the project, such as editors, browsers, and debuggers.

For operations engineers, Vagrant enables a disposable environment and consistent workflow for the development and testing infrastructure management scripts.

For designers, Vagrant provides the appropriate environment required, in order to make the designing process uncomplicated and easier.

Vagrant setup procedure and command line are extremely easy. The primary configuration location for any Vagrant environment is a VagrantFile. VagrantFile is a Ruby syntax file, which includes any configuration option for the development environment. It provides the possibility to install all the necessary files and modify the initial configuration.

Vagrant is able to define and manage multiple guest machines for each Vagrant-File created. So it can be used to create multiple machine environments.

Vagrant Boxes are the package format in order to create Vagrant environments. Boxes can be used by anyone and on any platform, supported by Vagrant, to create an identical virtual environment. There is a publicly available catalog of Vagrant boxes but there is the possibility to add other customized boxes. Boxes can also support versioning, so with the use of Vagrant each box can be updated and necessary fixes can be pushed.

Some interesting features introduced in Vagrant are:

1. The existence of large amount of image files that can be used.
2. The ability to back up a current machine to a Vagrant box file in order to be able to eventually share it back and reproduce it.
3. The ability to adjust settings on the virtual machine.
4. The introduction of VagrantFiles, which provides the possibility of installing the necessary packages and modifying configuration in the initial setup.
5. The integration with configuration management tools such as Puppet, Chef, and Ansible.

Configuration management tools, such as Puppet, Chef, and Ansible, are used for the configuration of the created environments. These CM tools are assimilated with Vagrant by HashiCorp. In this particular chapter the configuration management tool that will be presented is Ansible.

Ansible is an IT automation engine that is used to automate cloud provisioning as well as configuration and service orchestration and makes application and system deployment easier. It is extremely easy to setup and use and uses a limited amount of overhead. Ansible is exceptionally powerful and is composed by agentless architecture. It uses open SSH and no agents are used to exploit or update.

First, as shown in Fig. 31.3, the administration client connects to a server via SSH. As mentioned before, Ansible is agentless and all that is necessary in order to start provisioning is python and open SSH. The client then gathers facts from the server, such as operating systems and installed packages. Finally, after the client collects such information, Ansible runs a Playbook.

Playbooks are simple configuration files, written in YAML, which are human and machine readable. These files contain and define all the steps for deployment configuration. Playbooks are modular, can contain variables, multiple tasks and use modules in order to orchestrate configuration steps to one or multiple machines. The client with the use of Playbooks can run tasks such as the procedure of copying file, use specific modules, substitute variables (to create a database, etc.), and assign roles (to install firewall, etc.).

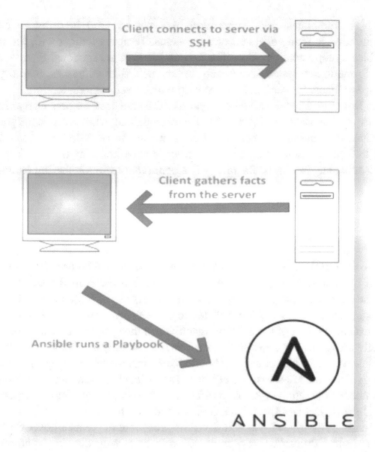

Fig. 31.3 Ansible initiation steps

Some of the most important attributes of Ansible are:

1. Host Inventory: Host inventory is a list of hosts or servers, which is assigned to manage and configure using Ansible. Using the host inventory the list of hosts and servers can be organized into groups and set ports.
2. Plays, Tasks, and Modules: A Playbook contains Plays. Plays are a set of tasks which consequently call Modules. Modules are used in order to alter or manage some configurations made on a server. The changes made using Modules are static. Modules can be used in order to install packages, run commands, mount drives, copy and template files, and manage services. There are many modules provided by Ansible; however, the developer also has the ability to create some.
3. Handlers: Handlers are specific tasks that get run after certain triggers. Handlers run at the end of each play and only run once. A handler can be used, for example, in order to restart configuration so that possible changes can take effect on the machine.

4. Variables, Templates, and Facts: Variables can be used in order to change configuration in many different environments. Templates allow with the use of variables to copy configuration files and update specific sections. Finally, Tasks are information collected about each server, such as IP addresses, memory, and disk space. Facts are used in order to help with server communication.

5. Roles: Roles are a special kind of Playbook. They are used for better organization and use of the tasks. Roles are fully self-contained with tasks, configuration templates, variables, and other files. Roles, in Ansible, include files and combine them to form reusable abstractions in order to save from any unnecessary steps which consume time. Ansible galaxy is a community site for sharing Roles.

31.4.2 Setup and Configuration

In order to create a Virtual Cloud Environment, as mentioned before, there are many tools and open-source programs that can be used. However, in the specific book chapter the cloud environment is represented with the use of Vagrant by HashiCorp. Vagrant can create many machines inside the same environment and each command typed is executed on all of the existing machines in the specific environment.

Initially a Vagrant box must be chosen from the publicly available list of Vagrant boxes. However, a developer can easily create one instead of choosing one from the preexisting list. There is a vast variety of Virtual Machines that can be used with a large number of operating systems in different versions such as Ubuntu, OpenSuse, CentOS, Fedora, Debian, OpenBSD, Solaris, and Kali Linux.

Regardless of the chosen Vagrant Box, the procedure to create a host or multiple hosts is the same (https://www.vagrantup.com/docs/). With the simple commands vagrant init and vagrant up the Virtual Machine is ready for use. In order to connect to the created machine, SSH is used and is represented with the command vagrant ssh.

With the production of the Virtual Machine, a VagrantFile is created. The entire necessary configuration takes place inside the specific file, for example, changing the machine's internal memory. In order to create interconnected virtual environment specific changes must take place in VagrantFile.

The following lines must be placed inside the configuration VagrantFile in order to create multiple machines, thus create the Virtual Cloud Environment (n represents the number of machines that will be created):

```
(1..n).each do |i|
  config.vm.define "node-#{i}" do |node|
    node.vm.hostname="node-#[i]"
      inline: "echo hello from node #{i}"
  end
end
```

Each host must be defined inside a host inventory file in order to be automatically picked up when the process of the configuration begins.

Vagrant integrates with configuration management tools such as Ansible. In order to create this connection, so that with the use of this automation tool the remote hosts will be secured, changes must be made inside VagrantFile. These changes are made in order to be able to use the created Ansible Playbook containing all the necessary configuration settings. Inside VagrantFile the line config.vm.provision "shell" must be uncommented and the following lines must be added:

```
config.vm.provision "ansible" do |ansible|
     ansible.playbook  ="playbook.yml"
     ansible.sudo= true
end
```

Once Ansible is installed, an Ansible Configuration File will be created defining the path to the host inventory, demonstrating all the created remote hosts.

With the procedure described, the virtual environment is created and multiple hosts are interconnected. With the use of the configuration management tool Ansible, the process of securing the specific hosts is initialized.

Although specific commands tend to differ between different operating systems, the basic steps in order to secure multiple hosts using Ansible are the same. Ansible uses Roles, which are a special kind of Playbooks containing tasks, configuration templates, variables, and other files. Roles, in Ansible, include files and combine them to form reusable abstractions in order to save from any unnecessary steps, thus organize the configuration better. Roles are simply created by typing ansible-galaxy init Rolename inside the created and chosen Roles directory.

Each Role created can contain many tasks which are similar to each other. Including task files allows breaking the configuration policy into smaller parts. However, there can be as many roles as possible in order to organize the procedure better. Roles allocation is objective and can be organized according to the developer's needs.

Once the Vagrant boxes are up and running, Ansible runs the Playbook that calls all the additional roles created in order to harden multiple machines created by Vagrant. At the following command line there is a simple Playbook example, which demonstrates how Playbooks execute all the additional roles.

```
-hosts: all
 become: true
 roles:
   -Role
```

Security is an on-going process. There are configuration steps that will make a server more secure, however, the risk factor remains. As mentioned before, although the command line for the installation and configuration of different tools using Ansible is different in each operating system, the basic configuration steps remain the same.

Securing Linux Server Distributions, basic configuration steps include:

1. Updates: Keeping system software updated is one of the biggest security precautions that can be taken, in order to harden any operating system. Software updates could be patches for major security vulnerabilities or even minor bug fixes. Since new versions are constantly released, computers are

no longer receiving all the necessary security patches. So system's update and upgrade is essential for security reasons. Using Ansible, a role must be created using Ansible Galaxy that runs update and upgrade commands for multiple machines, since Vagrant's and Ansible's host directories are interconnected and automatically all hosts are picked up. Once the machines are up and running the necessary commands will be run.

2. Creation and Configuration of Deploy User Accounts: Another important step that must be taken in order to ensure the safety of the server is to assign user accounts for each machine using encrypted passphrases. In order to assign users for multiple machines, an additional directory inside the main one must be created, which contains a variables file with every user including its hashed password. Inside Ansible's role directory assigned for the particular action there will be a task file including the necessary command line for the creation of users and encrypted passwords, as well as a variables file. The variables file contains the usernames and encrypted passwords which will be called by the specific task file.

3. Harden SSH Access: A cryptographic key-pair for the SSH access must be used for every account created, because it secures the system and makes brute-force attacks more challenging. SSH keys are a pair of cryptographic keys that are used to authenticate the access to a SSH server. Prior to the authentication process a private and public key-pair is created. The public key can be shared with anyone as opposed to private keys which are kept secret by the user. In order to configure the particular SSH key authentication the public keys must be placed inside the server in a special directory. When each created user connects to the server must use its private key in order to ensure that the client owns it. The server then allows the client to connect without a password. Setting up SSH key authentication allows disabling password-based authentication method. SSH keys include more bits of data ensuring in that way that a potential attack cannot brute-force attack in order to intercept the connection.

4. SSL Certificate Installation: The cryptographic key-pair used for application traffic encryption uses Secure Socket Layer (SSL) or Transport Layer Security (TLS) connection. A SSL Certificate is a method used in order to distribute a public key about a server and the organization responsible for it. Certificates can be digitally signed by a Certification Authority (CA), which is a trusted third party confirming that the information contained in the certificate is accurate [15]. A Certificate Installation must be made in all of the machines created by Vagrant in order to ensure a safe connection and encrypted application traffic. The generated certificates and keys must be placed inside a variables file for each account created and be called by the assigned role created for the particular step.

5. Installation and Configuration of a Firewall: There are many open-source security firewalls for Linux Systems such as IpTables, Ufw, IPCop Firewall, ShoreWall, and IpFire. The correct and more useful must be chosen, according to the used operating system, in order to protect the server from Dos attacks and unwanted system intrusion. Using a firewall to block unwanted inbound traffic provides a high level of security [16]. Using Ansible, the proper role

must be created including all the commands in order to secure the server using firewalls. Inside the created role nothing hardcoded is used, such as IP addresses or port numbers, but everything is being imported in a default variables file. The assigned role will call all the variables needed in order to run the specific security configuration step. Handlers are used in order to restart the service each time the machines are up and running.

6. Disable IPV6: A potential attacker could send malicious traffic via IPV6, since it is not always monitored by the administrators. Unless network configurations require IPV6, a proper configuration step in order to harden a server is to disable it. Using Ansible the proper command must be placed in the assigned role in order to secure each host from security intrusions.

7. Secure Shared Memory: Shared memory can be used in an attack against a running service, httpd or apache2, so a necessary configuration step is to secure it. Using Ansible a line must be added in fstab file, inside the assigned role, in order to secure the shared memory. The system then must be rebooted or the mount command must be run.

8. Determine Running Services: An important security step is to determine all the services running in each machine in order to be able to monitor the system. Each operating system uses different command line tools for monitoring running services such as netstat, tcpdump, top, and vmstat. The assigned role is being called by the Playbook, using Ansible, which contains the necessary commands for the particular configuration step.

9. Monitoring Tools: Applications create event files (LogFiles) to keep track of activities taking place at any given time. Monitoring tools must be installed and configured in each machine created by Vagrant, in order to monitor and analyze the event files and be able to prevent security issues. There is a vast variety of monitoring tools for each operating system such as Logwatch, Graylog 2, Logstash, and Logcheck.

10. Intrusion Detection System (IDS): Intrusion detection system is a software that monitors hosts or networks for security violations and malicious activity. Any activity detected is then reported to the system's administrator. The installation of such program is an important configuration step in order to prevent any potential threat. There is a vast variety of open-source intrusion detection tools such as Snort, AIDE, Suricata, Bro, and Kismet. According to the needs of a system's administrator the proper IDS is chosen. Using Ansible, the assigned role contains all the necessary task and variables files, in order to successfully install and configure the proper software.

11. Search for Rootkits: A rootkit is a malicious collection of computer programs which is designed to enable access to an unauthorized user to a computer or computer network. An attacker installs a rootkit on a computer by obtaining user-level access. This installation is made by exploiting a known vulnerability or by cracking a password. When the installation is complete it allows the attacker to gain privileged access to the system. There are many open-source tools that can be used for rootkit detection, for every remote host created by Vagrant using the automation tool Ansible, such as Chrootkit and Rkhunter.

31.5 Discussion

Using IT automation tools leads to an efficient and productive way to ensure cloud safety. This derives from the many benefits of the use of automation such as the elimination of repetitive tasks, the reduction of several errors that may occur in the configuration, the depravity of the configuration complexity, and the avoidance of outdated resources. The use of Vagrant and Ansible helps create and configure lightweight, portable environments avoiding maintenance and setup time, leading to the maximization of productivity and flexibility. Cloud hosts were configured in order to harden their systems to potential malicious attacks, with the use of the management configuration tool Ansible. Since security is an on-going process, we can never ensure absolute safety in the cloud. Although computer security's purpose is the solution of security issues potential adversaries leverage this knowledge and several security gaps in order to launch system attacks.

31.6 Conclusions

Cloud Computing is a model for enabling pervasive, convenient, on-demand network access to a shared pool of resources such as networks, servers, storage, applications, and services. These services regard hardware, software, data resources, and several applications which are stored in data centers and can be accessed through the internet. Over the recent years, Cloud Computing has grown in popularity, since many Information Technology organizations work and deliver services over cloud platforms. This expansive use of the cloud derives from the vast variety of services provided, as well as the many advantages of its use.

However, the major disadvantage of Cloud Computing is the security issues faced. The potential vulnerabilities can be exploited by malicious users in order to launch cloud-specific attacks. Furthermore, today economic aspects in information are considered [17]. In this particular book chapter, some of the major security issues were discussed and some of the main attacks were demonstrated. Due to the many benefits of IT automation, remote hosts were configured with the use of configuration management tools in order to be secured.

In order to create a Virtual Cloud Environment, open-source tool Vagrant was used. Vagrant is used to create and configure lightweight, portable development virtual environments which can easily be reproduced. It manages all the important configurations, in order to avoid all the unnecessary maintenance and setup time and is controlled by a single consistent workflow, in order to help maximize the productivity and flexibility. Finally, in order to automate the procedure of securing the created environment a configuration management tool must be used such as Ansible. Ansible is an IT automation engine that is used to automate cloud provisioning as well as configuration and service orchestration and makes application and system deployment easier.

Acknowledgment This work is supported under the framework of EU COST IC 1306: CRYP-TOACTION (Cryptography for Secure Digital Interaction) Project.

References

1. Mell, P., & Grance, T. (2011). *The NIST definition of cloud computing, recommendations of the National Institute of Standards and Technology* (NIST Special Publication 800-145. P 2,3).
2. Chauhan, M. Λ., Babar, M. A., & Benetallah, B. (2016). Architecting cloud-enabled systems: A systematic survey of challenges and solutions. p. 1. doi: 10.1002/spe.2409
3. Bulusu, S., & Sudia, K. (2012. *A study on cloud computing security challenges* (pp. 7–9). Dissertation, School of Computing Blekinge Institute of Technology, Sweden.
4. Popovic, K., & Hocenski, Z. (2010). Cloud computing security issues and challenges. In *MIPRO, 2010 Proceedings of the 33rd International Convention.*
5. Rastogi, N., Gloria, M. J. K., & Hendler, J. (2015). Security and privacy of performing data analytics in the cloud: A three-way handshake of technology, policy, and management. *Journal of Information Policy, 5*, 133–142.
6. Shrivastava, N., & Yadav, R. (2013). A review of cloud computing security issues. *International Journal of Engineering and Innovative Technology, 3*(1), 551.
7. Albeshri, A., Ahmad, A., & Caelli, W. (2010, September, 1–3). Mutual protection in a cloud computing environment. In *IEEE 12th international conference on high performance computing and communications (HPCC 2010)* (p. 641), Melbourne.
8. Swanson, M., & Guttman, B. (1996). *Computer security. generally accepted principles and practices for securing information technology systems* (NIST Special Publication 800-14). p. 3.
9. Grobauer, B., Walloschek, T., & Stocker, E. (2011). Understanding cloud computing vulnerabilities. *IEEE Security and Privacy Magazine, 9*(2), 50–57. doi:10.1109/MSP.2010.115. p. 3–6.
10. Mahmood, Z. (Ed.). (2014). *Cloud computing challenges, limitations and R&D solutions* (p. 5, 8–18). Heidelberg: Springer.
11. Antonopoulos, N., & Gillam, L. (Eds.). (2010). *Cloud computing principles, systems and applications* (p. 31). Heidelberg: Springer.
12. Song, D. X., Wanger, D., & Tian, X. (2001, August 13–17). Timing analysis of keystrokes and timing attacks on SSH. In *Proceedings of the10th USENIX security symposium* (p. 5). Washington, DC.
13. Sklavos, N. (2011, April 6–8). Cryptographic algorithms on a chip: Architectures, designs and implementation platforms. In *Proceedings of the 6th Design and Technology of Integrated Systems in Nano Era (DTIS'11)*, Greece.
14. Jensen, M., Schwenk, J., Gruschka, N., & Iacono, L. L. (2009, September 21–25). On technical security issues in cloud computing. In *Proceedings of the 2009 IEEE international conference on cloud computing(CLOUD '09)* (pp. 109–116), Bangalore.
15. Gohel, H. (2015). Design and development of combined algorithm computing technique to enhance web security. *International Journal of innovative and Emerging Research in Engineering, 2*(1), 77.
16. Selvi, V., Sankar, R., & Umarani, R. (2014). The design and implementation of on-line examination using firewall security. *IOSR Journal of Computer Engineering, 16*(6), 21.
17. Sklavos, N., & Souras, P. (2006). Economic models and approaches in information security for computer networks. *International Journal of Network Security, 2*(1), 14–20.

Chapter 32
A Survey and Comparison of Performance Evaluation in Intrusion Detection Systems

Jason Ernst, Tarfa Hamed, and Stefan Kremer

32.1 Introduction

Intrusion detection systems (IDS) are a rapidly changing field—out of necessity since security is often a cat-and-mouse game. The detection systems must constantly evolve in order to keep up with new attacks and vulnerabilities. Due to the rapidly changing nature of attacks and the IDSs created to detect these attacks, there is not much standardization in the field. While there have been some attempts in this direction, it is typically focused on standard datasets which catalog known attacks. Increasingly, to combat the rapidly changing nature of attacks and vulnerabilities, IDS are turning towards machine learning which provides some flexibility in detecting novel attacks. With the overview and comparison of the different IDSs provided in this chapter, the reader may then compare how new and existing systems compare with each other in the context of these previously known attacks, and their ability to adapt to new threats. The aim of this chapter is to collect and present a common framework for comparison of IDSs wherever possible, and also to highlight systems which are specialized towards specific attacks, and those systems which provide widespread and general coverage. In addition, where possible (based on data provided in the surveyed approaches), we aim to present comparison metrics so that the reader may try to gauge which IDS approaches may be the most effective in particular circumstances. First, we survey popular standardized datasets, and metrics which are commonly used to evaluate and compare IDSs and then we provide an in-depth survey and overview of particular IDSs.

J. Ernst (✉) • T. Hamed • S. Kremer
University of Guelph, Guelph, ON, N1G 2W1, Canada
e-mail: jason@left.io; tyaseen@uoguelph.ca; skremer@uoguelph.ca

© Springer International Publishing AG 2018   555
K. Daimi (ed.), *Computer and Network Security Essentials*,
DOI 10.1007/978-3-319-58424-9_32

32.2 Standardized Datasets for Benchmarking IDSs

One popular dataset is the KDD-99 dataset [1] which "contains a standard set of data to be audited, which includes a wide variety of intrusions simulated in a military network environment". The KDD-99 dataset covers four types of attacks: DOS (Denial of Service), R2L (Remote to Local), U2R (User to Root) and Probe [2]. Another well-known dataset for benchmarking IDS is the Lincoln Labs dataset from 1999 [3]. It also covers four types of attacks which can be classified into the same categories as the KDD-99 dataset. While both of these datasets are over 15 years old, they are still used in recent papers, particularly because they are so commonly used. However, modern intrusions require updated datasets including attacks such as those from botnets which recently caused significant security issues with the Internet [4]. A more modern dataset called CTU-13 [5] has been recognized by some researchers as the most valuable botnet dataset [6]. There are also many active discussions about which datasets should be used to replace KDD-99 and Lincoln Labs datasets, as well as suggestions for datasets for particular types of attacks, different training techniques and other situations by researchers [7, 8].

32.3 Metrics Used to Evaluate and Compare IDSs

In the majority of the papers surveyed, there are few common metrics which are used to evaluate the performance across all of the approaches. Most of the performance metrics are illustrated in Fig. 32.1. Typically the most common are [9]:

1. *False positive rate* which is the percentage of traffic identified as anomalies or threats but are actually legitimate traffic.
2. *False negative rate* which is the percentage of traffic which is actually anomalies or threats but are missed by the system.

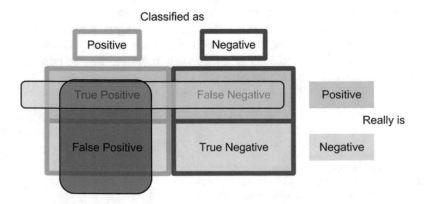

Fig. 32.1 Some common metrics illustrated (*precision red*, recall *yellow*) [8]

3. *True positive (or attack detection) rate* which is the percentage of actual attacks or anomalies which are correctly identified by the system.
4. *True negative rate* which is the percentage of traffic which is correctly identified as not an anomaly or threat.
5. *Precision*: is the number of true positives divided by the total of true positives and false positives.
6. *Sensitivity/recall*: is the number of true positives divided by the total of true positives and false negatives.
7. *Processing speed/capacity* which is the amount of time it takes an IDS to classify all of the attacks in a dataset.
8. *Training speed* which is the amount of time a system must be trained on a dataset to reach the effectiveness reported in the results.
9. *System utilization*: the amount of memory and CPU required to run the IDS [10]

32.4 Specification Method

The specification-based method is one where "manually specified program behavioral specifications are used as a basis to detect attack" [11]. Misuse detection, in contrast, is unable to detect attacks which have not been seen before, and anomaly detection often has high false positive rates. Specification methods instead focus on characterizing legitimate program or network behaviour. The biggest drawback would be every time new behaviour is required, it would need to be included into the specification method so as to not result in a false positive. For the specification-based method which was presented in [12], the experiment was directed to lower layer protocols (TCP and IP) which correspond to Denial of Service (DoS) and probing attacks in the Lincoln Labs data. The interesting points of the experiment were [12]:

1. A high rate of attack detection was achieved: The system proved its ability to detect all the attacks (100% rate of attack detection) that were supposed to be detected according to the prototype.
2. The system was able to detect IP-sweeps also which many anomaly detection systems fail to detect.
3. A low rate of false positives was achieved: The system proved to produce a very low false positive rate (about 5.5 false alarms/day).
4. The system exhibited an acceptable processing capacity: The system proved fast enough to be able to process an entire day's data within only 10 min.

For the IP machine, Table 32.1 explains the details of the detected attacks [12]. The experiment was performed on 43 attacks varied in type from DoS attacks, probing attacks and eavesdropping attacks. As shown from Table 32.1, the system was able to detect the whole number of given attacks.

Table 32.1 Attacks detected in 1999 Lincoln labs IDS evaluation data [12]

Attack name	Attacks present	Attacks detected	Description	Threat category
Apache2	2	2	DoS attack on Apache web server	DoS/comm.
Back	3	3	DoS attack on Apache web server	DoS/comm.
IP Sweep	6	6	Probe to identify potential victims	Access/IP sweep
Mailbomb	3	3	Large volume of mail to a server	DoS/comm.
Mscan	1	1	Attack tool	DoS/comm.
Neptune	3	3	SYN-flood attack	DoS/comm.
Ping-of-death	4	4	Over-sized ping packets	DoS/system
Smurf	3	3	ICMP echo-reply flood	DoS/comm.
Queso	3	3	Stealthy probe to identify victim OS	Access/ eavesdropping
Satan	2	2	Attack tool	Access/ eavesdropping
Portsweep	13	13	Probing to identify exploitable servers	DoS/comm.
Total	43	43		

32.5 Support Vector Machines

As opposed to manually coding behaviours, misuse or attacks, it is also possible to train systems to detect improper behaviour. Typically, Support Vector Machines (SVM)s, which are a type of machine learning, require a training stage along with labelled data which indicates normal traffic or attack traffic. The approach taken in [13] is a hybrid SVM which combines a trained SVM with a portion which can detect novel attacks so that it is flexible and has potentially better performance than completely untrained SVMs. The empirical evaluation phase of [13] involved three steps:

1. To verify the efficiency of the packet filtering approach, labelled datasets were used. A normal dataset consisted of collected normal packets during certain agreed times for the experiment. More than one dataset is used, the abnormal dataset #1 represents attack packets using TCP/IP covert channels, abnormal #2 represents DoS attacks and abnormal #3 contains malformed TCP/IP packets that disobey the published TCP/IP standard. The details of packet filtering using TCP/IP are given in Table 32.2, where the last column in the table represents the ratio of the number of filtered packets to the total number of packets [13].
2. Classification was performed using a soft-margin SVM, which showed high performance as expected, since the soft-margin SVM is an established, well-developed, time-proven supervised learning method. In spite of the above, it is

Table 32.2 Packet filtering results using passive TCP/IP finger printing [13]

	No. of packets	Inbound filtering	Outbound filtering	No. of filtered packets	Ratio (%)
Normal #1	1,369,134	1,249,924	39,804	1,289,728	94.20
Normal #2	1,574,274	1,435,838	46,374	1,482,212	94.20
Normal #3	1,571,598	1,439,880	44,071	1,483,951	94.40
Normal #4	1,783,444	1,589,337	65,035	1,654,372	92.80
Total	6,298,450	5,714,979	195,284	5,910,263	Ave: 93.84 STD: 0.74

not suitable for detecting novel attacks because of its supervised nature. The one-class SVM used an RBF kernel which showed excellent performance (94.65%), but the false positive rate was as high as well [13]. Since the enhanced SVM employs two kinds of machine learning features, it showed that it is very sensitive to feature mapping using a kernel function, since it incorporates two types of machine learning features. The enhanced SVM experiment with a sigmoid kernel outperformed in two areas: it produced similar detection performance as the soft-margin SVM, and it produced lower false positive and negative rates than the previous methods.

3. The third stage focused on comparing results with a real NIDS: In order to confirm the efficiency of the proposed approach, a comparison with a real NIDS was shown. Some well-known NIDSs (like Snort and Bro) are used for comparison. Those NIDSs are de facto standards used in the industry. The proposed framework is compared using the enhanced SVM method with NIDS systems based on real data.

Four kinds of test sets were used in each test experiment as follows:

1. Test #1 was an attack-free first source dataset not used during training
2. Test #2 contained ten kinds of attacks found in the data captured by the second source
3. Test #3 contained nine kinds of attack data from the DARPA IDS Test
4. Test #4 was comprised of real data collected by the second source.

The detailed results of the proposed framework with comparison to Snort and Bro are given in Table 32.3.

32.6 Machine Learning

In addition to SVMs there are also many other machine learning approaches. Several of the approaches based on classifier decision trees and decision rules are evaluated in [14]. The empirical evaluation was performed by comparing the performance of the several classification algorithms. The comparison involved calculating True

Table 32.3 Real-world test results [13]

		Test sets	Detection rate (%)	False positive rate (%)	False negative rate (%)
Enhanced SVM	TR1	Test#1—Normal	92.40	7.60	–
		Test#2—Attacks	68.70	–	31.30
		Test#3—Attacks	74.47	–	25.53
		Test#4—Real	99.80	0.20	–
	TR2	Test#1—Normal	94.53	5.46	
		Test#2—Attacks	66.90		33.10
		Test#3—Attacks	65.60		34.40
		Test#4—Real	99.93	0.07	
	TR3	Test#1—Normal	95.63	4.37	
		Test#2—Attacks	64.20		35.8
		Test#3—Attacks	57.20		42.80
		Test#4—Real	99.99	0.01	–
	Analysis	Test#1—Normal	Ave: 94.19, SD: 1.64	Ave: 5.81, SD: 1.64	Ave: 0.00, SD: 0.00
		Test#2—Attacks	Ave: 66.60, SD: 1.26	Ave: 0.00, SD: 0.00	Ave: 33.40, SD: 2.26
		Test#3—Attacks	Ave: 65.76, SD: 8.64	Ave: 0.00, SD: 0.00	Ave: 34.24, SD: 8.63
		Test#4—Real	Ave: 99.90, SD: 0.10	Ave: 0.09, SD: 0.09	Ave: 0.00, SD: 0.00
Snort		Test#1—Normal	94.77	5.23	–
		Test#2—Attacks	80.00	–	20.00
		Test#3—Attacks	88.88	–	11.12
		Test#4—Real	93.62	6.38	–
Bro		Test#1—Normal	96.56	3.44	–
		Test#2—Attacks	70.00	–	30.00
		Test#3—Attacks	77.78		22.22
		Test#4—Real	97.29	2.71	–

While TR1, TR2, TR3 mean training set 1,2 and 3, respectively

Positive (TP), False Positive (FP) and Prediction Accuracy (PA). An evaluation was made for these criteria to determine the best algorithm for the given attack. Table 32.4 shows the evaluation criteria for decision rules used in [14]. For the first column, CCI means Correctly Classified Instances and ICCI means Incorrectly Classified Instances [14]. This evaluation revealed that different algorithms that should be used to process different types of network attacks since some algorithms have an outstanding detection performance on specific types of attacks, compared with others [14].

Table 32.4 Evaluation criteria for decision rules [14]

Evaluation criteria	Classifier rules			
	JRip	Decision table	PART	OneR
CCI	63,810	63,250	63,979	56,909
ICCI	1724	2284	1555	8625
PA (%)	97.36	96.51	97.62	86.83

32.7   Behaviour-Based Approaches

Behaviour-based approaches focus on trying to capture, represent and encode the behaviour of malware, or an attack in a standardized format. An example of such a system is Malware Instruction Set (MIST) [15]. The biggest benefit of focusing on representing the behaviour is that it works well with other classification approaches. Good representation can enhance classification. Furthermore, due to the standardized and efficient encoding, it is possible to reduce the size of the reports generated on systems running IDSs. For behaviour-based analysis which was presented in [15], the empirical evaluation phase involved using 500 malware binaries. Then, the system used six independent antivirus software applications to assign the samples to five different classes. The assigned malwares were then executed under control of CWSandbox to produce 500 behaviour reports. That resulting distance matrix for the MIST showed high classification accuracy by the dark colour of the matrix diagonal (the darker the better). The results showed that MIST performed better than the XML representation especially in classifying Allaple and Looper malwares. In addition, the size of the resulting MIST report was very small as compared with XML, which leads to a reduction of analysis run-time as a result [15].

32.8   Mobile Agents

While many of the previous approaches apply on a single node, or even in a centralized manner, mobile agents allow the network to co-operatively determine if the network is being attacked. Since the task is divided, one of the benefits is often decreased load in terms of CPU and memory, however, it often comes at the cost of increased overhead in terms of network communications in order to coordinate with each other. The empirical evaluation phase of using a mobile agent with intrusion detection was a little bit different. In [10], the empirical evaluation phase consisted of studying the system resource utilization and capturing data statistics. For the resource utilization, results showed that the IDS has very low occupancy of the system resources (memory and CPU). For capturing data statistics, results showed that the IDS is very sensitive so that it can detect the "ping command"—which is a command used for checking a network connection—from the sharp change of the

ICMP package [10]. In [16], which proposes an intrusion detection model of mobile agent based on Aglets (IDMAA) for detecting intrusions, the empirical evaluation can be illustrated via two properties [16]:

1. Strength of IDMAA: The system consisted of some detection agents, and these agents were independent from each other, so that if one agent was stopped, it will affect the detection rate on that node only. That confirms that the system is fault-tolerant and robust.
2. Dynamic adaptability: If a host generated an agent and sent them to another host, there is no need to restart the system. The same applies when a host disposes of an agent; that does not affect the other detection agents. IDMAA can increase or decrease the detection agents according to the need to enhance detection performance. That is the idea of dynamic adaptability.

32.9   Genetic Network Programming

Similar to the goals of the hybrid SVM approach in [13], the Genetic Network Programming (GNP) approach in [17] tries to balance the ability to detect novel attacks with having too high of a false positive rate. The empirical evaluation phase of the approach presented in [17]—which focuses on using GNP for intrusion detection—consisted of two parts: choosing the sample space and checking the detection rates. After selecting the training sample space in the data collecting phase, the testing sample space was specified by dividing the database into two parts: 748 unlabelled normal connections and 320 unlabelled intrusions (Neptune, smurf, portsweep and nmap) connections. For checking detection rates, the results are given in Table 32.5 [17].

In order to evaluate the testing results, some criteria were calculated such as: Positive False Rate (PFR), Negative False Rate (NFR) and accuracy. The PFR corresponds to the false positive rate, the NFR corresponds to false negative rate and the accuracy corresponds to detection rate, respectively. Those criteria were calculated for the above table as follows [17]:

PFR $= (4+6)/748 = 1.34\%$
NFR $= (0+24)/320 = 7.5\%$
Accuracy $= (738+170+55)/1068 = 90.16\%$

Table 32.5 Simulation results of GNP intrusion detection [17]

	Normal (T)	Known intrusion (T)	Unknown intrusion (T)	Total (T)
Normal (R)	738	4	6	748
Known intrusion(R)	0	170	70	240
Unknown intrusion (R)	24	1	55	80
Total	762	175	131	1068

Table 32.6 Detection results for testing dataset of DOS, Probe, R2L and U2R with Clustering, C4.5 Rules, RIPPER and FILMID [18]

Attack class		Clustering	C4.5 Rules	RIPPER	FILMID
DoS	DR (%)	83.64	91.89	92.34	92.37
	FPR (%)	3.13	2.41	2.36	2.38
Probe	DR (%)	71.45	81.29	81.42	81.41
	FPR (%)	0.68	0.54	0.47	0.46
R2L	DR (%)	68.26	78.61	78.65	78.71
	FPR (%)	1.52	1.12	1.14	1.13
U2R	DR (%)	11.85	17.53	17.56	17.49
	FPR (%)	0.47	0.31	0.35	0.35

32.10 Fast Inductive Learning

"Given a series of positive examples and negative examples about one concept, the task of inductive learning is to induce a common concept description from these examples". In [18], which used fast inductive learning for intrusion detection (FILMD), the empirical evaluation phase began after training FILMD, RIPPER and C4.5 rules on the training datasets (in the detection phase), the system was tested on the specified dataset for the four attack classes. Next, some of the criteria were calculated for the three algorithms and clustering-based anomaly detection against all the attack classes. These criteria include: Detection Rate (DR) and False Positive Rate (FPR). The obtained results are shown in Table 32.6 [18]. The FILMD approach is similar to other machine learning approaches [13, 14], but the focus is on fast detection time while still maintaining high performance. The RIPPER and C4.5 Rules approaches are other competing inductive learning algorithms used in [18] to evaluate the performance of FILMD

It can be concluded that the FILMID algorithm which was based on double profiles has given better results than the other three algorithms (Clustering, C4.5Rules and RIPPER). In addition, the detection time was also calculated for both the RIPPER and the FILMID algorithms using the same computer and operating system. The detection time for the FILMID algorithm was better than that of RIPPER since the FILMID algorithm generates fewer rules than the RIPPER algorithm [18].

32.11 Situational Awareness

Situation awareness tries to capture sequential patterns in attack data, rather than focusing on classifying a single event. The idea is that a chain of suspicious events makes it much more likely that an attack has occurred. Since it determines which events are likely related, the number of attack events can be greatly reduced since many anomalous traffic events may be part of the same attack. A different empirical evaluation phase is presented in [19], which used situational awareness for intrusion

Table 32.7 Comparison of the detection results between SVM and Back propagation [18]

Algorithm	Total number/the number of error categorization	Detection accuracy
SVM	350/331	94.57%
Back propagation	350/307	87.71%

detection. Sensors provide input for the events in the IDS. After reporting the records of the sensors during all the stages, the alert events generated from security sensors will be exposed to simplifying, filtering, fusing and correlating. This led to decreasing warning events from 64,481 to 6,164. A risk value was also calculated which could be used to know the route of the attack [19].

32.12 Back Propagation

To some extent, the empirical evaluation phase in [18] was similar to what was used in the previously mentioned machine learning approaches. The KDD-Cup99 dataset of MIT Lincoln Lab was used in the experiment. Four SVMs were used to generate the detection model. The training dataset was divided into 300 samples of normal data and 200 intrusion samples from DoS, U2R, R2L and Probing datasets. The testing dataset was divided into 200 normal samples of normal data and 150 intrusion samples from DoS, U2R, R2L and Probing datasets. The detection rates given by the system based on SVM and back propagation network are shown in Table 32.7 [18]. In [20], the model was evaluated by calculating two performance measures: Detection Rate and FAR after applying back-propagation algorithm on the KDD cup99 dataset.

32.13 Fuzzy Logic

Fuzzy logic is a relatively new approach in intrusion detection. The benefit of a fuzzy system is that training of the system (in a machine learning sense) is not required. On the other hand, it is not as flexible to adapt to attacks which are novel. For instance, consider an attack that is nothing like an attack that is defined within the fuzzy rules—it will likely be misclassified. Consider a second attack that is similar to one of the approaches, but not quite the same—it may be misclassified by the fuzzy rules. The empirical evaluation used in [21], which was based on fuzzy logic, was specified for calculating the overall accuracy. Calculating the overall accuracy was based on some criteria such as the False Positive (FP), False Negative (FN), True Positive (TP) and True Negative (TN) rates. After calculating all the above criteria for the system, the classification performance of the system was noticeably improved so that it gave more than 90% accuracy for the four types of attacks [21].

Table 32.8 Metrics used

Metric	Papers
True Positive (TP)	[14, 22–24]
False Positive (FP)	[2, 14, 17, 22–28]
False Negative (FN)	[2, 17, 22–24, 26, 28]
True Negative (TN)	[22–24]
Recall	[22, 23, 29]
F-measure	[22, 23, 29, 30]
Precision	[22, 23, 29]
Overall accuracy	[20, 22–24, 28–33]
Prediction Accuracy (PA) or Detection Rate(DR)	[2, 14, 17, 20, 25–27, 30, 33–37]
False Alarm Rate (FAR)	[2, 12, 20, 23, 27, 29–31, 33, 35–37]
Correctly Classified Instances (CCI)	[14]
Incorrectly Classified Instances (ICCI)	[14]

32.14   Comparisons of IDSs by Metrics Used

Generally, researchers use different kinds of empirical evaluation metrics to evaluate their IDS. To summarize, Table 32.8 above shows the most used metrics and the papers that used them.

32.15   Conclusion

In conclusion, in this chapter has provided a review of the contemporary intrusion detection systems literature structured around a metrics and performance evaluation criteria. The aim of the chapter is to provide the reader with a different and new review and taxonomy of the IDSs spanning a variety of approaches, with a particular emphasis on approaches which are flexible and perform well with respect to the metrics outlined. First, we outline the typical datasets used in intrusion detection system performance evaluation. Then, the metrics used to evaluate the performance of IDSs were also defined and explained. The chapter then gives an overview of a large variety of approaches which are used to detect anomalies, intrusions and improper behaviour on the network. Lastly, this chapter explained the empirical evaluation of the IDS by discussing multiple metrics used in these kinds of systems including the standard ones and custom ones. The value of this chapter lies not only on its treatment of the source papers discusses, but also in its novel style in presenting the information about IDSs to the reader. The chapter's concept is to make the reader flows with the stream of the data from the input through the internal processing to the final output decision. This manner gives researchers a comprehensive knowledge about ID and what has been done in this field until now in terms of pre-processing, classifier algorithms and the achieved results. In addition, this style helps the reader to find which feature can be used in detecting certain kind of intrusions and which papers have used that. Another benefit of this approach is that it can reveal which papers have used training and testing or testing data only. It is hoped that this chapter will have a significant impact on future research in the IDS area by providing readers new to this area which a "jumping-off point" into the source literature. In addition, the structure of the review should provide some perspective of how researchers can investigate specific aspects of IDS and what solutions have been previously explored within each aspect. In addition, the review conducted important comparisons and provided some critiques after each component of IDS supported by some tables to give the reader a better perspective about that particular component. Intrusion detection will remain an interesting research topic for as long as there are intruders trying to gain illicit access to a network. The discipline represents a perpetual arms-race between those attempting to gain unauthorized control and those trying to prevent them. We hope that this chapter has provided an overview of this fascinating field and a starting point for future study.

References

1. Cup, K. (1999). *Dataset.* Available at the following website http://kdd.ics.uci.edu/databases/kddcup99/kddcup99.html
2. Sharma, V., & Nema, A. (2013). Innovative genetic approach for intrusion detection by using decision tree. In *2013 international conference on communication systems and network technologies* (pp. 418–422).
3. Lippmann, R., Haines, J. W., Fried, D. J., Korba, J., & Das, K. (2000). The 1999 DARPA off-line intrusion detection evaluation. *Computer Networks, 34*(4), 579–595.
4. J. G. Elevate Communications (2016). Terabit-scale multi-vector DDoS attacks to become the new normal in 2017, Predict DDoS Experts, *Business Wire.*
5. García, S., Grill, M., Stiborek, J., & Zunino, A. (2014). An empirical comparison of botnet detection methods. *Computers & Security, 45,* 100–123.
6. Małowidzki, M., Berezinski, P., & Mazur, M. (2015). Network intrusion detection: half a kingdom for a good dataset. In *Proceedings of NATO STO SAS-139 Workshop.* Portugal.
7. Scully, P. (2016). Where can I get the latest dataset for a network intrusion detection system?. *Quora* [Online]. Available: https://www.quora.com/Where-can-I-get-the-latest-dataset-for-a-network-intrusion-detection-system. Accessed January 12, 2017.
8. ubershmekel (2012). Precision, recall, sensitivity and specificity. *Ubershmekel's Uberpython Pythonlog* [Online]. Available: https://uberpython.wordpress.com/2012/01/01/precision-recall-sensitivity-and-specificity/. Accessed February 09, 2017.
9. Natesan, P., Balasubramanie, P., & Gowrison, G. (2012). Improving the attack detection rate in network intrusion detection using adaboost algorithm. *Journal of Computer Science, 8*(7), 1041–1048.
10. Mo, Y., Ma, Y., & Xu, L. (2008). Design and implementation of intrusion detection based on mobile agents. In *2008 IEEE international symposium on IT in medicine and education* (pp. 278–281).
11. Uppuluri, P., & Sekar, R. (2001). Experiences with specification-based intrusion detection. In *Recent advances in intrusion detection* (pp. 172–189).

12. Sekar, R. et al. (2002). Specification-based anomaly detection: A new approach for detecting network intrusions. In *Proceedings of the 9th ACM conference on computer and communications security* (pp. 265–274). Washington, DC, USA.
13. Shon, T., & Moon, J. (2007). A hybrid machine learning approach to network anomaly detection. *Information Science, 177*(18), 3799–3821.
14. MeeraGandhi, G., Appavoo, K., & Srivasta, S. (2010). Effective network intrusion detection using classifiers decision trees and decision rules. *International Journal Advanced network and Application, 2*(3), 686–692.
15. Trinius, P., Willems, C., Holz, T., & Rieck, K. (2009). A malware instruction set for behavior-based analysis. Tech. Rep. TR-2009-07, University of Mannheim.
16. Xu, J., & Wu, S. (2010). Intrusion detection model of mobile agent based on Aglets. In *2010 international conference on computer application and system modeling (ICCASM 2010)* (Vol. 4, pp. V4-347–V4-350).
17. Gong, Y., Mabu, S., Chen, C., Wang, Y., & Hirasawa, K. (2009). Intrusion detection system combining misuse detection and anomaly detection using Genetic Network Programming. *ICCAS-SICE, 2009.*
18. Yang, W., Wan, W., Guo, L., & Zhang L. J. (2007). An efficient intrusion detection model based on fast inductive learning. In *2007 international conference on machine learning and cybernetics* (Vol. 6, pp. 3249–3254).
19. Lan, F., Chunlei, W., & Guoqing, M. (2010). A framework for network security situation awareness based on knowledge discovery. In *2nd international conference on computer engineering and technology* (Vol. 1, pp. V1-226–V1-231).
20. Jaiganesh, V., Sumathi, P., & Mangayarkarasi, S. (2013). An analysis of intrusion detection system using back propagation neural network. In *2013 international conference on information communication and embedded systems (ICICES)* (pp. 232–236).
21. Shanmugavadivu, R., & Nagarajan, N. (2011). Network intrusion detection system using fuzzy logic. *Indian Journal of Computer Science and Engineering (IJCSE), 2*(1), 101–111.
22. Sen, J. (2010). Efficient routing anomaly detection in wireless mesh networks. In *2010 first international conference on integrated intelligent computing* (pp. 302–307).
23. Aggarwal, P., & Sharma, S. K. (2015). An empirical comparison of classifiers to analyze intrusion detection. In *2015 fifth international conference on advanced computing communication technologies* (pp. 446–450).
24. Vyas, T., Prajapati, P., & Gadhwal, S. (2015). A survey and evaluation of supervised machine learning techniques for spam e-mail filtering. In *2015 IEEE international conference on electrical, computer and communication technologies (ICECCT)* (pp. 1–7).
25. Rieck, K., Schwenk, G., Limmer, T., Holz, T., & Laskov, P. (2010). Botzilla: Detecting the phoning home of malicious software. In *Proceedings of the 2010 ACM symposium on applied computing* (pp. 1978–1984).
26. Lane, T. (2006). A decision-theoretic, semi-supervised model for intrusion detection. In M. A. Maloof (Ed.), *Machine learning and data mining for computer security* (pp. 157–177). London: Springer.
27. Warrender, C., Forrest, S., & Pearlmutter, B. (1999). Detecting intrusions using system calls: alternative data models. In *Proceedings of the 1999 IEEE symposium on security and privacy (Cat. No.99CB36344)* (pp. 133–145).
28. Joo, D., Hong, T., & Han, I. (2003). The neural network models for IDS based on the asymmetric costs of false negative errors and false positive errors. *Expert System with Applications, 25*(1), 69–75.
29. Kolias, C., Kambourakis, G., Stavrou, A., & Gritzalis, S. (2016). Intrusion detection in 802.11 networks: Empirical evaluation of threats and a public dataset. *IEEE Communications Surveys Tutorials, 18*(1), 184–208.
30. Subramanian, U., & Ong, H. S. (2014). Analysis of the effect of clustering the training data in Naive Bayes classifier for anomaly network intrusion detection. *Journal of Advances in Computer Networks, 2*(1), 91–94.

31. Casas, P., Mazel, J., & Owezarski, P. (2012). Unsupervised network intrusion detection systems: Detecting the unknown without knowledge. *Computer Communications, 35*(7), 772–783.
32. Muzammil, M. J., Qazi, S., & Ali, T. (2013). Comparative analysis of classification algorithms performance for statistical based intrusion detection system. In *3rd IEEE international conference on computer, control and communication (IC4)* (pp. 1–6).
33. Tan, Z., Jamdagni, A., He, X., Nanda, P., Liu, R. P., & Hu, J. (2015). Detection of denial-of-service attacks based on computer vision techniques. *IEEE Transactions Computers, 64*(9), 2519–2533.
34. Bhuse, V., & Gupta, A. (2006). Anomaly intrusion detection in wireless sensor networks. *Journal of High Speed Networks, 15*(1), 33–51.
35. Zhao, Y. J., Wei, M. J., & Wang, J. (2013). Realization of intrusion detection system based on the improved data mining technology. In *8th international conference on Computer Science and Education*. Colombo, Sri Lanka.
36. Mahoney, M. V., & Chan, P. K. (2001). *PHAD: Packet header anomaly detection for identifying hostile network traffic* (Tech. Rep. CS-2001-4). Melbourne, FL: Florida Institute of Technology.
37. Sedjelmaci, H., & Senouci, S. M. (2015). An accurate and efficient collaborative intrusion detection framework to secure vehicular networks. *Computers and Electrical Engineering, 43*, 33–47.

Chapter 33
Accountability for Federated Clouds

Thiago Gomes Rodrigues, Patricia Takako Endo, David W.S.C. Beserra, Djamel Sadok, and Judith Kelner

33.1 Introduction

The computational service delivery has evolved according to users' necessities, increasing the rigorousness of the requirements along the years; it can be divided into the following distinct eras (Fig. 33.1): monolithic, client-server, web, Service-Oriented Architecture (SOA), and cloud computing [3, 8, 9]. Focusing on the last era, cloud computing has brought many advantages for both provider and customer. From the provider perspective, clouds facilitate the infrastructure management, providing resource control mechanisms (dynamic allocation, elasticity) and at the same time, minimizing the costs to a new infrastructure investment [15]. On the other hand, from the customer perspective, cloud computing represents a good and easy model to rent computational resources, offering on-demand self-service with a pay-as-you-go model.

However, cloud computing suffers from many weaknesses, and according to the NIST, *"security, interoperability, and portability [...] are the major barriers to broader adoption"* of the clouds [18]. Beyond that, evidence distributed among different machines with different hardware architectures, Operational Systems (OS), and infrastructures also increases the complexity in performing accountability properly.

T.G. Rodrigues (✉) • D. Sadok • J. Kelner
Federal University of Pernambuco, Recife, Brazil
e-mail: trodrigues@gprt.ufpe.br; jamel@gprt.ufpe.br; jk@gprt.ufpe.br

P.T. Endo
University of Pernambuco, Caruaru, Brazil
e-mail: patricia.endo@upe.br

D.W.S.C. Beserra
Centre de Recherche en Informatique, Université Paris 1 Panthéon-Sorbonne, Paris, France
e-mail: David.Beserra@malix.univ-paris1.fr

© Springer International Publishing AG 2018
K. Daimi (ed.), *Computer and Network Security Essentials*,
DOI 10.1007/978-3-319-58424-9_33

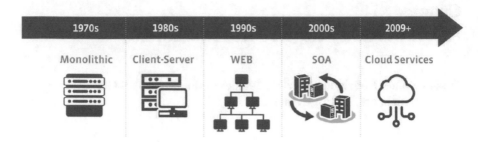

Fig. 33.1 Computational service delivery eras

Considering clouds interconnection, some authors named it as cloud federation or inter-cloud. The first uses the providers' interface, while inter-cloud is based on standards and open interfaces [23]. Nevertheless, both approaches aim to obtain interoperability [3] by offering new services that combine components from different clouds; in this way, we decided to use cloud federation and inter-cloud nomination interchangeably in this work.

Cloud federation provides scalability, hardware heterogeneity [5], more availability when compared against traditional clouds, geographic distribution to deal with disaster recovery and low latency access, avoiding vendor lock-in, as well as cost efficiency and energy savings [1, 3, 6]. However, interoperability between different clouds may be affected by integration problems or security breaches. In addition to the security concerns inherent to virtualized environments, cloud interoperation raises more security challenges, such as trust, authorization and identity management, and policy and interoperability control [23].

Furthermore, in federated scenarios, it is necessary to provide transparency between operations in order to guarantee that applications are fulfilling the Service Level Agreements (SLAs), that services are being billed properly and that security routines are being applied. For that, performing accountability is crucial but hard at the same time, because the evidences are spread across different servers, infrastructures, and applications, and the federation members may have different infrastructures with their own and specific security procedures, systems, and controls.

Considering accountability as *"the acknowledgement or assumption of responsibility for actions and decisions of persons or organizations that affects others"* [16], detail "when, where, what, who, and how" is one of the key functions for a proper accountability mechanism. The other key function of a proper accountability mechanism is store and retrieve the evidences, and increase the liability, transparency, responsiveness, responsibility, and controllability. Therefore, if the cloud computing environment allows accountable routines, it will be considered more trustworthy and, consequently, it will attract more customers.

Nonetheless, the current approaches do not support the audit process, infrastructure management, planning and billing at the same time. To properly support these four procedures, the information from infrastructure, virtualization and application layers must be collected. The related works of the state of the art that provide accountability routines [4, 6, 19, 20] collect information at one layer, supporting only billing process. They do not consider legal aspects related to the registry's safeguard, neither are able to provide different configurations according to the user needs. They also do not provide alerts routine when detect some SLA metric violation, contractual clauses, or nonstandard activities.

Considering the aforementioned motivations, the main objective of this chapter is to present relevant concepts about security and accountability in federated cloud, discuss main challenges in this research area, and propose a Cloud-based Accountability Framework for Federated Clouds, named CloudAcc, capable to enable audit process, infrastructure management, planning, and billing collecting the evidences dispersed through whole infrastructure.

This chapter is organized as follows: Sect. 33.2 discusses about cloud environments. Section 33.3 presents why accountability is important for federated clouds. After that, Sect. 33.4 describes the CloudAcc Framework thoroughly each framework module. Section 33.5 presents the CloudAcc implementation in a real federated cloud infrastructure. Finally, Sect. 33.6 will present the final considerations and future remarks.

33.2 Cloud Environments

Cloud computing has changed the way IT services are consumed. The computing utilities become consumed like other utility services available in the contemporaneity society. The main idea behind the IT services provided by a cloud provider is that the users must to pay only by the consumed resources. Furthermore, consumers should not expend money building a new and complex IT infrastructure to support their business needs. Currently, in order to expand the IT resources and provide more availability for their costumers, some cloud providers are operating in cooperation with other providers, composing what is called as cloud federation.

According to [23], cloud federation is the practice of interconnecting different cloud infrastructures, in order to provide scalability and hardware heterogeneity [6]; other key features include availability and disaster recovery, geographic distribution and low latency access, interoperability and avoiding vendor lock-in, legal issues and meeting regulations, as well as cost efficiency and energy savings [1, 3].

Cloud interoperability can be obtained by interface standardization or brokering. In an interface standardization approach all providers adopt the same interface. Nevertheless, developing a set of standards is difficult and hard to be adopted by all providers. The broker approach uses a service broker to translate messages among cloud interfaces making them interoperate with each other. A combination of these two approaches aforementioned often occurs in practice.

Despite the advantages provided in cloud federation considering the security aspects, the federation increases the security concern. Different access and authorization policies should compromise the access control management and communication problems may affect the services' availability. In addition, problems related to identity and access management, data security and trust and assurance are closely linked to these type of environment.

These security issues are even more critical when we consider other current scenarios integrated with cloud computing, such as fog computing and Internet of Things (IoT) communication. Both approaches have as main goal the idea to bring the computing more close to the users, when the process at the cloud is not viable (because it can take a long time to get the process result, for instance). In this case, due to several and different components existing in the environment, the accountability mechanism performs a crucial role, but even more complex too.

33.2.1   Accountability for Federated Clouds

Accountability is a concept that has different definitions. In [13], the author proposes five important aspects of accountability: transparency, liability, controllability, responsibility, and responsiveness. The author considers transparency and liability as the main important foundations of the accountability's concept. Transparency provides the way needed to assess organizational performance results. Liability means that organizational members should be held liable for their actions.

Another accountability concept that is closely connected to cloud computing environments is presented in [24]: *"accountability is a concept to make the system accountable and trustworthy by biding each activity to the identity of its actor. Such biding should be achieved under circumstance that all actors within the system are semi-trusted."*

In [7], the European Union Agency for Network and Information Security (ENISA) defines that accountability offers three capabilities: validation, attribution, and evidence. Validation allows users to verify if everything works as expected. Attribution allows, in the case of fault, that the responsibility can be assigned. Evidence is an artifact used to convince when a dispute arises. They identified *"falsifying record collection," "tampering with records,"* and *"destroying or suppressing the transmission of records"* as main threats against integrity of the collection, storage, and transmission of accountability registers.

These threats are also present in cloud computing and federated cloud environments, and compromise the forensic investigation [10]. Furthermore, an accountable infrastructure (Fig. 33.2) must provide sufficient evidences for: an audit process, infrastructure management, planning, and billing. Users should be able to trust that their contract will be fulfilled following the SLA, once their records can be checked.

Mixing the concepts aforementioned, we consider accountability as a mechanism that provides more transparency, responsiveness, responsibility, controllability, and liability through attribution or responsibility, evidence collection and validation, giving support for auditing, billing, management, and planning process.

Fig. 33.2 Accountable infrastructure

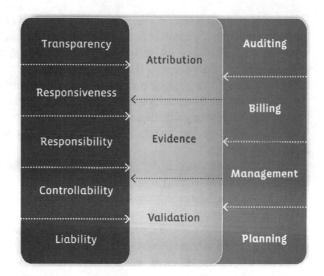

33.3   Why Accountability Is Important for Federated Clouds?

Despite existing effective countermeasures that mitigate and/or solve cloud security threats, there are many barriers to adopt cloud computing as deployment model. Why cloud computing and inter-cloud computing are not being used doubtless as infrastructure to support new applications and services? The answer for this question is that stakeholders do not trust that the cloud providers apply properly the control needed to provide security. Then we can conclude that applying security controls is not enough if the stakeholders do not trust that they are effectively applied.

Trust in cloud is a term that involves the confidence about confidentiality, integrity, availability, accountability, and auditability routines provided by a cloud provider [10]. This is the main concern considered in the decision to adopt or not a cloud computing environment. In this case, cloud consumers must trust in cloud providers to grant proper control mechanisms in order to avoid losing data confidentiality, integrity, and availability with accuracy in accounting [2]. A proper accountability routine increases the trust because it allows correctness billing, transparency and tractability, and increments the auditability support.

In a federated scenario, cloud providers must trust each other. In addition, the reputation of each federated member directly affects the global reputation [11, 23]. The establishment of trust in interconnected clouds is a complex and non-trivial activity because each federation member has its own security policies and procedures [12]. Federation exposes users' assets to new security concerns that were avoided when it was in an internal infrastructure. These risks include users' rights, transitive trust issues, and different system security requirements [15].

Public-key Infrastructure (PKI) is the common trust model adopted in cloud environments to provide user access. In federated cloud environments, the trust-worthiness is established by adding the Root Certified Authority (CA) certificate in chain-of-trust certificate for all federation members.

In federated clouds, despite the aforementioned security concerns, accountability is used as detective control, increasing the infrastructure security and trust, since the actions can be properly identified and logged. Therefore, strong actions identification and proper log generation and storage are the key requirements for a suitable accounting system.

An appropriate accountability mechanism in a federated scenario must be capable to increase security and confidence among the members of the federation. An accountable infrastructure must provide sufficient evidences for an audit process, infrastructure management, planning, and billing.

The existing projects that try to solve interconnection problems did not consider accountability in whole federated platform. Furthermore, an improper evidence collection affects the infrastructure trustworthiness, transparency, and liability. This work proposes a Cloud-based Accountability Framework for Federated Clouds, named **CloudAcc**, that provides cloud microservices to enable audit process, infrastructure management, planning, and billing collecting the evidences dispersed through whole infrastructure, increasing the trustworthiness, transparency, and liability. CloudAcc considers a multilayer evidence collection performing copy of logs to another infrastructure that will process and store them. Why the log processing is done in another infrastructure? The answer for this question is, because we want to avoid *"falsifying record collection," "tampering with records,"* and *"destroying or suppressing the transmission of records"* problems. In addition, the collection of evidence distributed for whole infrastructure may provide sufficient information, e.g., to detect malicious activities, misconfigurations, and correctness in billing process and SLA fulfillment.

Table 33.1 summarizes the acting areas for existing works focused in cloud federation that collect some accountability information with CloudAcc framework acting areas.

Considering the acting areas summarized by Table 33.1, we recognized that VM resources, services, and infrastructures must be accountable to increase the trustworthiness and provide proper evidence collection for a whole platform.

Table 33.1 Comparison with existing accountability frameworks

Framework	VM resources	Service	Infrastructure	Legal
RESERVOIR	x			
Cloudbus InterCloud	x			
OpenCirrus	x			
STRATOS		x		
CloudAcc	x	x	x	x

Considering the previous motivations, we propose the CloudAcc framework to increase the trustworthiness, transparency, and liability in cloud federation environments, improving the audit process, infrastructure management, planning and billing. In addition, the CloudAcc framework provides the security mechanisms needed to properly manage and store the collected evidences, solving or mitigating the security concerns related to accountability in federated environments.

33.4 CloudAcc Framework

Frequently, the integration between different clouds occurs through cross-signing the root certificate. This solution avoids the need of creating a new public-key hierarchy. In addition to this, the use of proxies is commonly adopted to provide locality transparency. In this work, we consider the use of proxies to provide cloud interconnection, as illustrated in Fig. 33.3.

Figure 33.3 depicts the peer-to-peer federation or interconnection between clouds C1 and C2. In this approach each federation member interconnects its own proxies through a secure tunnel (such as IPSec tunnel) that redirects the incoming requests to appropriate service.

The CloudAcc framework was designed to provide less impact in federation functioning or compromising the available resources from each federation member. In this way, CloudAcc can also be implemented in other scenarios, such as fog computing and IoT devices. Considering this requirement, it is divided into **CloudAcc Agent** and **CloudAcc Core**. Considering the federated scenario depicted in Fig. 33.3, CloudAcc Agent collects the evidences from the federation members C1 and C2, and sends them to another cloud that runs the CloudAcc Core.

The CloudAcc Agent collects the evidences in each federation member (clouds C1 and C2) and sends them through a secure channel for our accountability cloud that runs the CloudAcc Core. The evidences are collected considering three layers: (1) the infrastructure layer; (2) the virtualization layer; and (3) the system layer. The evidences in infrastructure layer can be collected using the Simple Network Management Protocol (SNMP). The administrator must configure the Agent with the credentials in order to properly collect the evidences in infrastructure layer. In virtualization layer, the CloudAcc Agent connects on hypervisor and collects the evidences in this layer. Lastly, in the system layer the CloudAcc Agent collects the system and applications logs.

On the other hand, the CloudAcc Core runs in a different cloud from the accounted federation. The main idea behind this is that providing all processing routines in another infrastructure (accountability cloud), we do not consume the available resources for each federation member. Beyond that, security problems, such as falsifying record collection, tampering with records and destroying or suppressing the transmission of records, can be avoided because we send all logs to other infrastructure, preventing that anybody modifies logs. This requirement is even more relevant if we consider fog computing scenario or IoT communication,

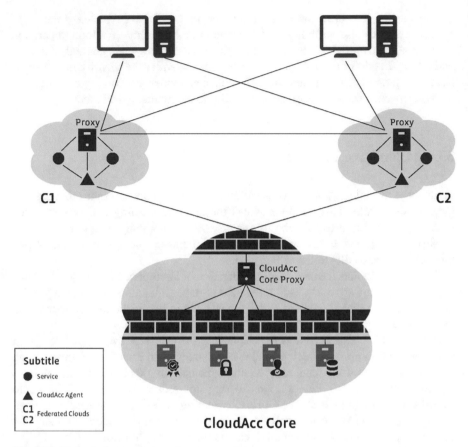

Fig. 33.3 Federated scenario overview

since the components, such as sensors, have less available resources. In addition, the registries stored in a trusted third-party increase the difficult of attacks against accountability because the attacker must corrupt the registries in federation and the registries in accountability cloud.

The CloudAcc Core was designed considering microservices architecture, that supports the security requirements, accountability routines, and legal issues. Moreover, each functionality should be configured according to the users' need. Each microservice has a local firewall that accepts only connections from other CloudAcc microservice or from the CloudAcc Core proxy. The CloudAcc Core is composed of two layers (Fig. 33.4): (1) Supporting Layer, and (2) Service Layer.

Fig. 33.4 General overview
of CloudAcc Core

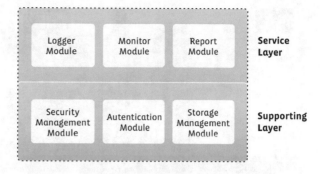

Fig. 33.5 SecMM
conceptual model

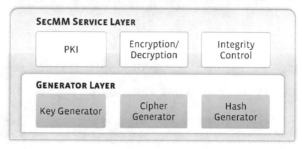

33.4.1 Supporting Layer

The supporting layer is composed of three modules: Security Management Module (SecMM), Authentication Module (AM), and Storage Management Module (SMM). Each module is independent but can consume the services from other modules. In addition, the supporting layer is responsible to perform security and access control, and store and retrieve all collected evidences.

The SecMM is a set of APIs that implements security routines interfacing hardware for secure computing as Trusted Platform Module (TPM) and Smart Cards to provide an appropriate security management for service infrastructure and client's data. The SecMM implements services, such as Public-key Infrastructure, Encryption/Decryption, and Integrity Control to support the security requirements needed to provide strong communication, confidentiality, integrity, authentication, and non-repudiation. This module is responsible to manage the symmetric and asymmetric keys, hash functions, digital signature, and Message Authentication Code (MAC) to support the security procedures needed to securely manipulate and store the evidences (Fig. 33.5).

The AM provides authentication and access control. We consider two authentication routines: infrastructure authentication and user authentication. In both approaches strong authentication is a mandatory requirement. Infrastructure authentication occurs to grant that all CloudAcc members are properly identified and accountable. CloudAcc considers two strong authentication methods for users: (1) digital certificate and smart card, and (2) a pair username/password with 2-

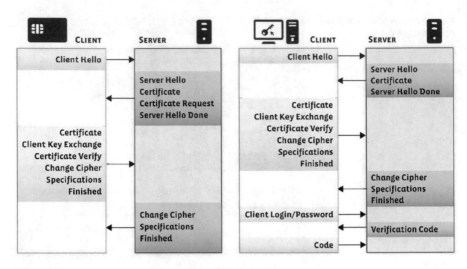

Fig. 33.6 Both client authentication approaches supported

Fig. 33.7 Secure storage
sequence diagram

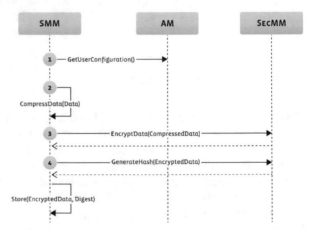

step verification. Each authorized user can configure his/her profile with personal settings that will be used to define who will access the user's data, and which cryptography will be used to protect the data (Fig. 33.6).

The SMM is responsible for providing secure storage of users' data based on settings established by themselves. It uses the services provided by SecMM and the configurations set in AM to execute security routines according to the user needs respecting the local legal issues, such as the Brazilian law #12965 or the National Security Systems (NSS) [17].

In Fig. 33.7, the sequence performed to grant the privacy in users' data is depicted. Firstly, the user configuration is retrieved from AM. The second step performs the data compression. After that, SMM uses SecMM to encrypt the compressed data following the configuration set by user. In the fourth step, to perform the integrity control, the encrypted data digest is generated. In the fifth step, SMM stores the data and its digest. The steps 1–4 are enough to grant the user's data privacy.

33.4.2   Service Layer

Service layer implements the core functionalities supported by CloudAcc. It consumes the services implemented in supporting layer to manipulate properly the users' data through interfaces and provides a set of functionalities that should be accessed through a RESTful interface. It is composed of Logger Module (LM), Monitor Module (MM), and Report Module (RM). This layer considers the configuration updated by each user in order to manipulate the evidences, monitoring the performance, and generating the reports according to the users' needs.

The LM is responsible for manipulating all distributed log collected on federated members executing merge/split operations. LM uses the modules presented in supporting layer (Sect. 33.4.1) to perform integrity control, confidentiality, authentication, and security storage.

The MM (Fig. 33.8) is responsible for checking if the SLA is being fulfilled, and if the security routines are in accordance with regulatory compliance. The RM (Fig. 33.9) provides reports about resource consumption, and alerts when the SLA is not fulfilled or some security configuration does not respect the regulatory compliance.

33.5   CloudAcc Implementation in a Real Federated Cloud

The CloudAcc was implemented in two distinct infrastructures, as shown in Fig. 33.10: at Google Cloud Platform (the CloudAcc Core) and at a real federated infrastructure (the CloudAcc Agent). In this example, C1 and C2 are the private clouds that federate their infrastructures.

Fig. 33.8 Monitor Module conceptual model

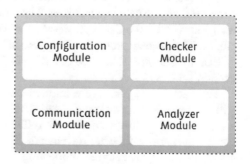

Fig. 33.9 Report Module
conceptual model

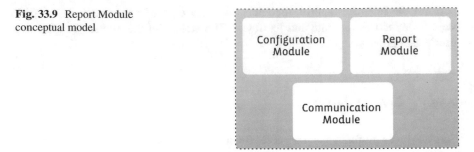

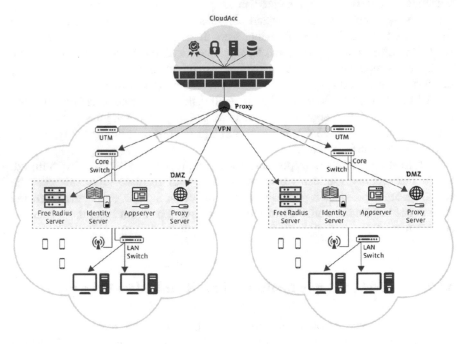

Fig. 33.10 CloudAcc implementation in a real scenario

The CloudAcc Core was implemented in the Google Cloud Platform, that was composed of two sets of machines. The first group is a Google f1-micro instance with one vCPU, 0.6 GB type RAM, 10 GB HD, Debian 8 and MySQL version 5.6.31. The second group is composed of Google g1-small instancies with one vCPU, 1.8 GB type RAM, 10 GB HD, Debian 8, Java SE Runtime Environment version 1.8.0, and Tomcat 8.

The CloudAcc agents were implemented in a real federated infrastructure. We will not provide infrastructure detail that we run the CloudAcc agent because we signed a Non-Disclosure Agreement (NDA). Considering this, Fig. 33.10 overviews our test scenario with generic elements. The black triangle in Fig. 33.10 represents a CloudAcc Agent running, collecting the evidences and sending them to the CloudAcc Core.

Agents collect evidences from three elements: the core switch, the web proxy, and the network authentication server. The web proxy is an instance with 4 vCPUs, 4 GB type RAM, 10 GB type HD, Debian 8, NGINX 1.9. The network authentication server is an instance with 2 vCPUs, 1 GB type RAM, 10 GB type HD, Ubuntu server 14.04, and Freeradius v3. The CloudAcc agent is a Java 1.8 implementation that includes the snmp4j version 2.4.3, jDOM version 2.0.6, libvit version 0.5, and JSON.simple API version 1.1.1.

We performed two types of experiments in our testbed: the performance and the proof-of-concept tests. The performance tests were executed in the Google Cloud Platform; and the second set of tests were performed in the real federated infrastructure. However, due to page limitation, we do not present these results here. For experimental results, please see [21].

33.5.1 CloudAcc Implementation Challenges

Despite the federation members still under the same timezone, we faced clock synchronization problems. In order to properly mount the users' activities across the servers the servers' clocks must still be synchronized. Furthermore, clock synchronization problems may compromise all accountability routines of CloudAcc. Clock synchronization may affect the CloudAcc agent, because the evidence collection in infrastructure and virtualization layers considers the local timestamp as key index of the collected evidence. To overcome these problems, we run a local Network Time Protocol (NTP) server instance that synchronized all the servers' clocks. Despite the security problems involving NTP, the time is the fundamental part for applications that use cryptographic routines (DNSSec, bitcoin, Time Stamping Authority (TSA), etc.) and we decided to adopt NTP in our framework considering the security concerns listed in [14] and their countermeasures.

Regarding the implementation challenges, we consider the CloudAcc core functionalities in microservices, and supporting the Brazilian "Marco Civil" requirements as the main challenging implementation problems. Modeling the CloudAcc in microservices architecture allowed scalability and reuse generated integration and communication problems. When the CloudAcc was deployed in Google Cloud platform, we faced several integration problems. As a workaround to solve the communication and integration problems, we configured internal static IPs for each machine running the CloudAcc microservices.

At last, there was the auditing process established in "Marco Civil" that specifies the log safeguard and the access. Considering the safeguard requirements, CloudAcc has encrypting routines to properly store the clients' information.

33.6 Final Considerations

A proper accountability system should support audit process, infrastructure management, planning, and billing. To support these functionalities, the accountability systems must collect evidences from physical, virtualization, and application layers. However, existing approaches in the literature do not support the four requirements, because they collect evidences in one layer only.

In this work, we proposed and implemented an accountability framework for federated cloud environment, named CloudAcc. Our main objective was implementing a framework to support audit process, infrastructure management, planning, and billing in federated environments according to the users' needs in compliance with the Brazilian "Marco Civil."

Another important aspect about security is its cost to the provider. According to [22], *"although deployment of such technologies may reduce security vulnerabilities and losses from security breaches, it is not clear to organizations how much they must invest in information security."* In this way, as a future work, we plan to analyze what is the cost to implement and maintain the CloudAcc framework and estimate the return of security by using the CloudAcc, as proposed by Sklavos and Souras, [22].

We also want to support accounting in Network Function Virtualization (NFV), that is an initiative that aims to implement network functions supporting interoperation from different hardware vendors. Furthermore, NFV organizes the network functions into service boxes that can be connected to create novel services. The NFV functionalities increase the network capabilities because it enables the creation, for instance, of security services using firewalls, and antispam, reducing the costs of maintaining the infrastructure and improving the use of the resources.

References

1. Aoyama, T., & Sakai, H. (2011). Inter-cloud-computing. *Wirtschaftsinformatik, 53*(3), 171–175.
2. Ardagna, C. A., Asal, R., Damiani, E., & Vu, Q. H. (2015). From security to assurance in the cloud: A survey. *ACM Computing Surveys (CSUR), 48*(1), 2.
3. Armbrust, M., Fox, A., Griffith, R., Joseph, A. D., Katz, R., Konwinski, A., Lee, G., Patterson, D., Rabkin, A., Stoica, I., et al. (2010). A view of cloud computing. *Communications of the ACM, 53*(4), 50–58.
4. Avetisyan, A. I., Campbell, R., Lai, K., Lyons, M., Milojicic, D. S., Lee, H. Y., Soh, Y. C., Ming, N. K., Luke, J. -Y., & Namgoong, H. et al. (2010). Open cirrus: A global cloud computing testbed. *IEE Computer Society, 43*(4), 35–43.
5. Barreto, L., Fraga, J., & Siqueira, F. (2015). Cloud federations and security attributes. In *2015 XXXIII Brazilian Symposium on Computer Networks and Distributed Systems (SBRC)* (pp. 140–149). New York: IEEE.
6. Buyya, R., Ranjan, R., & Calheiros, R. N. (2010). Intercloud: Utility-oriented federation of cloud computing environments for scaling of application services. In *Algorithms and architectures for parallel processing* (pp. 13–31). Heidelberg: Springer.

7. Castelluccia, C., Druschel, P., Hübner, S., Pasic, A., Preneel, B., & Tschofenig, H. (2011). Privacy, accountability and trust-challenges and opportunities. ENISA [Online]. Available: http://www.enisa.europa.eu/activities/identity-and-trust/library/deliverables/pat-study/atdownload/fullReport.

8. Dagger, D., O'Connor, A., Lawless, S., Walsh, E., & Wade, V. P. (2007). Service-oriented e-learning platforms: From monolithic systems to flexible services. *Internet Computing, IEEE, 11*(3), 28–35.

9. Erl, T. (2008). *Soa: Principles of service design* (Vol. 1). Upper Saddle River: Prentice Hall.

10. Farina, J., Scanlon, M., Le-Khac, N. -A., Kechadi, M., et al. (2015). Overview of the forensic investigation of cloud services. In *2015 10th International Conference on Availability, Reliability and Security (ARES)* (pp. 556–565). New York: IEEE.

11. Fernandes, D. A .B., Soares, L. F. B., Gomes, J. V., Freire, M. M., Inácio, P. R. M. (2014). Security issues in cloud environments: A survey. *International Journal of Information Security, 13*(2), 113–170.

12. Fernandez, E. B., Monge, R., & Hashizume, K. (2016). Building a security reference architecture for cloud systems. *Requirements Engineering, 21*(2), 225–249.

13. Koppell, J. G. S. (2005). Pathologies of accountability: Icann and the challenge of "multiple accountabilities disorder". *Public Administration Review, 65*(1), 94–108.

14. Malhotra, A., Van Gundy, M., Varia, M., Kennedy, H., Gardner, J., & Goldberg, S. (2016). The security of NTP's datagram protocol. Cryptology ePrint Archive, Report 2016/055. http://eprint.iacr.org/2016/055.

15. Mell, P., & Grance, T. (2011). The NIST definition of cloud computing [online]. Available: http://csrc.nist.gov/publications/nistpubs/800-145/SP800-145.pdf.

16. Nakahara, S., & Ishimoto, H. (2010). A study on the requirements of accountable cloud services and log management. In *2010 8th Asia-Pacific Symposium on Information and Telecommunication Technologies (APSITT)* (pp. 1–6). New York: IEEE.

17. National Security Agency and Central Security Service. (2016). Information assurance directorate. https://cryptome.org/2016/01/CNSA-Suite-and-Quantum-Computing-FAQ.pdf, Accessed: 2016-09-27.

18. NIST. (2010). Cloud computing. https://www.nist.gov/itl/cloud-computing. Accessed: 2016-05-27.

19. Pawluk, P., Simmons, B., Smit, M., Litoiu, M., & Mankovski, S. (2012). Introducing stratos: A cloud broker service. In *2012 IEEE Fifth International Conference on Cloud Computing* (pp. 891–898). New York: IEEE.

20. Rochwerger, B., Breitgand, D., Levy, E., Galis, A., Nagin, K., Llorente, I. M., Montero, R., Wolfsthal, Y., Elmroth, E., Caceres, J., et al. (2009). The reservoir model and architecture for open federated cloud computing. *IBM Journal of Research and Development, 53*(4), 4–1.

21. Rodrigues, T. G. (2016). *Cloudacc: A cloud-based accountability framework for federated cloud.* PhD Thesis.

22. Sklavos, N., & Souras, P. (2006). Economic models & approaches in information security for computer networks. *IJ Network Security, 2*(1), 14–20.

23. Toosi, A. N., Calheiros, R. N., Buyya R. (2014). Interconnected cloud computing environments: Challenges, taxonomy, and survey. *ACM Computing Surveys (CSUR), 47*(1), 7.

24. Yao, J., Chen, S., Wang, C., Levy, D., & Zic, J. (2010). Accountability as a service for the cloud. In *2010 IEEE International Conference on Services Computing (SCC)* (pp. 81–88). New York: IEEE.

Chapter 34
A Cognitive and Concurrent Cyber Kill Chain Model

Muhammad Salman Khan, Sana Siddiqui, and Ken Ferens

34.1 Stages of Cyber-Attack Kill Chain

In 2011, Lockheed Martin adapted the military notion of *kill chain*, which models the structure of a military attack for cyber security and intrusion in a computer network. They developed a kill chain to define the different stages of a cyber-attack and proposed an intelligence driven framework around this kill chain for the analysis, detection, and prevention of cyber-attacks and intrusions [1]. As shown in Fig. 34.1, there are seven stages of the traditional kill chain model: (1) Reconnaissance (R), (2) Weaponization (W), (3) Delivery (D), (4) Exploit (E), (5) Installation (I), (6) Command and Control (CnC), and (7) Actions on Objectives (A). This framework lays out a sequential model or process of how an adversary would carry out its malicious objectives. The model is based on the premise that threat actors attempt to infiltrate computer networks in a sequential, incremental, and progressive way. The model is structured so that if any stage of the kill chain is blocked then the attack will not be successful. Cyber security experts aim to detect a threat as early in the kill chain as possible to minimize losses. Nowadays, Security Operation Centers (SOC), Incident Response (IR), and Threat Intelligence (TI) teams in organizations use the cyber kill chain model as a mental process guide to analyze the data captured from attacks and take preventive measures at each step of the kill chain [2]. These seven sequential steps in a cyber kill chain provide information about the tactics, techniques, and procedures (TTP) of the adversary and are defined below.

M.S. Khan (✉) • S. Siddiqui • K. Ferens
Electrical and Computer Engineering, University of Manitoba, Winnipeg, MB, Canada
e-mail: muhammadsalman.khan@umanitoba.ca; siddiqu5@myumanitoba.ca;
ken.ferens@umanitoba.ca

© Springer International Publishing AG 2018
K. Daimi (ed.), *Computer and Network Security Essentials*,
DOI 10.1007/978-3-319-58424-9_34

Fig. 34.1 The Lockheed Martin cyber kill chain model

34.1.1 Reconnaissance (R)

Reconnaissance is the first step in the kill chain, during which a threat actor collects network and/or endpoint information about the potential target, who could be a single person, an organization, or a hardware/software part of the target network. At this stage, a threat actor performs covert investigation about the target entity, identifies the potential methods, and ways to break into the network. Also, research at this stage provides information on what type of malware objects could be deployed into the target network without getting detected by the cyber security defenses (vulnerabilities) [3]. Further, backdoors in the target network are found out [1]. Moreover, at this stage the adversary determines an appropriate set of infiltration objectives to be carried out into the target network. If the objective is to steal private information, then the threat actor must research, identify, and find a way to create a two way link, so that it can first enter the network, find the information of interest, and then steal it out of the network. If the objective is to destroy or disturb the network, then a one way link can do this task. Regardless, the threat actor aims to find a weakness of the network or the users to exploit and enter illegally. An example of a one way link is spam phishing emails, which are being evolved into delivering malware [4], in addition to their traditional objectives of stealing credentials, and are sent to a specific user with a malware file as an attachment; if the user opens that file, then system will be compromised and damaged. This example demonstrates how the threat actor exploits the email medium to attack the network. An example of a two way link is finding open ports using a port scanner; upon finding an open port, an authorized two way telnet communications session is created.

34.1.2 Weaponization (W)

Weaponization is the second stage, during which the threat actor develops the deliverable payload. An adversary uses the information collected at the reconnaissance stage (target vulnerabilities and backdoors) to prepare and plan *what* should be delivered, to exploit the vulnerability, and *how* should it be delivered, to utilize the discovered backdoor [5]. There are two types of payload that can be delivered at this stage: (1) malware that does not require any communication with the adversary, e.g., viruses and worms, or (2) malware that requires communication with the adversary to get command and control signals and/or send the stolen information back to the adversary. The latter are known as Remote Access Trojans (RAT). A RAT requires both client and the command and control server. The RAT client is the actual payload of the deliverable that is also configured on how to communicate back with the command and control server, which resides on the internet and is controlled by the adversary. For example, from the reconnaissance stage an adversary learned that at a certain university, the email system does not allow *.exe file but does allow *.pdf file attachments in the emails. Also, the adversary learned that professors routinely open *.pdf files of emails from students seeking registration to the university's graduate programs. Accordingly, the adversary creates a RAT malware file, with the capability of communicating with a command and control server, and embeds this malware file within a portable document file (pdf) named myCV.pdf, which is to be sent as an attachment via a phishing email. Another example is the credential harvesting mechanism to lure/force the target to visit a well-known, but counterfeited and cloned website and gather privacy information that will be exploited against victims.

34.1.3   Delivery (D)

After the malware payload has been developed and the backdoor to deliver the payload has been identified, the delivery stage is executed. The malware can be delivered either by luring or forcing the user to interact with the malware exploit or it can be delivered automatically by exploiting the weaknesses of the protocols and/or software. For example, the phishing email can be used to deliver the malware payload in a file attachment by duping the user to think that the email is genuine and thus download the malicious attachment; or, they are duped to input their privacy information in a typical phishing email that will be transmitted back to the attacker. Delivery is a critical part to ensure a successful attack by remaining undetected by existing security mechanisms. Therefore, adversaries design their attacks in such a way that although the traces of their attack may not be removed, however, those traces should masquerade the attack source from the security and forensic experts. Further, the threat actors utilize multiple delivery methods to increase their rate of success. The malware exploits which do not require user interaction are most difficult to catch, because they utilize an inherent flaw in the protocol, program, or software to deliver the payload or execute a piece of software for illegitimate purpose [6]. This inherent flaw is called a vulnerability of the software and requires software patching. For example, a malicious Java script within a Flash software file can be exploited to deliver the malware to the target computer as demonstrated in [7].

Following is a summary of various categories of malware delivery.

34.1.3.1   User Interaction Required

Emails Asking Users to Submit Private Data or Download an Attachment [8, 9]

Threat actors craft seemingly legitimate but phishing emails, which are targeted to many users, with the intention that some of these users will respond by following the instructions given in these emails. Typically, these emails pretend to be coming from a trusted organization, e.g., bank and telecommunications company, asking the users to share their account credentials. Also, phishing emails are crafted with malware attachments and lure the users to download those attachments pretending that they are downloading a valid and legitimate file. However, such files are either a well-known file type with the malware file embedded inside the file, or they are a malware saved in different file format other than an executable.

Malvertising [10]

Malicious advertisements are used to either exploit a vulnerability of web browsers, steal user information, or redirect the users towards infected websites to expose the malware directly to victim's computer. Malvertising redirects the user to the malicious web page which finds vulnerability in the user browser and delivers the malicious payload that could be a key logger to steal user passwords, to steal cookie information having user private data, or to install a malicious package [11].

Insider Threat [12]

Insiders are considered all the employees or trusted partners/members of an organization who are entrusted with network assets and know insider information of a network. A disgruntled employee could become a source of delivering a malicious payload inside a network using administrative/elevated accounts. Further, they could steal or reveal network credentials, thus exploiting the trust of the network.

34.1.3.2   No User Interaction Required

Web Exploits [13]

Web exploits are the vulnerabilities of web browsers/applications which an adversary can use to infiltrate/attack a computer or a network. Web browsers are used to connect to a server to download a web page or a file. For example, a multimedia web page requires content to be downloaded from different websites, and if there is a vulnerability that could redirect the web browser to the adversary website to download a malicious payload, then it could cause the user computer to execute the malware without user knowledge [14, 15].

Silent Downloaders [16]

These downloaders are also called the drive-by downloaders and are used to download malware at the target computer. These downloader software enter into the target computer by vulnerable websites or servers without the knowledge of the target. Typically, open source legitimate websites are exploited to transfer these downloaders into the target computers.

Code Injection [17]

Dynamic web technologies are based on processing strings of commands and data together. For example, when a web browser accesses a HTML webpage having JavaScript, then at the user computer, the web page runs the HTML code and JavaScript together, which is further processed by the web server. Cross site scripting (XSS) is a web technology to run different scripts inside a browser controlled by various web servers. An attacker can use XSS to run malicious code inside a web browser to execute a command from malicious server [18]. Also, in case of searching on SQL based web servers, SQL scripts are used behind the search button of web pages and can be exploited by attackers to inject malicious SQL code to perform malevolent activities.

Protocol Misuse [9]

In this type of attack, attackers use inherent flaws of communication protocols to carry out malicious objectives. For example, HTTP protocol uses referrer field in HTTP packet header to log information about the packet route. Legitimate websites use this information to optimize web pages on search engines. Attackers use HTTP referrer information of a legitimate website to increase ranking of abusive websites in search engine optimization and mislead the users to browse their websites, which can be further used to either generate spam emails or download malicious code.

Bot Attacks [19]

Bot is an internet software to perform certain web tasks automatically. Web crawlers are a type of bots that are used to index web sites on a search engine for optimization. However, bots are also used to deliver a malicious payload to the target computer by finding the open internet services automatically and creating internet connection or secure tunnels. These connections are used by the attackers to launch another bot to enter the network, either by exhaustive search of the admin account credentials, open further backdoors, or launch denial of service attacks [20].

34.1.4 Exploitation (E)

After the successful delivery of the malware payload to the target computer, the exploitation stage is initiated by installing the malware inside the target computer. As mentioned in [21], following conditions should be fulfilled to initiate a malware installation:

a. The malware should have the required access rights to be installed in the target computer.
b. The operating system or software of the target computer should be able to install the malware without additional requirements. For example, malware compiled for the Linux operating system cannot be installed on a Microsoft Windows operating system.
c. The antimalware defenses of the target computer should not be able to detect the malware otherwise the attack will fail causing cyber kill chain to be broken.

The exploitation stage does not actually perform the installation; rather, it prepares the environment to launch the installation stage of the kill chain. However, this stage is connected closely with the installation phase since all the requirements of the installation stage should be fulfilled by the exploitation phase. In order for the delivered payload to be installed, there must be a software or hardware bug that the payload can exploit for either installation or execution. These bugs are called Common Vulnerabilities and Exposure (CVE). A public database of the vulnerabilities is available at [22] which is updated as soon as a vulnerability is found out. Further, researchers and developers use this database to improve their software and provide patches. However, this database is not complete and there are unknown vulnerabilities which can be utilized by the threat actors to execute the exploit stage. Further, attacks can be a combination of the above categories, e.g., a web based ransomware attack is launched after exploiting both software vulnerabilities and network vulnerabilities [23]. CVEs can be categorized as follows [24].

34.1.4.1 Software Vulnerabilities

Vulnerabilities in this category can be exploited through a weakness in the operating system or the software installed in an operating system. Although, primarily an operating system is responsible for ensuring security through administrative access controls, there are some software, which can be misused for malware exploitation [25]. For an operating system exploit, an admin access or kernel level access is required. For example, buffer overflow is a software bug that requires access to privileged part of the memory that sometimes is only accessible by the kernel [26].

34.1.4.2 Network Vulnerabilities

Network vulnerabilities are also software vulnerabilities but exist in network devices, device drivers, or protocols which either switch or route internet traffic. Some examples of such vulnerabilities are HTTP protocol referrer misuse, Heartbleed bug in the OpenSSL protocol of transport layer [27], and router based split horizon route advertisement.

34.1.4.3 Embedded/Firmware/Hardware Vulnerabilities

Hardware programming is also called embedded or firmware programming. Controllers, ASICs, FPGAs, ASSPs, and System-on-Chips (SOC) are some examples of hardware where specialized firmware programming languages (Verilog, VHDL, Assembly) are used to develop dedicated operating systems or software. This category of devices is also vulnerable to cyber threats likewise the software and network vulnerabilities [28, 29]. These attacks have been used primarily to steal intellectual property or counterfeit the device. However, threat actors have started utilizing commodity hardware/firmware vulnerabilities to launch sophisticated threats using specialized Internet of Things (IoT) devices [30].

34.1.5 Installation (I)

A computer infection starts at the installation stage. If the malware is an executable file or the malicious activity is based on code injection or an insider threat, then the installation stage is not required. However, if the malware needs to be installed in the target computer, then the delivery stage should have delivered the dropper or downloader in the target computer and the exploit stage has already been completed by disabling the security defenses and finding a hook in the operating system to start the installation of the malware. At this stage, malware is installed and the installed files either use the libraries and support files of the operating system or acquire those files from the downloader or dropper packages [31]. Further, the malware installation updates the file access mechanism of the operating system using the privileged access rights. In addition, the malware changes the appearance of its files either by changing the format of the file or hiding those files from user access. Also, advanced malware are able to change their memory footprints to evade detection by sandbox algorithms or behavior based antimalware systems. These malware are also called as either polymorphic malware or metamorphic malware, respectively [32]. Polymorphic malware does not change the payload of the malware but changes the header so that it looks like a normal file in the memory. Metamorphic malware changes the pattern of the payload in the memory as well.

Installation stage not only installs the backdoor inside the target victim but also ensures that the threat actors are able to communicate with the victim computer persistently [33]. It is noted that this stage does not start communication with the command and control mechanism for the malware activity. Also, this stage differs from exploitation stage due to locality of the activity, i.e., exploitation stage ascertains that the malicious package is ready for installation and all requirements of exploitation stage are fulfilled while during installation stage, the actual payload starts setting up its foothold inside the victim computer locally. Also, a persistent connection with the server is established to initiate command and control communication. For example, a malicious web shell is installed on the victim computer at this stage which ensures persistence and avoidance from detection mechanisms using either new signatures or changing heuristics of its behavior by introducing polymorphism or metamorphism.

34.1.6 Command and Control (CnC)

Command and control networks can be servers, peer-to-peer networks, or social media servers. Further, there can be multiple levels of command and control networks (e.g., bot networks) to evade detection. Command and control stage is present in those malicious attacks whose objectives are as follows:

a. Steal the information (passwords, financial data, intellectual property, etc.) from the target computer, e.g., key loggers, Zeus, and Trojan.Coinbitclip [34].
b. Send instructions to the malware in the target computer to spread the malware to other parts of the network connected with the target computer, execute the malware, or activate the encryption for ransomware activities.

It is worthwhile to mention that for cyber defenders, CnC stage is the last stage to block the malicious activity. CnC stage takes place similar to normal internet communication. In addition to endpoint defenses, network perimeter monitoring plays an important role in detecting an illegitimate network connection. There are two major types of command and control servers based on communication:

a. Servers having meta information about the compromised nodes through beacons or heartbeat messages.
b. Servers communicating actively with the target nodes by issuing commands to control the victim nodes and performing more malicious actions like data exfiltration.

Further, command and controls servers can be divided into direct and indirect communication categories as follows:

a. In direct communication, malware at the victim node contains a list of various command and control server IPs so that if a certain IP is blocked, the malware hops for another IP to maintain the communication. This trait is also called persistence.

b. In indirect communication, threat actors utilize legitimate intermediary nodes to maintain communication. A group of legitimate nodes is compromised for the purpose of establishing communication link whereas the source of the link remains hidden from the victim [35]. Therefore, a botnet is created to re-route the communication from victim to the source.

Attackers may use different methods to establish an outbound connection. They may use email protocols to deliver the payload but then use single or multiple HTTP connections to establish an outbound link. Further, they may use compression mechanisms for siphoning data. Attacker would diversify their tactics and techniques to evade detection. However, the common denominator is network traffic and if an endpoint security mechanism cannot detect the presence of a communicating malware, network defenses can do the job.

34.1.7 Actions on Objectives (A)

This is the last stage of cyber kill chain and is responsible for executing the objectives of the attack. At this stage, if the malware is deployed in the target computer, then it starts performing the programmed function either through the instructions from command and control server or independently. This stage is also known as the detonation stage where the cyber kill chain is completed successfully.

Following are the major categories of actions that fall in this stage:

a. Data exfiltration: stealing private and intellectual data from the network.
b. Ransomware: Threat actors hold victim data hostage by encrypting the victim's data, change credentials or blocks network resources using various encryption methods, and demand ransom payment.
c. Cyber terrorism: threat actors inflict damage by erasing data or corrupting the files completely.

34.2 Critique of Cyber Kill Chain

As the internet has evolved from a traditional web model to a cloud based model, and with the prevalence of social networking and mobile technologies, traditional security measures including, but not limited to, firewall, antivirus, intrusion detection/prevention systems, and access control lists are proving to be futile and insufficient. Even with multiple layers of security defenses, threat actors are able to infiltrate a network and render persistence in their tactical measures and techniques. Persistence is a characteristic of threats whereby threat actors launch sophisticated attacks in such a way that the behavior and signature of those attacks evolve dynamically to remain undetected by the security defenses and the duration of attacks spans a large time interval [37]. These threats are typically orchestrated by organized entities and the threat actors are resourceful, patient, and aim to exfiltrate confidential information, damage the network's operational capabilities (e.g., taking down an online shopping website), and ultimately strike at the organization's credibility in the eyes of their clients and customers. For example [36], reported that Yahoo has declared hacking of more than 1 billion user account IDs and passwords from the Yahoo email servers in 2013 and 2014. This attack is known as the largest known attack in the history of internet thus far. This attack also involves stealing of user private information including date of birth and security questions. Also, these attacks include stealing Yahoo proprietary source code that is equivalent to stealing intellectual property information. This news of attack was disclosed when Yahoo was negotiating with Verizon Inc. to sell its Yahoo email business at a price of US$4.8 billion. This news has reportedly hurt this deal as well. According to [37], persistent threats should be analyzed using adaptive approaches and the network should be scanned continuously for new and old anomalies. Further, along with traditional security approaches, defense-in-depth and layered security should also be considered to stop Advanced Persistent Threats (APT). These defenses include sandboxes, honeypots, and machine learning engines to analyze existing data for a possible command and control activity. Existing cyber kill chain stages remain valid but the factor of persistence should be taken into account for those chains which are not completed yet. According to [38], a cyber kill chain attack contains one intrusion or one attempt of an intrusion while APT attack is composed of multiple kill chains. Each kill chain contains one of many indicators of attacks and an APT attack could be detected by evaluating and analyzing all such indicators together. For example, an APT attack starts with initial reconnaissance, skips weaponization, finds an exploit to escalate the privileges, performs an internal reconnaissance, moves laterally to the high valued computing nodes, waits and performs internal reconnaissance again, adapts the behavior according to the normal behavior of the network to avoid detection, ultimately infiltrates the target computer(s), and establishes command and control. As can be seen in Fig. 34.2, there are multiple reconnaissance stages and various stages of the kill chain are skipped. Further, there are many permutations of each cyber kill chain stage that could be taken by the APT actors to infiltrate a network successfully [39]. Also, it can be considered as reusing different stages of cyber kill chain multiple times to achieve the final malicious objective.

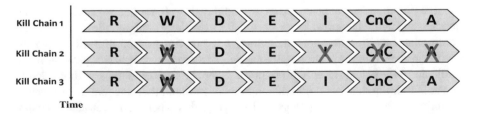

Fig. 34.2 Example of an APT attack with multiple cyber kill chains

Fig. 34.3 Marc Liliberte [41]—modified cyber kill chain

It is of paramount importance that security research should not be focused on conventional single kill chain; however, a holistic idea of multiple kill chains (persistence) should be taken into consideration when beefing up against threats. Further, it should be important to understand the motives of the attacker, rather than looking at what the attack is doing. For example, if the motivation is to steal financial information from servers, then there will be a command and control communication going outbound the network, and it should be visible to detect and stop. In order to curb APTs, there are various modifications proposed in the literature as follows:

a. Author in [40] suggested a modification to look at internal cyber kill chain stages. However, this model lacks the critical factor of analyzing the network defenses in a layered approach. Network perimeter security does an important job of holding back plethora of network based threats, such as illegitimate SSL connections.
b. Author in [41] proposed an updated cyber kill chain as shown in Fig. 34.3, where the weaponization stage is removed and the lateral movement is introduced after the command and controls stage, which emphasizes the horizontal movement of the attack to gain access on the targeted assets with the use of intermediary nodes. However, many advanced threats perform lateral movement using multiple command and control communications [40]; therefore, this model lacks a holistic approach in addressing the variety of threats altogether.
c. Authors in [42] discussed a holistic model covering the policy making, technical and legal aspects of an attack simultaneously. Their model performs better in generating a consensus among various stakeholders about an attack and deciding the course of action to thwart it. However, their work is more focused towards addressing the shared inference concerns of a threat faced by an organization and helping the stakeholders in converging towards a combined consensus.

Similarly there are varieties of other models that emphasize one or more aspects of the cyber kill chain and either add or modify the kill chain stages to reinforce the defenses.

34.3   Cognitive Cyber Security

Cognitive cyber security model is an adaptive method of analyzing data using machine learning, natural language processing, and artificial intelligence by mimicking how human brain functions and learns through processing data to detect cyber threats [20, 43]. Every internet connected organization relies on cyber security professionals to thwart attacks. These professionals are typically categorized as cyber incident response (IR) and threat intelligence (TI) teams [44]. Some organizations rely on outsourcing the cyber security tasks to dedicated cyber security organizations to look after their network while other organizations employee local teams to do the security job (security operation centers). Further, some organizations rely on both outsourcing and local cyber security experts and include banks, insurance companies, and government offices. This methodology is called as collaborative cyber security whereby different cyber security expert entities share their expertise to stop unavoidable cyber threats. With the evolution of machine learning, cloud based data analytics, and big data tools, cognitive cyber security has gained prominence to detect ever increasing sophisticated threats. Cognitive cyber security based first application is attributed to IBM which developed the first cognitive computing infrastructure called IBM Watson [45]. With the evolution of open source Hadoop infrastructure, big data analysis has become easier, faster, and near-real-time making possible the mining of large amount of data using machine learning to find correlation among various components of the network data including but not limited to network packet captures, firewall log and operating system logs etc. [43].

Machine learning, natural language processing, and artificial intelligence are a few fields of the broader spectrum of cognitive informatics and computing which is an active area of research [46]. Cyber security is a complex research domain and involves various heterogeneous aspects of the data and network including the user behavior and the analysis of mental model of the threat actors. It is necessary to incorporate such complexity in the cognitive data analysis for the detection of threats autonomously, adaptively, and dynamically. It is required to understand the intellectual model of the threat actors and how they bypass existing security defenses. Expertise of cyber professional experts is required to be incorporated in the cognitive analysis of threats. Although, traditional cyber kill chain is an abstract model but in the current era of high speed evolution of cyber threats, a modification is required to incorporate additional details of cognitive aspects. The authors argue that various stages of cyber kill chain should be combined into following four categories:

1. External and internal reconnaissance for exploitation of security weakness (**R**).
2. Delivery of the payloads (**D**).
3. Develop persistence to hide below the security radar using polymorphic and metamorphic behavior (**P**).
4. Command and control communications from the network and lateral movement within the network using endpoints/computing nodes (**CnC**).

As shown in Fig. 34.4, a cognitive time series of the proposed model is shown conceptually. If the time series is divided into N time steps, then each step constitutes a four phase cycle of the proposed four stages of cyber kill chain. An object could be a network trace, server logs, and packet captures. Moreover, the same object can be analyzed at multiple time steps but all the four stages are considered concurrently at each time step. Threat actors are capable of incorporating polymorphism and

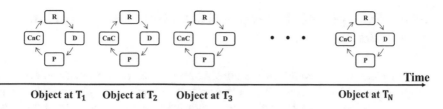

Fig. 34.4 Proposed cognitive analytical kill chain model for simultaneous analysis of data

metamorphism [32] in their attack techniques at each of the above four stages and therefore, any cognitive security system should also be able to focus on these categories concurrently [47]. In order to analyze a threat object, an analyst will not only look at the current indicators of compromise and indicators of attacks but will also analyze how a threat entered the network and what network defenses are compromised. Further, if a threat intelligence expert analyzes possibility of a future attack, then not only the objectives and actions of the threats are analyzed but the method of threat delivery and the techniques of those delivery mechanisms will be investigated as well. Compounding the analysis efforts, threat actors are also changing their techniques by making them low and slow [37]. This is being done to evade cyber defense mechanisms which correlate data anomalies against normal data. In addition, threats are able to spread laterally and thus pose extra challenge of analyzing infection both locally and globally. Therefore, the proposed model depicts an efficient and fast method of analyzing objects which is close in human thought process as well. In the next section, two examples are provided to validate the effectiveness of the proposed model.

It is noted here that the four stages of the proposed cyber kill chain model, R, D, P, and CnC, are not necessarily evaluated or considered in the order as shown in Fig. 34.4, however, a security expert concurrently uses these four stages to analyze an incident. With the traditional cyber kill chain model, it is observed that forensic analysis should begin with the first stage of reconnaissance which is not possible without knowing the details of the attack. Further, other stages cannot be analyzed until the malicious activity is analyzed first. It may be argued that traditional cyber kill chain model can be modified for forensic analysis by switching stages according to priority and available evidences. Nevertheless, there are two problems with this approach: (1) with state-of-the-art available tools for threat activities and the rise of attacks having targeted motivation (political or economic), all the seven stages cannot be differentiated and analyzed separately. Adversaries execute reconnaissance stage keeping in view their weaponization and exploitation capabilities. Therefore, in the proposed model, reconnaissance, weaponization, and exploitation stages of traditional stages are grouped in one reconnaissance stage, (2) traditional cyber kill chain treats threat analysis as a sequential analysis model while humans analyze attacks concurrently for all the stages of a cyber kill chain. Therefore, with concurrency, the proposed model closely complies with human mental model of analysis.

34.4 Examples of the Proposed Cognitive Analytical Kill Chain

In this section, two examples are considered to analyze the effectiveness of the proposed cognitive kill chain. These examples are based on mapping cognitive experience of Incident Response (IR) teams with the proposed model. In almost every organization having cyber security team, there are multiple layers of technologies deployed for cyber security. For example, in a hypothetical ecommerce organization, there are network defenses and endpoint defenses. Further, this organization follows an automated mechanism of patching the operating systems and software. There are multiple firewalls deployed for network defenses. Anti-phishing software is also installed over email gateways. There are antimalware systems installed on endpoints with strict domain level access controls. There are various data orchestration and analysis tools which ingests the data coming out of these defenses and provides an aggregated and human comprehendible view for IR team. As almost all the tools work on known signatures or traces, threat actors look for new ways to enter this organization network consistently to perpetrate a zero-day.

34.4.1 Example 1: Phishing Emails

Threat actors craft and send a very deceptive phishing email with malware attachment disguised as a marketing brochure to the finance department of this organization alluring them to download this attachment. It is noted that finance department can be an easy prey for such phishing campaigns as being non-cyber security professional. Let us assume that anti-phishing tool cannot find a known detectable pattern (signature of known regular expressions) and the email reaches to the users' inbox. Unless, a user is intelligent enough to find this email suspicious and convey this to the IR team or the signatures are updated by the security vendors, there is no way that the IR team could know about it. Therefore, this is a zero-day attack. In either case, as soon as the IR team come to know the existence of such emails, they start performing standard preventive actions as follows:

a. They quarantine the infected computers for further forensic analysis.
b. They issue a warning message to everyone in the organization to avoid being victim of such emails.

It is an established practice in Incident Response team to identify, analyze, and respond to the incidence in a timely manner to minimize the damage and reduce the associated cost [48]. Therefore, it is of paramount importance for IR teams to monitor the network traffic, computer logs, endpoint activities, and other related data to look for attacks. This includes observing traces of reconnaissance, e.g., port scanning activities, DNS traffic, user behavior analytics, etc. In the event of an attack incidence, IR team performs forensic analysis of the victimized computers [49], and does not only look for traces of malicious patterns in the quarantined computer, but also analyzes why the email went through their defenses undetected. At this stage, they do want to ensure that such emails should not go through their defenses again; therefore, they also find a way to update firewall rules. Further, based on the type of analysis, experts do want to see how the infected computer is trying to establish communication with outside servers including legitimate and unknown servers. Also, it is investigated to find if the infected computer has communicated internally with other computers and servers of the organization. If this process is mapped with the traditional kill chain model, the experts would be required to look for all the seven stages of kill chain sequentially, which, however, is not a cognitive approach. In sum, IR team concurrently investigates an attack for:

a. the malicious payload (D),
b. traces of malicious communication with other computers and servers (CnC),
c. analyzing the possible ways of how the phishing email reaches successfully to the finance department bypassing all the defenses (P),
d. what was the email content that attackers used to lure the users (R).

34.4.2   Example 2: DNS Denial of Service (DoS) Attack

In the hypothetical ecommerce organization, cyber experts find out an abnormally high DNS traffic and receive complaints from their users of slow browsing. After initial analysis, it is discovered that the organization's DNS servers are facing enormous traffic from legitimate internet servers. It could be a signal of a botnet based DoS attack [35]. However, further analysis may be required since the servers host online shopping websites and are connected to the organization for daily business operations. Typically, DNS based rate limiters are applied to reduce anomalously higher traffic, however, it may hurt the business operations if the rise in traffic is due to legitimate reasons, e.g., legitimate increase in traffic due to online shopping during holiday season. Therefore, security analysts first acquire the traces of the DNS traffic packet capture files/logs, look at the payload and IP addresses (CnC), and try to figure out if the traces conform to normal DNS communication (D), i.e., is there any repeated resolution of DNS name. Further, using IP address, they look for the source of traffic and look for any anomalous pattern. For example, if the DNS requests do not have legitimate payload consistently, then the traffic is a DoS attack because attackers may have used garbage payload or no payload for such DoS attacks. In addition, they will look for potential weaknesses of the system which can be exploited for this DoS attack, e.g., if DNS servers are updated with DNSSEC extension to ensure that valid DNS communication should be passed through the server (R). Further, they will also evaluate if the attack is showing any sign of persistence (P), e.g., if the DNS based DoS attack keeps changing the bot servers since this will be very hard to stop attack immediately using firewall black lists, etc.

34.5 Conclusion

This chapter provides a descriptive analysis of the existing cyber kill chain model to detect intrusion and threats. Different types of threats are also categorized with each stage of cyber kill chain. Further, a critique of the cyber kill chain model is discussed with modifications proposed in the context of evolving cyber challenges and cognitive mental model of cyber security experts. A new kill chain model is proposed that not only reduces the kill chain stages into four levels but also argues a simultaneous use of the four stages of the proposed model to analyze threats which is described as being close to human cognitive intelligence. Also, two practical examples are provided to describe the effectiveness of the proposed cyber kill chain model.

References

1. Hutchins, E., Cloppert, M., & Amin, R. (2011). Intelligence-driven computer network defense informed by analysis of adversary campaigns and intrusion kill chains. In *Proceedings of leading issues in information warfare and security research*.
2. NTT Group Security. (2016). *2016 NTT Group–Global threat intelligence report* (Online). Available: https://www.nttgroupsecurity.com
3. Achleitner, S., Porta, T. L., McDaniel, P., Sugrim, S., Krishnamurthy, S. V., & Chadha, R. (2016, October). Cyber deception: Virtual networks to defend insider reconnaissance. In *Proceedings of the 8th ACM CCS international workshop on managing insider security threats*.
4. PhishMe. (2016, June 4). *Q1 2016 sees 93% of phishing emails contain ransomware* (Online). Available: https://phishme.com
5. Cobb, S., & Lee, A. (2014, October). Malware is called malicious for a reason: The risks of weaponizing code. In *Proceedings of 6th IEEE international conference on cyber conflict (CyCon 2014)*.
6. Harbison, C. (2016, March 29). New ransomware installers can infect computers without users clicking anything, say researchers. *iDigitalTimes* (Online). Available: http://www.idigitaltimes.com
7. Brandt, A. (2016, April 25). *Android Towelroot Exploit used to deliver "Dogspectus" ransomware* (Online). Available: https://www.bluecoat.com
8. Rivlin, A., Mehra, D., Uyeno, H., & Pidathala, V. (2016, June). System and method of detecting delivery of malware using cross-customer data. U.S. Patent US9363280-B1, 7.
9. Mansoori, M., Hirose, Y., Welch, I., & Choo, K.-K. R. (2016, March). Empirical analysis of impact of HTTP referer on malicious website behaviour and delivery. In *Proceedings of IEEE 30th international conference on advanced information networking and applications (AINA)*.
10. Taylor, T., Xin, H., Wang, T., Jang, J., Stoecklin, M. P., Monrose, F., & Sailer, R. (2016, March). Detecting malicious exploit kits using tree-based similarity searches. In *Proceedings of the 6th ACM conference on data and application security and privacy (CODASPY)*.
11. Sood, A. K., & Enbody, R. J. (2011). Malvertising–exploiting web advertising. *Computer Fraud and Security, 2011*(4), 11–16.
12. Sanzgiri, A., & Dasgupta, D. (2016). Classification of insider threat detection techniques. In *Proceedings of the 11th annual cyber and information security research conference*.
13. Fang, Y., & Tung, Y.-Y. (2014, January). Patcher: An online service for detecting, viewing and patching web. In *Proceedings of IEEE 47th Hawaii international conference on system science*.

14. Salas, M. I. P., & Martins, E. (2015). A black-box approach to detect vulnerabilities in web services using penetration testing. *IEEE Latin America Transactions, 13*(3), 707–712.
15. University of Maryland. (2015, October 28). *Researchers find vulnerabilities in use of certificates for web security: Study finds website admins not revoking certificates, browsers not checking certificate revocation status* (Online). Available: www.sciencedaily.com
16. Kwon, B. J., Srinivas, V., Deshpande, A., & Dumitras, T. (2017, November). Catching worms, Trojan horses and PUPs: Unsupervised detection of silent delivery campaigns. In *Proceedings of network and distributed system security symposium.*
17. Taylor, T., Snow, K. Z., Otterness, N., & Monrose, F. (2016, February). Cache, trigger, impersonate: Enabling context-sensitive honeyclient analysis on-the-wire. In *Proceedings of network and distributed system security symposium (NDSS).*
18. Jin, X., Xunchao, H., Ying, K., Wenliang, D., & Yin, H. (2014, November). Code injection attacks on HTML5-based mobile Apps: Characterization, detection and mitigation. In *Proceedings of the 2014 ACM SIGSAC conference on computer and communications security.*
19. Stringhini, G., Hohlfeld, O., Kruegel, C., & Vigna, G. (2014, June). The Harvester, the Botmaster, and the Spammer: On the relations between the different actors in the spam landscape. In *Proceedings of the 9th ACM symposium on information, computer and communications security.*
20. Khan, M. S., Ferens, K., & Kinsner, W. (2015). Multifractal singularity spectrum for cognitive cyber defence in internet time series. *International Journal of Software Science and Computational Intelligence, 7*(3), 17–45.
21. Yadav, T., & Mallari, R. A. (2016, June). Technical aspects of cyber kill chain. In *Proceedings of international symposium on security in computing and communication.*
22. NIST. *National Vulnerability Database, DHS/NCCIC/US-CERT* (Online). Available: https://nvd.nist.gov/. Accessed 29 December 2016.
23. Cabaj, K., & Mazurczyk, W. (2016). Using software-defined networking for ransomware mitigation: The case of CryptoWall. *IEEE Network, 30*(6), 14–20.
24. Na, S., Kim, T., & Kim, H. (2016, November). A study on the classification of common vulnerabilities and exposures using Naive Bayes. In *Proceedings of the international conference on broadband and wireless computing, communication and applications.*
25. Zhang, N., Yuan, K., Naveed, M., Zhou, X., & Wang, X. (2015, May). Leave me alone: App-level protection against runtime information gathering on android. In *Proceedings of 2015 IEEE symposium on security and privacy.*
26. Muthuramalingam, S., Thangavel, M., & Sridhar, S. (2016). A review on digital sphere threats and vulnerabilities. *Combating Security Breaches and Criminal Activity in the Digital Sphere, 1*(21).
27. Durumeric, Z., Kasten, J., Adrian, D., Halderman, A. J., Bailey, M., Li, F., Weaver, N., Amann, J., Beekman, J., Payer, M., & Paxson, V. (2014, November). The matter of Heartbleed. In *Proceedings of the 2014 conference on internet measurement conference.*
28. Lee, R. P., Markantonakis, K., & Akram, R. N. (2016, May). Binding hardware and software to prevent firmware modification and device counterfeiting. In *Proceedings of the 2nd ACM international workshop on cyber-physical system security.*
29. Novotny, M. (2016, June). Cryptanalytical attacks on cyber-physical systems. In *Proceedings of 5th IEEE mediterranean conference on embedded computing (MECO).*
30. Ho, G., Leung, D., Mishra, P., Hosseini, A., Song, D., & Wagner, D. (2016, June). Smart locks: Lessons for securing commodity internet of things devices. In *Proceedings of the 11th ACM on Asia conference on computer and communications security.*
31. Xue, Y. L. (2014, March). Systems and methods for pre-installation detection of malware on mobile devices. Patent US9256738-B2.
32. Fraley, J. B., & Cannady, J. (2016, October). Enhanced detection of advanced malicious software. In *Proceedings of IEEE annual conference on ubiquitous computing, electronics and mobile communication conference (UEMCON).*
33. Alert Logic. (2016, December 30). *The cyber kill chain: Understanding advanced persistent threats* (Online). Available: https://www.alertlogic.com

34. Payet, L. (2014, February 9). *Hearthstone add-ons, cheating tools come with data-stealing malware*. Symantec Corporation (Online). Available: https://www.symantec.com
35. Khan, M. S., Ferens, K., & Kinsner, W. (2015, July). A cognitive multifractal approach to characterize complexity of non-stationary and malicious DNS data traffic using adaptive sliding window. In *Proceedings of IEEE 14th international conference on cognitive informatics and cognitive computing (ICCI*CC)*.
36. Goel, V., & Perlroth, N. (2016, December). *Yahoo says 1 billion user accounts were hacked* (Online). Available: http://www.nytimes.com
37. Siddiqui, S., Khan, M. S., Ferens, K., & Kinsner, W. (2016). Detecting advanced persistent threats using fractal dimension based machine learning classification. In *Proceedings of the 2016 ACM on international workshop on security and privacy analytics*.
38. Ussath, M., Jaeger, D., Cheng, F., & Meinel, C. (2016, March). Advanced persistent threats: Behind the scenes. In *Proceedings of IEEE 2016 annual conference on information science and systems (CISS)*.
39. Dell Secureworks. (2014). *Understand the threat* (Online). Available: http://www.secureworks.com/
40. Greene, T. (2016, August). *Why the 'cyber kill chain' needs an upgrade* (Online). Available: http://www.networkworld.com
41. Laliberte, M. (2016, September). *A new take on the cyber kill chain* (Online). Available: https://www.secplicity.org
42. Happa, J., & Fairclough, G. (2016). A model to facilitate discussions about cyber attacks. In M. Taddeo & L. Glorioso (Eds.), *Ethics and policies for cyber operations* (Vol. 124, pp. 169–185).
43. Grahn, K., Westerlund, M., & Pulkkis, G. (2017). Analytics for network security: A survey and taxonomy. In I. M. Alsmadi, G. Karabatis, & A. Aleroud (Eds.), *Information fusion for cyber-security analytics* (Vol. 691, pp. 175–193). New York: Springer International Publishing.
44. Jasper, S. E. (2016, November). U.S. cyber threat intelligence sharing frameworks. *International Journal of Intelligence and CounterIntelligence, 30*, 53–65.
45. Rashid, F. Y. (2016, November). *How IBM's Watson will change cybersecurity* (Online). Available: http://www.infoworld.com
46. Wang, Y., Widrow, B., Zadeh, L. A., Howard, N., Wood, S., Bhavsar, V. C., Budin, G., Chan, C., Fiorini, R. A., Gavrilova, M. L., & Shell, D. F. (2016). Cognitive intelligence: Deep learning, thinking, and reasoning by brain-inspired systems. *International Journal of Cognitive Informatics and Natural Intelligence (IJCINI), 10*(4), 1–20.
47. Thuraisingham, B., Kantarcioglu, M., Hamlen, K., Khan, L., Finin, T., Joshi, A., Oates, T., & Bertino, E. (2016, July). A data driven approach for the science of cyber security: Challenges and directions. In *Proceedings of IEEE 17th international conference on information reuse and integration*.
48. Ruefle, R., Dorofee, A., & Mundie, D. (2014). Computer security incident response team development and evolution. *IEEE Security and Privacy, 12*(5), 16–26.
49. Sivaprasad, A., & Jangale, S. (2012, March). A complete study on tools and techniques for digital forensic analysis. In *Proceedings of 2012 IEEE international conference on computing, electronics and electrical technologies (ICCEET)*.

Chapter 35
Defense Methods Against Social Engineering Attacks

Jibran Saleem and Mohammad Hammoudeh

35.1 Defense Against Social Engineering Attacks

Social Engineers have the potential to cause some serious damage to their victims. This damage can be social, economical, or reputational. It is now, more than ever, vital to understand what precautions can be taken to prevent, alleviate, and contain the devastation that can be potentially caused as a result of a Social Engineering attack. This section thus outlines the common Social Engineering mitigation strategies that companies and individuals can employ to protect themselves from potential Social Engineering attacks.

35.1.1 Physical Security

For any security-conscious business, a strong physical security must be enforced throughout the organization, without exception. If security is lax, attackers will have little trouble accessing the stations they need in order to launch their digital attack. In addition, once clear and concise security policies are established and implemented, they should be periodically tested in order to determine the state of security awareness among staff members. This is imperative to identify and resolve any potential gaps. It is equally important for members of staff to be continually reminded that the possibility of an attack is indeed real; it can occur at any time, without warning. Commenting on this issue, Kevin Mitnick asserts:

J. Saleem (✉) • M. Hammoudeh
Manchester Metropolitan University, Manchester, UK
e-mail: jibransaleem@gmail.com; m.hammoudeh@mmu.ac.uk

© Springer International Publishing AG 2018
K. Daimi (ed.), *Computer and Network Security Essentials*,
DOI 10.1007/978-3-319-58424-9_35

> People generally don't expect to be manipulated and deceived, so they get caught off guard by a Social Engineering attack. [1]

It is good practice to maintain signs throughout the premises, reminding employees not to plug-in any USB drives or any other digital device they find around the premises. Instead, they should submit them to the relevant department for expert analysis. In addition, they should be vigilant and report any suspicious behaviour to security, no matter how minor they perceive it to be. It is also a good idea to have employees acknowledge and sign a 'reminder of best security practices' each month.

Physical security could be bolstered with comprehensive CCTV coverage, coupled with a clearly defined human perimeter defence space on the premises. Installation of protective physical barriers, security lightings, alarms, motion detection systems, and the use of biometrics to identify employees could go a long way in protecting a business from a potential attack. Michael Erbschole states the following on physical security:

> The bottom line here is, no matter how good cyber security is, if an individual can walk in to a facility and gain access to systems, that individual has in effect circumvented cyber security defences. [2]

With sufficient physical controls in place, it may be possible for a company to repel a substantial Social Engineering attack. However, without implementation of strict physical security protocols, the company is effectively keeping their doors open to unauthorized visitors with malicious intent. They have free reign to visit and intrude the premises, offload malwares, Trojans, spywares, and circumvent the controls to access the desired data.

35.1.2   Internal/Digital Security

Another logical step that should be taken in the fight against Social Engineering is the rolling out of a series of digital protective services and software tools. This should be implemented to negate the risk of attacks. It is also worth mentioning that although the use of digital security services may be effective in combatting certain types of Social Engineering attacks, they may turn out to be completely useless in other types of attacks. For example, a reliable spam protection guard with an updated blacklist, compounded with an antivirus/malware protection and a good firewall, may go a long way in protecting a company from phishing attacks.

With the above being said, these measures will prove to be completely inadequate against physical baiting or tailgating. This does not necessarily mean that enterprises should not invest in software protection mechanisms, because they provide partial protection nonetheless. In protecting digital data and assets, the more security measures are undertaken, the better. Explaining the severity of complications that may occur if businesses are not using digital protection mechanisms, Charles elaborates:

... some TEISME's (Technology Enabled Information Small Medium Enterprises) enable intruders to gain 'system administrator status', download sensitive files such as passwords, implant 'sniffers' (what is dubbed here as Internet dogs or spyware), to copy transactions, insert 'trap doors' to permit easy return, or implant programs that can be activated later for a variety of purposes. [3]

To negate some of the risks listed above, utilizing sandboxing mechanisms can be very productive. Sandboxing is the creation of an isolated virtual machine, use of which will protect the network from propagative malwares. It has a tendency to spread itself over the domain, even if an employee inadvertently plugs in a compromised USB flash drive into their computer. Use of sandboxing against some visual deception attacks is so effective that some popular browsers (Chromium [4], Firefox [5]) have built in sandboxing technologies in order to prevent exploitation through internet browsers.

In 2010, Long Lu and his colleagues developed and tested an interesting browser-independent concept OS. The system named BLADE [6], which stands for 'Block All Drive-By Exploits' focused on preventing automatic unauthorized execution of binary files on the system. Drive-by downloads occur when a direct connection to the compromised website takes place, resulting in the installation of a malware without web user's authorization.

By taking the unconsented execution prevention approach, the author of the OS developed BLADE as a kernel driver. This kernel extension allowed the system to enforce a rule that barred any executable files on the system that did not have explicit user consent. Any downloads that do occur are directed to a sandbox where they are held and await further instruction from the authorized user. During the initial evaluation stage, the system performed at 100% efficiency, prohibiting all 18,896 drive-by download attempts from compromised websites. There are no further updates on the project since the last preliminary evaluation. This indicates that the project did not materialize, possibly due to lack of funds or resources. Nevertheless, it remains an excellent concept; if it was made into an open source and was developed further to cover all executables, not just the ones downloaded from browsers, it could serve as an excellent tool to protect the system from technical manipulative tools used by Social Engineers.

Other dedicated measures can prove to be very effective mitigation strategy against Social Engineering attacks. Such measures include proactive monitoring, aggressive user authentication/accounting, and use of targeted machine-learning and analysis algorithms. Normal system behaviours can be observed, and it can self-educate to distinguish between legitimate and illegitimate user actions and data/packet inconsistencies. Machine and behavioural learning systems in particular have become so efficient that they are capable of detecting and stopping sophisticated Social Engineering attacks such as Spear phishing.

A group of tech enthusiasts led by Gianluca and Olivier [7] developed a vector machine-based learning system, which has the ability to identify and block spear phishing email. The authors describe the system as working by monitoring user habits and developing user profiles. The profile is based on the user's writing style, use of punctuations, character recognition, word frequency, inbox email content, usual times of email receipt and delivery, and other parameters. Once the profile is developed, it is updated every time an email is sent or received. When the algorithm reaches a prime state, it takes over and blocks every email it deems to be a spear phishing attack. The authors claimed to have achieved a false positive detection rate lower than 0.05% in the final evaluation stage, which is a remarkable achievement considering the diversity of content that can appear in a spear phishing attack email.

The internal security mechanisms described above, as well as many other security solutions available through online specialist vendors, can serve as a powerful shield that can be used to protect businesses from Social Engineering attacks. Upon implementation, these solutions may require continuous manual monitoring. An example of this may be daily, weekly, or monthly analysis of the detected and blocked attacks. Such procedures are necessary to ensure legitimate connections are not being unnecessarily stopped.

These digital protective measures may block the first few attempts made by Social Engineers. However, businesses must understand that Social Engineers and hackers are devoted to finding exploits, often dedicating their full time 'occupation' to doing so. This is especially the case if they have identified a good motivation to hack a particular company. The system may be able to block certain number of attempts, but then the attacker might gain an upper hand and find a technical exploit, allowing them the access they require. By continually analysing attack attempts and upgrading the infrastructure accordingly, businesses can better protect themselves from these attacks.

35.1.3 Implementation of Efficient Security Policy and Procedures

Due to the ever-changing dynamics in today's IT world, it is crucial that the managers and employees alike are aware of their company's current security policies and procedures. The security policy contains procedures and guidelines that dictate data and asset protection methods of an organization. It is imperative to have a concise and clearly defined set of rules for maximum efficacy, and these rules should be available to all employees, irrespective of rank. That being said, the policies should also be protected from unauthorized access that could help the attackers gain insight into the inner workings of a company. The lack of a clear security policy can, in effect, become the cause of overwhelming non-compliance among employees, leading to successful attacks and fines from authorities.

Mitnick has written comprehensively on the utility of a well-researched security policy. He has an extensive and dedicated chapter in his book, the art of deception [8], aimed at policy writers and researchers. On the importance of having an organized and coherent security policy, Mitnick notes:

Designed at lowering exposure to semantic attacks, well-maintained policy and organizational procedures help to mitigate and significantly lower the risk of a potential exploit occurring, without relying on the technical capabilities of users. [8]

The above statement makes clear that not only are the security policies important for a company's survival, they are an integral tool in protecting the employees from any potential harm. Therefore, it is of paramount importance for the managers to be aware of any change in the company's security policy. It is also their responsibility to ensure that the changes are communicated to their employees, and that they are implemented consistently across the board. In a survey paper published in December 2015, Ryan and George write:

Policy and procedures need to be flexible to unknown and unforeseen attacks and, therefore, appropriate to the changing threat landscape. Fixed guidelines can quickly become out of date as new attack methods are constantly being developed. [9]

We have thus learned that one of the greatest benefits of enforcing security policies and procedures is that it protects the company not only from intruder attacks, but also from potential lawsuits. Examples of which include policy on data protection, prohibition of business related information on social media, and policies on the use of BYOD (Bring Your Own Device). Such procedures can prevent lawsuits that may arise in case of a successful attack and crackdown from local authorities due to business non-compliance.

A well-maintained and regularly updated policy is the end result of compressive research, updated laws, and lessons learned from previous attacks. It is derived from policies of other successful businesses in the same industry, and can result in greatly reduced security risks.

Implementing security policies is directly related to computer use at work. An employee willfully accessing a compromised website, or a victim of a phishing attack, will put the enterprise at risk due to their workstation being connected to the network. Potent and effective computer access and authorization policies, along with competent firewall and robust and reliable enterprise antivirus, should be sufficient to put a stop to any inadvertent exposure to potential harm to the company's IT infrastructure.

35.1.4 Penetration Testing

When a company has employed enough security measures and feels confident that it has protected itself from an attack, it is still a good idea to search for a second opinion from an established and professional penetration tester. The primary purpose of a penetration test is to determine technical vulnerabilities and weaknesses in the network, systems, and applications being used by the business. As well as testing the resilience of the company's digital assets, many firms that test penetration also offer their services to determine the security outlook of business employees.

By employing the same tactics as a malicious Social Engineer, but with the company's consent, an official penetration tester will attempt to access the system by human manipulation, direct hacking, or use of other tricks. Such tricks range from telephone pretexting and phishing, to bating, tailgating, and other browser-based exploitation attacks. Once the simulated attack is completed, the firm leading the attack presents the employer with a report detailing the vulnerabilities identified, probable causes of weaknesses, and remedial strategies. The business can follow up on the feedback to patch up the identified fragility.

If the focus of simulated attack was internal employees as well as infrastructure, then the company may also discover which human manipulation technique was used to gain access to the desired information. The information obtained can be very useful in hardening the network and employees in preparation for a real life attack. Commenting on the importance of penetration testing, Steve notes:

> Its not enough to secure often and update often, though these two items certainly go a long way towards ensuring a secure environment. Another basic point of security in depth is to test often. Testing ensures that the security policies are being enforced and the implementation of those security policies is successful. [10]

In today's age, there is an unprecedented complexity and frequency of attacks targeting businesses, and with the exponential growth of cyber-led criminal activity, it is ever more important for businesses to take every security precaution available to them. Steve's abovementioned statement clearly emphasizes the utility of having an updated and secure system. It is also clear from the remark that penetration-testing allows companies to identify weaknesses in the day-to-day implementation of the security policies.

It is reported by Navigant that the average cost of security breaches in 2013 was $6,200,200 [11]. Navigant also reports that Cenzic Security's testing performed in the same year led to the discovery of technical flaws in 96% of the cases. An average loss of $6,200,200 is a substantial amount, whereas security testing would only cost a fraction of this amount. These incredible statistics provide every reason for security-conscious businesses to develop the habit of undergoing regular penetration tests.

Further research indicates that more than half of all UK businesses have been hit by a ransomware attack in 2015 [12], a malicious program that is commonly transmitted through phishing attacks. A separate study shows that one in five UK businesses hit by ransomware attacks is forced to close [13]. This is due to a variety of reasons, ranging from high ransom demands, to loss of data, negative publicity, and lawsuits.

To defeat the cancer of cybercrime, companies must go above and beyond normal business practices to stay on top of the game. The security challenges in today's digital world are dynamic, daunting, and convoluted to say the least. Therefore, robust cyber security and continual testing of infrastructure and employees should be a company's top priority. A holistic and comprehensive strategy that deals with risk management, cyber security will help businesses go a long way in protecting themselves from the dangers of cybercrime and Social Engineering attacks. With the aid of automated technology, vital security gaps can be identified and dealt with accordingly.

35.1.5   User Training and Security Awareness

People are more easily accessible and exploitable than machines, and thus the human element in businesses remains most vulnerable to Social Engineers. Policies that ensure strong passwords, two-factor authentications for work login, top of the range firewalls, and IDS are all made redundant if employees do not appreciate the importance of maintaining the safety of their pin, passwords, and access cards. A company's security is only as strong as their weakest link, which in this case is the employee.

Since the inception of modern technology, Social Engineers and hackers have understood that the human link in any technological equation is always the most exploitable element. Humans are the mouldable key that can be easily manipulated to gain entry to any network, system, or data. As such, the trend to access targets by 'technology only' is changing. Obtaining information from someone under false pretences, manipulation, deceit, and coercion is now conventional. The following quote aptly summarizes the rationale behind the increased number of attacks on employees as opposed to infrastructure:

Why waste your efforts on cracking passwords when you can ask for it - Unknown

In essence, the most effective mitigation strategy when dealing with Social Engineering is education. With periodic and systematic security training and frequent reminders urging the need to stay on guard and staying vigilant against suspicious behaviour, businesses can effectively turn their weakest link in to the strongest. It is vital for employees to understand the significance of protecting sensitive information, as well as the importance of knowing how a Social Engineer might strike. With a greater awareness, they can develop the knowledge of various attack vectors and establish the capability to differentiate between a dispersed or a direct attack. Employees can learn that a Social Engineer will not directly ask for a code; they will not blurt: "Give me access code for the server room, please?" Rather, they will tie little pieces of information they have acquired over time, decipher cues and signals given to them by multiple employees, and then connect the pieces of the jigsaw puzzle to unearth the information they have been after.

The single most important measure that can protect the company from a Social Engineering attack is a *continued* awareness program on information security. The word "continued" is purposefully stressed; a recent study [14] found that after attending a business training session, employees in general tend to forget 50% of the information in an hour, 70% in 24 h, and 90% in a week. So although preparatory work for training as well as the actual delivery itself can be manually intensive and costly, it is nevertheless the necessary plunge that companies must take if they wish to fortify themselves against Social Engineering attacks.

Guido Robling, a respected name in the field of academia with over a 100 publications to this date, holding numerous academic awards, presents this comment on the significance of security awareness:

> Only two things really help against Social Engineering: awareness and vigilance. Users need to know about Social Engineering, how it works, and be on alert when "strange" phone calls or emails occur. [15]

The message Guido is trying to deliver could not be anymore clearer; absolute security can never be guaranteed, but by playing smart and educating employees on security awareness, companies can turn their ignorant workers into educated and resourceful watchmen. In essence, employees are turn from liabilities into assets.

35.2 Analysis of Mitigation Strategies

In the section above, the five different strategies that firms can employ to protect themselves from Social Engineering attacks have been presented and discussed. In the following, the most effective and useful approaches that can truly turn the tables on attackers are elaborated.

35.2.1 The Most Potent Approach: Security Awareness

As the internet world is expanding, so is the horizon of knowledge of those who are curious. Previously, people would have to make a concentrated effort to learn hacking and social engineering. Now, with the internet within easy reach (and so full of information) learning exploitation techniques have become much simpler. Accessible tutorials and the availability of dedicated online social engineering tutoring websites means that 'spare time and dedication' is all that is needed for one to master the art of social engineering.

The need for businesses to be wary of this ever-growing threat is now fundamental. A lack of wariness will eventually result in catastrophe. Therefore, out of the many actions a company can take, security awareness is perhaps the most effective against social engineering attacks. As mentioned repeatedly in this chapter, businesses can take every single security measure available to them, but if their employees are not educated on the risks of disclosing internal sensitive information to strangers, all existing security measure are meaningless.

It is also essential to understand that security awareness is not just for employees who use phones and computers. From high-profile managers to security guards, cleaners and catering staff, everyone within an organization must have a solid understanding of risks arising from social engineering attacks. By involving all staff members in a security training (including non-IT staff) not only helps them understand the need to remain vigilant, but also ensures that they embrace the security program as a whole, which will consequently improve the security outlook of the entire organization.

Gragg [16] talks extensively in his research about the need to have a well-established security awareness amongst all workers. He suggests that each organization must have a specific security policy addressing social engineering. He goes on to suggest that every employee must complete security awareness training, while those who are easily manipulated should also go through resistance training.

Sarah Granger, a media innovator and author, states that:

Combat strategies... require action on both the physical and psychological levels. Employee training is essential. The mistake many corporations make is to only plan for attack on the physical side. That leaves them wide open from the social-psychological angle. [17]

This is an apt observation. Adding to this statement, Martin asserts that the case for information security in businesses is very strong. He claims that if physical security is the engine, staff awareness is the oil that drives this system forward [18]. Expressing his thoughts, Shuhaili states that with the ever-changing security landscape and people's increasing adoption of technology, the need to maintain an up-to-date levels of awareness is imperative [19]. Similarly, the European Union Agency for Network and Information Security (ENISA) claims that educated employees will help enhance the consistency and effectiveness of existing information security controls, and potentially stimulate the adoption of cost-effective controls [20]. In essence, a comprehensive training program will gradually reduce expenditure on IT security.

The real-world benefits arising from educating employees on internet security practices are unending. Not only will companies save money due to a reduction in security breaches (and resulting fines), they will also protect themselves from having to respond to any negative press and intrusive scrutiny from authorities, which often occurs after a breach. Further, a sterling reputation amongst company clients for being competent and strict on security through periodic penetration testing means that the prospects of growth in clientele could be endless.

In comparison, take, for example, the case of Talk Talk. This organization firmly established itself as a budget broadband provider and a leader in fibre-optics over a relatively short period of time. Nonetheless, they started gathering negative attention from the media and public after three consecutive high profile security breaches in a single year. These breaches resulted in a loss of 101,000 customers, and a financial loss of an estimated £60 million. [21].

To further support our view that security awareness among employees is an effective strategy in combating social engineering attacks, we will devote the next section of this chapter to practical case studies. We will evaluate the security improvements before and after the employees attended a security awareness course. Analysis of these cases will demonstrate that security awareness is the crucial and most effective tool in the fight against social engineering attacks and, therefore, is an indispensable component of a healthy business.

35.2.2   Case Studies

35.2.2.1   Company A

A small financial institution [22]. Company A had been aware of targeted phishing
and spear phishing attempts aimed at SMEs. However, they were unable to train their
employees in security awareness, except for some key staff in their IT department.
As part of a new initiative, some of the recently employed staff had been given
very limited and basic exposure to IT security. Thereafter, the company decided
to make security training mandatory for its entire workforce, contracting with an
IT security training provider. As part of the training process, phishing tests were
conducted before and after the training was delivered. According to the report, initial
tests indicated that 39% of the company employees are highly likely to click on a
phishing email, which could result in a major security breach.

In response to the recommendations, the company introduced a mandatory training session for managers lasting 40 min, with a condensed 15-min version tailored
for the other employees. After all staff members received their security awareness
training, another test was conducted to determine how employees would respond to
phishing emails. The report revealed that not a single one of the employees clicked
on a phishing link. The probability of employees becoming victim to a phishing
attack fell from 39% to 0%. Reportedly, the company averaged 1.2% over the next
12 months in subsequent simulated phishing attacks—a considerable improvement.

35.2.2.2   Company B

A shipping and logistics business [23]. Company B had over 3000 employees,
most of whom were issued with company-supplied PDAs and laptops. After a
new security manager took charge of his office, he noted a prevalence of poor IT
practices amongst employees. For example: misuse of user access rights; passwords
being shared openly between employees; sharing access credentials; use of simple
passwords (e.g., 123456123456); staff members leaving computers unlocked when
away from desks, and unauthorized disclosures made to third parties. An audit also
discovered that in most cases it was an employee's ignorance or unintentional error
that led to the incident. This had been the standard of IT security for years, so the
company decided to act and began working on a large-scale IT security awareness
campaign. After consultation, the company implemented the PDCA (Plan, Do,
Check, Act) standard for information security management, prescribed in the ISO
27110:2005 [24].

After implementing mandatory security training sessions (lasting 120 min per
module), the company saw notable improvement in employee's attitude towards
information security. In the sessions, trainers actively encouraged employees to
use more sensible and strong passwords. Pre-training assessment figures reveal that
57.9% of staff members were using simple passwords, which were cracked by the

penetration testers in around 2 h. The audit commissioned soon after the training shows that use of simple passwords fell immediately to 20%. Overall, after security training, the company noticed considerable improvement among staff in terms of compliance to security policy. After the company introduced a continued security awareness program for its entire workforce, the rates of unintentional security breaches, unauthorized disclosure, and bad IT practices fell significantly.

35.2.2.3   Company C

A large global manufacturing company [25]. Company C had over 5000 employees across the globe and been in the manufacturing business for decades. Despite robust authentication and filtering systems, the company began noticing malware attacks on its infrastructure—mostly through phishing attacks and browser infections. There was no employee awareness on IT security at all, and the company had neither a policy nor a plan in place to educate users on the ill-effects of thoughtlessly clicking on a URL. It was estimated that these infections were costing the firm in excess of $700,000 annually in repair costs alone.

Fearing the worst, the company decided to suppress the growing number of malware infections. They contracted an online security awareness training provider that offered the course in multiple languages. Since the majority of the workforce had been using the company's computers for emails and internet browsing, the company focused its efforts on increasing security awareness on three key areas: email security, safer web browsing, and URL training. With close collaboration with the security course provider, the company managed to train 95% of their employees in 12 months. It is reported that prior to the training program, the company was dealing with 72 malware infections per day. In a review undertaken 4 months after the program commenced, the company noted a reduction of 46% in malware infections globally. This resulted in substantial savings, which would have previously been spent on strenuous system repairs and recovery.

35.2.3   Review of Case Studies

All three case studies listed above have one thing in common: the businesses had no effective security awareness plan in place. This resulted in IT malpractices, infections, and attacks on their infrastructure. We then notice that significant reduction in the IT related problems was observed once the institutions implemented an effective security-training program. It is also evident that all three businesses received quick returns on the investment they made in the awareness course delivery. This was true in terms of overall savings on the cost of remedial actions. Finally, their staff also developed a healthy sense of suspicion against cyber attacks, which in itself is everything a sensible and smart employer should encourage and expect from their employees.

What should also be understood here is that the review of these case studies had only one main focus, namely the overall impact after the delivery of awareness courses. If employers also begin integrating other defence methods described in this chapter, the benefits arising from that decision would be positively far reaching and its effects would be long-term. The protection achieved through a comprehensive multi-level and prolonged defence strategy could potentially bring businesses to near immunity against cyber and Social Engineering attacks.

35.2.4   Methods to Improve User Awareness

Social Engineers are on constant search for new technical and psychological vulnerabilities so they can continue exploiting their targets. Unfortunately, uneducated and naive workers make the task of manipulation easier for the Social Engineers. Unwittingly, uninformed workers extend a helping hand to malicious Social Engineers and end up becoming part of skirmish, which brings enduring hardship to the business that trusted them.

However, as argued thoroughly in this chapter, there are numerous measures, which businesses can take to prevent themselves from becoming victim of Social Engineering attacks. One of which is security awareness, which can be delivered in a number of ways. This sections lists different approaches that are available to employers, should they choose to convert uneducated workers into knowledgeable and security aware employees.

1. Onsite Training
     An arrangement can be made to prepare an internal staff member who can conduct regular in-house coaching to in turn educate other staff members on security awareness. Alternatively, external trainers can also be hired for the same purpose. The key here is that these sessions should not be lengthy; they should be delivered in small, bite-sized sessions with regular breaks. That way, the message will be easily absorbed by the audience, and they will not suffer from training fatigue.
     Another important factor to consider is that the sessions must not contain technical jargon. Employees who are not involved in a technical role are not required to understand how to operate a firewall, or how malware containment programs work. The training must be delivered in simple, easy to understand language with clear objectives and focus on spotting and preventing Social Engineering from occurring.

2. Intranet
     A company's intranet can be very resourceful in facilitating security awareness programs. For example, a company can integrate a security course, prepared locally or externally by accredited personnel, and list the program as learning guide in a prominent section of the intranet. The managers must then encourage the workers to review the content on recurring basis, so that the information is

engraved in the minds of workers. The intranet is also a good medium to circulate security notifications to workers regarding recent security risks, with instructions on how to deal with the threat and who to report the incident to.

3. Screensavers

Screensavers can play a big part in promoting security awareness among employees. They can be used to display short reminders on topics such as keeping the password safe, disallowing tailgating, challenging anyone without a company badge/pass, reporting any suspicious behaviour to relevant departments, and so on. Efforts must be made to ensure bigger, bolder fonts and appropriate and relevant imagery are used in order for the content to be viewed and understood from a reasonable distance.

4. Posters

Displaying bright and vibrant posters with big fonts can be an effective attention-grabber. Putting brief and targeted messages on security issues concerning the business can act as an effective strategy in creating awareness among employees. General security reminders on posters should be rotated routinely, which will provide employees the opportunity to digest multiple security messages with ease and convenience. However, posters with more important and specific reminders can be placed in a prominent part of the workplace on a semi-permanent basis.

5. Manual Reminders

Concise and direct reminders can also be delivered to the workforce through printed sheets. In cases where staff intranet or other resources are not available, this could be an affordable model to keep the employees informed about the risks associated with Social Engineering. Managers could also implement a system where these manual/physical reminders are circulated in the workplace with a staff name list and date. That way, everyone who has read and understood the content can sign the form acknowledging that they have reviewed security awareness reminders, and those who have not can be re-approached with the reminders.

6. Online Courses

Employers also have an option to choose from one of the many online security training providers. Online courses not only allow self-paced learning and flexibility, but some providers offer intranet integration and specialist software in the package as well. Managers can thus track the progress of their employees from their own computers. Although many online training websites charge a fee for supplying courses, there are some excellent and free resources available online too such as www.cybrary.it. These websites can be very effective in developing a worker's knowledge on risks related to Social Engineering, and has an added benefit of zero cost to employers, proving to be very beneficial for cash-strapped businesses.

The reality is, with the presence and availability of such a variety of training methods as well as many more ingenious ways of awareness development, businesses have no excuse to leave their workers uneducated on the hazards of Social

Engineering. Once training is finalized and the work force is adequately aware of the risks posed by attackers, the employer automatically gains an upper hand in this battle; the business is less likely to suffer from an attack due to their trained staff exercising due diligence to protect the company.

35.3 Chapter Summary

This chapter contains detailed analyses of potential Social Engineering mitigation techniques used by companies to protect themselves from attacks. In addition, it has also been concluded in this chapter, after rigorous consultation of published papers on the topic of Social Engineering prevention and reviews of various case studies, that security awareness is the most significant tool in the combat against Social Engineering. The last section of this chapter lists and examines various approaches that are available to deliver security awareness coaching and reminders to employees in an office environment.

The steps outlined in this chapter are by no means exhaustive; it would be more beneficial to combine all the defence measures listed in this report to achieve maximum protection. A multi-layered defence program will undoubtedly be more effective against Social Engineering attacks compared to a single defence method. To become competent in defence, employees must understand exploitation methods used by Social Engineers. It is often the case that the sole reason attackers manage to gain entry to a target is because they are successful in exploiting the weaknesses found in employees. Therefore, companies must spend their time and effort to ensure that their workforce truly understands and appreciates the threat of Social Engineering. By recognizing the general exploitation methods that Social Engineers use to execute attacks, workers can play a huge part in the defence, namely by taking preventative measures.

Using creativity in their own refined methods, businesses can also trigger various behavioural defence instincts in their workers. An excellent way to achieve this is by conducting regular brainstorming sessions, so that employees can present new defence ideas and learn from each other's experiences. The unfortunate reality though is that there is no such thing as absolute "fool-proof" security. However, if all the defence methods outlined in this chapter are implemented with efficiency and sincerity, those security measures will make it much more difficult for a Social Engineer to successfully penetrate a company.

References

1. Mitnick, K. (2005). *Art of intrusion C: The real stories behind the exploits of hackers, intruders and deceivers* (1st ed.). Princeton: Wiley.
2. Erbschloe, M. (2004). *Physical security for IT* (1st ed.). Dorset: Digital Press.

3. Shoniregun, C. A. (2014). *Impacts and risk assessment of Technology for Internet Security (Advances in Information Security)* (1st ed.). New York: Springer.
4. The Chromium Projects. (Unknown). *Sandbox FAQ.* Available at: https://www.chromium.org/ developers/design-documents/sandbox/Sandbox-FAQ. Accessed 11 July 2016.
5. Mozilla Wiki. (Unknown). *Security/Sandbox.* Available at: https://wiki.mozilla.org/Security/ Sandbox. Accessed 11 July 2016.
6. Lu, L., Yegneswaran, V., Porras, P., & Lee, W. (2010). *BLADE: An attack-agnostic approach for preventing drive-by malware infections.* Available at: http://ants.iis.sinica.edu.tw/ 3bkmj9ltewxtsrrvnoknfdxrm3zfwrr/17/BLADE-ACM-CCS-2010.pdf. Accessed 11 July 2016.
7. Stringhini, G., & Thonnard, O. (2015). *That ain't you: Blocking spearphishing through behavioral modelling.* Available at: http://www0.cs.ucl.ac.uk/staff/G.Stringhini/ papers/spearphishing-dimva2015.pdf. Accessed 11 July 2016.
8. Mitnick, K., & Simon, W. (2002). *The art of deception.* Indianapolis: Wiley.
9. Heartfield, R., & Loukas, G. (2015). A taxonomy of attacks and a survey of defence mechanisms for semantic social engineering attacks. *ACM Computing Surveys, 48*(3), 37.
10. Suehring, S. (2015). *Linux firewalls: Enhancing security with Nftables and beyond* (4th ed.). Boston: Addison Wesley.
11. Navigant. (2014). *Cyber security trends for 2014—Part 1.* Available at: http:// www.navigant.com/insights/hot-topics/technology-solutions-experts-corner/cyber-security-trends-2014-part-1/. Accessed 12 August 2016.
12. Mendelsohn, T. (2016). *More than half of UK firms have been hit by ransomware—Report.* Available at: http://arstechnica.co.uk/security/2016/08/more-than-half-of-uk-firms-have-been-hit-by-ransomware-report/. Accessed 12 August 2016.
13. Ashford, W. (2016). *One in five businesses hit by ransomware are forced to close, study shows.* Available at: http://www.computerweekly.com/news/450301845/One-in-five-businesses-hit-by-ransomware-are-forced-to-close-study-shows. Accessed 12 August 2016.
14. Kohn, A. (2014). *Brain science: The forgetting curve–the dirty secret of corporate training.* Available at: http://www.learningsolutionsmag.com/articles/1379/brain-science-the-forgetting-curvethe-dirty-secret-of-corporate-training. Accessed 13 August 2016.
15. Robling, G., & Muller, M. (2009). *Social engineering: a serious underestimated problem.* Available at: https://www.researchgate.net/publication/220807213_Social_engineering _a_serious_underestimated_problem. Accessed 13 August 2016.
16. Gragg, D. (2002). *A multi-level defense against social engineering.* Available at: https:// www.sans.org/reading-room/whitepapers/engineering/multi-level-defense-social-engineering-920. Accessed 15 August 2016.
17. Granger, S. (2002). *Social engineering fundamentals, part II: Combat strategies.* Available at: http://www.symantec.com/connect/articles/social-engineering-fundamentals-part-ii-combat-strategies. Accessed 15 August 2016.
18. Smith, M. (2006). *The importance of employee awareness to information security.* Available at: http://digital-library.theiet.org/content/conferences/10.1049/ic_20060320. Accessed 16 August 2016.
19. Talib, S., Clarke, N. L., & Furnell, S. M. (2010). *An analysis of information security awareness within home and work environments.* Available at: http://ieeexplore.ieee.org/xpl/ login.jsp?tp=&arnumber=5438096&url=http%3A%2F%2Fieeexplore.ieee.org%2Fxpls%2 Fabs _all.jsp%3Farnumber%3D5438096. Accessed 16 August 2016.
20. ENISA. (2010). *The new users' guide: How to raise information security awareness (EN).* Available at: https://www.enisa.europa.eu/publications/archive/copy_of_new-users-guide. Accessed 16 August 2016.
21. Palmer, K., & McGoogan, C. (2016). *TalkTalk loses 101,000 customers after hack.* Available at: http://www.telegraph.co.uk/technology/2016/02/02/talktalk-loses-101000-customers-after-hack/. Accessed 17 August 2016.
22. KnowBe4. (2016). *CASE STUDY Financial Institution.* Available at: https://cdn2.hubspot.net/ hubfs/241394/Knowbe4-May2015-PDF/CaseStudy_Financials.pdf?t=1471185563903. Accessed 17 August 2016.

23. Eminagaoglu, M., Ucar, E., & Eren, S. (2010). *The positive outcomes of information security awareness training in companies e A case study.* Available at: http://www.csb.uncw.edu/people/cummingsj/classes/mis534/articles/Ch5UserTraining.pdf. Accessed 17 August 2016.
24. ISO.   (2013).   *ISO/IEC   27001:2005.*   Available   at: http://www.iso.org/iso/catalogue_detail?csnumber=42103. Accessed 17 August 2016.
25. Wombat. (2016). *Global manufacturing company reduces malware infections by 46%.* Available   at: https://info.wombatsecurity.com/hs-fs/hub/372792/file-2557238064-pdf/WombatSecurity_CaseStudy_Manufacturing_46PercentMalwareReduction_090815.pdf? submissionGuid=ffd67461-ca8b-4466-9d8b-a4ad57a5d9df. Accessed 18 August 2016.

Printed in the United States
By Bookmasters